FROM THE LIBRARY OF
David L. Nelson

The Clostridia and Biotechnology

BIOTECHNOLOGY SERIES

1.	R. Saliwanchik	*Legal Protection for Microbiological and Genetic Engineering Inventions*
2.	L. Vining (editor)	*Biochemistry and Genetic Regulation of Commercially Important Antibiotics*
3.	K. Herrmann and R. Somerville (editors)	*Amino Acids: Biosynthesis and Genetic Regulation*
4.	D. Wise (editor)	*Organic Chemicals from Biomass*
5.	A. Laskin (editor)	*Enzymes and Immobilized Cells in Biotechnology*
6.	A. Demain and N. Solomon (editors)	*Biology of Industrial Microorganisms*
7.	Z. Vaněk and Z. Hošťálek (editors)	*Overproduction of Microbial Metabolites: Strain Improvement and Process Control Strategies*
8.	W. Reznikoff and L. Gold (editors)	*Maximizing Gene Expression*
9.	W. Thilly (editor)	*Mammalian Cell Technology*
10.	R. Rodriguez and D. Denhardt (editors)	*Vectors: A Survey of Molecular Cloning Vectors and Their Uses*
11.	S.-D. Kung and C. Arntzen (editors)	*Plant Biotechnology*
12.	D. Wise (editor)	*Applied Biosensors*
13.	P. Barr, A. Brake, and P. Valenzuela (editors)	*Yeast Genetic Engineering*

14. S. Narang (editor)	*Protein Engineering: Approaches to the Manipulation of Protein Folding*
15. L. Ginzburg (editor)	*Assessing Ecological Risks of Biotechnology*
16. N. First and F. Haseltine (editors)	*Transgenic Animals*
17. C. Ho and D. Wang (editors)	*Animal Cell Bioreactors*
18. I. Goldberg and J.S. Rokem (editors)	*Biology of Methylotrophs*
19. J. Goldstein (editor)	*Biotechnology of Blood*
20. R. Ellis (editor)	*Vaccines: New Approaches to Immunological Problems*
21. D. Finkelstein and C. Ball (editors)	*Biotechnology of Filamentous Fungi*
22. R. Doi and M. McGloughlin (editors)	*Biology of Bacilli: Applications to Industry*
23. J. Bennett and M. Klich (editors)	Aspergillus: *Biology and Industrial Applications*
24. J. Davies and W. Reznikoff (editors)	*Milestones in Biotechnology: Classic Papers on Genetic Engineering*
25. D.R. Woods (editor)	*The Clostridia and Biotechnology*

The Clostridia and Biotechnology

Edited by

D.R. Woods
Department of Microbiology
University of Cape Town
Rondebosch, South Africa

Butterworth–Heinemann
Boston London Oxford Singapore Sydney Toronto Wellington

Recognizing the importance of preserving what has been written, it is the policy of Butterworth–Heinemann to have the books it publishes printed on acid-free paper, and we exert our best efforts to that end.

Library of Congress Cataloging-in-Publication Data
The Clostridia and biotechnology / edited by D.R. Woods.
p. cm.—(Biotechnology series ; 25)
Includes bibliographical references and index.
ISBN 0-7506-9004-6 (acid-free paper)
1. *Clostridium*. 2. Microbial biotechnology. I. Woods, D.R. (David R.) II. Series: Biotechnology series (Reading, Mass.) ; 25.
TP248.27.M53C56 1993
660′.62—dc20 93-16943
CIP

British Library Cataloguing-in-Publication Data
A catalogue record for this book is available from the British Library.

Butterworth–Heinemann
80 Montvale Avenue
Stoneham, MA 02180

10 9 8 7 6 5 4 3 2 1

Printed in the United States of America

CONTRIBUTORS

Hubert Bahl
Institut für Mikrobiologie
Georg-August-Universität
Göttingen, Germany

George N. Bennett
Department of Biochemistry and Cell Biology
Rice University
Houston, Texas

Nicola Bodsworth
Molecular Genetics Group
Centre for Applied Microbiology and Research
Porton Down, United Kingdom

John K. Brehm
Molecular Genetics Group
Centre for Applied Microbiology and Research
Porton Down, United Kingdom

Karin Bronnenmeier
Institute for Microbiology
Technical University Munich
Munich, Germany

Francesco Canganella
Present address:
Department of Agrobiology and Agrochemistry
University of Tuscia
Viterbo, Italy

Jiann-Shin Chen
Department of Anaerobic Microbiology
Virginia Polytechnic Institute and State University
Blacksburg, Virginia

Peter Dürre
Institut für Mikrobiologie
Georg-August-Universität
Göttingen, Germany

H.J. Gilbert
Department of Agricultural Biochemistry and Nutrition
University of Newcastle upon Tyne
Newcastle upon Tyne, United Kingdom

Gerhard Gottschalk
Institut für Mikrobiologie
Georg-August-Universität
Göttingen, Germany

N.A. Gutierrez
Biotechnology Department
Massey University
Palmerston North, New Zealand

G.P. Hazlewood
Department of Biochemistry
Agricultural and Food Research Council
Institute of Animal Physiology and Genetics Research
Cambridge, United Kingdom

John L. Johnson
Department of Anaerobic Microbiology
Virginia Polytechnic Institute and State University
Blacksburg, Virginia

David T. Jones
Department of Microbiology
University of Otago
Dunedin, New Zealand

I.S. Maddox
Biotechnology Department
Massey University
Palmerston North, New Zealand

Margaret L. Mauchline
Molecular Genetics Group
Centre for Applied Microbiology and Research
Porton Down, United Kingdom

Nigel P. Minton
Molecular Genetics Group
Centre for Applied Microbiology and Research
Porton Down, United Kingdom

J.G. Morris
Department of Biological Sciences
University of Wales
Aberystwyth, United Kingdom

John D. Oultram
Molecular Genetics Group
Centre for Applied Microbiology and Research
Porton Down, United Kingdom

E.T. Papoutsakis
Department of Chemical Engineering
Northwestern University
Evanston, Illinois

N. Qureshi
Biotechnology Department
Massey University
Palmerston North, New Zealand

Gilles Reysset
Unité des Anaérobies
Institut Pasteur
Paris, France

Palmer Rogers
Department of Microbiology
Medical School
University of Minnesota
Minneapolis, Minnesota

J. Santangelo
Department of Microbiology
University of Cape Town
Rondebosch, South Africa

Madeleine Sebald
Unité des Anaérobies
Institut Pasteur
Paris, France

Walter L. Staudenbauer
Institute for Microbiology
Technical University Munich
Munich, Germany

Tracy-Jane Swinfield
Molecular Genetics Group
Centre for Applied Microbiology and Research
Porton Down, United Kingdom

Sarah M. Whelan
Molecular Genetics Group
Centre for Applied Microbiology and Research
Porton Down, United Kingdom

Juergen Wiegel
Department of Microbiology and Center for Biological Resource Recovery
University of Georgia
Athens, Georgia

D.R. Woods
Department of Microbiology
University of Cape Town
Rondebosch, South Africa

Michael Young
Department of Biological Sciences
University of Wales
Aberystwyth, United Kingdom

CONTENTS

PREFACE

The genus *Clostridium* comprises a heterogenous group of approximately 100 anaerobic species, which are able to degrade polysaccharides and proteins, and produce a variety of industrially important products via branched fermentation pathways. The importance of the clostridia for biotechnology was recognized at the beginning of this century when *Clostridium acetobutylicum* was used to produce acetone, butanol, and ethanol in the industrial acetone, butanol, and ethanol (ABE) fermentation pioneered by Chaim Weizmann, Manchester, England. The development of chemical processes for the production of acetone and butanol from oil, and the increase in the cost of starch and molasses substrates for the fermentation process resulted in the demise of the ABE fermentation in the Western world. The ultimate requirement for the production of solvents from renewable resources, and the fluctuating price and availability of oil has triggered a renewed interest in the study of clostridia for solvent and organic acid production. Evidence of this interest was demonstrated by the successful *1st International Workshop on the Genetics and Physiology of Saccharolytic Clostridia* held in Salisbury, England, August 1990.

Key factors in the eventual exploitation of the clostridia for biotechnology involve an understanding of the biochemistry, genetics, and regulation of the various metabolic pathways, and the development of genetic systems for the genetic manipulation of the clostridia. Significant advances have recently been made in these areas and the groundwork is being laid for the genetic manipulation of the clostridia for industrial purposes. This book, containing contributions from the world's leading groups studying the molecular biology of the clostridia, documents these exciting developments, which have the potential of converting our reliance on extractive technologies to those based on renewable resources.

D.R. Woods

CHAPTER
1

History and Future Potential of the Clostridia in Biotechnology

J.G. Morris

The *Vibrion butyrique*, which so intrigued Pasteur in 1861 with its ability to grow in the absence of air and its inhibition on exposure to air, was more than likely a butyric clostridium. In 1863, Trécul coined this word to describe the swollen spore-forming rods of the butyric fermentation. The establishment of the genus *Clostridium*, however, to accommodate all such anaerobic fermentative spore-forming bacteria is credited to Prazmowski (1880). The aerointolerance, which at the outset had been the most remarkable feature of these organisms, was also to prove in those early years the greatest impediment to their isolation in pure culture and subsequent study. In considering their place in nature, Pasteur instinctively assigned the role of rendering the habitats oxygen-free to accompanying aerobic microbes. This led to the widespread adoption of methods of growing such anaerobes in mixed culture with aerobes, as Roux (1887) described. Meanwhile, Pasteur, with his colleague Joubert, had in 1877 described a sepsis vibrio (probably *C. septicum*) whose appearance was that of a sporulating bacterium that would only grow in a vacuum or an atmosphere of pure CO_2. Kitasato (1889) noted that Nicolaier's club-shaped organism, which was thought to be associated with tetanic infection (Nicolaier 1884), was capable of growth on coagulated serum, but only and always in company with other microbes. Applying the methodology that he had learned in Koch's laboratory, Kitasato attempted to isolate a pure culture of what he rightly divined would prove to be an anaerobic organism. To this end, he subjected the mixed cultures to preliminary heat treatment at 80°C for 45 to 60 min and obtained colonial growth in plate cultures incubated under hydrogen (though no growth was obtained under pure CO_2). The purified Gram-positive rods produced endospores that were killed by steaming at 100°C for 5 min though they survived 10 h immersion in 5% carbolic acid. Winogradsky made similar observations when he pursued the nature of the bacterium, which, in 1893, he had found to be an agent of N_2 fixation in fertile soils from St. Petersburg, Russia. At first, he was surprised by its being an anaerobic rod, but he successfully isolated it on slices of carrot held under vacuum in sealed glass tubes. When seeded as a pure culture into a thin layer of a sugary liquid medium, this isolate, which he designated as *C. pastorianum* (Winogradsky 1895), refused to grow.

> However, if other bacilli or even a common mold were seeded at the same time fermentation and growth of the specific rod started immediately. These are just the conditions that this rod, which is very common in soil, finds everywhere, and this explains why it is possible for these anaerobes to grow in such a well-aerated medium as soil.

With considerable percipience, having determined that the chief products of its fermentation of glucose were butyric acid, CO_2 and hydrogen, he further speculated that,

> The chemical mechanism of the nitrogen-fixing phenomenon in this case seems to be due to the effect of nascent hydrogen on gaseous nitrogen in the midst of the living protoplasm. If this hypothesis is correct, ammonia would be the immediate result of this synthesis.

Almost simultaneously in Delft, Beijerinck (1893) was reporting that among the anaerobic spore-forming microbes of the butylic fermentation there could be distinguished both butyric-acid and butyl-alcohol producers. Several years earlier, Fitz had described a *Bacillus butylicum* that gave butyl alcohol as a minor product alongside butyric acid. Grimbert had described a *B. orthobutylicus* but Beijerinck was of the opinion that his organisms differed from both of these. He recorded, as early as 1886, that certain cereal grains (especially barley), after having been ground and soaked in boiling water, readily entered into a gassy fermentation whose products included butyl alcohol. Starting from other samples however, butyric acid was obtained as the most prominent product. He went on to describe how the butylic organisms could routinely be isolated by applying his accumulation principle (Anhäufung) that formed the basis of the enrichment culture procedure which he had earlier applied to the isolation of sulphate reducers. Thereafter, he applied it to so many distinctive classes of bacteria that the technique became virtually a trademark of his laboratory. Being struck by the apparently close relationship between the butylic isolate, two of his butyric types and Prazmowski's facultatively anaerobic *Bacillus polymyxa* he placed all of these as species of the one genus *Granulobacter*. These were so named because under appropriate growth conditions all gave rise to cells swollen by the deposits of granules within them of a starch-like polysaccharide which in turn was called *granulose*. In his full description of his *G. butylicum*, Beijerinck mistakenly reported that *n*-propyl alcohol was a coproduct with *n*-butyl alcohol of its fermentative metabolism. When many years later van der Lek (1930) revived *G. butylicum* from a dried spore preparation made by Beijerinck in 1893, he showed that, in fact, it was *iso*-propyl alcohol that was formed alongside *n*-butyl alcohol. Prophetically, alongside his description of his butylic anaerobe, Beijerinck forecast that if butyl alcohol were ever to become a product of technical importance it could easily and cheaply be prepared by the fermentation method. In 1896, Beijerinck described an enrichment method for isolating a butyric fermenter, namely *G. saccharobutyricum*, and further extended the list of species within his newly contrived genus when with his junior colleague H. van Delden he studied the newly introduced Belgian warm water procedure for retting flax (Beijerinck and van Delden 1904). His previous botanical training was undoubtedly helpful to him as, from an anatomical study of the changes wrought in the flax plant during retting, it was obvious to him that the process consisted of a degradative fermentation of pectin that caused the plant fibers to become separated from the surrounding tissue. Again by an enrichment procedure, he isolated a *G. pectinovorum* that was seemingly similar to or even identical with the *Plectridium pectinovorum* described a year earlier by Störmer. He was possibly unaware of the fact that in 1895 Winogradsky with Fribes had published a full report on the anaerobic retting process and the microbial agent chiefly responsible for the pectin digestion therein.

Meanwhile, human disease-causing clostridia were also being identified. Following Kitasato's successful isolation of *C. tetani* in 1889, Welch and Nuttall (1892) isolated *C. perfringens* from a cadaver, Finney (1893) obtained *C. difficile* from a case of post-operative infection, and van Ermengen (1896) discovered *C. botulinum* in a sample of ham. Thus, by the turn of the century, the genus *Clostridium* accommodated both pathogenic and nonpathogenic species, the members of the former group being notable for their production of potent toxins while those in the latter group displayed properties (e.g., capacity for N_2 fixation, solventogenesis, secretion of polymer-degrading enzymes) that were discerned to be of potential technical importance (the term *biotechnology* had yet to be invented). Even so, the problems that had beset the initial cultivation of these anaerobes remained a disincentive to their more widespread study, though their production of durable endospores facilitated the enrichment culture procedures employed for their isolation and the long-term storage of pure viable cultures once these had been obtained.

Interestingly, it was the 1st World War that was to provide the impetus for a more intensive study of the properties of various *Clostridium* spp. and, more generally, of the media and other conditions that favored their growth. Thus, it was possible for Willis (1990) to say

> Before the war of 1914–1918 the study of the spore-bearing anaerobes had been undertaken fitfully and by imperfect methods with much attention to their pathogenicity but little to their general biology . . . It was not until the exigencies of war rendered an intensive study of the anaerobes necessary, and till the introduction of the McIntosh and Fildes jar (1916) made it relatively easy to obtain pure cultures, that the obscurity surrounding this group was dispersed.

The provocation was in the first place the terrible incidence of gangrenous infections of traumatic wounds that provoked studies of the invasiveness of the infectious organisms. (*C. perfringens*, *C. novyi*, and *C. septicum*) and of their predeliction for anoxic niches of low redox potential. Simultaneously, as we shall see, in the quest for supplies of acetone for cordite manufacture the first experience was being gained of batch growth of a *Clostridium* on an industrial scale, with resources being devoted to the rapid development of this fermentation that would never have been made available apart from the urgency imparted by the war. In the inter-war years, research was continued on the subject of the reducing potential of growing cultures of clostridia and on how addition of reducing agents to their culture media could accelerate growth from small inocula (e.g., Hall 1929; Hewitt 1931). The clostridia also figured large in the experiments that persuaded McLeod and Gordon (1923) that the basis of the hypertoxicity of oxygen to obligate anaerobes was their lack of catalase. Though this unitary theory of anaerobiosis was later demonstrated to be flawed, it held sway for several decades until it became evident that oxygen toxicity was multifactorial in its basis and that moderate anaerobes such as the clostridia were not wholly devoid of protective mechanisms (Morris 1988). The 2nd World War predictably revived concerns regarding the treatment of gangrenous wounds (Willis 1991) though now radical surgery could be supported by serotherapy (using a polyvalent antiserum containing antitoxins to the three major clostridial species). It was in this period that Macfarlane and Knight (1941) showed that the α-toxin of *C. perfringens* was actually an enzyme, namely lecithinase C. This was an opportune discovery that was the prelude to the recognition that several other of the clostridial toxins were also enzymes (MacLennan 1962). With the arrival of penicillin, there was a gradual decline in the fatality rate of the war-wounded and the subsequent chance-discovery (Shinn 1962) of the potency versus anaerobic bacteria of metronidazole so revolutionized the treatment of invasive histotoxic and systemic clostridial infections that it rendered superfluous the marketing of a human anti-gangrene vaccine—though the prophylactic anti-tetanus (toxoid vaccine) is still of service.

While the imperatives of War may have supplied the major incentives for the isolation of several pure cultures of *Clostridium* spp. and the growth of these under controlled conditions so as to produce maximal yields of desired products, e.g., solvents or toxins/enzymes, during the years of peace species of *Clostridium* played what, in retrospect, appears to be a disproportionately prominent role in the exploration of metabolic pathways.

1.1 CONTRIBUTIONS MADE BY *CLOSTRIDIUM* SPECIES TO KNOWLEDGE OF METABOLIC BIOCHEMISTRY

Particularly during that period (1925–1965) when the attention of so many biochemists was devoted to discovering and unravelling the routes of cellular anabolism and catabolism and analyzing the intermediates and enzymic catalysts thereof, clostridia turned out to be among the most rewarding of experimental subjects. It is relevant in the present context briefly to recollect some of their contributions to this era of dynamic biochemistry (Baldwin 1947) since the findings made at this time were to suggest the ways in which these organisms might most profitably be exploited for biotechnological purposes.

In a far from exhaustive list of significant and influential findings made with *Clostridium* spp. during this period, I would include the following:

1. The discovery of the Stickland reaction in proteolytic clostridia (Stickland 1934; Seto 1980) and the identification of the oxidative and reductive branches of these amino acid fermentations among whose enzymes several proved to be of interest as agents of possibly useful biotransformations (Bader et al. 1982).
2. The isolation by H.A. Barker in 1939 of *C. kluyveri* (Barker and Taha 1942), whose fermentative conversion of ethanol plus acetate to butyrate and caproate, demonstrated that fermentation, though exergonic, was not necessarily degradative in character. More importantly, soluble cell-free extracts of *C. kluyveri* could accomplish this exergonic fatty acid synthesis and in the presence of oxygen could oxidize these fatty acids. Accordingly, in 1951, Barker proposed a mechanism for the oxidation of butyrate in which the thiolytic cleavage of acetoacetyl-CoA was the final step. It was not until a year later that a soluble enzyme system capable of oxidizing fatty acids in the presence of coenzyme A was obtained from animal tissues (Stadtman 1976). Furthermore, formation of acetyl-CoA by the *C. kluyveri* extract as the result of NAD- and CoA-dependent oxidation of acetaldehyde, established the principle that oxidation of aldehydes could be coupled with the esterification of CoA to yield acyl-CoAs, whose "energy-rich" nature was again evidenced in the *C. kluyveri* extract by the ability therein of acetyl-CoA to acetylate imidazole, cyanide, and mercaptans.

   Interestingly, the necessary involvement of an acyl-carrier protein in the biosynthesis of longer chain-length fatty acids (with malonyl-CoA serving as the C_2 unit donor) was also first discovered with *C. kluyveri*, as a by-product of studies of the course of oxidation of propionate in soluble cell-free extracts (Vagelos and Earl 1959).

3. The discovery of the coenzyme forms of vitamin B_{12} and of related corrinoid compounds. These were first encountered and identified during studies of the mechanism of glutamate fermentation by *C. tetanomorphum* (Weissbach et al. 1959).
4. The C1-carrier role of tetrahydrofolate and the possibility of free energy conservation associated with the formyltetrahydrofolate synthetase reaction were highlighted by the key involvement of this cofactor in the fermentative metabolism of the purinolytic clostridia (e.g., Rabinowitz and Pricer 1956). Both corrinoid and tetrahydrofolate were found to play crucial roles in the biosynthesis of acetic acid from hydrogen and CO_2 (Poston et al. 1966; Braun et al. 1981; Hu et al. 1982; Ljungdahl et al. 1989) by homoacetogens such as *C. aceticum* (Wieringa 1941) and *C. thermoaceticum* (Fontaine et al. 1942). Also central to this acetogenesis was the operation of CO dehydrogenase (Diekert and Thauer 1978).
5. The discovery of novel low potential (high energy) electron donors. Thus, bacterial ferredoxin (Fd) was discovered during an investigation of the means of production of hydrogen by extracts of *C. pasteurianum* (Mortenson et al. 1962). Rubredoxin (Lovenberg and Sobel 1965) and flavodoxin (Knight and Hardy 1966) were also isolated from *C. pasteurianum*. The potency of reduced Fd as an electron donor was additionally demonstrated by its ability to serve as reductant in the carboxylative generation of pyruvate from acetyl-CoA in *C. acidiurici* (Raeburn and Rabinowitz 1965) and *C. kluyveri* (Andrew and Morris 1965).
6. The first demonstration of N_2 fixation by a cell-free bacterial extract, namely of *C. pasteurianum* (Carnahan et al. 1960). This led to an ongoing detailed examination of the structure of the components of the nitrogenase of this organism that has considerably advanced knowledge of metalloprotein (bio)chemistry.
7. The demonstration of the catalytic versatility of metalloproteins has been more strikingly made with clostridia than with any other group of organisms. For example, among selenoproteins alone can be numbered: the xanthine dehydrogenase of purinolytic clostridia, which is a selenium and molybdenum, iron sulfur protein with flavin adenine dinucleotide (FAD) as a prosthetic group; the nicotinate hydroxylase of *C. barkeri*, which is another selenomolybdo-iron sulfur flavoprotein; the glycine reductases of purinolytic and proteolytic clostridia; the formate-NADP oxidoreductase of *C. thermoaceticum*, which is a selenotungsten protein. The fascinating hexameric CO dehydrogenase of *C. thermoaceticum* contains, per molecular unit, six Ni atoms, three Zn or Mg atoms, and at least six $[Fe_4S_4]$ clusters with additional Fe atoms (Ragsdale et al. 1983).
8. The demonstration of the presence in saccharolytic clostridia of a NADH-Fd reductase (Jungermann et al. 1971, 1973), which inter alia could (1) account for the ability of NADH to serve as an electron donor

in N_2 fixation by *C. pasteurianum*; (2) by coupling with hydrogenase provide a means of disposing of excess reducing equivalents; (3) in collaboration with a NADPH-Fd oxidoreductase effect a cytoplasmic reduction of NADP by NADH. The discovery of this NADH-Fd oxidoreductase served as a timely reminder that when considering the bioenergetics of metabolism one should be concerned primarily not with changes in the modified standard state ($\Delta G^{o\prime}$ values) but with the real activities of products and reactants (ΔG values). Thus, while under standard state conditions at pH 7, reduction of Fd by NADH could not be spontaneously accomplished ($\Delta E^{o\prime} = -80$ mV), under the intracellular conditions that would prevail when the clostridium was actively evolving hydrogen, the ΔE value would be positive and therefore the ΔG value negative for oxidation of NADH by Fd.

A second lesson learned from the manner in which NADH could be reoxidized via hydrogen generation by the saccharolytic clostridia, was that in the fermentative metabolism of these organisms there was controlled purposive coupling between electron flow and free energy conservation. This stemmed from the fact that the acetyl-CoA produced in the course of glucose fermentation could be further metabolized by either of two competing routes namely (1) conversion to acetate with associated production of 1 mol ATP per mol acetyl-CoA used; (2) reduction to yield butyrate with associated production of 0.5 mol ATP per mol acetyl-CoA so reduced. In practice, the available acetyl-CoA is partitioned between these competing routes in a proportion determined by the activity of the NADH-Fd reductase, for it is this which dictates what fraction of the fermentatively generated NADH remains available for butyrate formation. The key discovery (Thauer et al. 1971) was that the activity of the NADH-Fd reductase was subject to regulation by the intracellular ratio of acetyl-CoA (activator) to CoA (inhibitor). In this manner, the fates of NADH and acetyl-CoA are intertwined and the efficiency of the fermentation-associated free energy conservation is regulated.

While in these and other laboratory-scale studies clostridia had yielded much information of general relevance, knowledge had simultaneously been accruing as to how several of these organisms could be cultivated on a much larger industrial scale.

1.2 COMMERCIAL EXPLOITATION OF CLOSTRIDIAL FERMENTATIONS

Although predated by the traditional (though unwitting) exploitation of pectinolytic clostridia in the retting of flax, and although large-scale growth of various pathogenic clostridia has been undertaken for the preparation of the semi-purified toxins used as the basis of protective toxoid manufacture, it is the acetone-butanol (AB) fermentation that dominates the history of commercial cultivation of *Clostridium* spp. The fascinating story of the origins and operation of the AB fermentation using strains of *C. acetobutylicum* growing on starchy grains or molasses has been told in articles written at intervals throughout the years in which the fermentation was operated in the United Kingdom, North America, and South Africa (e.g., Killeffer 1927; Gabriel and Crawford 1930; Beesch 1953; McCutchan and Hickey 1954; Ross 1961; Weizmann 1963; Hastings 1971; Spivey 1978). The articles often described the current industrial practice and speculated as to what improvements might reasonably be made. Of the more recent reviews, that by Jones and Woods (1986) is particularly readable, giving a full account of the development and decline of the AB fermentation on an industrial scale and linking this with a survey of subsequent research on the physiology of *C. acetobutylicum* and prospects for improving the economics of the process. The startlingly abrupt rise in petroleum prices in the 1970s led to a renewed interest in the feasibility of obtaining fuels or chemical feedstocks by microbial conversion of renewable biomass. The AB fermentation naturally came in for fresh scrutiny, especially since in its heyday it ranked second in importance to the ethanol fermentation that is undertaken with yeast and when, even today, The People's Republic of China relies on the AB fermentation for about one-half of its supply of butanol. This forced regeneration of governmental interest when translated into funds for fresh research in turn promoted further reviews of the nature of the fermentation and the prospects for its revival on a large scale (e.g., Lenz and Moreira 1980; Gibbs 1983; Moreira 1983; Haggström 1985; McNeil and Kristiansen 1986; Ennis et al. 1986; Marlatt and Datta 1986; Awang et al. 1988; Jones and Woods 1989; Ladisch 1991).

In the light of these accounts, it would not be sensible to recount again the course of the rise and demise of the AB fermentation industry in the Western world. Yet several important lessons can be learned from this history by those who might hope to revive the fermentation or operate a similar anaerobic fermentation. In the first phase, "classical" microbiology had to be undertaken to isolate the strains(s) of *C. acetobutylicum* that could best use the available substrate and produce maximal specific yields of solvents in the most desirable proportions. Thus, the early isolate of Fernbach that was capable of fermenting potatoes was quickly discarded in favor of Weizmann's BY strain that was able to ferment a wider range of starchy substrates to give an improved yield of butanol. When war and blockade greatly restricted the availability of starchy grains in the United Kingdom, the fermentation process was transferred first to Canada and then to the Midwest Corn Belt of the USA. After the war, when the emphasis switched to butanol production, the fermentation continued to prosper so that by the end of 1927 Commercial Solvents Corp. operated 96 fermentors at Peoria, IL and 52 fermentors in its Terre Haute, IN plant. Though at this time it was sensible to locate the fermentation plants close to the supply of maize grain, i.e., in Illinois and

Indiana, later when it became desirable to replace maize by molasses their situation was no longer advantageous, with substantial costs being incurred in the transport of the molasses to the inland plants. In this new situation, it was worthwhile in 1935 to build an entirely new fermentation plant in Bromborough near Liverpool, United Kingdom to which molasses could be bulk transported by sea. At the same time, a search was instituted among existing and new isolates of *C. acetobutylicum* for strains that were able to ferment sucrose and molasses at high rates and maximum specific productivity. Unfortunately, the productivity of the fermentation process was constrained by the toxicity of the accumulated butanol that restricted the sugar concentration to about 6% (assuming 30% w/w conversion to solvents, and recognizing that about 1.35% butanol was sufficient to halt further fermentation). With the Weizmann strain in batch culture in 90,000-l fermentors, 5850 kg of fermentable sugar at 6–6.5% initial concentration gave 1053 kg butanol with 526 kg acetone, 175 kg ethanol, 2900 kg CO_2, 117 kg H_2, and biomass during a 30- to 34-h fermentation (Spivey 1978).

When the economic competitiveness of the fermentation was threatened by the lower cost of alternative chemical syntheses of acetone (from propylene) and butanol (from ethylene) steps were taken to maximize the profitability of the microbial process. During the 1st World War, sale of acetone had to bear the whole cost of production, for at that time little or no market existed for the butanol that was cogenerated as 60% of the total solvent yield. After the war it was butanol (and butyl esters) that were in greater demand and profitability was likely to be increased if use could be made of strains that gave proportionately more butanol. At the same time, attempts were made to operate an integrated fermentation plant in which the solvents were removed by distillation and the evolved gases were collected and separated, with the CO_2 being marketed as the compressed gas or as dry ice and the hydrogen being used for the manufacture of synthetic methanol or hydrogenation of edible oils in margarine production. The residual solids from the batch fermentation were found to be rich in riboflavin and were separated and sold as an animal feed supplement. Yet, although the AB fermentation process underwent a temporary revival during the 2nd World War, it was judged unlikely that it could thereafter compete under normal circumstances with petroleum-based solvent synthesis. The three chief reasons which led to this conclusion were:

1. The dependence of the fermentative process on unreliable supplies of carbohydrate feedstock whose price was liable to fluctuate with the true harvest cost and market demand but was also subject to governmental intervention.
2. The problem posed by the increasing cost of disposing of very large volumes of waste effluent in an environmentally nonpolluting manner: this being one consequence of the ability to produce only relatively weak "beers", i.e., of low solvent concentration.

3. The fact that in a petrochemical plant most operations occur in the gaseous phase with the products being removed by condensation in a continuous operation: this contrasts with the batch fermentation process wherein the products had to be recovered by costly distillation.

Thus, it is not just the microbiology of the process that must be perfected if it is to be made competitive, there is a role too for improvements in the biochemical engineering. Indeed in the earliest years of the attempted scale-up of the AB fermentation, the greatest obstacle to success was lack of experience of large-scale axenic cultivation of an anaerobic bacterium. As Hastings (1971) put it, "when the British Government established a team under Weizmann to develop the manufacture of acetone by the AB process, there was literally not a single fermentation vessel in Britain suitable for the purpose." The commandeering of breweries to undertake the fermentation was not particularly helpful since British beers of the time were largely made using top fermenting yeasts with the yeast "cream" being skimmed off and used under conditions of no more than adequate hygiene as a large inoculum for a subsequent brew. Such a procedure was inappropriate for the AB fermentation and new equipment had to be developed and new practices learned. Had it not been for the urgency imparted by the wartime government it might on commercial grounds have been deemed unprofitable to confront these problems. Yet by 1916, production of acetone in the United Kingdom had risen to more than 2000 lb per week (Gabriel 1928) though much more rapid progress in plant development occurred after transfer of the process to North America. The lessons then learned concerning sterilization of pressurized vessels and their attendant pipework and valves formed the basis for the fermenter design that was undertaken during the next World War at the start of the antibiotic production industry.

The contamination that was to plague the operators of the early AB fermentation plants was bacteriophage infection. This was countered by the selection of mutant strains of the producer organism that were specifically resistant to infection by the prevalent phage. A collection of these phage-resistant (immunized) strains of *C. acetobutylicum* was maintained sometimes at plant level but certainly centrally by such institutions as the Northern Regional Research Laboratory (NRRL) of the U.S. Department of Agriculture, so that a change to a different immunized strain could be made if a batch developed a phage infection. Biochemical engineering was, however, to play a further important role in ensuring efficient downstream processing of the butyl beers resulting from the fermentation. Thus, when during the 2nd World War productivity at Bromborough exceeded the capacity of the batch distillation plant, multiple column continuous distillation units were so successfully employed that this remained the favored method of solvent recovery.

A real, if old, lesson that could also have been learned from the AB fermentation is that sentiment has little place in business. The United Kingdom firm of Strange and Graham Ltd. had initiated the investigation of the feasibility of producing butanol or isoamylalcohol by fermentation and had engaged the services of Fernbach and Schoen from the Pasteur Institute and of Perkin and Weizmann from the University of Manchester, only then in 1916 to have its plant at Kings Lynn commandeered by the British Government. When eventually, and this was not until 1923, the factory was returned to the company and it had recommenced acetone and butanol production therein, it was successfully sued in the High Court in London for infringement of the Weizmann 1915 British Patent the rights to which had in the interim been acquired by the Commercial Solvents Corp. of Maryland, USA. With production already affected by an explosion at the Kings Lynn plant, Strange and Graham Ltd. went into liquidation. Today such companies would doubtless be better advised by their patent lawyers.

In summary then, the history of the only clostridial fermentation to have been operated commercially on a massive scale teaches us that good microbiology and the development of a high productivity process are not of themselves sufficient; they need to be backed up by sound biochemical engineering which extends to the downstream processing of the fermented liquors, and an eye to economic competitiveness that perceives the need to adjust practices in line with supply costs and the demands of the marketplace. The chance of successful development of a commercially viable clostridial fermentation would, therefore, be enhanced if progress were to be made along the following lines.

1. Select the best-available strain of the producer species; if necessary, maintain a collection of strains which between them are sufficiently versatile to cope well with a variety of diverse feedstocks. If a number of products were simultaneously accumulated, strains that form these products in different ratios could be employed to match market needs. If the fermentation is subject to product inhibition the selection of hyperresistant strains could be attempted as also would be the development of bacteriophage-resistant mutants.
2. From detailed knowledge of the physiological behavior of the producer strains (gained in laboratory-scale studies) establish the appropriate conditions wherein the fermentation will most efficiently and rapidly convert substrate to desired product.
3. Pay attention to the design of the fermentation equipment. If possible, operate the fermentation in a continuous flow mode to minimize the unprofitable "downtime" that is a feature of batch operation. Yet continuous flow operation could throw up new problems, including deterioration of the producer strain. This is the case with some strains of *C. acetobutylicum* which in continuous culture give rise to mutant derivatives that are asporogenous and nonsolventogenic. Other strains give asporogenous mutants that have retained their solvent-forming ability and it has been suggested that these might prove useful, when immobilized on var-

ious support materials, in continuous flow solvent manufacture (e.g., Largier et al. 1985).

4. Optimize the downstream process/recovery of the desired products. In the AB fermentation, it has at different times been proposed that the danger of accumulation of butanol to potentially toxic concentrations could be alleviated by continuous removal of the solvent products by on-line vacuum distillation, or by adsorption on silicalite, continuous dialysis/ultrafiltration or partition into a recoverable nontoxic organic liquid.
5. Construct and operate an integrated fermentation plant with a commercial outlet for as many as possible of the "incidental" products of the fermentation. As previously noted, it should be sufficiently versatile to be able to accept different substrates for the fermentation so that it may benefit from any marked fluctuations in commodity prices and short-term surpluses. It would be best of all if the fermentation could be geared to the utilization of waste material which is in search of a market, e.g., lactose whey, sulfite liquor, and straw lignocellulose.

1.3 FUTURE PROSPECTS FOR BIOTECHNOLOGICAL EXPLOITATION OF CLOSTRIDIA

1.3.1 Fermentations

Anaerobic fermentations start with the major advantage that, unlike aerobic processes, they do not incur the cost penalties associated with the need to sustain high rates of culture aeration, agitation, and cooling (Morris 1983).

Under pH-controlled conditions, growth of cultures of many clostridia can proceed to relatively high cell densities and, in many instances, the release of fermentation gases satisfies the requirements for mixing. Among *Clostridium* spp. are to be found many ethanol-forming organisms several of which ferment and grow at much higher temperatures than do strains of yeast. Thus, *C. thermohydrosulfuricum*, which is widely distributed in nature and ferments a wide range of carbohydrates including lactose and cellobiose, during growth on starch at pH 6.8 and 68°C produces about 1.6 mol of ethanol per mol glucose equivalent (Parkinnen 1986). Another thermophilic species, *C. thermosaccharolyticum*, ferments cellobiose producing as its major products acetate, ethanol, lactate, and smaller amounts of butyrate. By mutagenesis with nitrosoguanidine followed by selection of fluoroacetate-resistant clones, at least one mutant strain was obtained in which ethanol production was doubled at the expense of much decreased acetate formation (Rothstein 1986).

Several reasons have been advanced for employing such thermophilic organisms in fuel alcohol or solvent production including their supposedly rapid growth, the diminished risk of contamination and improved prospects for on-line stripping of product(s) from the still growing culture. Furthermore, many of the species of greatest interest are able to hydrolyze and use polysaccharides such as starch, pectin, inulin, and xylan as well as free sugars. Thus, both thermophilic and mesophilic *Clostridium* spp. have been isolated by reason of their ability to degrade cellulose. The thermophile *C. thermocellum* is of especial interest, since during its anaerobic growth at 60°C it digests cellulosic substrates to yield ethanol plus acetate (Lynd et al. 1989). Although its in situ production of cellulase is high (Lynd 1989), its ethanol yield is relatively low. It has, therefore, been proposed that better results might be obtained with mixed cultures of *C. thermocellum* with either *C. thermosaccharolyticum* or *C. thermohydrosulfuricum* (Rogers 1986). Even so, as stated by Jones and Woods (1989),

> at present, anaerobic fermentation processes for the production of fuels and chemicals suffer from a number of serious limitations including low yields, low productivity and low final concentrations. Unless some of these limitations can be overcome it is unlikely that the fermentative route will become competitive.

Since the cost of the fermentation substrate is generally a substantial proportion of the total process costs, if truly cheap waste materials, e.g., lignocellulose, could be employed then the bioprocess would immediately be more attractive. It is debatable, however, whether the answer lies in the use of a potent cellulolytic organism in pure or mixed culture, or even in the implantation of cellulase genes into a noncellulolytic fermenter, for low cost, high efficiency prehydrolysis of lignocellulose by chemical means can be adopted. For example, in the United Kingdom, ICI plc have developed a process based on catalyzed acid hydrolysis that from straw lignocellulase (wood chips or bagasse) can produce: (1) a pentose-rich liquor derived from the hemicellulose component, with such low contents of furfural and hydroxymethylfurfural that it is eminently suitable for clostridial fermentation; (2) a pure glucose stream derived from cellulose; and (3) lignin (solid) that can be used as a fuel, for production of aromatic compounds including phenol and benzene, or other purposes. A large increase in butanol production would be required if this alcohol were to be used as a "clean" motor fuel additive or as a co-agent in micellar flooding for tertiary oil recovery (Ladisch 1991). The butanol used for these purposes could be less pure, and thus cheaper to produce than any that was to be used as a chemical feedstock. In particular, because butanol shows a lower affinity for water and a higher solubility in diesel fuel than either methanol or ethanol, it might be particularly useful as a diesel additive. If butanol were to be produced by clostridial fermentation, it might be desirable to turn to a species or strain that produces proportionately more of the alcohol than did the commercial AB fermentation strains. For example, *C. puniceum* in 4% glucose minimal medium at 25°C and pH 5.5 grew until all the glucose had been consumed (50 h) when the supernatant contained 173 mM butanol, only 17 mM each of ethanol and acetone and no acetic or butyric acids. This represented a high efficiency of conversion of glucose to butanol (32 g butanol per 100 g glucose used). This organism rapidly macerates raw potato pieces with the production of high concentrations of butanol and is also able to ferment pentose sugars (Holt et al. 1988). The process would also be much benefited if the maximal butanol concentrations currently attainable in clostridial fermentations (namely, about 1.5%) could be increased, possibly by the development of mutant strains that display enhanced tolerance to butanol. If only the final butanol concentration could be increased to 4%, this would represent a 50% saving in the energy required for distillation (Ladisch 1991). As was also pointed out by Ladisch (1991) modified bioreactor configurations, such as trickle bed or hollow fiber fermentor-extractor combinations could also improve productivity.

Interestingly, *C. acetobutylicum* has also been recommended for possible in situ use in microbial-enhanced oil recovery (Jang et al. 1983). The particular features that commend its application in such a process are firstly that it could be applied as a suspension of its spores that pass easily through porous rock and whose strength and rigidity make them tolerant to high shear forces and to pressure. Then when germinated and growing the organism would liberate gas plus oil mobility-enhancing metabolites, e.g., acids and solvents. The process would occur in three stages. In the first, the spores would be injected along with nutrients into a water-flooded oil reservoir. During the second state, both the injection and production wells would be sealed off for several months giving an opportunity for migration, fermentation, and reproduction of the clostridia with attendant production of gas and other metabolic products. Finally, after completion of the in situ incubation, the production well would be opened and a waterflood used to displace the released oil.

Apart from the AB fermentation, the clostridial fermentation that holds out most promise of commercial exploitation for synthesis of a bulk, low unit cost product is that performed by homoacetogenic species (Hugenholtz and Ljungdahl 1990). Greatest attention has been given to the thermophile *C. thermoaceticum*, which can homoferment both glucose or xylose, with 1 mol glucose yielding 3 mol of acetic acid. It is intriguing to note reflections of the experiences gained in the AB fermentation in the strategy planned to make the *C. thermoaceticum* fermentation an attractive alternative to the chemical synthesis of acetic acid. Thus, it was recognized early that a prime requirement would be the availability of a cheap fermentation substrate and a cost-effective minimal growth medium (Lundie and Drake 1984). High-productivity rates were also a target of early experimentation, with Wang and Wang (1984) choosing to make incremental additions of glucose in batch culture so as to minimize any substrate inhibition of the fermentation. By this means, and with the culture maintained at pH 6.9, they achieved a final acetic acid concentration of 56 g l^{-1} though substantial cell lysis occurred. When the culture pH was not controlled it fell to pH 5.4 and only 15.3 g acetic acid l^{-1} was produced. Reed and Bogdan (1985) achieved higher production rates, though only a concentration of 13.4 g acetic acid l^{-1}, by use of a cell-recycling procedure together with immobilization of the clostridial cells on corn cob granules and activated carbon. With a tube fermenter fitted with internal rotating discs that provided a large surface area for clostridial cell attachment and for cell-substrate interaction, a production rate of 11 g acetic acid l^{-1} h^{-1} was sustained in a medium of pH 6.6 (see Ljungdahl et al. 1989).

As in the AB fermentation, recovery of the product (acetic acid) is costly in energetic and hence financial terms when it is performed in the most obvious manner of distillation following acidification of the medium. Accordingly, attention has been paid to more economical ways of downstream processing including extraction into a recoverable organic liquid after acidification of the medium with CO_2 under pressure, or use of selective membrane permeation with or without electrodialysis. Again reminiscent of the toxicity of butanol to *C. acetobutylicum* is the toxicity of the product acetic acid to the homoacetogenic clostridia. At pH values lower than 6, this toxicity is attributable to the undissociated acid so that it becomes more marked as the pH of the medium falls. Schwartz and Keller (1982) surmised that to have a chance of being commercially competitive an acetogenic fermentation, for reasons having to do with the conventional means of product recovery, should be undertaken at a pH as low as 4.5 and should proceed with a productivity rate of at least 5 g l^{-1} h^{-1} that might be accomplished were the mass doubling time to be about 7 h and the final acetic acid concentration about 50 g l^{-1}. Accordingly, they instituted a search for mutant strains of *C. thermoaceticum* that displayed enhanced tolerance of undissociated acetic acid. Starting with a mutant strain capable of growth at pH 6 in media initially containing 10 g acetic acid l^{-1}, they sought to increase its tolerance of acetic acid by serial passage through pH-controlled fermentations operated in the fed batch mode with progressive decrease in pH and increase in acetic acid concentration providing the selective pressures. The outcome was a strain of *C. thermoaceticum*, namely ATCC 31490, which was capable of growth and acetic acid production at pH 4.5 although it disappointed its selectors by growing only very slowly at that pH (doubling time of 36 h) and producing only 4.5 g acetic acid l^{-1} (see also Parekh and Cheryan 1990). *C. thermoautotrophicum* growing on methanol or glucose in minimal medium could form 1.8–2.7 g acetic acid l^{-1} (Savage and Drake 1986) and other and better homoacetogens have doubtless yet to be isolated. Yet even currently available strains of *C. thermoaceticum* could prove to be a competitive source of acetic acid, with this prospect being made much more likely if a very large increase in production was to be demanded by use of calcium magnesium acetate (CMA) as a highway deicer, as has been proposed in the USA. Thus, in 1986, it was calculated that CMA could be produced by fermentation of hydrolyzed corn starch at a price close to that being charged for the chemically synthesized compound (see Ljungdahl et al. 1989). There may additionally be the prospect of some premium price being chargeable for organic acids of natural biological origin as opposed to synthetic chemical origin. Thus butyric acid might be produced by an organism such as *C. thermobutyricum* that accomplishes an almost homobutyric fermentation (Wiegel et al. 1989), or even by other butyric clostridia, e.g., *C. butyricum*, *C. pasteurianum*, or *C. tyrobutyricum*, once the appropriate culture conditions have been determined (Michel-Savin et al. 1990).

The saccharolytic fermentation accomplished by these species, especially *C. butyricum*, has been considered as a possibly useful source of hydrogen, and H_2 generation has been studied with growing cultures and suspensions of cells (Breure et al. 1986; Heyndrickx et al. 1987) and also with immobilized cells (Karube et al. 1982). *C. perfringens* was also employed to generate hydrogen from natural materials such as cucumber or banana pulps (Porter 1974). Over an 8-h period in a 10-l fermenter the organism produced 45 l of gaseous hydrogen. Yet it would hardly be sensible to consider H_2 generation by fermentation of organic matter as anything like an efficient process for fuel production. As was pointed out by Hungate (1974), even the best H_2-forming bacterial fermentation produces only 4 mol H_2 mol^{-1} carbohydrate used, which represents only one-third of the energy of combustion of the substrate. On the other hand, liberated H_2 is easily collected and purified and, as in the AB fermentation, it can be a valuable by-product of any fermentation whose sale could help to offset some of the cost of the process.

Particularly when the fermentation follows a branched pathway, changes in environmental conditions can redirect the flow of intermediates through competing branches and can even open up new outlets leading to the accumulation of unusual end products. This is evident in the *C. acetobutylicum* fermentation wherein the preliminary acidogenic phase gives way to the solvent-producing phase only when the medium has achieved a suitably low pH and contains an adequate concentration of butyric acid. Diversion of the normal electron flow in this fermentation is provoked when glycerol is co-metabolized to 1,3-propanediol (Forsberg 1987) or when added propionate, caproate, or 4-hydroxybutyrate are reduced to their respective alcohols (Jewell et al. 1986). An interesting attempt to distort the normally propionate-yielding fermentation of lactate by *C. propionicum* so that acrylate is accumulated, was made by Sinskey et al. (1981). In normal circumstances, lactyl-CoA is converted to propionyl-CoA via acrylyl-CoA as intermediate and free acrylate does not make an appearance. When a washed suspension of *C. propionicum*, however, was supplied with lactate (25 mM) and excess propionate (200 mM) in the presence of oxygen and methylene blue to serve as an alternative electron sink to acrylyl-CoA, then some acrylate (18 mM) accumulated. Although unpromising as a source of acrylate, the strategy employed might yet find some application in channelling fermentations to yield unusual products. Certainly it is worthwhile to discover whether environmentally imposed constraints could serve this end. For example, under conditions of phosphate limitation *C. sphenoides* forms both R(−)-1,2-propanediol and D(−)-lactate from glucose (Tran-Dinh and Gottschalk 1985), and although *C. thermosaccharolyticum* is better known for its ability to produce ethanol, under certain conditions it can form substantial quantities of R(−)-1,2-propanediol and acetol from various sugars, as a consequence of the routing of intermediates via a methylglyoxal bypass (Cameron and

Cooney 1986). As pointed out by Culberson and Donaldson (1982), because chemical feedstocks typically command higher unit prices than do fuels, the economics of processes for producing chemicals from biomass could result in an incentive structure that favors the use of biomass for chemical rather than for fuel production. In that event, some of the more unusual products of normal or distorted fermentation processes could well prove to be particularly desirable as "synthons."

1.3.2 Biotransformations

Clostridia in growing cultures, as washed suspensions of whole or semi-permeabilized cells or as immobilized cells, and also enzymes derived therefrom, can be employed to effect specific structural changes in a wide range of chemical compounds. Unlike the bulk fermentations that yield large quantities of low unit cost products, such targeted bioconversions can form small amounts of high value-added compounds and yet be operated profitably. As previously noted, it was recognized early in the history of their discovery that many species of *Clostridium* secreted enzymes that served to digest substrate polymers and, in some disease-causing organisms, served as invasive virulence factors. Included among such lytic enzymes are various amylases, pectinases and cellulases, proteases, collagenase, and aminopeptidase (Saha et al. 1989). Those enzymes produced by thermophilic species can be particularly useful because of their enhanced thermostability, e.g., the pectinolytic enzymes of *C. thermosulfurogenes* (Schink and Zeikus 1983) or the pullulanase of *C. thermohydrosulfuricum* (Melasniemi 1987). An even more notable feature of the anaerobic clostridia is, of course, their highly reducing character that enables them to develop and sustain a low E_h value in the culture medium and effect specific reductions of alien oxidants when these are added to their cultures or cell suspensions. This was noted in the case of redox-indicator dyes, but the phenomenon is much more widespread and important to the well being of the clostridium, for reduction of an alien-electron acceptor can enable it (1) to ferment more highly reduced substrates and (2) accumulate higher-than-normal proportions of the more highly oxidized end products of fermentation (a circumstance that is generally associated with a greater-than-usual specific yield of ATP). Reductive enzymes displaying broad substrate specificity might indeed have been evolved by clostridia (and other obligate anaerobes) to facilitate the disposal of excess electrons. This could be the explanation of the wide variety of hydrogenations and reductive elimination reactions (dehydroxylations, dehalogenations, and deaminations) that can be accomplished by clostridia. The clostridial bioconversions of bile acids and steroids (Morris 1989) illustrate the variety of reactions that can be catalyzed by clostridia, but it is for the purpose of regio- and stereo-specific reductions that clostridia have chiefly been employed.

Though many bioreductions can be catalyzed in vitro using purified clostridial oxidoreductases (Saha et al. 1989) when the progress of the reaction depends on the recycling of the prime electron donor [usually NAD(P)H or reduced Fd/flavodoxin] the same biotransformation can often be carried out more simply and cheaply by using whole organisms (semi-permeabilized if necessary) in suspension or immobilized in or on some suitable support material. Immobilized cells of a *Clostridium* sp. lend themselves well to continuous flow processes, as for example in the stereospecific Δ^4-3-keto-steroid reduction by *C. paraputrificum* (Abramov et al. 1990). Hydrogen, used via a hydrogenase, can frequently serve as the primary reductant, otherwise the requisite electron donor is generated by the cell suspension from a supplied fermentation substrate. Enantioselective reduction of ketones to form chiral secondary alcohols, reduction of 2-oxoacids, biohydrogenation of unsaturated compounds, and reductive carboxylation of acyl-CoA esters are among the reactions so accomplished (Morris 1989). The manner of pregrowth of the organism can determine not only its specific reductive activity but also the outcome of a reduction. For example, cells of *C. tyrobutyricum* La1 harvested after growth on crotonate and then supplied in suspension with H_2 and methylviologen, reduced the ketone sulcatone (6-methylhept-5-ene-2-one) predominantly to the S(+) form of the alcohol product sulcatol. Cells of the same organism pregrown on glucose medium gave R(−)sulcatol under the same test conditions (Belan et al. 1987). There may be surprises, but knowledge of the physiology of an organism can suggest ways in which its reductive ability can be fully harnessed or even improved. Thus, the same *C. tyrobutyricum* La1 is unable to grow on mannitol (a substrate that is more reduced than glucose) unless it is simultaneously supplied with an ancillary electron acceptor. Acetate generally serves this role (being reduced to butyrate). But by supplying pentan-2-one to a chemostat culture of *C. tyrobutyricum* growing on excess mannitol with limiting acetate, it was possible to force the election of a mutant strain that was much enhanced in the specific rate at which it reduced not just pentan-2-one but also several other ketones and aldehydes (Tidswell et al. 1991).

Simon and his colleagues (Simon et al. 1985) undertook a most rewarding study of the enoate reductases of *C. kluyveri* and *C. tyrobutyricum* that are of broad substrate specificity and catalyze the reduction of a variety of non-activated, Δ^2- unsaturated carboxylic acids and aldehydes. One of the most intriguing features of this enoate reductase is that although it is normally NADH-dependent it can accept electrons from reduced methylviologen. Since methylviologen can serve as oxidant for H_2 in the reaction catalyzed by the organism's hydrogenase, it is not surprising that addition of methylviologen to cell suspensions of *C. tyrobutyricum* held under H_2 was helpful to the enoate reductase reaction. Furthermore, since methylviologen can be reduced at the cathode of an electrochemical cell it can mediate the electromicrobial reduction of 2-enoate derivatives by a cell suspension of *C. tyrobutyricum*. Lovitt et al. (1987) with suspensions of permeabilized cells of *C. sporogenes* supplied with methylviologen as mediator accomplished the synthesis of several 2-oxo acids by bioelectrical reductive carboxylation and developed polarographic assays for the 2-oxo acid synthase, MV-NAD reductase and proline reductase of the organism. Thus, the possibility exists of using clostridial cell suspensions as the catalytic agents in electromicrobial reductions, though present procedures which employ mercury as the metallic cathode are not very satisfactory.

The particularly low redox potential developed by the homoacetogen *C. formicoaceticum* when this was supplied with either CO or formate as reductant, was sufficient to support the reduction by this organism of nonactivated carboxylates (Fraisse and Simon 1988). Extracts of the organism when supplied with CO reduced unbranched, branched, saturated, and unsaturated carboxylates to the corresponding alcohols. Suspensions of cells of *C. thermoaceticum* given CO or formate and mediators such as methylviologen or copper sepulchrate, also undertook the reduction of nonactivated carboxylates and 2-enoates (Simon et al. 1987).

Such findings present us with glimpses of what is probably, in breadth and depth, a much greater repertoire of potentially exploitable skills resident in clostridia. Certainly too few species have been studied in any depth. Only recently has it become possible to effect gene transfer by procedures that hold out the prospect of being able to eradicate unhelpful traits and enhance desirable properties. Even small changes could prove helpful. For example, to obtain a useful concentration of ethanol from the fermentation of starch by *C. thermohydrosulfuricum* the organism's tolerance of ethanol should be increased and the fermentation should proceed under conditions in which little maltose is produced as this leads to incomplete fermentation of the starch. As pointed out by Parkinnen (1986), a clone possessing higher α-glucosidase activity could thus prove especially useful.

The amplified production of useful clostridial proteins can be accomplished by cloning of the requisite genes. Whereas it might have been imagined that another Gram-positive organism such as *Bacillus subtilis* might have made the most appropriate host, in fact, in spite of its phylogenetic unrelatedness to the clostridia, *Escherichia coli* has proved to be an excellent recipient for clostridial genes isolated from recombinant DNA libraries constructed in *E. coli* (Young et al. 1989). As a host, *E. coli* has the advantage of being able to be grown either anaerobically or aerobically, and alien clostridial genes are stably maintained and well expressed. Therefore, one is encouraged to believe that there will prove to be no insuperable obstacle to the successful cloning of all of the clostridial enzymes likely to be of practical value in effecting desirable biotransformations including some derived from thermophilic species that have the advantage of increased thermostability (e.g., Haeckel and Bahl 1989). Examples are listed in Young et al. (1989).

1.4 CONCLUSION

The once successful operation of the acetone-butanol fermentation should encourage us in the conviction that clostridia could again be exploited for industrial-scale syntheses of chemical feedstocks from renewable plant materials. The balance of economic advantage will not perpetually favor the petrochemical industry and the prospect of exhaustion of fossil fuels will sooner or later force a reappraisal of the "biorefining of biomass" as a route to fuels and useful chemicals. Even modest improvements in strain performance or efficiency of product recovery and effluent disposal could hasten the day when such processes will find favor in the eyes of accountants as well as biochemical engineers. As a result, plants are likely to be large and integrated, operating in such a way that by a combination of chemical and biological pre- and posttreatments an optimal mixture of multiple saleable products will be generated. Meanwhile, along with other microbes, clostridia will supply a diverse range of commercially useful enzymes and, as whole organisms, will be employable as catalysts of stereospecific chemical transformations. They are well suited to continuous flow processes of the upward flow sludge blanket type or to sustained percolation through beds of microbial cells immobilized on a suitable support material, and their particular utility as agents of specific bioreductions though recognized has, to date, scarcely been explored.

According to Best (1988) "the recent upsurge of interest in biological systems as potent agents of chemical change will significantly affect the shape and outlook of the chemical industry over the next few decades." Even on the evidence of past successes, we can expect species of *Clostridium*, including some that have yet to be isolated, to play prominent roles in this new chemical biotechnology.

REFERENCES

Abramov, S., Aharonowitz, Y., Harnick, M., Lamed, R., and Freeman, A. (1990) *Enzymol. Microb. Technol.* 12, 982–988.

Andrew, I.G., and Morris, J.G. (1965) *Biochim. Biophys. Acta* 97, 176–179.

Awang, G.M., Jones, G.A., and Ingledew, W.M. (1988) *Crit. Rev. Microbiol.* 15 (suppl. 1), S33–S67.

Bader, J., Rauschenbach, P., and Simon, H. (1982) *FEBS Lett.* 140, 67–72.

Baldwin, E. (1947) *Dynamic Aspects of Biochemistry*, Cambridge University Press, Cambridge.

Barker, H.A., and Taha, S.M. (1942) *J. Bacteriol.* 43, 347–363.

Beesch, S.C. (1953) *Appl. Microbiol.* 1, 85–95.

Beijerinck, M.W. (1893) *Verh. K. Akad. Wet. Amsterdam*, 2de Sect. 1, no. 10.

Beijerinck, M.W. and van Delden, H. (1904) *Arch. Neerl. Sci. Exactes Nat.* Ser. 2 9, 418–441.

Belan A., Bolte, J., Fauve, A., Gourcy, J.G., and Veschambre, H. (1987) *J. Org. Chem.* 52, 256–260.

Best, D.J. (1988) in *Molecular Biology and Biotechnology* (Walker, J.M., and Gingold, E.B., eds.), 2nd ed., pp. 259–293, Royal Society of Chemistry, London.

Braun, M., Mayer, F., and Gottschalk, G. (1981) *Archiv. Microbiol.* 128, 288–293.

Breure, A.M., Duetz, W.A., Zoutberg, G.R., Mulder, R., and Van Andel, J.G. (1986) *FEMS Microbiol. Lett.* 33, 289–292.

Cameron, D.C., and Cooney, C.L. (1986) *Bio/Technology* 4, 651–654.

Carnahan, J.E., Mortenson, L.E., Mower, H.F., and Castle, J.E.(1960) *Biochim. Biophys. Acta* 38, 188–189.
Culberson, O.L., and Donaldson, T.L. (1982) *Biotechnol. Bioeng. Symp.* 12, 291–296.
Diekert, G.B., and Thauer, R.K. (1978) *J. Bacteriol.* 136, 597–606.
Ennis, B.M., Gutierrez, N.A., and Maddox, I.S. (1986) *Process Biochem.* 21 (5), 131–147.
Finney, J.M.T. (1893) *Bull. Johns Hopkins Hosp.* 4, 53–62.
Fontaine, F.E., Peterson, W.H., McCoy, E., and Johnson, M.J. (1942) *J. Bacteriol.* 43, 701–715.
Forsberg, C.W. (1987) *Appl. Environ. Microbiol.* 53, 639–643.
Fraisse, L., and Simon, H. (1988) *Archiv. Microbiol.* 150, 381–386.
Gabriel, C.L. (1928) *Ind. Eng. Chem.* 20, 1063–1067.
Gabriel, C.L., and Crawford, F.M. (1930) *Ind. Eng. Chem.* 22, 1163–1165.
Gibbs, D.F. (1983) *Trends Biotechnol.* 1, 12–15.
Haeckel, K., and Bahl, H. (1989) *FEMS Microbiol. Lett.* 60, 333–337.
Haggström, L. (1985) *Biotechnol. Adv.* 3, 13–28.
Hall, I.C. (1929) *J. Bacteriol.* 17, 225–301.
Hastings, J.H.J. (1971) *Adv. Appl. Microbiol.* 14, 1–45.
Hewitt, L.F. (1931) *Oxidation-Reduction Potentials in Bacteriology and Biochemistry*, London County Council, London.
Heyndrickx, M., De Vos, P., Thiban, B., Stevens, P., and De Ley, J. (1987) *Syst. Appl. Microbiol.* 9, 163–168.
Holt, R.A., Cairns, A.J., and Morris, J.G. (1988) *Appl. Microbiol. Biotechnol.* 27, 319–324.
Hu, S-I., Drake, H.L., and Wood, H.G. (1982) *J. Bacteriol.* 136, 597–606.
Hugenholtz, J., and Ljungdahl, L.G. (1990) *FEMS Microbiol. Rev.* 87, 383–389.
Hungate, R.E. (1974) *Am. Soc. Microbiol. News* 40, 833–838.
Jang, L-K., Chang, P.W., Findley, J.E., and Yen, T.F. (1983) *Appl. Environ. Microbiol.* 46, 1066–1072.
Jewell, J.B., Coutinho, J.B., and Kropinski, A.W. (1986) *Curr. Microbiol.* 13, 215–220.
Jones, D.T., and Woods, D.R. (1986) *Microbiol. Rev.* 50, 484–524.
Jones, D.T., and Woods, D.R. (1989) in *Clostridia* (Minton, N.P., and Clarke, D.J., eds.), pp. 105–144, Plenum Press, New York.
Jungermann, K., Leimenstoll, G., Rupprecht, E., and Thauer, R.K. (1971) *Archiv. Mikrobiol.* 80, 370–372.
Jungermann, K., Thauer, R.K., Leimenstoll, G., and Decker, K. (1973) *Biochim. Biophys. Acta* 305, 268–280.
Karube, I., Vrano, N., Matsunaga, T., and Suzuki, S. (1982) *Eur. J. Appl. Microbiol. Biotechnol.* 16, 5–9.
Killeffer, D.H. (1927) *Ind. Eng. Chem.* 19, 46–50.
Kitasato, S. (1889) *Z. Hyg.* 6, 105–116.
Knight, E., Jr., and Hardy, R.W.F. (1966) *J. Biol. Chem.* 241, 2752–2756.
Ladisch, M.R. (1991) *Enzymol. Microb. Technol.* 13, 280–283.
Largier, S.T., Long, S., Santangelo, J.D., Jones, D.T., and Woods, D.R. (1985) *Appl. Environ. Microbiol.* 50, 477–481.
Lenz, T.G., and Moreira, A.R. (1980) *Ind. Eng. Chem. Prod. Res. Dev.* 19, 478–483.

Ljungdahl, L.G., Hugenholtz, J., and Wiegel, J. (1989) in *Clostridia* (Minton, N.P., and Clarke, D.J., eds.), pp. 145–191, Plenum Press, New York.
Lovenberg, W., and Sobel, B.E. (1965) *Proc. Natl. Acad. Sci. USA* 54, 193–199.
Lovitt, R.W., James, E.W., Kell, D.B., and Morris, J.G. (1987) in *Bioreactors and Biotransformations* (Moody, G.W., and Baker, P.B., eds.), pp. 263–278, Elsevier, Amsterdam.
Lundie, L.L. Jr., and Drake, H.L. (1984) *J. Bacteriol.* 159, 700–703.
Lynd, L.R. (1989) *Adv. Biochem. Eng. Biotechnol.* 38, 1–52.
Lynd, L.R., Grethlein, H.E., and Wolkin, R.H. (1989) *Appl. Environ. Microbiol.* 55, 3131–3139.
Macfarlane, M.G., and Knight, B.C.J.G. (1941) *Biochem. J.* 35, 884–896.
MacLennan, J.D. (1962) *Bacteriol. Rev.* 26, 177–276.
McCutchan, W.N., and Hickey, R.J. (1954) *Ind. Ferment.* 1, 347–388.
McIntosh, J., and Fildes, P. (1916) *Lancet* 1, 768.
McLeod, J.W., and Gordon, J. (1923) *J. Pathol. Bacteriol.* 26, 332–343.
McNeil, B., and Kristiansen, B. (1986) *Adv. Appl. Microbiol.* 31, 61–92.
Marlatt, J.A., and Datta, R. (1986) *Biotechnol. Prog.* 2, 23–28.
Melasniemi, H. (1987) *Biochem. J.* 246, 193–197.
Michel-Savin, D., Marchal, R., and Vandecasteele, J.P. (1990) *Appl. Microbiol. Biotechnol.* 32, 387–392.
Moreira, A.R. (1983) in *Organic Chemicals from Biomass* (Wise, D.L., ed.), pp. 385–406, Benjamin/Cummings, Menlo Park, CA.
Morris, J.G. (1983) in *Biotechnology* (Phelps, C.F., and Clarke, P.H., eds.), pp. 147–172, Biochemical Society, London.
Morris, J.G. (1988) in *Homeostatic Mechanisms in Microorganisms* (Whittenbury, R., et al., eds.), pp. 84–98, Bath University Press, Bath, England.
Morris, J.G. (1989) in *Clostridia* (Minton, N.P., and Clarke, D.J., eds.), pp. 193–225, Plenum Press, New York.
Mortensen, L.E., Valentine, R.C., and Carnahan, J.E. (1962) *Biochem. Biophys. Res. Commun.* 7, 448–453.
Nicolaier, A. (1884) *Dtsch. Med. Wochenschr.*10, 842–844.
Parekh, S.R., and Cheryan, M. (1990) *Process Biochem. Int.* 25 (4), 117–121.
Parkkinen, E. (1986) *Appl. Microbiol. Biotechnol.* 25, 213–219.
Pasteur, L. (1861) *Comptes Rendus* 52, 344–347.
Pasteur, L., and Joubert, M. (1877) *Comptes Rendus* 85, 900–906.
Porter, J.R. (1974) *Am. Soc. Microbiol. News* 40, 813–825.
Poston, J.M., Kuratomi, K., and Stadtman, E.R. (1966) *J. Biol. Chem.* 263, 16000–16006.
Prazmowski, A. (1880) *Untersuchungen über die Entwicklungsgeschichte und Fermentwirkung einiger Bacterien Arten*, Hugo Voigt, Leipzig.
Rabinowitz, J.C., and Pricer, W.E., Jr., (1956) *J. Am. Chem. Soc.* 78, 4176–4178 and 5702-5704.
Raeburn, S., and Rabinowitz, J.C. (1965) *Biochem. Biophys. Res. Commun.* 18, 303–314.
Ragsdale, S.W., Clark, J.E., Ljungdahl L.G., Lundie, L.L., and Drake, H.L. (1983) *J. Biol. Chem.* 258, 2364–2369.
Reed, W.M., and Bogdan, M.E. (1985) *Biotechnol. Bioeng. Symp.* 15, 641–647.
Rogers, P. (1986) *Adv. Appl. Microbiol.* 31, 1–60.
Ross, D. (1961) *Prog. Ind. Microbiol.* 3, 73–85.

Rothstein, D.M. (1986) *J. Bacteriol.* 165, 319–320.
Roux, E. (1887) *Ann. Inst. Pasteur* (*Paris*) 1, 49–60.
Saha, B.C., Lamed, R., and Zeikus, J.G. (1989) in *Clostridia* (Minton, N.P., and Clarke, D.J., eds.), pp. 227–263, Plenum Press, New York.
Savage, M.D., and Drake, H.L. (1986) *J. Bacteriol.* 165, 315–318.
Schink, B., and Zeikus, J.G. (1983) *FEMS Microbiol. Lett.* 17, 295–298.
Schwartz, R.D., and Keller, F.A., Jr. (1982) *Appl. Environ. Microbiol.* 43, 117–123 and 1385–1392.
Seto, B. (1980) in *Diversity of Bacterial Respiratory Systems* (Knowles, C.J., ed.), vol. 2, pp. 50–64, CRC Press, Boca Raton, FL.
Shinn, D.L.S. (1962) *Lancet* 1, 1191.
Simon, H., Bader, J., Günther, H., Neumann, S., and Thanos, J. (1985) *Angew. Chem. Int. Ed. Engl.* 24, 539–555.
Simon, H., White, H., Lebertz, H., and Thanos, J. (1987) *Ang. Chem. Int. Ed. Engl.* 26, 785–787.
Sinskey, A.J., Akedo, M., and Cooney, C.L. (1981) in *Trends in the Biology of Fermentations for Fuels and Chemicals* (Hollaender, A., et al., eds.), pp. 473–492, Plenum Press, New York.
Spivey, M.J. (1978) *Process Biochem.* 13 (11), 2–4 and 25.
Stadtman, E.R. (1976) in *Reflections on Biochemistry* (Kornberg, A., et al., eds.), pp. 161–172, Pergamon Press, Oxford.
Stickland, L.H. (1934) *Biochem. J.* 28, 1746–1759.
Thauer, R.K., Rupprecht, E., Ohrloff, C., Jungermann, K., and Decker, K. (1971) *J. Biol. Chem.* 246, 954–959.
Tidswell, E.C., Thompson, A.N., and Morris, J.G. (1991) *J. Appl. Microbiol. Biotechnol.*, 35, 317–322.
Tran-Dinh, K., and Gottschalk, G. (1985) *Archiv. Microbiol.* 142, 87–92.
Vagelos, P.R., and Earl, J.M. (1959) *J. Biol. Chem.* 234, 2272–2279.
van der Lek, J.B. (1930) *Onderzoekingen over de butylalkoholgisting*. Diss. Technical University, Delft, the Netherlands.
van Ermengen, E. (1896) *Cent. Bakt. Parasit. Infekt. Abt.* 1, 19, 442–444.
Wang, G., and Wang, D. (1984) *Appl. Environ. Microbiol.* 47, 294–298.
Wiegel, J., Kuk, S., and Kohring, G.W. (1989) *Int. J. Syst. Bacteriol.* 39, 199–204.
Weissbach, H., Toohey, J.I., and Barker, H.A. (1959) *Proc. Natl. Acad. Sci. USA* 45, 521–525.
Weizmann, C. (1963) *Trial and Error*, Hamish Hamilton, London.
Welch, W.H., and Nuttall, G.H.F. (1892) *Johns Hopkins Hosp. Bull.* 3, 81–91.
Wieringa, K.T. (1941) *Brennstoff-Chemie* 22, 161–164.
Willis, A.T. (1990) in *Topley & Wilson's Principles of Bacteriology, Virology and Immunity*, 8th ed., vol. 2, pp. 212–246, Edward Arnold, London.
Willis, A.T. (1991) in *Anaerobes in Human Disease* (Duerden, B.I., and Drasar, B.S., eds.), pp. 299–323, Edward Arnold, London.
Winogradsky, S. (1895) in *Microbiologie du Sol* (1949), pp. 367–402, Masson et Cie, Paris.
Winogradsky, S., and Fribes, M.V. (1895) see Winogradsky, S. (1949) *Microbiologie du Sol*, pp. 405–409, Masson et Cie, Paris.
Young, M., Staudenbauer, W.L., and Minton, N.P. (1989) in *Clostridia* (Minton, N.P., and Clarke, D.J., eds.), pp. 63–103, Plenum Press, New York.

CHAPTER
2

Biochemistry and Regulation of Acid and Solvent Production in Clostridia

Palmer Rogers
Gerhard Gottschalk

The clostridia are anaerobic spore-forming bacteria that have a Gram-positive type of cell wall and are unable to carry out a dissimilatory sulfate reduction (Cato et al. 1986; Hippe et al. 1991). Because of these four rather general criteria many bacterial species have been assigned to the genus *Clostridium* which comprises now approximately 100 species. The organisms grouped together into this genus are extremely heterogeneous, since the G + C content of their DNA ranges from 21–54 mol%. Within the clostridia are included psychrophilic, mesophilic, and thermophilic species. The major role of these organisms in nature is to degrade organic material, notably polymers such as starch, cellulose, hemicelluloses, and proteins. Therefore, many clostridial species secrete enzymes such as amylases, cellulases, proteinases, and collagenases, some of which are of biotechnological interest (Saha et al. 1989).

The biochemistry of the various pathways by which the substrates are fermented has been carefully studied over the last 20 years in a few key species of *Clostridium*. As a result of these studies, the enzymic reactions of the basic clostridial fermentations are now known. These are: the homoacetate fermentation;

the butyrate-acetate fermentation; the acetone-butanol fermentation; and the ethanol fermentation. They will be discussed in this chapter as well as several other fermentations such as the homofermentative lactic acid fermentation as carried out by *C. thermolacticum*, the propionate fermentation by *C. propion-*

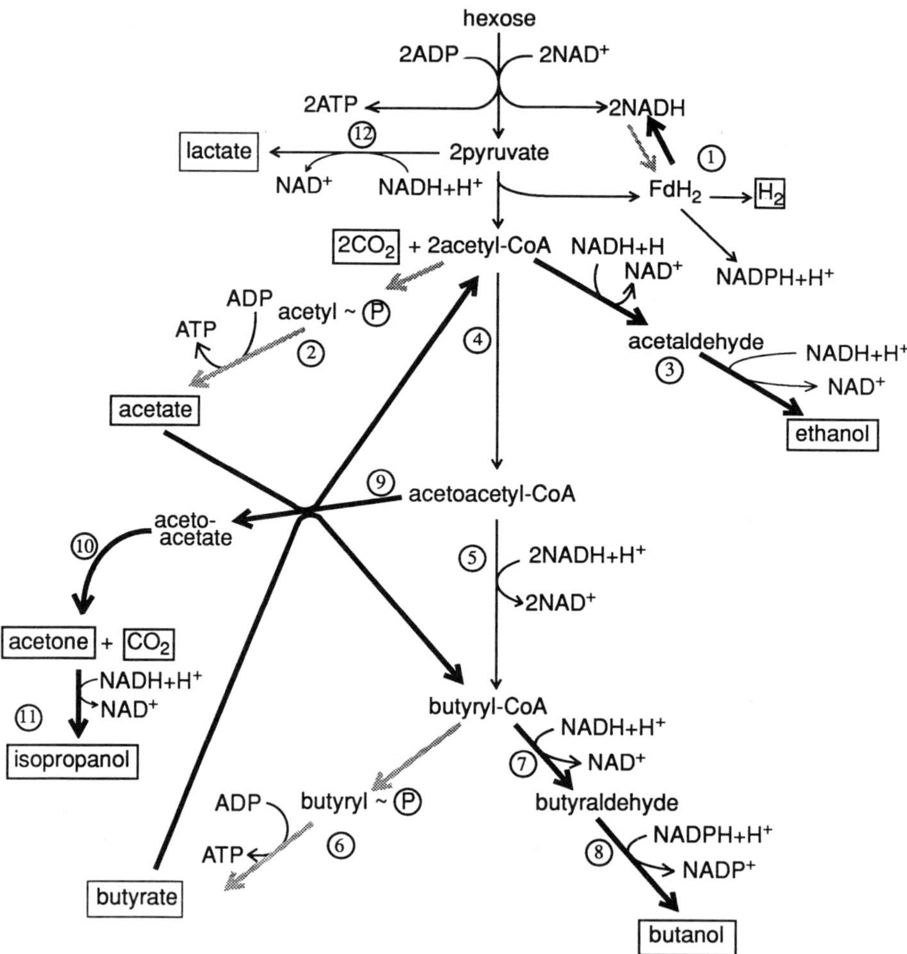

FIGURE 2–1 The butanol fermentations. The number indicates enzymes as follows: (1) three enzymes that exchange electrons with reduced ferredoxin (FdH_2): NAD^+: and $NADP^+$: ferredoxin oxidoreductases, and hydrogenase; (2) phosphotransacetylase and acetate kinase; (3) acetaldehyde dehydrogenase and ethanol dehydrogenase; (4) acetyl CoA acetyltransferase (or thiolase); (5) three enzymes producing butyryl-CoA; (6) phosphotransbutyrylase and butyrate kinase; (7) butyraldehyde dehydrogenase; (8) butanol dehydrogenase; (9) acetoacetyl-CoA: acyl-CoA transferase; (10) acetoacetate decarboxylase; (11) isopropanol dehydrogenase; (12) lactate dehydrogenase. The thick gray arrows indicate reactions that predominate during acidogenesis and the thick dark arrows indicate reactions that predominate during solventogenesis.

icum, and the *C. kluyveri* fermentations. The important clostridial fermentations of N-containing compounds such as amino acids, purines, and pyrimidines that led to the discovery of several novel pathways will not be covered here and the reader is referred to a recent review (Andreesen et al. 1989).

There is a special problem with clostridial fermentations that is relevant whenever these fermentations are considered as a basis of a biotechnological process. Except for the homoacetate fermentation (where only acetate is produced), clostridia usually produce a variety of products via branched fermentation pathways (Thauer et al. 1977; Rogers 1986). Figure 2–1 represents an example of this principle, showing that *C. acetobutylicum* produces five organic fermentation products from acetyl-CoA as a result of three branch points in its fermentation pathway. Although this variety of pathways enjoyed by the clostridia is vital to the physiology and adaptability of these bacteria to different growth conditions, the usefulness of these organisms for production of commodity chemicals will depend upon directing these fermentations toward one or perhaps two major products. Thus, present-day research is designed to elucidate the physiological signals and details of the molecular sensing and responding mechanisms in these bacteria that ultimately control the direction of flow of these fermentations toward specific products. Only if the clostridial fermentations are better understood and production strains can be generated, will these organisms play a role in biomass conversion to chemicals sometime in the 21st century, particularly as crude petroleum becomes more expensive as a starting material.

The organisms to be covered in the chapter are listed in Table 2–1 along with the major products of fermentation. As compared to the large number of

TABLE 2–1 Some Clostridial Species and Their Typical Fermentation Products

Organism	*Substrates*	*Products*
C. thermoaceticum	Sugars, H_2 + CO_2	Acetate
C. thermoautotrophicum	CO	
C. aceticum		
C. thermolacticum	Sugars	Lactate
C. propionicum	Lactate	Propionate
C. tyrobutyricum	Sugars	Butyrate, acetate, CO_2 + H_2
C. butyricum		
C. kluyveri	Ethanol, acetate	Butyrate, caproate, H_2
C. acetobutylicum	Sugars, starch	Acetone, butanol, ethanol, isopropanol, CO_2, H_2, acetate, butyrate
C. beijerinckii		
C. thermohydrosulfuricum	Starch, sugars	Ethanol, CO_2, H_2, acetate
C. thermocellum	Cellulose, sugars	
C. thermosaccharolyticum	Starch, sugars	
C. sphenoides	Starch, sugars	

known clostridial species this is a rather small collection, but it comprises those species that are of biotechnological interest because of the products formed.

2.1 FORMATION OF ORGANIC ACIDS

2.1.1 Homoacetate Fermentation

Microorganisms able to carry out a homoacetate fermentation are not only found in the genus *Clostridium*, but also in other genera such as: *Acetobacterium*, *Acetogenium*, and *Sporomusa* (Fuchs 1986). The three species mentioned in Table 2–1 are typical saccharolytic clostridia which, in addition to growth on sugars, are able to grow chemolithotrophically at the expense of H_2 + CO_2 or CO with the formation of acetate (Ljungdahl et al. 1989). The pathway used for acetate formation is summarized in Figure 2–2.

Breakdown of sugars to pyruvate proceeds via the Embden-Meyerhof-Parnas pathway. From the two pyruvate molecules formed, two acetates are produced that involve the enzymes pyruvate:ferredoxin (Fd)-oxidoreductase, phosphotransacetylase, and acetate kinase. This is typical of the reactions found in most clostridial species. These organisms are unique among the clostridia, in that CO_2 acts as the hydrogen acceptor for the reducing equivalents generated during sugar breakdown. To provide a hydrogen acceptor, CO_2 is reduced to formate and then linked to tetrahydrofolate (H_4F) in the formyl-H_4F synthetase reaction and it is reduced stepwise to methyl-H_4F. The final reactions resulting in the formation of a third acetate are complex and require the enzyme CO-dehydrogenase, which has three active sites (Figure 2–3). The methyl group is transferred from methyl-H_4F via a corrinoid enzyme to the CO-dehydrogenase. Independently, the second CO_2 molecule from glycolysis is reduced to yield enzyme-bound CO. An acetyl enzyme is formed which subsequently is cleaved by the action of coenzyme A generating acetyl-CoA, which again can be converted into acetate via acetyl phosphate. Thus, three molecules of acetate are formed from one molecule of hexose, and there are practically no by-products.

The heterotrophic version of the homoacetate pathway produces the maximum ATP yield that can be expected for a fermentation process. Activation of sugar requires two ATPs (or equivalent), while a third ATP is consumed in the formyl-H_4F synthetase reaction. At least seven ATPs are synthesized (four during glycolysis and three in the course of acetate formation). As a consequence, growth yields are high, e.g., in the order of 50 g cells (dry weight) mol^{-1} glucose fermented for *C. thermoaceticum* (Andreesen et al. 1973). The formation of 3 mol of acetate from 1 mol of glucose is associated with a free energy change of $\Delta G^{o\prime} = -310.9$ kJ mol^{-1} glucose. Since the energy of ATP hydrolysis amounts to $\Delta G^{o\prime} = -31.8$ kJ mol^{-1} the thermodynamic efficiency under standard conditions is in the range of 41%, which is acceptable.

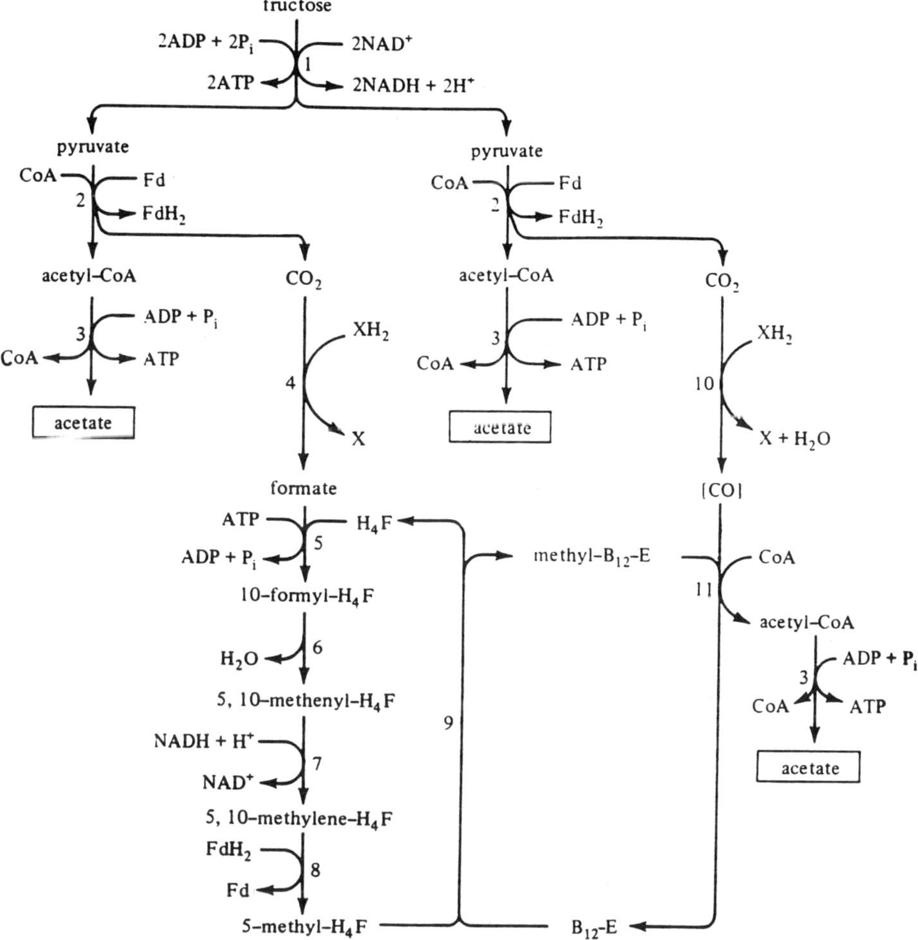

FIGURE 2–2 Pathway of the acetate fermentation. (1) Degradation of fructose via the Embden-Meyerhof-Parnas pathway; (2) pyruvate-ferredoxin (Fd) oxidoreductase; (3) phosphotransacetylase plus acetate kinase; (4) formate dehydrogenase; (5) formyltetrahydrofolate (H_4F) synthetase; (6) methenyl-H_4F cyclohydrolase; (7) methylene-H_4F dehydrogenase; (8) methylene-H_4F reductase; (9) H_4F: B_{12} methyltransferase; (10) CO dehydrogenase; (11) acetyl-CoA-synthesizing coenzyme; (CO), enzyme-bound. From Gottschalk (1986).

The energetics of growth and acetate formation from H_2 + CO_2 is much more complex. In this case, one ATP is consumed in the formyl-H_4F synthetase reaction and only one ATP is produced by substrate-level phosphorylation during acetate kinase-catalyzed acetate formation. Therefore, it has been proposed that an electron transport system is functioning in these microorganisms that generates an electrochemical proton gradient which is subse-

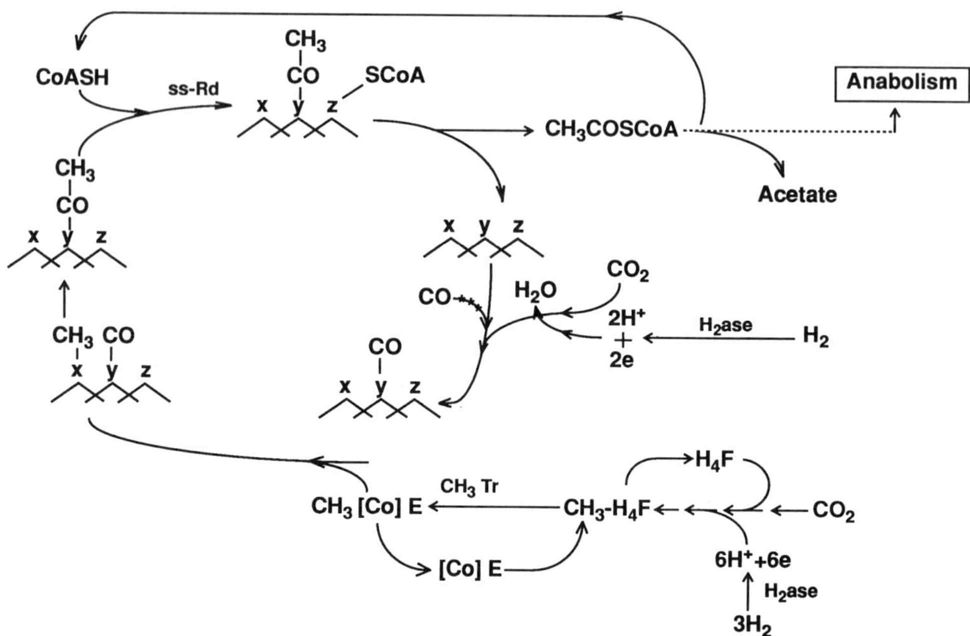

FIGURE 2–3 The acetyl-CoA pathway for autotrophic growth by acetogenic clostridia. H_4F, tetrahydrofolate; CH_3Tr, methyl-transferase; CoE, corrinoid enzyme; /\/\/\, CO dehydrogenase with three subsites, X, Y, Z; SS-Rd. CO dehydrogenase disulfide reductase; H_2ase, hydrogenase. The broken arrow indicates anabolic reactions. From Wood et al. (1986).

quently used by the F_1F_0-ATPase for ATP synthesis (Fuchs 1986; Das et al. 1989). Electron carriers such as cytochromes and menaquinone have been shown to be present in *C. thermoaceticum* (Gottwald et al. 1975). These carriers appear to be involved in establishing the proton gradient across the cytoplasmic membrane that can be used by the proton-translocating ATPase (Mayer et al. 1986). During chemoorganotrophic growth on sugars, NADH is the most likely electron donor, while under chemolithotrophic conditions H_2 (through hydrogenase) or CO (through CO-dehydrogenase) serve as electron donors (Figure 2–4). The most likely physiologic electron acceptor is

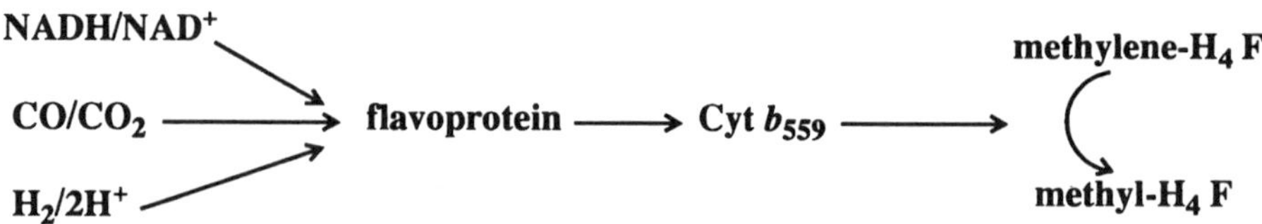

FIGURE 2–4 Electron transport chain as operative in acetogenic clostridia. Reproduced with permission from Das et al. (1989).

methylene-H_4F; its reduction to methyl-H_4F with H_2 is associated with a free energy change of $\Delta G^{o\prime} = -39$ kJ mol^{-1}. The responsible enzyme, methylene-H_4F reductase is membrane-associated (Hugenholtz et al. 1987). Some other acetogens, e.g., *Acetobacterium woodii* lack cytochromes but here a sodium requirement for acetogenesis could be demonstrated (Heise et al. 1989); thus instead of proton gradients, Na^+-gradients might be built up and taken advantage of for energy conservation.

The homoacetate fermentation is of interest from a biotechnological standpoint because it allows the production of 100% acetate from sugars without any other by-products. The acetogenic clostridia listed in Table 2–1 also carry out the synthesis of acetate from gases such a CO_2 + H_2 or CO. The drawback of this fermentation as compared to acetic acid production by aerobic acetic acid bacteria is that clostridia responsible for the anaerobic process are fairly pH sensitive. At pH 5 or below acetic acid functions as an uncoupler, and growth and acetogenesis is no longer possible (Baronofsky et al. 1984; Wang and Wang 1984). In contrast to the aerobic acetic acid bacteria, the anaerobes apparently are not able to pump out protons rapidly enough to allow growth in the presence of high concentrations of acetic acid (Menzel and Gottschalk 1985). Thus, the homoacetate fermentation can only be taken advantage of to produce acetate if the pH is controlled. Processes have been developed for *C. thermoaceticum*, *C. thermoautotrophicum*, and *C. formicoaceticum* that work at pH values around 6.5 (Ljungdahl et al. 1985; Tang et al. 1988; Parekh and Cheryan 1990). Using milled dolomite to control the pH, the production of calcium magnesium acetate (CMA) from sugars is possible (Ljungdahl et al. 1986). The CMA has been proposed for use as an environmental safe road deicer.

2.1.2 Lactate and Propionate Fermentation

Only a few clostridial species are known which produce lactate in a lactate/sugar ratio comparable to the one of homofermentative lactic acid bacteria. A species to be mentioned in this context is *C. thermolacticum* (Le Ruyet et al. 1985). It ferments a great variety of sugars including starch, xylan, lactose, and various pentoses. Its temperature optimum lies between 60 and 65°C. Since the pH-optimum of growth is between 6.8 and 7.4, this organism probably does not have a biotechnological potential.

C. propionicum converts lactate to propionate, acetate and CO_2 according to the following equation (Cardon and Barker 1947):

$$3 \text{ lactate} \longrightarrow 2 \text{ propionate} + 1 \text{ acetate} + 1\ CO_2$$

This fermentation is probably not of biotechnological importance although some acrylate accumulation was observed (Akedo et al. 1983). It is interesting, because lactate is directly reduced to propionate (and not via succinate as in the case of the propionibacteria). The difficult reaction of this pathway is the removal of water from lactate to yield acrylate. This dehy-

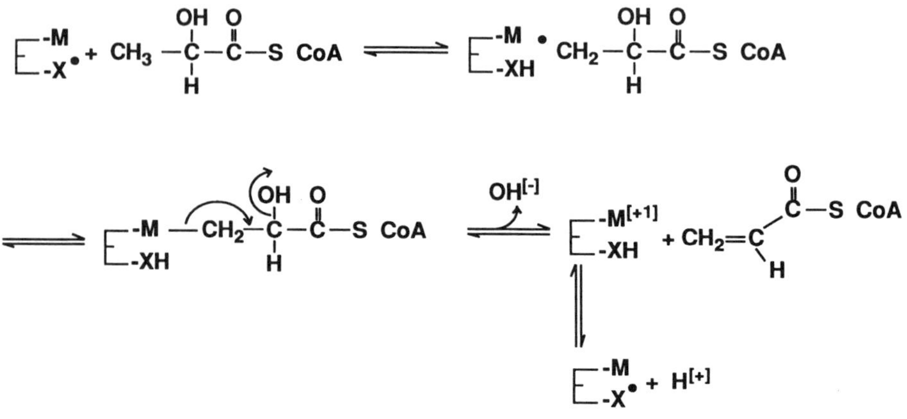

FIGURE 2–5 Proposed mechanism for the dehydration of lactyl-CoA to acryl-CoA. Reproduced with permission from Kuchta and Abeles (1985).

dration has to proceed against the rule of Markownikoff and requires the conversion of lactate to lactyl-CoA which then is dehydrated to acrylyl-CoA. Analogous dehydrations occur when 2-hydroxyglutaryl-CoA is converted into glutaconyl-CoA or 4-hydroxybutyryl-CoA into crotonyl-CoA (Schweiger et al. 1987; Willadsen and Buckel 1990). Two enzyme proteins and ATP are required for the dehydration. According to Kuchta and Abeles (1985), the reaction is initiated by removing a β-hydrogen radical followed by binding of the carbon radical to the enzyme (Figure 2–5).

2.1.3 *Clostridium kluyveri* Fermentation

C. kluyveri was shown some time ago to carry out an interesting fermentation: the conversion of ethanol and acetate to butyrate, caproate, and H_2. A typical fermentation balance is the following (Bornstein and Barker 1948):

$$6 \text{ ethanol} + 3 \text{ acetate} \longrightarrow 3 \text{ butyrate} + 1 \text{ caproate} + 2H_2 + 4H_2O + H^+$$

The energy metabolism of *C. kluyveri* is very curious: per mol of H_2 evolved, 0.5 mol acetyl-CoA becomes available to the cells for conversion to acetate via acetyl phosphate (Thauer et al. 1977; Gottschalk 1986). Thus, the more H_2 evolved, the more ATP becomes available to the organisms for growth. How the pathways of electrons and carbon are regulated in these organisms to control the quantities of butyrate, caproate, and H_2 formed, remains unknown.

One interesting *C. kluyveri* fermentation became known only recently and that is the fermentation of succinate and ethanol according to the following equation (Kenealy and Waselefsky 1985):

$$2 \text{ succinate} + 3 \text{ ethanol} \longrightarrow 2 \text{ butyrate} + 3 \text{ acetate}$$

Evidence has been presented that succinate is converted to crotonyl-CoA via 4-hydroxybutyrate (Figure 2–6). Thus, this fermentation comprises a re-

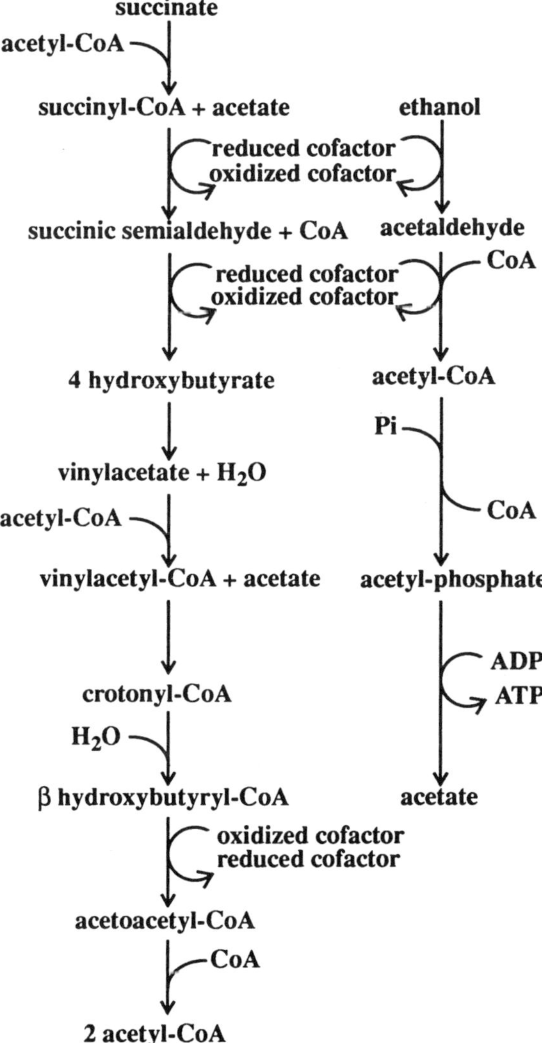

FIGURE 2–6 Hypothetical scheme for the succinate-ethanol fermentation by *C. kluyveri*. From Kenealy and Waselefsky (1985).

duction of a dicarboxylic acid to a hydroxy-monocarboxylic acid that is interesting and may have a biotechnological potential.

2.1.4 Butyrate-acetate Fermentation

Clostridia are primarily known for their ability to ferment sugars to a mixture of butyrate, acetate, CO_2 + H_2. Several species produce additional products such as ethanol or lactate, but *C. butyricum* and *C. tyrobutyricum* are purely acidogenic, and they are of interest for the production of butyrate. The fermentation proceeds according to the following equation (Wood 1961):

$$1 \text{ glucose} \longrightarrow 0.8 \text{ butyrate} + 0.4 \text{ acetate} + 2CO_2 + 2.4\ H_2$$

The pathway of sugar breakdown in these organisms is similar to the one used by the homoacetate fermenters or the solvent producers. But they contain an active hydrogenase so that the reduced Fd produced at the pyruvate: Fd-oxidoreductase step is reoxidized with H_2-evolution. Apparently, these species of clostridia do not contain alcohol dehydrogenase and acetaldehyde dehydrogenase so that only acids are produced. The branch point to acetate and butyrate is subject to regulation. Under a low partial pressure of H_2 the butyrate/acetate ratio decreases, more acetate is produced accompanied by an increase of the ATP yield of this fermentation (Jungermann et al. 1973). Thus, in a biotechnological process designed to produce butyrate this has to be taken into account (van Andel et al. 1985; Michel-Savin et al. 1990).

2.2 SOLVENT FORMATION

The clostridia are perhaps best known for their role in the large-scale commercial production of the solvents, butanol, acetone, and in some cases isopropanol, earlier in this century from inexpensive starting materials such as molasses and starches (Prescott and Dunn 1959; Bahl and Gottschalk 1988). Other thermophilic clostridia have been shown to produce ethanol from various sugars as well as cellulosic materials at rather high temperatures (55–65°C) in experimental scale fermentations. The clostridia can also be induced to produce other solvents such as 1,3-propanediol, 1,2-propanediol, and propanol under specific conditions. Aside from disastrous virus infections that plague all bacterial fermentations, the employment of the clostridia for commercial production of solvents in the future has two important disadvantages: (1) low tolerance of the bacteria to accumulated solvents making their recovery from the beers a high-cost operation and (2) branched fermentation pathways resulting in mixtures with acids and thus lower solvent yields (Linden et al. 1983; Phillips and Humphrey 1983). These specific problems as well as experimental attempts to deal with them for each type of fermentation are addressed as follows.

2.2.1 Ethanol Fermentations

The three thermophilic clostridia that produce large amounts of ethanol and are the best candidates for future bioconversion of biomass to this solvent are *C. thermocellum*, *C. thermohydrosulfuricum*, and *C. thermosaccharolyticum* (Table 2–1). In connection with the utilization of cellulose as feedstock, the centerpiece of research and development of an ethanolic fermentation by the clostridia is *C. thermocellum* (Lynd 1989; Slapack et al. 1987; Wiegel and Ljungdahl 1986). Strains of *C. thermocellum* hydrolyze cellulosic biomass; and all three clostridial species ferment cellobiose and other sugars at 55–70°C forming ethanol, acetate, CO_2, and H_2 (Figure 2–7). During the ethanolic fermentation most of the pyruvate is converted to acetyl-CoA, reduced Fd (FdH_2), and CO_2 with a significant amount reduced to lactate by lactate dehydrogenase. In one branch of the fermentation acetyl-CoA is reduced to acetaldehyde and then to ethanol catalyzed by NAD^+-linked dehydrogenases. (Figure 2–7, reaction 4 and 5). Acetate is formed via another

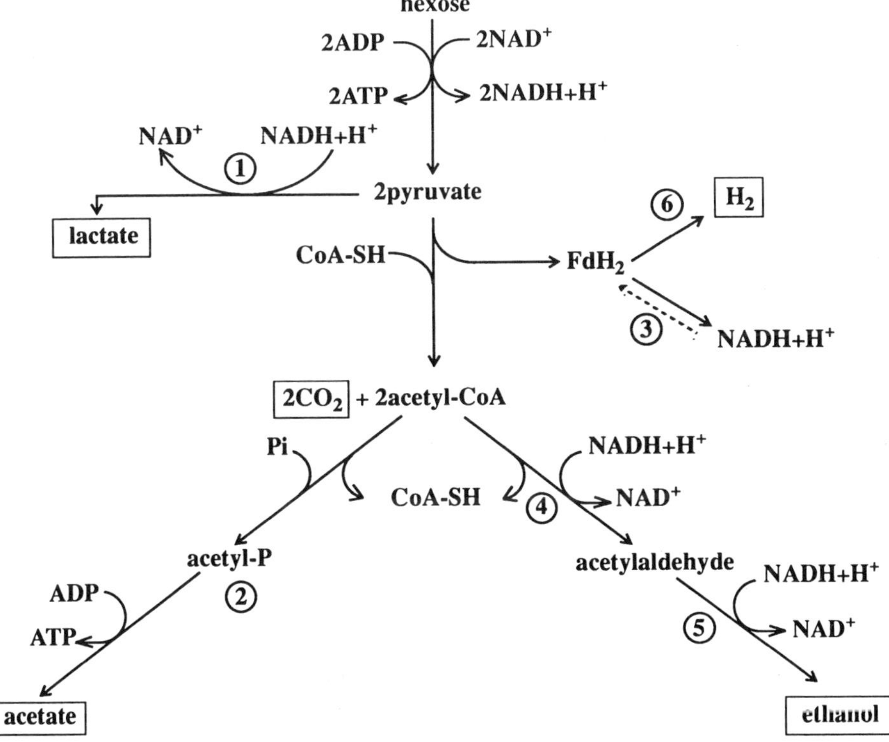

FIGURE 2–7 Clostridial ethanol fermentation. The numbers indicate enzymes as follows: (1) lactate dehydrogenase; (2) phosphotransacetylase + acetate kinase; (3) NADH: ferredoxin oxidoreductase; (4) acetaldehyde dehydrogenase; (5) ethanol dehydrogenase, and (6) hydrogenase.

branch yielding stoichiometric amounts of ATP as with other clostridial fermentations (Figure 2–7, reaction 2). Using this triple-branched pathway, *C. thermocellum* and *C. thermosaccharolyticum* produce about 0.6–1.0 mol of ethanol per mol of hexose, while in both *C. thermohydrosulfuricum* and *C. saccharolyticum* fermentations, 1.8–1.9 mol of ethanol are produced per mol of hexose, classing these latter two strains as excellent ethanol producers.

The biochemical basis for the different product ratios observed for the different species using the same branched pathway (Figure 2–6), remains unknown, but has been studied indirectly. For example, during the first 24–48 h in a batch fermentation of cellulose by *C. thermocellum*, ethanol is the major product while more acetate and lactate accumulate from 2–5 days. Also a higher ethanol/lactate ratio is favored by pH 7.2 over pH 6.5 (Tailliez et al. 1989). Thus, in a batch fermentation there is both an effect of the pH of the fermentation and the time of sampling on ethanol production. When the ethanologens make excess ethanol (ethanol >1) the organism must transfer a significant quantity of electrons from reduced Fd via Fd:NAD^+ oxidoreductase to $NADH^+$ (Figure 2–6, reaction 3), which permits added ethanol formation and a reduction in the H_2 yield per mol hexose. The role of this electron-balancing function was shown when strain LQR1 or *C. thermocellum*, that fermented cellobiose to exactly 1:1 ethanol/acetate, was found to be devoid of detectable Fd/NAD^+ oxidoreductase in crude extracts (Lamed and Zeikus 1980). In contrast, strain AS39 and *Thermoanaerobium brockii* contained this enzyme and were capable of producing excess ethanol and low H_2 in experimental fermentations. Because of effective degradation of various crystalline cellulose materials by *C. thermocellum*, the cellulase enzyme complexes or "cellulosomes" they produce have been extensively studied (Bayer and Lamed 1986; Kobayashi et al. 1990; Kohring et al. 1990; see Chapter 13, this volume). In order to take advantage of this capability for cellulose utilization, several mutant strains of *C. thermocellum* have been developed that somewhat improve the ratio of ethanol to acetate and lactate (see reviews, Doung et al. 1983; Rogers 1986). *C. thermocellum* does not produce more than 5 g of ethanol l^{-1} from cellulose, therefore strain improvement is aimed at selecting ethanol-tolerant ethanol hyperproducing strains. A mutant strain that produces 14.5 g ethanol l^{-1} of medium has been recently reported (Tailliez et al. 1989). Both batch and continuous culture conditions are being optimized for cellulosic biomass fermentation (Lynd et al. 1989). Thermophilic conversion of carbohydrates to ethanol by *C. thermosaccharolyticum* suffers from the same unfavorable low ratio of ethanol to acids as *C. thermocellum*. Improved mutant strains are being developed that are deficient in acetate production and produce two-times more ethanol than the parent strain (Rothstein 1986). Since *C. thermohydrosulfuricum* produces better than a 9:1 ratio of ethanol to acetate and lactate during fermentation of sugars, it would be a better candidate for high yield and high temperature conversion of sugars to ethanol. However, *C. thermohydrosulfuricum* cannot degrade cellulose. The biochemical or regulatory basis for the high ethanol and low acid product ratio of this organism remains unknown. Molecular genetic techniques for combining the favorable ethanol/acid ratio of *C. thermohydrosulfuricum* with the cellulose degradation capability by the cellulosome of *C. thermocellum* into one super organism are becoming available (see Chapters 5, 6, and 7, this volume). A second approach is also available, which uses the principle of symbiotic coculture. This method has been developed and used on an experimental level with cocultures of *C. thermocellum—C. thermohydrosulfuricum* (Ng et al. 1981), *C. thermocellum—C. thermosaccharolyticum* (Avgerinos 1982), and *C. thermocellum—Thermoanaerobacter ethanolicus* (Wiegel and Dykstra 1984). In all cases, the coculture produces a substantially higher yield of ethanol from cellulose than a monoculture of *C. thermocellum*. A metabolic explanation for the stable coculture of *C. thermocellum* and *C. thermohydrosulfuricum* was reported (Ng and Zeikus 1982). A balance develops between the two bacteria growing in the coculture. *C. thermohydrosulfuricum* depends upon the products, cellobiose and glucose, from the cellulase-complex degradation of cellulose that is produced by *C. thermocellum*. Glucose utilization by *C. thermocellum* is inhibited by cellobiose, and both cellobiose and glucose are more rapidly used by *C. thermohydrosulfuricum*. This results in a product yield distorted towards a high ratio of ethanol/acetate plus lactate. Recent example cocultures of this type (see Figure 2–8) yielded end-product ratios of ethanol/acetate of 10.5–4.3 compared to 1.8 or 1.9 for the monocultures (Mori 1990). It should be pointed out that starch is also an

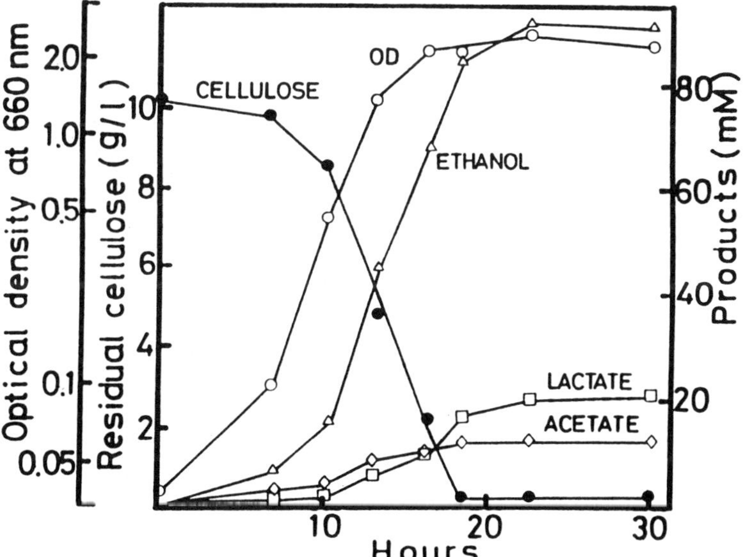

FIGURE 2–8 Fermentation profile of coculture of *C. thermohydrosulfuricum* YM3 and *C. thermocellum* YM4 on 1% Avicel. Reproduced with permission from Mori (1990).

important feed-stock for alcohol production in a thermophilic process. Here, cocultures are not necessary because *C. thermohydrosulfuricum* and related organisms (e.g., *C. thermosulfurogenes* secrete potent amylases and pullulanases (Antranikian 1990).

2.2.2 Butanol Fermentations

Only a few strains of the butyric acid-forming clostridia are able to produce *n*-butanol as a major fermentation product. These include: *C. acetobutylicum*, *C. aurantibutyricum*, *C. beijerinckii*, (syn. *C. butylicum*), and *C. tetanomorphum*. In addition to *n*-butanol, all of these strains produce butyrate, acetate, ethanol, and acetone or isopropanol, except that *C. tetanomorphum* cannot produce acetone or isopropanol but rather produces equal amounts of butanol and ethanol (Gottwald et al. 1984).There has been a rather remarkable increase in basic research effort focused on these bacteria since 1980, particularly using strains of *C. acetobutylicum* and *C. beijerinckii*. The major thrust of this work has been directed toward: (1) understanding the controls and functions of various branches and parts of the fermentation network; (2) development of genetic systems for DNA transfer and gene cloning; and (3) understanding of the molecular signals and mechanisms that initiate solvent formation. The early work up to 1988 on the biochemistry, genetics, and development of the fermentation process has been reviewed (Rogers 1986; Jones and Woods 1986; Bahl and Gottschalk 1988). The major metabolic pathways for the acetone-butanol fermentation have been pretty well formulated, and important physiologic controls of the switch to solvent formation are now tentatively identified. Figure 2–1 presents a summary of the major fermentation pathways used by *C. acetobutylicum* and *C. beijerinckii*.

During exponential growth of *C. acetobutylicum* on a variety of sugars and starch at pH values greater than about 5.6 (depending upon the strain) the major fermentation products are butyrate, acetate, CO_2, and H_2 (Figure 2–1, thick gray arrows). During this acidogenic phase these organisms are carrying out the butyrate-acetate fermentation described above. As in that case, the excess H_2 formed (usually 2.3–2.4 mol H_2 per mol glucose fermented) is derived from transfer of electrons from NADH through Fd:NADH oxidoreductase and hydrogenase to H_2 (Figure 2–1, reaction 1).

As acids accumulate in a batch culture, growth becomes linear and gradually stops. When the pH slips down to pH 4.5–4.0 (depending upon the strain) a classical shift in the fermentation occurs and solvent formation is triggered. Solvent formation requires the induction of new enzyme pathways in these cells (Figure 2–1, thick dark arrows) catalyzing the formation of butanol, acetone, and ethanol. The connection between accumulation of organic acids with a pH drop, the slowing of growth, and induction of solvent-forming enzymes will be addressed in the next section. At this same time, there is an uptake of butyrate and acetate from the broth back into the cells and their recycling and conversion to butanol and ethanol (Figure 2–1, thick dark arrows, reaction 9). These same events occur with asporogenous mutants of *C. acetobutylicum* grown in phosphate-limited two-stage chemostats where pH control is shifted to pH 4.3 to trigger solvent formation. Under these conditions these asporogenous mutants produce a yield of 0.34 g solvent per g glucose used, which duplicates the best batch cultures using sporulating strains (Bahl et al. 1982a). *C. beijerinckii* appears to follow the same general pathway of events leading to the onset of solvent production except that isopropanol is produced from the reduction of acetone (Yan et al. 1988).

Dependent upon the strain studied both of the previous clostridia use one or two aldehyde dehydrogenases and probably one or two alcohol dehydrogenases for butanol, ethanol, and isopropanol formation during solvent production (Hiu et al. 1987; Dürre et al. 1987; Palosaari and Rogers 1988; Andersch et al. 1983; Yan and Chen 1990; Bertram et al. 1990). Some of these dehydrogenases have been shown to be induced 40- to 200-fold at the fermentation shift due to new protein synthesis (e.g., Palosaari and Rogers 1988; Andersch et al. 1983) and they are responsible for alcohol production, (Figure 2–1, reactions 3, 5, and 6). It is clear from Figure 2–1 that the shift from net acetate and butyrate production to net butanol and ethanol formation results in an additional three changes: First, a reduction in acid production results in a lower ATP yield per glucose molecule fermented, since net butyrate and the acetate no longer accumulate which allowed production of an additional ATP from butyryl-phosphate and acetyl-phosphate during exponential growth (Figure 2–1, reactions 2 and 6); secondly, induction of the alcohol and aldehyde dehydrogenases requires 2 mol additional NADH oxidation per mol of alcohol produced; and thus thirdly, the amount of H_2 produced is reduced to about 1.4 mol H_2 per mol glucose used since additional electrons must be redirected from reduced ferredoxin to form NADH or NADPH (Figure 2–1, reaction 1). These changes in the fermentation energetics may be involved in the regulatory signals that promote the shift from acid to solvent production. A possible mechanism is covered in the next section.

It appears that the major pathway for acetone (and eventually isopropanol) synthesis requires two enzymes: Acetoacetyl-CoA:acyl-CoA transferase (ACT) and acetoacetate decarboxylase (Figure 2–1, reactions 9 and 10). These enzymes are also induced 50- to 100-fold in response to the shift in pH in *C. actetobutylicum* and *C. beijerinckii*; and they have been purified and characterized from both organisms (Wieseborn et al. 1989b; Yan et al. 1988).

What is the physiologic role of this pathway for these bacteria? Our best guess is that acetone (or isopropanol) is a by-product of a vital system for recycling of accumulated acids to form alcohols and thus detoxifying the surrounding environment of the organism. After these enzymes are induced, acetate and butyrate are taken up by the bacteria and ACT transfers the CoA from acetoacetyl-CoA to recycle the acids as acyl-CoA intermediates (Figure 2–1, reaction 9). The decarboxylase pulls the reaction 9 by forming CO_2 and acetone which then are excreted by the bacteria. Since the time courses of many fermentations show continued acetone synthesis both during a reduction in acid concentration and after there is no net change in acids in the growth medium, it appears that recycling of acids may continue throughout solvent formation. Consistent with this view is that both phosphotransbutyrylase (PTB) and butyrate kinase (BK) (Figure 2–1, reaction 6) are maintained at high activity during both acid- and solvent-phases of the fermentation (Wiesenborn 1989a; Hüsemann and Papoutsakis 1989; Andersch et al. 1983; Ballongue et al. 1986). There is 20 to 40 times the amount of PTB over butyraldehyde dehydrogenase (BAD) at maximal induction levels and the Km's for butyryl-CoA for these enzymes are about the same (0.045 mM for BAD and 0.065 or 0.11 mM for PTB) (Wiesenborn et al. 1989b; Palosaari and Rogers 1988). Apparently, butyric acid synthesis continues and organic acids in the medium are recycled to their respective alcohols throughout solvent formation (Hüsemann and Papoutsakis 1990; Jewell et al. 1986). Also, ^{14}C-butyrate is recycled to ^{14}C-butanol throughout solvent formation at a significant rate in *C. acetobutylicum*. This recycling most likely depends upon ACT activity, since *C. tetanomorphum*, which contains no ACT, and produces net butyrate throughout butanol formation, does not cycle ^{14}C-butyrate to ^{14}C-butanol (Palosaari 1990). It has been suggested that some recycling of acids to alcohols may occur via reversal of PTB and BK (e.g., Hüsemann and Papoutsakis 1990).

Using classical mutagenesis and selection techniques, several mutant strains of *C. acetobutylicum* have been reported that appear to show some promise for future strain development (Rogers 1986; Jones and Woods 1986; Bowring and Morris 1985). Noteworthy is the development of allyl alcohol-resistant mutants with an altered defective butanol dehydrogenase that produces significant butyraldehyde (bp 75°C) and lower amounts of butanol (bp 118°C). Production of butyraldehyde in place of butanol should be an advantage for a more cost-effective solvent recovery (Rogers and Palosaari 1987). Other mutants that produce no butanol (Dürre et al. 1986), little acetone (Junelles et al. 1987; Clark et al. 1989), or high levels of solvents with lower amounts of acid have also been reported (Rogers and Palosaari 1987; Cueto and Mendez 1990).

2.2.3 Other Solvent Products

The clostridia that form butanol, acetone, and isopropanol appear to be quite versatile in that dependent upon the carbon source and pH of the medium, they can produce several alternative solvents. For example, glycerol is fermented with production of 1,3-propanediol by 17 of 21 solventogenic strains of clostridia tested. During growth of *C. beijerinckii* (*C. butylicum*) in a chemostat culture at pH 6.5–6.6, 61% of the glycerol is converted to 1,3-propanediol (Forsberg 1987). *C. acetobutylicum* and *C. beijerinckii*, grown in a defined medium with a CO_2 gas phase and 0.2% Na HCO_3 convert rhamnose to propionic acid, 1,2-propanediol, and propanol as the major fermentation products (Forsberg et al. 1987). The biochemical mechanism for these conversions in clostridia have been proposed but not demonstrated (Forsberg 1987). Three strains of *C. thermosaccharolyticum* were found to produce up to 7.9 g l^{-1} of R(−)-1,2-propanediol with a best yield of 0.27 g g^{-1} glucose (Cameron and Cooney 1986). *C. sphenoides* normally produces ethanol, acetate, and lactate from fermentation of glucose. However, during extreme phosphate-limitation this fermentation yields as much as 34 mM D(−)-1,2-propanediol 100 mM^{-1} glucose (Tran-Din and Gottschalk 1985). It appears that a new pathway, the methylglyoxal by-pass, is induced by this organism during phosphate limitation.

A second strategy for unusual solvent production, using the principle of enzyme recruitment, is illustrated by the finding that *C. acetobutylicum* will convert several different organic acids through the recycle mechanism discussed above. For example, propionate, valerate, and 4-hydroxybutyrate were taken up and converted by this organism to their alcoholic derivitives, *n*-propanol, *n*-pentanol, and 1,4-butanediol (Jewell et al. 1986).

2.3 PHYSIOLOGIC CONTROL OF ACID AND SOLVENT FORMATION

To carry out an effective fermentation for solvents or organic acids, it is often only necessary to optimize the physiologic conditions in the process to ensure maximum yield of the desired products while preventing production of undesirable by-products. However, careful study and an understanding of the cellular regulatory networks governing the fermentation will increase our capability to develop new highly productive strains and control fermentation conditions more effectively. This approach is particularly important for developing many of the clostridial fermentations since branched fermentation pathways are typical, e.g., in *C. acetobutylicum* and *C. thermocellum*.

A batch culture of *C. acetobutylicum* in complex medium shifts from acid to solvent formation towards the end of growth. As mentioned earlier a continuous solvent-producing culture of *C. acetobutylicum* can be operated at low pH under phosphate limitation but also under other kinds of limitation with the substrate (glucose or another sugar) always in excess. Conditions that have been found to induce the shift are summarized in Table 2–2. It should be emphasized that the synthesis of the enzymes required for solvent formation is a prerequisite for the shift and that a shift does not occur if protein synthesis is inhibited.

It is apparent from Table 2–2 that solventogenesis can be induced: (1) by high intracellular concentrations of acids (production of acids with a concomittant decrease of the pH; addition of acids and uncouplers); or (2) by inhibition of H_2 evolution (addition of CO, H_2, or methylviologen). Under the condition of high intracellular acid concentrations the formation of both,

TABLE 2–2 Induction of Solventogenesis

Conditions				*Effects*	*References*
pH	Decrease during fermentation below 5.5			Increase of internal concentration of acids	Bahl et al. 1982
Butyrate	Addition	pH 7.0 pH 4.5	100 mM 20 mM	Increase of internal concentration of acids	Holt et al. 1984
Acetoacetate butyrate uncouplers	Addition	pH 4.5	8–12 mM 20 mM 5 μM	Induction (1–3 h after addition)	Hüsemann and Papoutsakis 1986
CO	Gassing	pH 4.5		Inhibition of hydrogenase, increase of butanol and ethanol yield	Kim et al. 1984
Methyl-viologen	Addition	pH 6.8 pH 5.0	0.1 g/l 0.1 g/l	Increase of ethanol production, induction, increase of acetone production	Rao and Mutharasan 1986
H_2	Increase of partial pressure			Fast induction	Doremus et al. 1985

acetone and butanol is switched on, whereas inhibition of H_2 evolution only results in butanol production (in addition to ethanol). The fastest shift is observed following the addition of methylviologen to a growing culture of *C. acetobutylicum* (Rao and Mutharasan 1986; Grupe 1991).

How the physiologic changes described above effect the generation of the signal for induction of solventogenic enzymes is not known. The transmembrane pH gradient or ΔpH (pH outside minus pH inside) is maintained during the shift (around 1 pH unit); it may even increase a little (Gottwald and Gottschalk 1985; Huang et al. 1986; Terraciano and Kashket 1986; Hüsemann and Papoutsakis 1988). Intracellular concentrations of acetic and butyric acids (undissociated forms) are in the order of 10–60 mM (Monot et al. 1984; Terraciano and Kashket 1986; Grupe 1991); a certain threshold concentration may be a prerequisite for the shift.

When the pH of a continuous culture of *C. acetobutylicum*, that is producing acetate and butyrate at pH 5.6, is allowed to fall to pH 4.3, there is a shift to acetone and butanol production. Just before solvent formation, the intracellular pool of ATP drops by one-half and the pool of NAD(P)H increases drastically (Grupe and Gottschalk 1992). However, adding methyl viologen to the continuous culture instead of a drop of the pH results in the rapid increase in the NAD(P)H pool and no change in ATP. The same result is noted after gasing with CO (Meyer and Papoutsakis 1989). Since these latter two treatments result in only butanol production but no acetone, a dual-signal model was proposed, in which the internal ATP level generates one signal and the NAD(P)H increase provides the other (Grupe and Gottschalk 1992).

2.4 REGULATORY MECHANISM FOR CLOSTRIDIAL FERMENTATIONS

As stated previously, for *C. acetobutylicum* the physiologic change that signals the shift to solventogenesis is most likely either a shift in the redox state of a component of the bioelectronic system of the bacterium, such as NADH or reduced Fd, or an increase of the intracellular concentration of butyric acid, butyrate, or H^+, or perhaps both. In this organism, the shift to solventogenesis is associated with at least seven rather striking events summarized in Table 2–3. Clearly, rather massive physiologic changes occur that involve induction and activation of several enzymes. At present, the complex of regulatory networks, proteins, and pathways that control these overlapping systems have not been worked out. For example, at this point, exactly which proteins receive the physiologic signals for triggering the above events in this bacterium are unknown. There are three general classes of intracellular molecular signaling systems in bacteria that appear to operate most regulatory networks required for stress control or survival under stress and adaptation

TABLE 2–3 Events Associated with Solventogenesis in *Clostridium acetobutylicum*[1]

Process	*Change with Switch to Solvent Formation*
Solvent-forming enzymes	20- to 100-fold induction of four or five enzymes
Acid recycling	Organic acids converted to alcohols
ATP generation	Reduced from about 3.3–2 mol per mol hexose
Growth: batch culture	Cell division stops, biomass increase slowed
Growth: continuous culture (phosphaphate limited)	Slow growth and cell division (asporogenous strain)
Granulose production (many strains)	Two enzymes induced
Clostridial stage	Cell becomes elongate and cigar shaped
Endospore formation	Morphogenesis program, may involve 100 induced/activated proteins

[1] The physiologic and mutant-strain evidence that links these processes has been summarized in reviews (Jones and Woods 1986; Rogers 1986).

to environmental change. These are: (1) signal transduction via protein phosphorylation (Stock et al. 1989); (2) small molecule effectors such as cAMP (Saier 1989) or ppGpp (Cashel and Rudd 1987); and (3) sigma-factor control of promoter specificity by RNA polymerase (Doi and Wang 1986). Some experimental studies briefly outlined here, begun primarily with *C. acetobutylicum*, encourages us to examine three possible models for how regulation of these fermentation pathways might occur.

2.4.1 Signal Transduction Pathway

In many bacterial species, adaptive responses to stress or nutritional shifts are controlled by a signal transduction system made up of two components. A protein kinase (PK), that serves as the sensor or modulator protein, autophosphorylates itself in response to an environmental signal (Stock et al. 1989). The phosphokinase can transfer the phosphate group to a response regulator protein (RR) that then can bind to the control regions of specific genes on the bacterial DNA. Specific sets of enzymes are synthesized and adaptation is complete. The simplest model of this type for switching from an acid fermentation to solvent production in response to an intracellular metabolite signal is shown by the following:

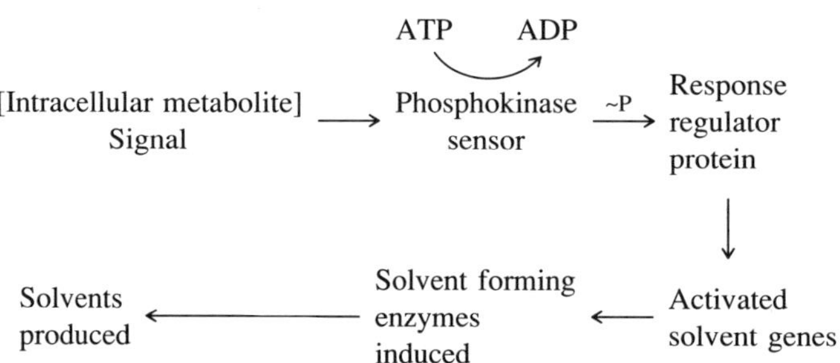

It has been found that in some signal transduction pathways there can be more than one signal and PKs and a series of RRs forming branches to more than one set of genes controlling different processes (Stock et al. 1989). For example, initiation of sporulation in *B. subtilis* is controlled by a multicomponent phosphorylation set of factors (Burbulys et al. 1991).

In support of this model, phosphorylation of proteins has been demonstrated in *C. sphenoides*, *C. thermohydrosulfuricum*, and *C. acetobutylicum* (Antranikian et al. 1985; Londesborough 1986; Balodimos et al. 1990). In the latter two organisms, phosphokinase activity that phosphorylates specific proteins was demonstrated. More specifically, purified clostridial DnaK, a stress protein, acts as a protein kinase that phosphorylates a single 50-kilodalton (kDa) clostridial protein (Balodimos et al. 1990). No specific sensor function or response regulator function have been assigned to any of these proteins. Also, the signal transduction pathway in *C. acetobutylicum* may be highly branched to accommodate control of the various processes responding with the switch of solventogenesis, as shown in Table 2–3.

2.4.2 A Small Molecule Effector

C. acetobutylicum accumulates diadenosine-5′,5‴-P^1-P^4-tetraphosphate (Ap_4A) and adenosine 5′-P^1,P^4-tetraphospho-5′guanosine (Ap_4G) in response to stress by temperature upshift or added butanol (Balodimos et al. 1988). Unfortunately, no change in these nucleotides occurred prior to solvent formation. These nucleotides may be intracellular stress signals for some stress situations in *E. coli*, but their role is unclear (Van Bogelen et al. 1987). Since initiation of sporulation of *B. subtilis* and *Saccharomyces cerevisiae* and differentiation of aerial mycelium in *Streptomyces* is associated with the decrease in the GTP pool, the role of GTP in differentiation and shift to solventogenesis of *C. acetobutylicum* was investigated (Santangelo et al. 1989). A dip in the pool of all nucleoside triphosphates was found at the pH break point in batch cultures; and a similar drop in ATP was detected in continuous culture during a controlled pH drop causing the shift to solvent formation (Grupe and Gottschalk 1990). Inhibitors causing a reduction in ribonucleotide level in other organisms also caused a significant increase in endospore formation. At this point, an association of specific nucleotides with signals for the solvent shift or morphogenesis in the clostridia is not clear, but this model for regulation may deserve further investigations.

2.4.3 A Stress-related Sigma Factor

A working model for how induction of solvent pathways is controlled and how this system interacts with induction of sporulation and other pathways (Table 2–3) can be based on recent research with stress-systems in *E. coli* and sporulation in *B. subtilis*. Different sigma factors (σ, the subunit of RNA polymerase that determines promotor specificity) allow expression of large families of genes in these bacteria, control adaptation to stress, nutrient deprivation, and trigger endospore formation (Doi and Wang 1986). For example, in response to heat or ethanol shock, *E. coli* induces synthesis of 17 proteins using a σ^{32}-RNA polymerase rather than the σ^{70}-RNA polymerase used for normal housekeeping functions (Neidhardt and van Bogelen 1987). *B. subtilis* uses an array of five temporally regulated sigma factors to signal expression of the 100 or so proteins needed for endospore formation (Helmann and Chamberlin 1988). If *C. acetobutylicum* switches to solvent formation as part of a global stress response, then proteins induced by heat-shock stress may overlap with proteins induced at the time the pH is low. *C. acetobutylicum* does produce 11 to 15 heat shock proteins (hsp) when exposed to a brief temperature upshift, exposure to 0.1% *n*-butanol, or to air (Terracciano et al. 1988; Pich et al. 1990). In continuous cultures shifted to a low pH five hsps appeared, including a DnaK-like and a GroEL-like protein, coincident with the change from acid to solvent formation (see Chapter 11, this volume for a detailed analysis). The relationship between the regulatory systems of endospore formation and solventogenesis in *C. acetobutylicum* has been derived from mutant selection studies. There are a set of sporulation-negative (Spo^-) mutants that are unable to produce spores, granulose, clostridial forms, or induce any of the terminal solvent-forming enzymes (Jones et al. 1982; Rogers and Palosaari 1987). However, some Spo^- mutants produce solvents normally and some mutants that produce no solvents are Spo^+. Thus, it is clear that although regulation of these two processes overlap they probably form two branches of a global regulatory network. The formation of transposon-induced regulatory mutants of *C. acetobutylicum* and investigation of the classes of these mutants will serve to identify the number and perhaps order of the regulatory elements controlling this network of physiologic processes listed in Table 2–3 (see Chapter 10, this volume for details of this work).

Although research is moving forward rather rapidly in our understanding of the regulatory events controlling the *C. acetobutylicum* fermentation, almost no data are available on the molecular systems controlling the branched pathways in the thermophilic ethanolic clostridia or the acid-forming clostridia. Future genetic and biochemical analysis of these organisms remains an essential part of development of strains for commercial fermentations.

REFERENCES

Akedo, M., Cooney, C.L., and Sinskey, A.J. (1983) *Bio/Technology* 2, 791–794.

Andersch, W., Bahl, H., and Gottschalk, G. (1983) *Appl. Microbiol. Biotechnol.* 18, 327–332.

Andreesen, J.R., Schaupp, A., Neurauter, C., Brown, A., and Ljungdahl, L.G. (1973) *J. Bacteriol.* 114, 743–751.

Andreesen, J.R., Bahl, H., and Gottschalk, G. (1989) in *Clostridia* (Minton, N.P., and Clarke, D.J., eds.), pp. 27–62, Plenum Press, New York.

Antranikian, G., Herzberg, C., and Gottschalk, G. (1985) *FEMS Microbiol. Lett.* 27, 135–138.

Antranikian, G. (1990) *FEMS Microbiol. Rev.* 75, 201–218.

Avgerinos, G.C. (1982) Ph.D. thesis, MIT, Cambridge, MA.

Bahl, H., Andersch, W., Braun, K., Gottschalk, G. (1982a) *Appl. Microbiol. Biotechnol.* 14, 17–20.

Bahl, H., Andersch, W., and Gottschalk, G. (1982b) *Appl. Microbiol. Biotechnol.* 15, 201–205.

Bahl, H., and Gottschalk, G. (1988) in *Biotechnology* (Rehm, H.-J. and Reed, G. eds.), vol. 6b. pp. 1–30, VCH Verlagsgesellschaft, Weinheim, Germany.

Ballongue, J., Amine, J., Petitdemange, H., and Gay, R. (1986) *FEMS Microbiol. Lett.* 35, 295–301.

Balodimos, I.A., Kashket, E.R., and Rapaport, E. (1988) *J. Bacteriol.* 170, 2301–2305.

Balodimos, I.A., Rapaport, E., and Kashket, E.R. (1990) *Appl. Environ. Microbiol.* 56, 2170–2173.

Baronofsky, J.J., Schreurs, W.J.A., and Kashket, E.R. (1984) *Appl. Environ Microbiol.* 48, 1134–1139.

Bayer, E.A., and Lamed, R. (1986) *J. Bacteriol.* 167, 828–836.

Bertram, J., Kuhn, A., and Dürre, P. (1990) *Arch. Microbiol.* 153, 373–377.

Bornstein, B.T., and Barker, H.A. (1948) *J. Biol. Chem.* 172, 659–669.

Bowring, S.N., and Morris, J.G. (1985) *J. Appl. Bacteriol.* 58, 577–584.

Burbulys, D., Trach, K.A., and Hoch, J.A. (1991) *Cell* 64, 545–552.

Cameron, D.C., and Cooney, C.L. (1986) *Bio/Technology* 4, 651–654.

Cardon, B.P., and Barker, H.A. (1947) *Arch. Biochem.* 12, 165–180.

Cashel, M., and Rudd, K.E. (1987) in *Escherichia coli and Salmonella typhimurium, Cellular and Molecular Biology* (Neidhardt, F.C., ed.), vol. 2, pp. 1410–1438, American Society for Microbiology, Washington, DC.

Cato, E.P., George, W.L., and Finegold, S.M. (1986) in *Bergey's Manual of Systematic Bacteriology* (Sneath, H.A., Mair, N.S., Sharpe, M.E., and Holt, J.G., eds.), vol. 2, pp. 1141–1200, The Williams & Wilkins Co., Baltimore.

Clark, S.W., Bennett, G.N., and Rudolph, F.B. (1989) *Appl. Environ. Microbiol.* 55, 970–976.

Cueto, P.H., and Mendez, B.S. (1990) *Appl. Environ. Microbiol.* 56, 578–580.

Das, A., Hugenholtz, J., Yan Halbeek, H., and Ljungdahl, L.G. (1989) *J. Bacteriol.* 171, 5823–5829.

Doi, R.H., and Wang, L.-F. (1986) *Microbiol. Rev.* 50, 227–243.

Doremus, M.G., Linen, J.C., and Moreira, A.R. (1985) *Biotechnol. Bioeng.* 27, 852–860.

Doung, T.V.C., Johnson, E.A., and Demain, A.L. (1983) in *Topics in Enzyme and Fermentation Biotechnology* (Weisman, A., ed.), vol. 7, pp. 156–195, John Wiley & Sons, New York.

Dürre, P., Kuhn, A., and Gottschalk, G. (1986) *FEMS Lett.* 36, 77–81.

Dürre, P., Kuhn, A., Gottwald, M., and Gottschalk, G. (1987) *Appl. Microbiol. Biotechnol.* 26, 268–272.

Forsberg, C.W. (1987) *Appl. Environ. Microbiol.* 53, 639–643.

Forsberg, C.W., Donaldson, L., and Gibbins, L.N. (1987) *Can. J. Microbiol.* 33, 21–26.

Fuchs, G. (1986) *FEMS Microbiol. Rev.* 39, 181–213.

Gottschalk, G. (1986) *Bacterial Metabolism,* pp. 208–282. Springer-Verlag, New York.

Gottwald, M., Andreesen, J.R., Le Gall, J., and Ljungdahl, L.G. (1975) *J. Bacteriol.* 122, 325–328.

Gottwald, M., and Gottschalk, G. (1985) *Arch. Microbiol.* 143, 42–46.

Gottwald, M., Hippe, H., and Gottschalk, G. (1984) *Appl. Environ. Microbiol.* 48, 573–576.

Grupe, H. (1991) Doctoral thesis, University of Göttingen, Göttingen, Germany.

Grupe, H., and Gottschalk, G. (1990) in *Proceedings of the Sixth International Symposium on Genetics of Industrial Microorganisms*, (Heslot, T., Davies, J., Foreat, J., et al. eds.), vol. 2, pp. 715–729, Sociéte Francaise de Microbiologie, Strasbourg.

Grupe, H., and Gottschalk, G. (1992) *Appl. Environ. Microbiol.* 58, 3896–3902.

Heise, R., Müller, V., and Gottschalk, G. (1989) *J. Bacteriol.* 171, 5473–5478.

Helmann, J.D., and Chamberlin, M.J. (1988) *Annu. Rev. Biochem.* 57, 839–872.

Hippe, H., Andreesen, J.R., and Gottschalk, G. (1991) in *The Prokaryotes*, vol. 2 (Balows, A., Harder, W., Schleifer, K.-H., and Trüper, H.G., eds.), pp. 1800–1866, Springer-Verlag, New York.

Hiu, S.F., Zhu, C.-F., Yan, R.-T., and Chen, J.-S. (1987) *Appl. Environ. Microbiol.* 53, 697–703.

Holt, R.A., Stephens, G.M., and Morris, J.G. (1984) *Appl. Environ. Microbiol.* 48, 1166–1170.

Huang, L., Forsberg, C.W., and Gibbins, L.N. (1986) *Appl. Environ. Microbiol.* 51, 1230–1234.

Hüsemann, M., and Papoutsakis, E.T. (1986) *Biotechnol. Lett.* 8, 37–42.

Hüsemann, M., and Papoutsaki, E.T. (1988) *Biotechnol. Bioeng.* 32, 843–852.

Hüsemann, M.H.W., and Papoutsakis, E.T. (1989) *Appl. Microbiol. Biotechnol.* 30, 585–595.

Hüsemann, M.H.W., and Papoutsakis, E.T. (1990) *Appl. Environ. Microbiol.* 56, 1497–1500.

Hugenholtz, J., Ivey, D.M., and Ljungdahl, L.G. (1987) *J. Bacteriol.* 169, 5845–5847.

Jewell, J.B., Coutinho, J.B., and Kropinski, A.M. (1986) *Curr. Microbiol.* 13, 215–219.

Jones, D.T., van der Westhuizen, A., Long, S., Allcock, E.R., Reid, S.R., and Woods, D.R. (1982) *Appl. Environ. Microbiol.* 43, 1434–1439.

Jones, D.T., and Woods, D.R. (1986) *Microbiol. Rev.* 50, 484–524.

Junelles, A.-M., Janati-Idrissi, R., Kanouni, A.E., Petitdemange, H., and Gay, R. (1987) *Biotechnol. Lett.* 9, 175–178.

Jungermann, K., Thauer, R.K., Leimenstoll, G., and Decker, K.L. (1973) *Biochim. Biophys. Acta* 305, 268–280.

Kenealy, W.R., and Waselefsky, D.M. (1985) *Arch. Microbiol.* 141, 187–194.

Kim, B.H., Bellows, P., Datta, R., and Zeikus, J.G. (1984) *Appl. Environ. Microbiol.* 48, 764–770.

Kobayashi, T., Romaniec, M.P.M., Fauth, V., and Demain, A.L. (1990) *Appl. Environ. Microbiol.* 56, 3040–3046.

Kohring, S., Wiegel, J., and Mayer, F. (1990) *Appl. Environ. Microbiol.* 56, 3798–3804.

Kuchta, R.D., and Abeles, R.H. (1985) *J. Biol. Chem.* 260, 13181–13189.

Lamed, R., and Zeikus, J.G. (1980) *J. Bacteriol.* 144, 569–578.

Le Ruyet, P., Dubourguier, H.C., Albagnac, G., and Prensier, G. (1985) *Syst. Appl. Microbiol.* 6, 196–202.

Linden, J.C., and Moreira, A.R. (1983) in *Biological Basis of New Developments in Biotechnology* (Hollaender, A., Laskin, A.I., and Rogers, P., eds.), pp. 377–404, Plenum Press, New York.

Ljungdahl, L.G., Carreira, L.H., Garrison, R.J., Rabek, N.E., and Wiegel, J. (1985) *Biotechnol. Bioeng. Symp.* 15, 207–223.

Ljungdahl, L.G., Carreira, L.H., Garrison, R.J. et al. (1986) Federal Highway Administration Rep. no. FHWA/RD-86/117, Natl. Tech. Infor. Serv., Springfield, VA.

Ljungdahl, L.G., Hugenholtz, J., and Wiegel, J. (1989) in *Clostridia* (Minton, N.P., and Clarke, D.J., eds.), pp. 145–191, Plenum Press, New York.

Londesborough, J. (1986) *J. Bacteriol.* 165, 595–601.

Lynd, L.R. (1989) in *Advances in Biochemical Engineering/Biotechnology* (Fiechter, A., ed.), vol. 38, pp. 1–52, Springer-Verlag KG, Berlin.

Lynd, L.R., Grethlein, H.E., and Wolkin, R.H. (1989) *Appl. Environ. Microbiol.* 55, 3131–3139.
Mayer, F., Ivey, D.M., and Ljungdahl, L.G. (1986) *J. Bacteriol.* 166, 1128–1130.
Menzel, U., and Gottschalk, G. (1985) *Arch. Microbiol.* 143, 47–51.
Meyer, C.L., and Papoutsakis, E.T. (1989) *Appl. Microbiol. Biotechnol.* 30, 450–459.
Michel-Savin, D., Marchal, R., and Vandecasteele, J.P. (1990) *Appl. Microbiol. Biotechnol.* 33, 127–131.
Monot, F., Engasser, J.M., and Petitdemange, H. (1984) *Appl. Microbiol. Biotechnol.* 19, 422–426.
Mori, Y. (1990) *Appl. Environ. Microbiol.* 56, 37–42.
Neidhardt, F.C., and van Bogelen, R.A. (1987) in *Escherichia coli and Salmonella typhimurium, Cellular and Molecular Biology* (Neidhardt, F.C., ed.), vol. 2, pp. 1334–1345, American Society for Microbiology, Washington, DC.
Ng, T.K., Ben-Bassat, A., and Zeikus, J.G. (1981) *Appl. Environ. Microbiol.* 41, 1337–1343.
Ng, T.K., and Zeikus, J.G. (1982) *J. Bacteriol.* 150, 1391–1399.
Palosaari, N.R. (1990) Ph.D. thesis, University of Minnesota, Minneapolis.
Palosaari, N.R., and Rogers, P. (1988) *J. Bacteriol.* 170, 2971–2976.
Parekh, S.R., and Cheryan, M. (1990) *Process Biochem. Int.* 25, 117–121.
Phillips, J.A., and Humphrey, A.E. (1983) in *Organic Chemicals from Biomass* (Wise, D.L., ed.), pp. 249–403, Benjamin/Cummings, Menlo Park, CA.
Pich, A., Narberhaus, F., and Bahl, H. (1990) *Appl. Microbiol. Biotechnol.* 33, 697–704.
Prescott, S.C., and Dunn, C.G. (1959) *Industrial Microbiology*, 3rd ed., pp. 240–284, McGraw-Hill Book Co., New York.
Rao, G., and Mutharasan, R. (1986) *Biotechnol. Lett.* 8, 893–896.
Rogers, P. (1986) *Adv. Appl. Microbiol.* 31, 1–60.
Rogers, P., and Palosaari, N.R. (1987) *Appl. Environ. Microbiol.* 53, 2761–2766.
Rothstein, D.M. (1986) *J. Bacteriol.* 165, 319–320.
Saier, M.H. (1989) *Microbiol. Rev.* 53, 109–120.
Saha, B.C., Lamed, R., and Zeikus, J.G. (1989) in *Clostridia* (Minton, N.P., and Clarke, D.J., eds.), pp. 227–263, Plenum Press, New York.
Santangelo, J.D., Jones, D.T., and Woods, D.R. (1989) *J. Gen. Microbiol.* 135, 711–719.
Schweiger, G., Dutscho, R., and Buckel, W. (1987) *Eur. J. Biochem.* 169, 441–448.
Slapack, G.E., Russell, I., and Stewart, G.G. (1987) *Thermophilic Microbes in Ethanol Production*, CRC Press, Boca Raton, FL.
Stock, J.B., Ninfa, A.J., and Stock, A.M. (1989) *Microbiol. Rev.* 53, 450–490.
Tailliez, P., Gireard, H., Millet, J., and Beguin, P. (1989) *Appl. Environ. Microbiol.* 55, 207–211.
Tang, I.-C., Yang, S.-T., and Okos, M.R. (1988) *Appl. Microbiol. Biotechnol.* 28, 138–143.
Terracciano, J.S., and Kashket, E.R. (1986) *Appl. Environ. Microbiol.* 52, 86–91.
Terracciano, J.S., Rapaport, F., and Kashket, E.R. (1988) *Appl. Environ. Microbiol.* 54, 1989–1995.
Thauer, R.K., Jungermann, K., and Decker, K. (1977) *Bacteriol. Rev.* 41, 100–180.
Tran-Din, K., and Gottschalk, G. (1985) *Arch. Microbiol.* 142, 87–92.
van Andel, J.G., Zoutberg, G.R., Crabbendam, P.M., and Breure, A.M.C. (1985) *Appl. Microbiol. Biotechnol.* 23, 21–26.

van Bogelen, R.A., Kelley, P.M., and Neidhardt, F.C. (1987) *J. Bacteriol.* 169, 26–32.

Wang, G., and Wang, D.J.C. (1984) *Appl. Environ. Microbiol.* 47, 294–298.

Wiegel, J., and Dykstra, M. (1984) *Appl. Environ. Biotechnol.* 20, 59–65.

Wiegel, J., and Ljungdahl, L.G. (1986) *Crit. Rev. Biotechnol.* 3, 39–107.

Wiesenborn, D.P., Rudolph, F.B., and Papoutsakis, E.T. (1989a) *Appl. Environ. Microbiol.* 55, 317–322.

Wiesenborn, D.P., Rudolph, F.B., and Papoutsakis, E.T. (1989b) *Appl. Environ. Microbiol.* 55, 323–329.

Willadsen, P., and Buckel, W. (1990) *FEMS Microbiol. Lett.* 70, 187–192.

Wood, H.G., Ragsdale, S.W., and Pezacka, E. (1986) *FEMS Microbiol. Rev.* 39, 345–362.

Wood, W.A. (1961) in *The Bacteria: A Treatise on Structure and Function* (Gunsalus, I.C., and Stanier, R.Y., eds.), vol. 2, pp. 59–150, Academic Press, New York.

Yan, R.-T., and Chen, J.-S. (1990) *Appl. Environ. Microbiol.* 56, 2591–2599.

Yan, R.-T., Zhu, C.-X., Golemboski, C., and Chen, J.-S. (1988) *Appl. Environ. Microbiol.* 54, 642–648.

CHAPTER 3

Properties of Acid- and Solvent-Forming Enzymes of Clostridia

Jiann-Shin Chen

Acetic and butyric acids are common products of anaerobic bacteria, and several distinct metabolic pathways are used by different groups of anaerobes for the synthesis of the two acids (Chen 1987). Within the genus *Clostridium*, species belonging to the butyric group (represented by *C. butyricum*) produce acetate and butyrate via reactions 1 through 9 of Figure 3–1. The solvent-producing clostridia also use these reactions for the synthesis of acetate and butyrate.

The production of ethanol, acetone, isopropanol, and butanol by clostridia requires the participation of three additional pathways (consisting of reactions 10 through 16 in Figure 3–1) which branch off from three metabolic intermediates (acetyl-CoA, acetoacetyl-CoA, and butyryl-CoA) that are also precursors for acetic and butyric acids. The acid- and solvent-producing pathways thus share a sequence of reactions (besides the glycolytic reactions) spanning between pyruvate and butyryl-CoA.

Research performed in the author's laboratory was supported by grants from the U.S. Department of Energy (Division of Energy Biosciences), NESTE Oy, and the Center for Innovative Technology of the Commonwealth of Virginia.

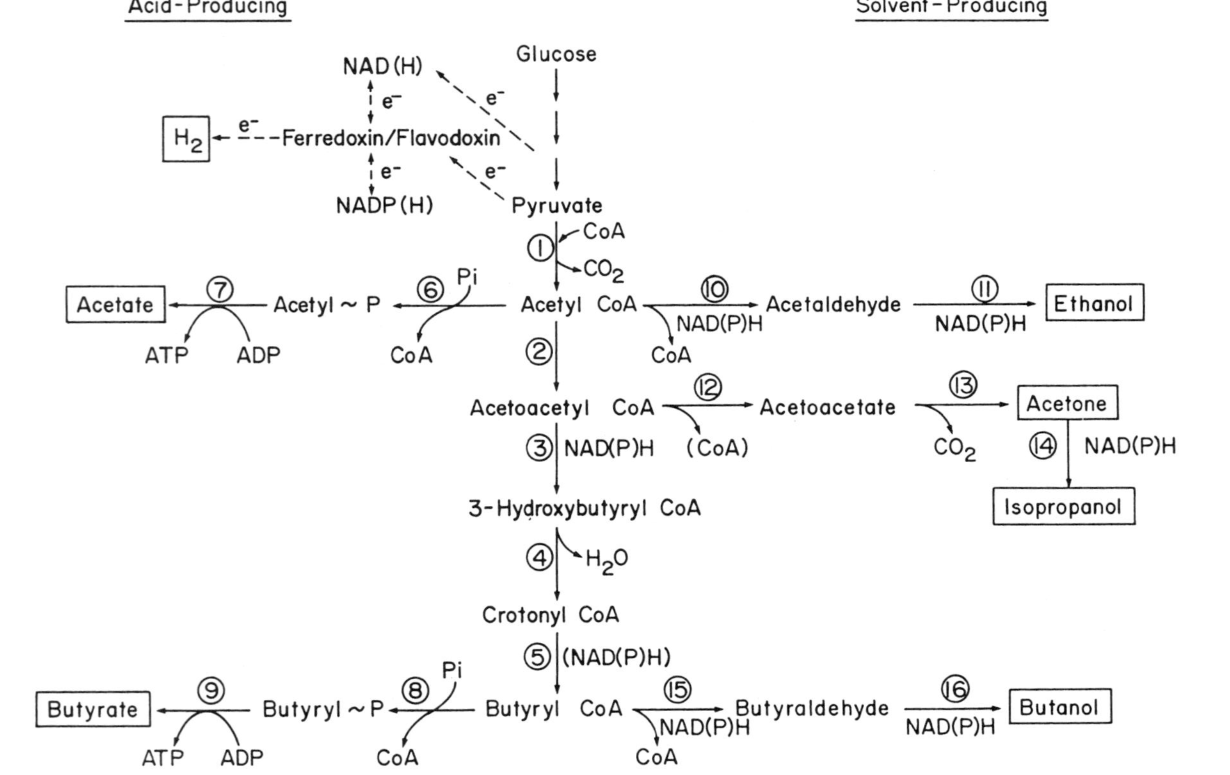

FIGURE 3–1 Metabolic pathway of acid and solvent production in clostridia. Enzymes catalyzing the numbered reactions are listed in Table 3–1. *C. beijerinckii* NRRL B593 carries out all of the reactions depicted in the figure. Other solvent-producing species may carry out all or some of the reactions.

For this chapter, the discussion begins with the enzyme catalyzing the oxidative decarboxylation of pyruvate (reaction 1 in Figure 3–1). The metabolic pathways (encompassing reactions 1 through 16) depicted in Figure 3–1 generally follow those outlined by Doelle (1975) and Gottschalk (1986). Most of these steps are now established through the measurement of pertinent enzyme activities (Andersch et al. 1983; Hartmanis and Gatenbeck 1984; Yan et al. 1988; Huesemann and Papoutsakis 1989) as well as the purification of enzymes catalyzing the reactions (see Sections 3.1–3.3). Nevertheless, a thorough understanding of each reaction and the interrelationship among the reactions will await further characterization of the enzymes involved and will also require genetic analyses of alternative enzymes and the control mechanisms. For instance, alternative alcohol dehydrogenases with different substrate and coenzyme specificities may be involved in the formation of butanol and ethanol. Biochemical and physiological studies with mutants defective in specific genes should indicate the relative importance of alternative enzymes, which will bear on the control of the product pattern via changes in the flow of carbon and reducing power.

The acid- and solvent-forming pathways are temporally regulated and selectively activated so that the product pattern can vary with growth stages or growth conditions (Jones and Woods 1986; Rogers 1986). The branched nature and the presence of alternative enzymes for analogous reactions of the pathways make the control of solvent production an interesting topic of physiological research. The results of such studies are of practical importance as the product pattern of industrial solvent fermentation can then be rationally manipulated to respond to market needs.

Manipulation of the product pattern and the yield of solvent fermentation requires amplification or suppression of enzymes catalyzing crucial reactions. An understanding of the properties of the acid- and solvent-forming enzymes leads to the elucidation of the control mechanisms for the reactions of the alternative pathways and the development of strategies for more precise control of the fermentation.

Enzymes catalyzing the acid- and solvent-forming reactions have been isolated from *C. acetobutylicum* or *C. beijerinckii* (including strains formerly known as "*C. butylicum*" and some strains formerly labeled as *C. butyricum*), and several of them have been highly purified over the past several years. Enzymes catalyzing similar reactions have also been isolated from other *Clostridium* spp. In this chapter, the focus will be on enzymes purified from *C. acetobutylicum* and *C. beijerinckii*. Table 3–1 is a listing of acid- and solvent-forming enzymes catalyzing reactions 1 through 16 of Figure 3–1. These enzymes will be discussed in the following order: (1) enzymes catalyzing the central reactions (1 through 5 in Figure 3–1) leading to both acid and solvent formation, (2) enzymes catalyzing specific acid-forming reactions (6 through 9 in Figure 3–1), and (3) enzymes catalyzing specific solvent-forming reactions (10 through 16 in Figure 3–1).

TABLE 3–1 Acid- and Solvent-Forming Enzymes from *C. acetobutylicum* (A) and *C. beijerinckii* (B)

Enzyme[1]	*Source*	*Subunit Mol. Wt.*	*Protein Purified (P) or Gene Cloned (G)*
Pyruvate: Fd oxidoreductase (1)	A	123,000	P
Thiolase (2)	A	44,000	P, G
3-OH-butyryl-CoA DH (3)	A	31,435	G
	B	30,800	P
Crotonase (4)	A	40,000	P
Butyryl-CoA DH (5)		ND[2]	
PTA (6)	B	ND	P
Acetate kinase (7)		ND	
PTB (8)	A	31,000	P, G
	B	33,000	P
Butyrate kinase (9)	A	39,000	P, G
CoA-transferase (12)	A	23,000; 25,000	P, G
	B	(23,000; 28,000)	P
AcAc decarboxylase (13)	A	27,519	P, G
ALDH (10, 15)	A	56,000	P
	B	55,000	P
ADH (11, 14, 16)	A	42,000 [NAD(P)H]	P
	A	43,274 [NADPH]	G
	B	38,000 [NADPH]	P, G
	B	40–43,500 [NAD(P)H]	P

[1] The number in parentheses refers to the reaction (see Figure 3–1) catalyzed by the enzyme. Abbreviation and key references are as follows: pyruvate:Fd oxidoreductase, pyruvate:ferredoxin oxidareductase (Meinecke et al. 1989); thiolase (Wiesenborn et al. 1988); 3-OH-butyryl-CoA DH, 3-hydroxybutyryl-CoA dehydrogenase (Youngleson et al. 1989a; Colby and Chen 1992); crotonase (Waterson et al. 1972); butyryl-CoA DH, butyryl-CoA dehydrogenase (Hartmanis and Gatenbeck 1984); PTA, phosphotransacetylase (Thompson and Chen 1990); acetate kinase (Huesemann and Papoutsakis 1989); PTB, phosphotransbutyrylase (Wiesenborn et al. 1989a; Thompson and Chen 1990); butyrate kinase (Hartmanis 1987); CoA-transferase, acetoacetyl-CoA:acetate/butyrate CoA-transferase (Wiesenborn et al. 1989b; Gerischer and Duerre 1990); AcAc decarboxylase, acetoacetate decarboxylase (Fridovich 1972; Gerischer and Duerre 1990); ALDH, aldehyde dehydrogenase (Palosaari and Rogers 1988; Yan and Chen 1990); ADH, alcohol dehydrogenase (Welch et al. 1989; Youngleson et al. 1988; Hiu et al. 1987; Yan, Zhu, and J.-S. Chen, unpublished observations).
[2] ND, not determined.

3.1 ENZYMES CATALYZING REACTIONS LINKED TO BOTH ACID AND SOLVENT FORMATION

Enzymes catalyzing reactions 1 through 5 (see Figure 3–1) are as follows: (1) pyruvate:ferredoxin (Fd) oxidoreductase, (2) thiolase, (3) 3-hydroxybutyryl-

CoA dehydrogenase, (4) crotonase, and (5) butyryl-CoA dehydrogenase. It is presumed that one and the same enzyme is responsible for each reaction under both acid- and solvent-producing conditions. However, it remains to be established whether the same form of an enzyme is involved under all conditions or whether an association between sequential enzymes (Srere 1987; Srivastava and Bernhard 1987), such as between enzymes for reactions 2 and 3 or between enzymes for reactions 2 and 12, occurs in the cell to regulate the flow of a metabolic intermediate into alternative pathways.

3.1.1 Pyruvate:Ferredoxin Oxidoreductase

Pyruvate:Fd oxidoreductae (EC 1.2.7.1) catalyzes the following reaction:

$$\text{Pyruvate} + \text{CoASH} + \text{Fd (oxidized)} \rightleftharpoons \text{acetyl-CoA} + CO_2 + \text{Fd (reduced)}$$

The reaction requires the participation of enzyme-bound thiamine pyrophosphate (TPP). The use of Fd or flavodoxin as the electron acceptor by the pyruvate-oxidizing enzyme is a distinct feature of anaerobes (Kerscher and Oesterhelt 1982). The coupling of pyruvate oxidation to the reduction of electron carriers of such low redox potentials enables the anaerobe to use hydrogenase to dispose of excess reducing power in the form of hydrogen gas (Adams et al. 1981). The use of protons as a terminal electron acceptor lowers the production of reduced compounds, such as alcohols, during the fermentation.

Pyruvate:Fd oxidoreductase has been purified from *C. acetobutylicum* DSM 1731 (Meinecke et al. 1989). The purification resulted in a 9.2-fold increase in specific activity with a yield of 5.2%. The purified enzyme can use Fd of *C. pasteurianum* as an electron acceptor. The apparent K_m for pyruvate and CoASH are, respectively, 322 and 3.7 μM. It contained 4.23 atoms of iron, 0.91 mol of inorganic sulfide, and 0.39 mol of TPP per subunit (M_r 123,000). The native molecular weight of the enzyme has not been determined, but a dimeric structure was inferred from a comparison with pyruvate oxidoreductases from other sources.

Like similar enzymes purified from other sources, the enzyme from *C. acetobutylicum* was sensitive to oxygen and dilution. About 80% of activity was lost during chromatography on Phenyl Sepharose, probably due to the loss of TPP and the iron-sulfur cluster from the enzyme. This significant activity loss points to the possibility that the purified enzyme contained inactive apoprotein and hence the fold of purification of the enzymic protein was underestimated (otherwise pyruvate oxidoreductase would constitute about 10% of total protein in crude extracts of *C. acetobutylicum*).

Pyruvate:Fd/flavodoxin oxidoreductases purified from *C. thermoaceticum* and *Klebsiella pneumoniae* are similar enzymes with a subunit molecular weight of 120,000–125,000 and contain about 8 atoms of iron and 5.5–6.5

mol of inorganic sulfide per subunit (Wahl and Orme-Johnson 1987). The pyruvate oxidoreductase of *K. pneumoniae* is encoded by the *nifJ* gene, and the deduced polypeptide from *nifJ* has a molecular weight of about 128,000 (Arnold et al. 1988; Cannon et al. 1988). That the 123,000-Da subunit of the *C. acetobutylicum* enzyme did not further dissociate (see Meinecke et al. 1989) suggests that the subunit of pyruvate:Fd/flavodoxin oxidoreductase is a conserved, relatively large polypeptide.

Ferredoxin and flavodoxin have been purified from several clostridia (Yasunobu and Tanaka 1973; Mayhew and Ludwig 1975), but a Fd has not yet been purified from *C. acetobutylicum*. Recently, the presence of flavodoxin in *C. acetobutylicum* was indicated (Santangelo et al. 1991). It remains to be shown whether the pyruvate oxidoreductase of *C. acetobutylicum* uses Fd or flavodoxin as the electron acceptor in vivo and whether the type of low potential electron carrier synthesized in this species is regulated by Fe concentrations in the growth medium (Mayhew and Ludwig 1975; Schoenheit et al. 1979).

The presence of Fd/flavodoxin-linked pyruvate oxidoreductase in *C. beijerinckii* NRRL B593 ("*C. butylicum*" strain 21 or B-593) is implicated by hydrogen gas production during pyruvate dissimilation (Koepsell and Johnson 1942) and the purification of hydrogenase from the organism (Peck and Gest 1957), which preceded the isolation of Fd from *C. pasteurianum* (Mortenson et al. 1963).

3.1.2 Thiolase

Thiolase (acetyl-CoA:acetyl-CoA *C*-acetyltransferase, EC 2.3.1.9) catalyzes the following reaction (Gehring and Lynen 1972):

$$2 \text{ acetyl-CoA} \rightleftharpoons \text{acetoacetyl-CoA} + \text{CoASH}$$

Both acetyl-CoA and acetoacetyl-CoA are situated at branch points leading to either acid formation or solvent formation (see Figure 3–1). The level of thiolase in *C. acetobutylicum* ATCC 824 cells reached a maximum at the end of the growth phase (Hartmanis and Gatenbeck 1984) or was higher under solvent-producing conditions (Wiesenborn et al. 1988); however, the enzyme level (specific activity, units mg^{-1}) in cells from continuous cultures did not always correlate with the specific production rate of butanol, acetone, and butyrate combined (Wiesenborn et al. 1988). For *C. beijerinckii*, the activity level of thiolase remained unchanged throughout growth in strain NRRL B592 or decreased with time in strain NRRL B593 (Yan et al. 1988).

Thiolase has been purified 70-fold to homogeneity from *C. acetobutylicum* ATCC 824 (Wiesenborn et al. 1988). It has a subunit molecular weight of 44,000, and the elution volume of the enzyme on a gel-filtration column corresponded to a native enzyme having four subunits of equal molecular weight (the native molecular weight was not reported). The native molecular weight of thiolase from *C. beijerinckii* NRRL B592 and NRRL B593 were,

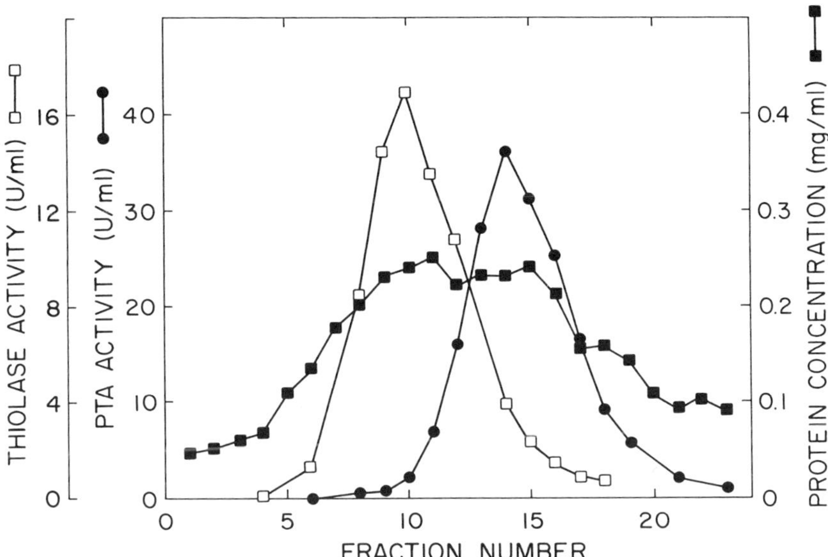

FIGURE 3–2 Elution of *C. beijerinckii* NRRL B593 thiolase and phosphotransacetylase (PTA) from a Sephacryl S-300 column. The applied sample was a fraction from a hydroxyapatite column on which a crude extract was resolved (see Figure 1 in Thompson and Chen 1990). The elution volume of thiolase and PTA corresponded, respectively, to molecular weights of 120,000 and 56,000 (D.K. Thompson, and Chen, J-S., previously unpublished observations).

respectively, 100,000 and 120,000 (Figure 3–2; D.K. Thompson, M. Walker, and J.-S. Chen, unpublished observations). The thiolase of *C. kluyveri* has a native molecular weight of 155,000–158,000 (Hartmanis and Stadtman 1982) and a subunit molecular weight of 41,000 (Sliwkowski and Hartmanis 1984).

The thiolase of *C. acetobutylicum* has a K_m of 0.27 mM for acetyl-CoA at 30°C and pH 7.4 (Wiesenborn et al. 1988). In the condensation (physiological) reaction, this thiolase was inhibited by micromolar levels of CoASH and even at low ratios of CoASH to acetyl-CoA. ATP (10 mM) or butyryl-CoA (1 mM) caused an inhibition of about 40%. The condensation reaction has a broad peak of activity between pH 5.5 and 7.9, which differs from the pH profile of the thiolase from *C. pasteurianum* (Berndt and Schlegel 1975). The thiolase of *C. acetobutylicum* was rapidly inactivated by iodoacetamide but more slowly by 5,5′-dithiobis(2-nitrobenzoic acid). In the thiolytic reaction, kinetic data are consistent with a ping-pong mechanism, which is similar to other thiolases. The K_m values for CoASH and acetoacetyl-CoA are, respectively, 4.7 and 32 μM at 30°C and pH 8.

Isoelectric focusing of purified thiolase from *C. pasteurianum* separated the enzyme into two species, with pI of 4.5 (major form) and 7.6 (minor form), but the two species had the same molecular weight of 158,000 (Berndt and Schlegel 1975). It is not known if the two species have different catalytic properties. Thiolase from solvent-producing clostridia has not been examined by isoelectric focusing for the possible presence of multiple forms.

3.1.3 3-Hydroxybutyryl-CoA Dehydrogenase

3-Hydroxybutyryl-CoA dehydrogenase (L-3-hydroxyacyl-CoA:NAD$^+$ oxidoreductase, EC 1.1.1.35) catalyzes the following reaction:

$$\text{Acetoacetyl-CoA} + \text{NAD(P)H} + \text{H}^+ \rightleftharpoons \text{3-hydroxybutyryl-CoA} + \text{NAD(P)}^+$$

NADH-linked activity of the enzyme has been measured in *C. acetobutylicum* ATCC 824 (Hartmanis and Gatenbeck 1984), and the enzyme has been purified from *C. beijerinckii* NRRL B593 (Colby and Chen 1992). The *C. beijerinckii* enzyme is active with either NADH or NADPH as a coenzyme, but the catalytic efficiency [k_{cat}/K_m] of the NADH-linked reaction is 60 times higher than that of the NADPH-linked reaction, suggesting that NADH is the physiological coenzyme. The K_m for acetoacetyl-CoA (with NADH) is 14 μM. The subunit and native molecular weights of the *C. beijerinckii* enzyme are, respectively, 30,800 and 213,000. When crude extracts were fractionated by successive gel filtration steps, active fractions with apparent molecular weights of 360,000; 235,000; and 120,000 were obtained, suggesting that the enzyme may exist as different multimers. However, the purified enzyme showed a constant molecular weight when samples between 3 and 0.05 mg protein per ml were examined. The physiologically relevant quaternary structure of the enzyme is thus not known.

The structural gene for the enzyme has been cloned from *C. acetobutylicum* P262, and it encodes a polypeptide with a molecular weight of 31,535 (Youngleson et al. 1989a). The deduced amino acid sequence of the gene showed homology with vertebrate 3-hydroxyacyl-CoA dehydrogenase, and a sequence characteristic of an NAD(H)-binding site motif is present near the amino-terminus of the deduced polypeptide. Of the 34 amino-terminal residues determined from the purified enzyme from *C. beijerinckii* NRRL B593, 32 are identical to those of the deduced sequence from *C. acetobutylicum* P262 (Colby and Chen 1992). The similar coenzyme specificity, amino-terminal sequence, and subunit size suggest that the enzyme may also be conserved within the two solvent-producing clostridial species.

In contrast, an NADP(H)-specific 3-hydroxybutyryl-CoA dehydrogenase (L-3-hydroxybutyryl-CoA:NADP$^+$ oxidoreductase, EC 1.1.1.157) has been purified from *C. kluyveri* (Madan et al. 1973; Sliwkowski and Hartmanis 1984). The native molecular weight is 215,000–220,000 (Maden et al. 1973), and the subunit molecular weight is 26,000 (Madan et al. 1973) or 33,500 (Sliwkowski and Hartmanis 1984). The NADP(H)-dependent 3-hydroxybutyryl-CoA dehydrogenase may be compared with the NAD(H)-dependent enzyme for structures conferring their different coenzyme specificities.

3.1.4 Crotonase

Crotonase (enoyl-CoA hydratase, EC 4.2.1.17) catalyzes the following reaction:

$$\text{3-hydroxybutyryl-CoA} \rightleftharpoons \text{crotonyl-CoA} + H_2O$$

In mitochondria, crotonase catalyzes the hydration of enoyl-CoA to 3-hydroxyacyl-CoA during beta-oxidation of saturated fatty acids. In butyrate- and butanol-forming clostridia, the enzyme catalyzes the dehydration of 3-hydroxybutyryl-CoA.

Crotonase has been purified 131-fold (67% yield) to homogeneity from *C. acetobutylicum* (probably strain NRRL B528; Waterson et al. 1972). The enzyme was active only toward C-4 and C-6 substrates, with a K_m of 30 μM for crotonyl-CoA. Crotonyl-CoA is an inhibitor at high concentrations in the hydration reaction, but hexenoyl CoA does not show substrate inhibition. Activity toward C-8 to C-16 enoyl-CoA substrates was found in crude extracts, which suggests that more than one hydratase is present in this organism. However, if the other hydratase(s) has a specificity for substrates longer than C-4, it is not expected to have a role in butyrate and butanol formation.

Crotonase of *C. acetobutylicum* has a native molecular weight of 158,000 and a subunit molecular weight of 40,000, indicating that the enzyme is a tetramer. The enzyme is not stable at 4°C or when frozen, but the lyophilized sample is stable.

3.1.5 Butyryl-CoA Dehydrogenase

Butyryl-CoA dehydrogenase (EC 1.3.99.2) catalyzes the following reaction:

$$\text{Crotonyl-CoA} + [\text{NAD(P)H} + H^+] \rightleftharpoons \text{butyryl-CoA} + [\text{NAD(P)}^+]$$

The enzyme has not been isolated from solvent-producing clostridia, and no NADH or NADPH linked activity has been demonstrated (Hartmanis and Gatenbeck 1984). Using dye-linked assays, activity of the enzyme was detected in this group of anaerobes (v. Hugo et al. 1972; Hartmanis and Gatenbeck 1984). The reduction of crotonyl-CoA to butyryl-CoA is the reverse of a reaction catalyzed by acyl-CoA dehydrogenase in mammalian tissues during the oxidation of saturated fatty acids. The mammalian enzyme has been difficult to purify and assay, and it requires an electron-transferring flavoprotein for the physiological reaction (Engel 1980).

Butyryl-CoA dehydrogenase from the butyrate-producing anaerobe *Megasphaera elsdenii* has been purified and extensively characterized (Fink et al. 1986; Van Berkel et al. 1988), and an electron-transferring flavoprotein (ETF) which mediates electron transfer between NAD(H) and the enzyme has also been purified and characterized (Whitfield and Mayhew 1974; Brockman and Wood 1975; Pace and Stankovich 1987). Although dye-linked assays have been developed to circumvent the need for ETF in measuring the activity of butyryl-CoA dehydrogenase, the catalytic assay is not routinely used to monitor the purification because of problems associated with variable contamination with ETF and thioesterase in impure preparations (Engel 1981). Difficulties associated with the dye-linked assays and instability of CoA-free butyryl-CoA dehydrogenase at pH values above and below 6 (Fink et al. 1986) may explain the extremely low activity obtained with *C. acetobutylicum* (Hartmanis and Gatenbeck 1984).

The presence of an ETF for butyryl-CoA dehydrogenase in solvent-producing clostridia may be speculated because of the isolation of NADH- or NADPH-specific ETFs from additional anaerobic bacteria (Stanton 1989; Dietrichs et al. 1990) and anaerobic mitochondria (Komuniecki et al. 1989).

3.2 ENZYMES CATALYZING SPECIFIC ACID-FORMING REACTIONS

Two analogous pathways lead to the formation of, respectively, acetic and butyric acids. Each pathway consists of two reactions: first, the conversion of the respective acyl-CoA into acyl-phosphate (reactions 6 and 8 in Figure 3–1), and second, the formation of the acid from acyl-phosphate with concomitant phosphorylation of ADP (reactions 7 and 9 in Figure 3–1). The two sequential reactions are catalyzed, respectively, by an acyltransferase and a kinase, but distinct enzymes are responsible for the analogous reactions of the parallel pathways.

3.2.1 Phosphotransacetylase

Phosphotransacetylase (PTA, acetyl-CoA:orthophosphate acetyltransferase, EC 2.3.1.8) catalyzes the following reaction:

$$\text{Acetyl-CoA} + \text{Pi} \rightleftharpoons \text{acetyl phosphate} + \text{CoASH}$$

PTA activity has been measured in cell extracts of *C. acetobutylicum* (Andersch et al. 1983; Hartmanis and Gatenbeck 1984; Huesemann and Papoutsakis 1989), and the enzyme has been partially purified from *C. beijerinckii* (Thompson and Chen 1990; D.K. Thompson, M. Walker, and J.-S. Chen, unpublished observations).

PTA of *C. beijerinckii* strains NRRL B592 and B593 had a molecular weight between 56,000 and 57,000 (see Figure 3–2; also D.K. Thompson, M. Walker, and J.-S. Chen, unpublished observations), which is significantly lower than the molecular weight of phosphotransbutyrylase (PTB) from the same strains (Thompson and Chen 1990). PTA from these strains showed no activity toward butyryl-CoA or acetoacetyl-CoA, and it required ammonium sulfate (0.2 M) for stabilization, which also distinguishes PTA from PTB (Thompson and Chen 1990; also D.K. Thompson and J.-S. Chen, unpublished observations).

PTA has been purified from *C. kluyveri* (Stadtman 1952) and later crystallized (Klotzsch 1969). The purified enzyme required K or ammonium ions for activity, whereas sodium or lithium ions inhibited (Stadtman 1952). The activity of PTA from *C. acetobutylicum* (Hartmanis and Gatenbeck 1984) and from *C. beijerinckii* (D.K. Thompson and J.-S. Chen, unpublished observations) was also enhanced by potassium or ammonium ions.

3.2.2 Acetate Kinase

Acetate kinase (EC 2.7.2.1) catalyzes the following reaction:

$$\text{Acetyl phosphate} + \text{ADP} \rightleftharpoons \text{acetate} + \text{ATP}$$

Activity of acetate kinase has been measured in *C. acetobutylicum* (Andersch et al. 1983; Hartmanis and Gatenbeck 1984; Huesemann and Papoutsakis 1989), but the enzyme has not been isolated.

Acetate kinase is a widely distributed bacterial enzyme, and its distribution parallels that of PTA, which reflects the physiological role of the two-enzyme sequence in the formation or activation of acetate (Rose 1962). Acetate kinase, with a molecular weight of 60,000, has been purified from *C. thermoaceticum* (Schaupp and Ljungdahl 1974). It displayed the Michaelis-Menten type of kinetics with acetate, but its activity with respect to varying concentrations of ATP was sigmoidal. The regulation of acetate kinase and butyrate kinase (along with PTA and PTB) in solvent-producing clostridia is an aspect pertinent to the understanding of the uptake and reutilization of acids during solvent formation (Huesemann and Papoutsakis 1989).

3.2.3 Phosphotransbutyrylase

Phosphotransbutyrylase (PTB, butanoyl-CoA:orthophosphate butanoyltransferase, EC 2.3.1.19) catalyzes the following reaction:

$$\text{Butyryl-CoA} + \text{Pi} \rightleftharpoons \text{butyryl phosphate} + \text{CoASH}$$

PTB has been prepared from *C. acetobutylicum* (Gavard et al. 1957; Wiesenborn et al. 1989a) and from *C. beijerinckii* (Valentine and Wolfe 1960; Thompson and Chen 1990). Table 3–2 is a comparison of the properties of the purified PTB from *C. acetobutylicum* ATCC 824 and from *C. beijerinckii* NRRL B593.

PTBs from *C. acetobutylicum* and *C. beijerinckii* are similar in their kinetic properties. In the butyryl phosphate-forming direction, PTB activity peaked between pH 7.5–8.0, and the activity dropped precipitously when the pH decreased from 7.5 to 6, which is likely a factor in the regulation of acid production during the cell's transition into solvent production (Wiesenborn et al. 1989a). The subunit molecular weight for PTB is 31,000 for *C. acetobutylicum* ATCC 824 (Cary et al. 1988; Wiesenborn et al. 1989a) and 33,000 for *C. beijerinckii* NRRL B593 (Thompson and Chen 1990), but there is a significant difference in their native molecular weights (264,000 versus 205,000). Because the native molecular weights were estimated in different

TABLE 3–2 Comparison of Phosphotransbutyrylase Purified from *C. acetobutylicum* ATCC 824 and from *C. beijerinckii* NRRL B593

Property	*Organism*	
	C. acetobutylicum[1]	C. beijerinckii
Native mol. wt.	264,000	205,000
Subunit mol. wt.	31,000	33,000
K_m value (mM) for		
Acetyl-CoA	6.20	3.33
Butyryl-CoA	0.11	0.04
Acetoacetyl-CoA	ND[2]	1.10
Phosphate		
With acetyl-CoA	610	ND
With butyryl-CoA	14	6.50

[1] Data for *C. acetobutylicum* PTB are from Wiesenborn et al. (1989a). Data for *C. beijerinckii* PTB are from Thompson and Chen (1990).

[2] ND, not determined.

buffers containing different additives, it remains to be shown if an inherent difference exists.

PTB of *C. beijerinckii* also reacts with acetoacetyl-CoA in the presence of phosphate, and CoASH is a product of the reaction (Thompson and Chen 1990). A postulated product of the reaction is acetoacetyl phosphate, which is yet to be synthesized and confirmed. Whether or not the acetoacetyl-CoA-linked activity of PTB has any physiological significance is still an unanswered question.

3.2.4 Butyrate Kinase

Butyrate kinase (EC 2.7.2.7) catalyzes the following reaction:

$$\text{Butyryl phosphate} + \text{ADP} \rightleftharpoons \text{butyrate} + \text{ATP}$$

The enzyme has been partially purified from *C. beijerinckii* ATCC 6014 (formerly "*C. butyricum* ATCC 6014," Twarog and Wolfe 1962) and from *C. tetanomorphum* ATCC 3606 (Twarog and Wolfe 1963), and it has now been purified from *C. acetobutylicum* ATCC 824 (Hartmanis 1987).

Butyrate kinase from *C. acetobutylicum* ATCC 824 has a native molecular weight of 85,000 (Hartmanis 1987) and a subunit molecular weight of 39,000 (Hartmanis 1987; Cary et al. 1988), suggesting that the enzyme is a dimer. The enzyme was reactive toward straight-chain saturated carboxylic acids with a chain length of 2 to 6, and butyrate gave the highest activity when all acids tested were at 700 mM (Hartmanis 1987). Acetate gave an activity equivalent to 6% of the butyrate-linked activity. K_m values for substrates of the reverse reaction (butyryl phosphate-forming) have been measured, but that for butyryl phosphate has not been reported (Twarog and Wolfe 1962; Hartmanis 1987).

3.3 ENZYMES CATALYZING SPECIFIC SOLVENT-FORMING REACTIONS

Enzymes specifically involved in the formation of ethanol, acetone, isopropanol, and butanol in solvent-producing clostridia (see Figure 3–1 for numbered reactions) fall into four groups: (1) the enzyme(s) converting acetoacetyl-CoA to acetoacetate (reaction 12), which is represented by acetoacetyl-CoA: acetate/butyrate CoA-transferase; (2) acetoacetate decarboxylase catalyzing reaction 13; (3) aldehyde dehydrogenase(s) catalyzing reactions 10 and 15; and (4) alcohol dehydrogenases catalyzing reactions 11, 14, and 16. Alcohol dehydrogenases with different structural and catalytic properties have been isolated from strains of *C. acetobutylicum* and *C. beijerinckii.* It is yet to be established whether additional, physiologically important forms also exist for the other solvent-forming enzymes.

3.3.1 Acetoacetyl-CoA: Acetate/Butyrate CoA-Transferase

The acetoacetyl-CoA: acetate/butyrate CoA-transferase (EC 2.8.3.9) catalyzes the following reaction:

$$\text{Acetoacetyl-CoA} + \text{acetate or butyrate} \rightleftharpoons \text{acetoacetate} + \text{acetyl-CoA or butyryl-CoA}$$

The reversible transfer of the CoA moiety between an acyl-CoA and acetoacetate is a reaction used by several bacteria either for the utilization of the carboxylic acids as an energy and carbon source (Nunn 1987) or because acetoacetate or acetoacetyl-CoA is a metabolic intermediate (Gottschalk 1986). Enzymes catalyzing the reaction belong to the class of CoA-transferase. Depending on the direction of the physiological reaction and the source organism, CoA-transferases differ in their substrate specificity, which is manifested by different relative activities toward carboxylic acids or toward thioesters of these acids.

Acetoacetyl-CoA: acetate/butyrate CoA-transferase (hereafter abbreviated as CoA-transferase) has been purified from *C. acetobutylicum* ATCC 824 (Wiesenborn et al. 1989b) and from *C. beijerinckii* NRRL B593 (G.D. Colby and J.-S. Chen, unpublished observations). The CoA-transferase of solvent-producing clostridia is an unstable enzyme. Wiesenborn et al. (1989b) successfully used high concentrations of ammonium sulfate (0.5–0.75 M) and glycerol (15–20% vol/vol) in 25 mM MOPS [3-(N-morpholino)-propanesulfonic acid] buffer at pH 7.0 to preserve enzyme activity. Similar conditions are also required for the purification of CoA-transferase from *C. beijerinckii.*

The purified *C. beijerinckii* CoA-transferase had a native molecular weight of 85,000, and two types of subunits with apparent molecular weights within the range of 28,000 and 23,000 were observed (G.D. Colby and J.-S. Chen, unpublished observations).

The CoA-transferase of *C. acetobutylicum* ATCC 824 has a native molecular weight of about 93,000 (Wiesenborn et al. 1989b) and consists of two types of subunit: the large (M_r 25,000 or 28,000) and the small (M_r 23,000 or 26,000) subunits (Wiesenborn et al. 1989b; Cary et al. 1990). Based on the amino acid sequence deduced from the gene cloned from *C. acetobutylicum* DSM 792 (= ATCC 824), Gerischer and Duerre (1990) found a molecular weight of 23,622 for the *large* subunit of CoA-transferase. It remains to be determined why higher molecular weights were obtained from electrophoresis on polyacrylamide gels in the presence of sodium dodecylsulfate.

The *C. acetobutylicum* CoA-transferase has unusually high K_m values for acetate (1.2 M) and butyrate (0.66 M), but its K_m values for acetoacetyl-CoA (21 μM with acetate and 56 μm with butyrate) are not unusual. The corresponding K_m values of the enzyme from *C. beijerinckii* NRRL B593 were: 500 mM for acetate, 10 mM for butyrate, and 30 μM (with butyrate) or 50 μM (with acetate) for acetoacetyl-CoA (G.D. Colby and J.-S. Chen, unpublished observations). The CoA-transferase from three strains of *C. beijerinckii* displayed a 6- to 15-fold higher activity with butyrate (at 20 mM) than with acetate (at 100 mM). Under identical assay conditions, the enzyme from *C. acetobutylicum* ATCC 824 gave a ratio of 0.045 (G.D. Colby and J.-S. Chen, unpublished observations), which is in general agreement with the results of Wiesenborn et al. (1989b). CoA-transferase is believed to be the sole enzyme for the re-utilization of acids during solvent production (Hartmanis et al. 1984). It is unclear how the opposite relative activities toward the two acids, as observed in the two species, are related to the cell's apparent preferential uptake of butyrate over acetate during solvent production. It was proposed that the high K_m values of the *C. acetobutylicum* enzyme for acetate and butyrate might allow the cell to have a "graded response" to the progressive toxic effects of increasing levels of these acids (Wiesenborn et al. 1989b).

From *C. beijerinckii*, two additional acetoacetyl-CoA-utilizing activities have been detected. PTB reacts with acetoacetyl-CoA in a phosphate-dependent reaction (see discussion in Section 3.2.3). In addition, acetoacetyl-CoA hydrolase activity is also present (Yan et al. 1988). The hydrolase activity, however, is low when compared with the CoA-transferase activity in crude extracts prepared in a buffer-containing ammonium sulfate and glycerol. Further studies will be needed to determine the significance of acetoacetyl-CoA or acyl-CoA hydrolase in clostridia.

A recent report (Nakamura et al. 1990) suggests that reactions of CoA-transferase and thiolase (see Section 3.1.2) are catalyzed by a single enzyme (molecular weight 200,000) in *C. acetobutylicum* ATCC 824. These workers used a coupled enzyme assay to measure acetate-linked activity only, instead of using a direct assay for measuring acetate- or butyrate-linked activity, which may explain the discrepancy between studies.

3.3.2 Acetoacetate Decarboxylase

Acetoacetate decarboxylase (EC 4.1.1.4) catalyzes the following reaction:

$$\text{Acetoacetate} \rightleftharpoons \text{acetone} + CO_2$$

Among solvent-forming enzymes, acetoacetate decarboxylase was the first to have been isolated from the cell and subjected to detailed mechanistic studies. The enzyme was at first partially purified from *C. acetobutylicum* strain BY (Davies 1943) and from "*Clostridium madisonii*" (strain A16 of E. McCoy at the University of Wisconsin, Madison) (Seeley and VanDemark 1950). Davies (1943) found the enzyme stable up to 70°C for 5 min.

Hamilton and Westheimer (1959a), in their investigation of the mechanism of enzymatic decarboxylation, prepared crystalline acetoacetate decarboxylase from *C. acetobutylicum* ATCC 862 (the strain has since been lost, but it is equivalent to NRRL B528; L.K. Nakamura, personal communication), which gave an activity two to three times higher than that obtained by Davies (1943). A molecular weight of about 300,000 was estimated with the crystalline enzyme (Hamilton and Westheimer 1959a).

Following the first crystallization of the enzyme, an exhaustive study of the structural and catalytic properties of the enzyme was conducted in two laboratories (Westheimer 1969; Fridovich 1972). Additional purification schemes were developed during this period: one being a modification of the earlier extraction procedure and without involving column chromatography (Fridovich 1963), and the other involving DEAE-cellulose column chromatography (Zerner et al. 1966). The latter procedure again gave a crystalline enzyme, but the specific activity was several times higher than that obtained previously (fold of purification not given). The procedure reported by Zerner et al. (1966) was successfully used in a recent purification of the enzyme (Gerischer and Duerre 1990), whereas the earlier procedure (Fridovich 1963) had to be modified to yield the pure enzyme (Petersen and Bennett 1990). A common first step for all purification schemes is the preparation of acetone powder from cell paste of *C. acetobutylicum*, which leads to efficient extraction of the enzyme from the cell (Fridovich 1972). The fold of purification for the homogeneous preparation, however, has not been reported, hence precluding an estimation of the level of the enzyme in the cell.

The molecular weight of acetoacetate decarboxylase of *C. acetobutylicum* has been reported as 260,000 (Lederer et al. 1966); 340,000 (Tagaki and Westheimer, 1968); or 330,000 (Gerischer and Duerre 1990). The subunit molecular weight was 29,000 (Tagaki and Westheimer 1968) or 28,000 (Gerischer and Duerre 1990) when measured with the protein, and the deduced amino acid sequence gave a molecular weight of 27,519 (Gerischer and Duerre

1990). With 2.5 M urea and at 10°C, Tagaki and Westheimer (1968) obtained a subunit dimer with a molecular weight of 62,000, from which 70% of the original enzyme activity was recovered (it is not clear if the subunit dimer itself is active). These results suggest that acetoacetate decarboxylase is composed of 8 or 12 subunits.

Acetoacetate decarboxylase has also been purified 370-fold to homogeneity from an acetone-producing isolate of *Bacillus polymyxa*, and sonication in the presence of Triton X-100 was necessary to extract the enzyme from lyophilized cells (Kimura et al. 1986). The *B. polymyxa* enzyme has a molecular weight of 280,000 and a subunit molecular weight of 27,000. Acetoacetate decarboxylase was isolated from wet cells of *C. beijerinckii* NRRL B592 and NRRL B593 without involving acetone powder or a detergent (C.G. Golemboski and J.-S. Chen, unpublished observations). Preliminary measurements using gel filtration on a Sephacryl S-300 column gave molecular weights in the range of 200,000–230,000. Further work will be needed to establish the subunit composition of the physiologically relevant form of acetoacetate decarboxylase in *C. beijerinckii*.

The K_m value for acetoacetate is 8 mM for the *C. acetobutylicum* enzyme (Fridovich 1972) and 0.94 mM for the *B. polymyxa* enzyme (Kimura et al. 1986). The optimal pH for activity is 5.9 for the enzyme from both organisms (Fridovich 1972; Kimura et al. 1986). The decarboxylation of acetoacetate by the *C. acetobutylicum* enzyme involves obligatory exchange of the carbonyl oxygen atom with the oxygen of the water, suggesting that a Schiff base is formed as an active intermediate between the enzyme and its substrate (Hamilton and Westheimer 1959b; Fridovich and Westheimer 1962). The active intermediate was later found to involve the lysine residue (with asterisk) in the sequence -Glu-Leu-Ser-Ala-Tyr-Pro-Lys*-Lys-Leu- (Warren et al. 1966; Laursen and Westheimer 1966), and this sequence corresponds to residue numbers 109 through 117 of the polypeptide (Gerischer and Duerre 1990).

Acetoacetate decarboxylase is remarkably resistant to thermal inactivation. The *C. acetobutylicum* enzyme can withstand 70°C for an hour in 0.1 M phosphate buffer at pH 5.9 (Neece and Fridovich 1967), and the *B. polymyxa* enzyme was stable at 60°C for 30 min (Kimura et al. 1986). One intriguing property of the crystalline enzyme is that its specific activity varied from one batch to another, and the enzyme activity could be irreversibly increased by heating at 50°C (Neece and Fridovich 1967). The molecular basis for this conversion and its biological implication remain unknown. Two apparently pertinent observations are that before heat activation only up to about 50% of the subunit could be labelled at the active site by acetoacetate (with borohydride) or acetopyruvate (Warren et al. 1966; Tagaki et al. 1968) and inactivating labeling of the active site had no effect on subsequent heat activation (Neece and Fridovich 1967), suggesting that the crystalline enzyme consisted of both active and inactive subunits. It should be important to elucidate the molecular mechanism for the activation and determine whether the property of the enzyme changes during purification.

3.3.3 Aldehyde Dehydrogenase

The CoA-acylating aldehyde dehydrogenase (EC 1.2.1.10) catalyzes the following reaction:

$$\text{Acyl-CoA} + \text{NAD(P)H} + \text{H}^+ \rightleftharpoons \text{aldehyde} + \text{CoASH} + \text{NAD(P)}^+$$

Coenzyme A-acylating aldehyde dehydrogenase (ALDH) has been purified from both *C. acetobutylicum* NRRL B543 (Palosaari and Rogers 1988) and from *C. beijerinckii* NRRL B592 (Yan and Chen 1990). The homogeneous preparation required only a 38-fold (from *C. beijerinckii*) or an 89-fold (from *C. acetobutylicum*) purification, indicating that the ALDH is a major protein in cells producing solvents. The result of amino acid sequence analysis of the N-terminal region of the *C. beijerinckii* ALDH further confirmed the purity of the protein (R.-T. Yan and J.-S. Chen, unpublished observations).

Table 3–3 compares important properties of the ALDH purified from *C. acetobutylicum* and from *C. beijerinckii*. The molecular size and subunit composition are similar for ALDH from the two sources. The purified ALDH can use either NAD(H) or NADP(H) as the coenzyme, but NADH is a more

TABLE 3–3 Comparison of Aldehyde Dehydrogenase Purified from *C. acetobutylicum* NRRL B643 and from *C. beijerinckii* NRRL B592

	Organism	
Property	C. acetobutylicum[1]	C. beijerinckii
Native mol. wt.	115,000	100,000
Subunit mol. wt.	56,000	55,000
K_m value (mM) and k_{cat}/K_m value (min^{-1} mM^{-1}; in parentheses) for		
NADH (with acetyl-CoA)	NA[2]	0.0082 (430)
(with butyryl-CoA)	0.003	0.0076 (3,100)
NADPH (with acetyl-CoA)	NA	0.21 (8.5)
(with butyryl-CoA)	0.040	0.067 (510)
Acetyl-CoA (with NADH)	0.055	0.15 (22)
(with NADPH)	NA	0.02 (89)
Butyryl-CoA (with NADH)	0.045	0.17 (140)
(with NADPH)	NA	0.072 (480)

[1] Data for *C. acetobutylicum* aldehyde dehydrogenase are from Palosaari and Rogers (1988). Data for *C. beijerinckii* aldehyde dehydrogenase are from Yan and Chen (1990).
[2] NA, Not available.

efficient coenzyme than NADPH as NADH gave a higher k_{cat}/K_m value than NADPH. Also, butyryl-CoA is a better substrate than acetyl-CoA for ALDH according to the k_{cat}/K_m value. The ratio of butyraldehyde-linked versus acetaldehyde-linked activities (NADH-dependent) remained constant during the purification of ALDH from *C. acetobutylicum* and *C. beijerinckii*, indicating that the purified ALDH is the predominant if not the sole ALDH. ALDH's higher activity with butryryl-CoA than with acetyl-CoA is consistent with the production of more butanol than ethanol by these organisms.

The coenzyme specificity of ALDH in cell-free extracts cannot be accurately determined if the enzyme activity is solely based on absorbance changes at 340 nm (oxidation or reduction of the coenzymes). Although the purified ALDH uses either NAD(H) or NADP(H) as the coenzyme, only NAD(H)-dependent activity (based on the oxidation or reduction of the coenzyme) was measured in cell-free extracts (Rogers 1986; Duerre et al. 1987; Yan et al. 1988). When the formation of butyryl-CoA, another product of the reaction, was monitored, both NAD^+- and $NADP^+$-linked activities of ALDH were readily detected in cell-free extracts of two strains of *C. beijerinckii* (Yan and Chen 1990). Even under anaerobic assay conditions, activities based on thioester formation were higher than activities based on NADH formation in cell-free extracts, indicating that reoxidation of NAD(P)H occurred and perhaps involved alcohol dehydrogenase(s) (Yan and Chen 1990). The presence in *C. acetobutylicum* DSM 792 (= ATCC 824) of a specific, NADH-dependent acetaldehyde dehydrogenase was recently suggested (Bertram et al. 1990); however, the enzyme activity was based on the oxidation of NADH in cell-free extracts.

The ALDH of *C. beijerinckii* NRRL B592 is O_2 sensitive, but it can be protected against O_2 inactivation by dithiothreitol (DTT). The O_2-inactivated ALDH can be reactivated by several systems (Yan and Chen 1990). The most effective system consists of CoASH plus DTT (giving an activity 5.7-fold of the control), which is followed by CoASH alone (2.7-fold). Neither DTT nor NAD^+ alone reactivates ALDH, but DTT and NAD^+ together are effective (1.8-fold). However, NAD^+ and CoASH together give a lower reactivation (1.6-fold) than CoASH alone (2.7-fold). Preincubation of the oxidized ALDH with DTT prevents the enzyme from activation by further incubation with CoASH.

The results of reactivation may be interpreted as follows: the reactivation of ALDH involves the regeneration of a sulfhydryl group(s) on the enzyme, and binding of CoASH to the enzyme puts the enzyme in a conformation most conducive to reductive activation by a thiol compound such as DTT. CoASH alone is effective because it is a thiol compound. NAD^+ may cause a similar but less-effective conformational change in the enzyme than CoASH. NAD^+ or DTT probably also binds to sites that affect the proper binding of CoASH and hence decreases the effectiveness of CoA for activation. These sites on ALDH may be involved in interactions with a nonprotein sulfhydryl group (shared by CoASH and DTT) and with an adenyl or ADP moiety (a similar structure between CoASH and NAD^+). NAD^+ may compete against CoASH. Indeed, NAD^+ or CoASH is inhibitory at high concentrations, but there is no inhibition when the two substrates are present at a constant ratio (Yan and Chen 1990). Shone and Fromm (1981) showed that NAD^+ was necessary, together with 2-mercaptoethanol, for the activation (and for the maintenance of the active conformation) of the *Escherichia coli* ALDH. These workers did not report whether CoASH could replace NAD^+ or NAD^+ plus 2-mercaptoethanol for activation. Because the measured activity of ALDH is easily affected by prior exposure to O_2 (in the presence of low levels of thiol compounds) and by the assay conditions, a comparison of ALDH activity from different sources must be done carefully.

3.3.4 Alcohol Dehydrogenase

Alcohol dehydrogenase (ADH; EC 1.1.1.1 [NAD^+]; 1.1.1.2 [$NADP^+$]; 1.1.1.71 [$NAD(P)^+$]) catalyzes the following reaction:

$$\text{Aldehyde or ketone} + \text{NAD(P)H} + H^+ \rightleftharpoons \text{primary or secondary alcohol} + \text{NAD(P)}^+$$

Multiple forms of ADH are often present in an organism, and the distribution of different ADH forms in eucaryotes may be organ- or organelle-specific and may also be developmentally regulated. In bacteria, the expression of ADH or different forms of ADH is usually regulated by growth conditions or growth stages. The presence of multiple forms of ADH within a bacterial species or within an individual strain complicates studies of the enzyme unless each form is separated from the others.

Individual strains of *C. acetobutylicum* and *C. beijerinckii* may contain one or more ADHs (Hiu et al. 1987; Duerre et al. 1987; Welch et al. 1989; Bertram et al. 1990), and a total of five distinct ADHs have been identified in two strains of *C. beijerinckii* (Hiu et al. 1987; C.-X. Zhu, R.-T. Yan, A. Ismaiel, and J.-S. Chen unpublished observations). These ADHs differ in structure and substrate and coenzyme specificities.

In an earlier study, an NAD^+-dependent butanol dehydrogenase was partially purified from *C. acetobutylicum* ATCC 824 (Petitdemange et al. 1968). ADH in cell-free extracts of *C. acetobutylicum* NCIB 8049 oxidized both primary and secondary alcohols with NAD^+ as the coenzyme (Fogarty and Ward 1970). However, a more recent study of ADHs from other strains of *C. acetobutylicum* detected no activity with secondary alcohols (Youngleson et al. 1988; Welch et al. 1989). Table 3–4 compares properties of ADHs from *C. acetobutylicum* and *C. beijerinckii*.

The ADH purified from *C. acetobutylicum* ATCC 824 is a primary ADH (Welch et al. 1989). It can use either NADH or NADPH as the coenzyme, with NADH giving a higher activity than NADPH. During its purification, about 50% of the NADPH-linked activity that was originally present in the cell-free extract was lost on the DEAE-cellulose column while there was no

TABLE 3–4 Comparison of Alcohol Dehydrogenase of *C. acetobutylicum* and *C. beijerinckii*

	Organism			
	C. acetobutylicum		C. beijerinckii	
Property	*ATCC 824*[1]	*P262*	*NRRL B592*	*NRRL B593*
Native mol. wt.	82,000	—[2]	80,000	100,000
Subunit mol. wt.	42,000	43,274	40–43,500	38,000
Coenzyme	NAD(P)H	NADPH	NAD(P)H	NADPH
K_m (mM) value and k_{cat}/K_m value (min^{-1} mM^{-1}; in parentheses) with				
Acetone	NA[3]	NA	NA	0.98 (8,500)
2-Butanone	—	—	—	1.5 (2,600)
2-Pentanone	—	—	—	8.5 (120)
Acetaldehyde	— (0.366)	—	—	11 (1,900)
Propionaldehyde	—	—	—	8.9 (3,400)
Butyraldehyde	16 (16.8)	—	—	34 (240)
NADH	0.18 (with butyraldehyde)	NA	—	NA
NADPH	—	—	—	0.022 (with acetone) 0.031 (with butyraldehyde)

[1] References: for *C. acetobutylicum* ATCC 824, Welch et al. (1989); for *C. acetobutylicum* P262, Youngleson et al. (1988, 1989b); for *C. beijerinckii* NRRL B592, R.-T. Yan, A. Ismaiel, and J.-S. Chen (unpublished observations); for *C. beijerinckii* NRRL B593, Hiu et al. (1987), A. Ismaiel, C.-X. Zhu, G.D. Colby, and J.-S. Chen (unpublished observations).
[2] —, data not available.
[3] NA, no measurable activity.

proportional loss of NADH-linked activity, suggesting the presence of an NADPH-specific ADH in this organism. The gene encoding an NADPH-specific primary ADH has been cloned from *C. acetobutylicum* P262 (Youngleson et al. 1988). While the NAD(P)H-ADH of *C. acetobutylicum* ATCC 824 required the presence of zinc in buffers during its purification, the deduced amino acid sequence of the NADPH-ADH of *C. acetobutylicum* P262 resembles that of an iron-activated ADH (Youngleson et al. 1989b). Recently, a gene cluster from *C. acetobutylicum* ATCC 824 was cloned and expressed in *Escherichia coli*, and it contains genes encoding two isozymes of NADH-dependent ADH and perhaps also an NADPH-dependent ADH (Petersen et al. 1991).

Bertram et al. (1990) suggested that in *C. acetobutylicum* DSM 792 (= ATCC 824) the NADPH-dependent ADH is responsible for ethanol production, whereas the NADH-dependent ADH is responsible for butanol formation. This may prove to be a significant advance in establishing the role of individual ADHs in *C. acetobutylicum*, although some uncertainties still remain. The assignment was based on mutants (AA2 and AA5) that lacked NADH-dependent ADH activity and did not produce butanol while they produced a low level of ethanol. However, the two mutants also lacked butyraldehyde dehydrogenase activity, which means butyraldehyde is not available for butanol formation whether or not the pertinent ADH is present. For example, Clark et al. (1989) obtained mutants (M3 and M5) of *C. acetobutylicum* ATCC 824 which still contained the NADH-ADH but did not produce butanol (the mutant lacked butyraldehyde dehydrogenase activity). Additional data, preferably with mutants defective in specific *adh* genes, can show more conclusively the role of individual ADHs.

During the purification of primary ADH from *C. beijerinckii* NRRL B592 (a strain producing acetone, butanol, and ethanol, but not isopropanol), two types of ADH with distinct coenzyme specificity and stability were clearly separated by chromatography on either Reactive Red 120- or Cibacron Blue 3GA-agarose (R.-T. Yan and J.-S. Chen, unpublished observations). One of them was NADPH-specific, whereas the other could use either NADH or NADPH as the coenzyme. The NADPH-specific ADH was more thermostable and was not inactivated by O_2. The NAD(P)H-ADH, which is the predominant ADH in solvent-producing cells, has been purified to a state that it contained only ADH (R.-T. Yan and J.-S. Chen, unpublished observations). The NAD(P)H-ADH had a native molecular weight of 80,000. The seemingly homogeneous NAD(P)H-ADH consistently gave two equally intense bands (Coomassie blue stained; molecular weights 40,000 and 43,500) after electrophoresis on polyacrylamide gel in the presence of sodium dodecylsulfate. Furthermore, three protein bands that all had NADH- and NADPH-linked ADH activities were observed when the seemingly homogeneous NAD(P)H-ADH was electrophoresed under nondenaturing conditions (A. Ismaiel and J.-S. Chen, unpublished observations). Results from subunit analysis and peptide mapping of the three electrophoretically separated forms of NAD(P)H-ADH suggest that the ADH in the upper, middle, and lower bands had, respectively, the following subunit composition: α_2, $\alpha\beta$, and β_2, with α and β designating subunits with molecular weights of 40,000 and 43,500, respectively (A. Ismaiel and J.-S. Chen, unpublished observations). Therefore, *C. beijerinckii* NRRL B592 contains at least four distinct primary ADHs.

Among solvent-producing clostridia, some strains of *C. beijerinckii* (including those previously known as "*C. butylicum*") have the distinct capacity to produce the secondary alcohol isopropanol, which is produced via the reduction of acetone (reaction 14 in Figure 3–1). A primary/secondary ADH, which is practically NADP(H)-specific (Hiu et al. 1987), has been isolated from each of four strains of butanol- and isopropanol-producing *C. beijerinckii* (C.-X. Zhu, M. Walker, and J.-S. Chen, unpublished observations). Primary/secondary ADH has been purified to homogeneity from two strains of *C. beijerinckii*, and the two purified ADHs have similar kinetic and structural properties, including the amino-terminal sequence (A. Ismaiel, C.-X. Zhu, G.D. Colby, and J.-S. Chen, unpublished observations). Table 3–4 is a summary of properties of the primary/secondary ADH purified from *C. beijerinckii* NRRL B593. The k_{cat}/K_m values were comparable between 2-butanone and propionaldehyde as well as between 2-pentanone and butyraldehyde, but the value for acetone was significantly higher than that for acetaldehyde. It suggests that the substrate-binding pocket of the primary/secondary ADH consists of two subsites: a smaller subsite that can accommodate either a hydrogen or a methyl group, and a larger subsite that can interact effectively with a C1-C3 alkyl group but not a hydrogen atom (A. Ismaiel and J.-S. Chen, unpublished observations).

The structural gene for the primary/secondary ADH of *C. beijerinckii* NRRL B593 has been cloned and sequenced (M. Rifaat and J.-S. Chen, unpublished observations). It encodes a polypeptide of 351 amino acid residues, which had a 75% identity with that of the thermostable ADH of *Thermoanaerobium brockii*, an anaerobe producing ethanol, but not butanol or isopropanol (Zeikus et al. 1979; Lamed and Zeikus 1981; Peretz and Burstein 1989). Figure 3–3 illustrates the striking similarity in the amino-terminal region, and the remaining sequences are equally similar, including the lack of a stretch of 18 amino acid residues that corresponds to the loop containing the four cysteinyl ligands for the structural zinc atom in the horse liver ADH (Peretz and Burstein 1989; M. Rifaat and J.-S. Chen, unpublished observations). Although the two ADHs have similar subunit molecular weights (about 38,000), they differ in their native molecular weights (100,000 for the *C. beijerinckii* primary/secondary ADH and 150,000 for the *T. brockii* ADH). Further research will be needed to compare the quaternary structure of the

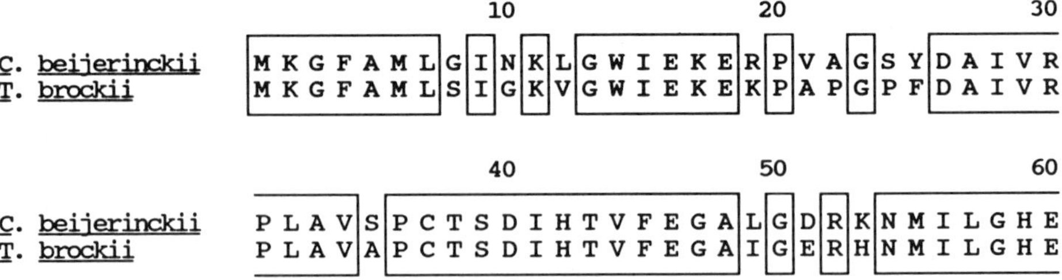

FIGURE 3–3 Comparison of the N-terminal amino acid sequence of alcohol dehydrogenases of *C. beijerinckii* NRRL B593 and *T. brockii*. Amino acid residues are represented by their single letter code. Conserved amino acid residues are boxed. The amino acid sequence of the *T. brockii* ADH was from Peretz and Burstein (1989).

two ADHs, to gain insights into the structure-function relationship in this type of ADH, and perhaps to gain new clues to the propagation of *adh* genes among alcohol-producing organisms.

3.4 CONCLUSION

The acid- and solvent-forming pathways of clostridia are interconnected through several shared metabolic intermediates. Enzymes catalyzing the reactions of the pathways have largely been identified, and the majority of them have been purified to a high degree of purity. In addition to the connection through the shared metabolic intermediates, solvent production is further linked to acid production through the reutilization of acetic and butyric acids. The re-entry of acetic and butyric acids into metabolism may involve different paths, hence different enzymes, under different growth conditions. Different strains of *C. acetobutylicum* (Woolley and Morris 1990) and *C. beijerinckii* (Chen and Hiu 1986; Hiu et al. 1987) encompass a range of physiological properties. The varied product pattern of different strains or of the same strain growing under different conditions suggests a flexible metabolic system that is amenable to manipulations for the adjustment of the product pattern. These possibilities point to a fermentation that can be regulated to respond to changing market needs.

A knowledge of the property of the acid- and solvent-forming enzymes will serve as the basis for genetic experiments to define the mechanism of regulation for the expression of individual enzymes. The presence of alternative enzymes, such as the alcohol dehydrogenases, for a key reaction further indicates a need for genetic studies to define the role or relative importance of individual enzymes. Recent advances in this area show the effectiveness of combining biochemical and genetic approaches for such investigations.

A comparison of respective acid- and solvent-forming enzymes between species or strains of the same species reveals significant similarities and also differences in their structural and catalytic properties. Differences in the size or amino acid sequence of homologous polypeptides may be attributed to genetic changes as a result of evolution. Differences in the size (the quaternary structure) of corresponding "native" enzymes that seem to share similar subunits are more difficult to explain, and it is even more difficult to understand the significance of size variations among different preparations of the same enzyme. It may be useful to consider the possible formation of different complexes between an enzyme and others to facilitate alternative reaction sequences or the occurrence of an enzyme with varied subunit compositions to confer different regulatory properties. Thus, the different forms of an enzyme may suggest expanded possibilities for metabolic control, or they may simply be artifacts of preparation, which must be differentiated. The study of acid- and solvent-forming enzymes of clostridia will continuously be challenged by a need to identify the physiologically relevant or practically useful forms of an enzyme and will be rewarded with a greater capability to further improve this fermentation.

REFERENCES

Adams, M.W.W., Mortenson, L.E., and Chen, J.-S. (1981) *Biochim. Biophys. Acta* 594, 105–176.

Andersch, W., Bahl, H., and Gottschalk, G. (1983) *Eur. J. Appl. Microbiol. Biotechnol.* 18, 327–332.

Arnold, W., Rump, A., Klipp, W., Priefer, U.B., and Puehler, A. (1988) *J. Mol. Biol.* 203, 715–738.

Berndt, H., and Schlegel, H.G. (1975) *Arch. Microbiol.* 103, 21–30.

Bertram, J., Kuhn, A., and Duerre, P. (1990) *Arch. Microbiol.* 153, 373–377.

Brockman, H.L., and Wood, W.A. (1975) *J. Bacteriol.* 124, 1447–1453.

Cannon, M., Cannon, F., Buchanon-Wollaston, V., et al. (1988) *Nucleic Acids Res.* 16, 11379.

Cary, J.W., Peterson, D.J., Papoutsakis, E.T., and Bennett, G.N. (1988) *J. Bacteriol.* 170, 4613–4618.

Cary, J.W., Peterson, D.J., Papoutsakis, E.T., and Bennett, G.N. (1990) *Appl. Environ. Microbiol.* 56, 1576–1583.

Chen, J.-S. (1987) in *Food Microbiology, Vol. 1, Concepts in Physiology and Metabolism* (Montville, T.J., ed.), pp. 61–101, CRC Press, Boca Raton, FL.

Chen, J.-S., and Hiu, S.F. (1986) *Biotechnol. Lett.* 8, 371–376.

Clark, S.W., Bennett, G.N., and Rudolph, F.B. (1989) *Appl. Environ. Microbiol.* 55, 970–976.

Colby, G.D., and Chen, J.-S. (1992) *Appl. Environ. Microbiol.* 58, 3297–3302.

Davies, R. (1943) *Biochem. J.* 37, 230–238.

Dietrichs, D., Meyer, M., Schmidt, B., Andreesen, J.R. (1990) *J. Bacteriol.* 172, 2088–2095.

Doelle, H.W. (1975) *Bacterial Metabolism*, 2nd ed., pp. 574–587, Academic Press, New York.

Duerre, P., Kuhn, A., Gottwald, M., and Gottschalk, G. (1987) *Appl. Microbiol. Biotechnol.* 26, 268–272.

Engel, P.C. (1980) in *Flavins and Flavoproteins* (Yagi, K., and Yamano, T., eds.), pp. 423–430, Japan Scientific Societies Press, Tokyo.

Engel, P.C. (1981) *Methods Enzymol.* 71, 359–366.

Fink, C.W., Stankovich, M.T., and Soltysik, S. (1986) *Biochemistry* 25, 6637–6643.

Fogarty, W.M., and Ward, J.A. (1970) *Proc. Biochem. Soc.* 119, 19P–20P.

Fridovich, I. (1963) *J. Biol. Chem.* 238, 592–598.

Fridovich, I. (1972) in *The Enzymes* (Boyer, P.D., ed.), 3rd ed., vol. 6, pp. 255–270, Academic Press, New York.

Fridovich, I., and Westheimer, F.H. (1962) *J. Am. Chem. Soc.* 84, 3208–3209.

Gavard, R., Hautecoeur, B., Descourtieux, H. (1957) *C.R. Acad. Sci.* Ser. D, 244, 2323–2326.

Gehring, U., and Lynen, F. (1972) in *Enzymes* (Boyer, P.D., ed.), 3rd ed., vol. 7, pp. 391–405, Academic Press, New York.

Gerischer, U., and Duerre, P. (1990) *J. Bacteriol.* 172, 6907–6918.

Gottschalk, G. (1986) *Bacterial Metabolism*, 2nd ed., Springer-Verlag, New York.

Hamilton, G.A., and Westheimer, F.H. (1959a) *J. Am. Chem. Soc.* 81, 2277.

Hamilton, G.A., and Westheimer, F.H. (1959b) *J. Am. Chem. Soc.* 81, 6332–6333.
Hartmanis, M.G.N. (1987) *J. Biol. Chem.* 262, 617–621.
Hartmanis, M.G.N., and Gatenbeck, S. (1984) *Appl. Environ. Microbiol.* 47, 1277–1283.
Hartmanis, M.G.N., Klason, T., and Gatenbeck, S. (1984) *Appl. Microbiol. Biotechnol.* 20, 66–71.
Hartmanis, M.G.N., and Stadtman, T.C. (1982) *Proc. Natl. Acad. Sci. USA* 79, 4912–4916.
Hiu, S.F., Zhu, C.-X., Yan, R.-T., and Chen, J.-S. (1987) *Appl. Environ. Microbiol.* 53, 697–703.
Huesemann, M.H.W., and Papoutsakis, E.T. (1989) *Appl. Microbiol. Biotechnol.* 30, 585–595.
Jones, D.T., and Woods, D.R. (1986) *Microbiol. Rev.* 50, 484–524.
Kerscher, L., and Oesterhelt, D. (1982) *Trends Biochem. Sci.* 7, 371–374.
Kimura, Y., Yasuda, N., Tanigaki-Nagae, H., et al. (1986) *Agric. Biol. Chem.* 50, 2509–2516.
Klotzsch, H.R. (1969) *Methods Enzymol.* 13, 381–386.
Koepsell, H.J., and Johnson, M.J. (1942) *J. Biol. Chem.* 145, 379–386.
Komuniecki, R., McCrury, J., Thissen, J., and Rubin, N. (1989) *Biochim. Biophys. Acta* 975, 127–131.
Lamed, R.J., and Zeikus, J.G. (1981) *Biochem. J.* 195, 183–190.
Laursen, R.A., and Westheimer, F.H. (1966) *J. Am. Chem. Soc.* 88, 3426–3430.
Lederer, F., Coutts, S.M., Laursen, R.A., and Westheimer, F.H. (1966) *Biochemistry* 5, 823–833.
Madan, V.K., Hillmer, P., and Gottschalk, G. (1973) *Eur. J. Biochem.* 32, 51–56.
Mayhew, S.G., and Ludwig, M.L. (1975) in *The Enzymes* (Boyer, P.D., ed.), 3rd ed., vol. 12, pp. 57–118, Academic Press, New York.
Meinecke, B., Bertram, J., and Gottschalk, G. (1989) *Arch. Microbiol.* 152, 244–250.
Mortenson, L.E., Valentine, R.C., and Carnahan, J.E. (1963) *J. Biol. Chem.* 238, 794–800.
Nakamura, T., Igarashi, Y., and Kodama, T. (1990) *Agric. Biol. Chem.* 54, 267–268.
Neece, M.S., and Fridovich, I. (1967) *J. Biol. Chem.* 242, 2939–2944.
Nunn, W.D. (1987) in *Escherichia coli and Salmonella typhimurium Cellular and Molecular Biology* (Neidhardt, F.C., Ingraham, J.L., Low, K.B., et al., eds.), pp. 285–301, American Society for Microbiology, Washington, DC.
Pace, C.P., and Stankovich, M.T. (1987) *Biochim. Biophys. Acta* 911, 267–276.
Palosaari, N.R., and Rogers, P. (1988) *J. Bacteriol.* 170, 2971–2976.
Peck, H.D., and Gest, H. (1957) *J. Bacteriol.* 73, 569–580.
Peretz, M., and Burstein, Y. (1989) *Biochemistry* 28, 6549–6555.
Petersen, D.J., and Bennett, G.N. (1990) *Appl. Environ. Microbiol.* 56, 3491–3498.
Petersen, D.J., Welch, R.W., Rudolph, F.B., and Bennett, G.N. (1991) *J. Bacteriol.* 173, 1831–1834.
Petitdemange, H., Desbordes, J., Berthelin, J., and Gay, R. (1968) *C.R. Acad. Sci.* Ser. D, 266, 1772–1774.
Rogers, P. (1986) *Adv. Appl. Microbiol.* 31, 1–60.
Rose, I.A. (1962) in *The Enzymes* (Boyer, P.D., Lardy, H., and Myrbaeck, K., eds.), 2nd ed., vol. 6, pp. 115–118, Academic Press, New York.
Santangelo, J.D., Jones, D.T., and Woods, D.R. (1991) *J. Bacteriol.* 173, 1088–1095.
Schaupp, A., and Ljungdahl, L.G. (1974) *Arch. Microbiol.* 100, 121–129.

Schoenheit, P., Brandis, A., and Thauer, R.K. (1979) *Arch. Microbiol.* 120, 73–76.
Seeley, H.W., and VanDemark, P.J. (1950) *J. Bacteriol.* 59, 381–386.
Shone, C.C., and Fromm, H.J. (1981) *Biochemistry* 20, 7494–7501.
Sliwkowski, M.X., and Hartmanis, M.G.N. (1984) *Anal. Biochem.* 141, 344–347.
Srere, P.A. (1987) *Annu. Rev. Biochem.* 56, 89–124.
Srivastava, D.K., and Bernhard, S.A. (1987) *Annu. Rev. Biophys. Biophys. Chem.* 16, 175–204.
Stadtman, E.R. (1952) *J. Biol. Chem.* 196, 527–534.
Stanton, T.B. (1989) *Appl. Environ. Microbiol.* 55, 2365–2371.
Tagaki, W., Guthrie, J.P., and Westheimer, F.H. (1968) *Biochemistry* 7, 905–913.
Tagaki, W., and Westheimer, F.H. (1968) *Biochemistry* 7, 895–900.
Thompson, D.K., and Chen, J.-S. (1990) *Appl. Environ. Microbiol.* 56, 607–613.
Twarog, R., and Wolfe, R.S. (1962) *J. Biol. Chem.* 237, 2474–2477.
Twarog, R., and Wolfe, R.S. (1963) *J. Bacteriol.* 86, 112–117.
Valentine, R.C., and Wolfe, R.S. (1960) *J. Biol. Chem.* 235, 1948–1952.
Van Berkel, W.J.H., Van Den Berg, W.A.M., and Mueller, F. (1988) *Eur. J. Biochem.* 178, 197–207.
v. Hugo, H., Schoberth, S., Madan, V.K., and Gottschalk, G. (1972) *Arch. Microbiol.* 87, 189–202.
Wahl, R.C., and Orme-Johnson, W.H. (1987) *J. Biol. Chem.* 262, 10489–10496.
Warren, S., Zerner, B., and Westheimer, F.H. (1966) *Biochemistry* 5, 817–823.
Waterson, R.M., Casterllina, F.J., Hass, G.M., and Hill, R.L. (1972) *J. Biol. Chem.* 247, 5266–5271.
Welch, R.W., Rudolph, F.B., and Papoutsakis, E.T. (1989) *Arch. Biochem. Biophys.* 273, 309–318.
Westheimer, F.H. (1969) *Methods Enzymol.* 14, 231–241.
Whitfield, C.D., and Mayhew, S.G. (1974) *J. Biol. Chem.* 249, 2801–2810.
Wiesenborn, D.P., Rudolph, F.B., and Papoutsakis, E.T. (1988) *Appl. Environ. Microbiol.* 54, 2717–2722.
Wiesenborn, D.P., Rudolph, F.B., and Papoutsakis, E.T. (1989a) *Appl. Environ. Microbiol.* 55, 317–322.
Wiesenborn, D.P., Rudolph, F.B., and Papoutsakis, E.T. (1989b) *Appl. Environ. Microbiol.* 55, 323–329.
Woolley, R.C., and Morris, J.G. (1990) *J. Appl. Bacteriol.* 69, 718–728.
Yan, R.-T., and Chen, J.-S. (1990) *Appl. Environ. Microbiol.* 56, 2591–2599.
Yan, R.-T., Zhu, C.-X., Golemboski, C., and Chen, J.-S. (1988) *Appl. Environ. Microbiol.* 54, 642–648.
Yasunobu, K.T., and Tanaka, M. (1973) in *Iron-Sulfur Proteins* (Lovenberg, W., ed.), vol. 2, pp. 29–130, Academic Press, New York.
Youngleson, J.S., Jones, D.T., and Woods, D.R. (1989a) *J. Bacteriol.* 171, 6800–6807.
Youngleson, J.S., Jones, W.A., Jones, D.T., and Woods, D.R. (1989b) *Gene* 78, 355–364.
Youngleson, J.S., Santangelo, J.D., Jones, D.T., and Woods, D.R. (1988) *Appl. Environ. Microbiol.* 54, 676–682.
Zeikus, J.G., Hegge, P.W., and Anderson, M.A. (1979) *Arch. Microbiol.* 122, 41–48.
Zerner, B., Coutts, S.M., Lederer, F., Waters, H.H., and Westheimer, F.H. (1966) *Biochemistry* 5, 813–816.

CHAPTER

4

Mutagenesis and Its Application in Biotechnology

David T. Jones

Until the advent of recombinant DNA technology, the use of genetic manipulation for improving strains of industrial bacteria, depended largely on the use of mutagenesis and selection. Methods for increasing the frequency of mutation in bacterial populations and selection techniques for the isolation of various types of mutants were first developed in the 1940s and 1950s (Drake and Baltz 1976) and played a major role in advancing bacterial genetics. These methods were later adapted for practical applications and used for the selection of strains of industrial microorganisms with superior or novel characteristics. Mutagenesis and selection played a key role in the success of the emerging antibiotic industry (Aharnowitz and Cohen 1981) and advances in the understanding of the biochemistry and regulation of metabolic pathways enabled similar strategies to be applied to several other fermentation processes (Hopwood 1981).

The mode of action of many chemical and physical mutagenic agents, and the nature of mutations which are produced, have been well characterized (Drake and Baltz 1976) and standardized procedures for their use have become established for many commonly used bacterial species (Carlton and Brown 1981). Such procedures employ either chemical or physical mutagenic agents, with known modes of action, to increase the frequency of the desired types of mutation. A wide variety of selection procedures have been employed to select for the presence of the mutations that arise at random. The isolation of mutant strains for use as genetic markers still remains a basic requirement for genetic analysis of any bacterial species. In addition, the study of mutant strains can also provide valuable information on gene function and regulation. The extent to which genetic analysis can be undertaken, however, tends to be limited unless genetic transfer from one strain to another is possible.

Although standard techniques for mutagenesis and selection suffer several limitations, they are relatively straight-forward procedures to apply. They still provide the simplest and most practical approach when dealing with poorly characterized species, where genetic techniques remain rudimentary, or where strain differences make adapting established techniques difficult.

4.1 SURVEY OF MUTANTS AND MUTAGENIC TECHNIQUES IN CLOSTRIDIAL SPECIES

Conventional mutagenic and selection techniques have been used for the isolation of a wide variety of mutants for several different clostridial species. As might be expected, studies involving mutant strains have concentrated on species that have clearly established medical importance or have been identified as having potential for biotechnological applications.

A few studies that report the isolation of mutants have made use of mutations which have arisen spontaneously. In the majority of studies, however, standard mutagenic agents have been used to enhance the mutation frequency prior to selection. The most widely used mutagenic agents have been methyl methanesulfonate (MMS), ethyl methane-sulfonate (EMS) *N*-methyl-*N′*-nitro-*N*-nitrosoguanidine (MNNG) and UV light. However, in most cases, little or no attempt has been made to evaluate the effectiveness of the mutagenic agents. Comprehensive investigations to determine the conditions required for enhanced mutagenesis and killing have only been reported for a few species.

The largest variety of mutants have been isolated from *Clostridium acetobutylicum.* A comprehensive study of conditions required for mutagenesis was conducted by Bowring and Morris (1985) who found that the direct mutagens EMS and MNNG, produced a significant increase in relative induced mutation frequency of auxotrophic and antibiotic resistant markers. In contrast, UV radiation and mitomycin C, which are indirect mutagens, were reported to be ineffective. Nalidixic acid, hydrogen peroxide, and metronidazole were also ineffective as mutagenic agents under the conditions tested. The optimal conditions for mutagenesis in *C. acetobutylicum* were also investigated by Lemmel (1985) and in this study MNNG, UV irradiation, and acridine half mustard ICR 191 were reported to be ineffective. In a similar study on *C. butyricum*, Carrasco and Soto (1987) reported that both MMS and UV irradiation were effective, while MNNG was only slightly mutagenic. Daldal (1985) also used EMS, MNNG and UV irradiation for the isolation of mutants from *C. pasteurianum*. This species was extremely sensitive to UV irradiation but the relative effectiveness of the various mutagenic agents was not determined.

It would appear that the mutagenic agents which act by direct mutagenic mechanisms such as MMS, EMS, and MNNG are more effective at bringing about an increase in relative induced mutation frequency in most clostridial species which have been investigated. Those that act as indirect mutagenic agents such as UV light, which depend on subsequent misrepair of damaged DNA have been reported to be less effective. This observation led to the suggestion that the clostridia may be deficient in error prone repair processes (Walker 1984; Bowring and Morris 1985). However, in some species at least, UV irradiation has been used successfully to enhance the mutation rate and in several instances MNNG has been found to be ineffective as a mutagenic agent, so that generalizations should be viewed with caution.

Methods using chemical and physical agents to enhance mutation frequency have been expanded and to some extent superseded by the use of biological agents. Transposons have been used to produce mutations by random insertion into genomic material in several different species of bacteria. The successful transfer of several transposons of streptococcal origin including Tn*916*, Tn*917*, Tn*925*, and Tn*1545* to *C. tetani* and *C. acetobutylicum* by direct conjugation or by way of composite conjugative transposons or plasmids has been achieved in several laboratories (Yu and Pearce 1986; Oultram et al. 1987; Davies et al. 1988; Volk et al. 1988; Bertram and Dürre 1989; Woolley et al. 1989; Bertram et al. 1990). In particular, the demonstration of high-frequency transfer and random integration into the chromosome of transposon Tn*916* in *C. acetobutylicum* has opened the way to the use of transposon mutagenesis in this species. The demonstration that some transposons which insert into multiple chromosomal sites, undergo precise deletions under the appropriate conditions, indicates that such systems will also provide a valuable method for the cloning of clostridial genes (see Chapter 10, this volume).

4.1.1 Use of Mutants as Genetic Markers

Auxotrophic mutations have been among the most common and extensively used genetic markers for many species of bacteria. The application of penicillin counterselection and replica-plating techniques makes the isolation of this type of mutant relatively easy from those species that possess extensive biosynthetic capabilities and relatively simple nutritional requirements. Chemically defined or minimal culture media have been developed for many species of clostridia (see reviews by Woods and Jones 1984; Rogers 1986). Among the saccharolytic clostridia auxotrophic mutants, which were defective in a variety of biosynthetic functions, including amino acid, purine, pyrimidine and vitamin synthesis, have been isolated from several species, including *C. acetobutylicum* (Bowring and Morris 1985; Jones et al. 1985), *C. butyricum* (Carrasco and Soto 1987) and *C. pasteurianum* (Daldal 1985). In addition, auxotrophic mutants have also been isolated from species with more complex growth requirements, such as *C. thermocellum* (Mendez and Gomez 1982) and *C. perfringens* (Sebald and Costilow 1975). Bowring and Morris (1985) reported that certain types of auxotrophic mutants were isolated at much higher frequency than others from mutagenized cultures of *C. acetobutylicum*.

Antibiotic resistant mutants constitute a second category of mutations that have proved to be relatively easy to isolate by direct selection techniques and have been widely used as genetic markers for many species of bacteria. However, reports of chromosomal encoded antibiotic resistance have been limited to a relatively small number of antibiotics for clostridial species. Ampicillin, erythromycin, and rifampicin are the antibiotics that have been most commonly used to generate antibiotic resistance mutants. The isolation of rifampicin-resistant mutants has also provided a useful method for obtaining pleiotropic mutants exhibiting defects in endospore formation, solvent production, and cell morphology in *C. acetobutylicum* (Jones et al. 1982; Long et al. 1984; Rogers and Palosaari 1987; Clark et al. 1989).

Although a variety of mutations suitable for use as genetic markers have been isolated for a few of the most extensively studied species, there have been few reports relating to the transfer and recombination of chromosomal markers among the clostridia, and little use has been made of these markers for gene analysis and mapping. Although a variety of procedures for transferring genetic material by transformation or conjugation have been reported for some clostridial species, these systems have been restricted almost entirely to the transfer of extrachromosomal elements and have not used chromosomal genes (Young et al. 1989). Protoplast fusion has been used to investigate genetic recombination in *C. acetobutylicum* following fusion between auxotrophic mutant strains (Jones et al. 1985). Results obtained from this study indicated that although recombinants occurred at a frequency comparable to aerobic species, analysis was complicated by the formation of various segregating biparental phenotypes. These results were similar to those obtained from protoplast fusion studies carried out on *Bacillus subtilis* (see review by Hotchkiss and Gabour 1985) and limits application for genetic analysis.

The availability of substantial numbers of cloned genes has facilitated their application as probes for the location of homologous genes on the chromosome of other species. This novel approach was used by Canard and Cole (1989) to produce a genomic map for *C. perfringens*. These workers used six rare-cutting endonucleases to cut the chromosome into fragments that were separated by pulsed-field gel electrophoresis. A physical map of the chromosome was constructed and the position of some 24 genes for which probes were available were then localized to produce a genetic map for this species. The availability of these newer methods will make the isolation and characterization of mutants as genetic markers less important.

4.1.2 Use of Mutants for Analysis of Gene Function

The isolation and characterization of mutants has also provided useful approaches for investigating various biochemical, physiological, and morphological processes in the bacterial cell and have provided valuable information on gene function and regulation. Even when methods for obtaining gene transfer are not available, mutations arising in individual genes can still provide useful information about genetic function, by comparison with the phenotypic characteristics of the parental strain.

Among the clostridia, the isolation and characterization of mutant strains exhibiting defects in endospore formation have been used to investigate the relationship between the onset of sporulation and other phenomena such as toxin and solvent production. Various selection procedures for the isolation of asporogenous mutants have been used, including alterations in colonial morphology, colony staining, and isolation of rifampicin-resistant colonies. Such techniques have been used to generate a variety of asporogenous and oligosporogenous mutants from several different species of clostridia, blocked at various stages of endospore development.

A variety of sporulation defective mutants isolated from *C. histolyticum* was used to establish a relationship between sporulation and toxin production (Sebald and Schaeffer 1965; Sebald 1968). Sporulation mutants blocked at stage II or later produced toxin in normal amounts whereas those blocked before stage II development produced little or no toxin. A similar relationship between the production of the enterotoxin responsible for food poisoning and endospore formation was reported to occur in strains of *C. perfringens* (Sebald and Cassier 1969; Duncan et al. 1972). Mutant strains of *C. perfringens* defective in toxin and hemagglutinin production have also been reported by Tatsuki et al. (1981).

A link between the induction of endospore formation and alterations in the energy-generating metabolic pathways has been reported to occur in several different clostridial species (see review by Woods and Jones 1984). A relationship between solvent formation and sporulation was noted by early workers, among solvent-producing strains of clostridia used for the industrial production of acetone and butanol. This led to the practice of selecting high solvent-producing strains by subjecting cultures to cycles of sporulation, heat activation, and outgrowth to generate vigorous production strains with good solvent-producing ability (Beesch 1952). Most strains of *C. acetobutylicum* exhibit distinct morphological changes during the fermentation associated with the shift from the acidogenic phase to the solventogenic phase. Morphological changes, associated with the reduction in growth, include a loss of motility, accumulation of granulose as a storage product, and may include the formation of swollen cigar-shaped clostridial forms. These changes are normally followed by the initiation of forespore septation that, under the appropriate conditions, leads to the production of mature endospores (see reviews by Jones and Woods 1986; Rogers 1986; Jones and Woods 1989a).

The isolation and characterization of mutants blocked at various stages of morphological development have been used to investigate the relationship between solventogenesis and the onset of spore formation, as well as factors involved in triggering the shift to solvent production in *C. acetobutylicum*. Mutants unable to form the clostridial stage also failed to undergo the shift to solvent production, did not produce granulose, and failed to sporulate. However, asporogenous mutants that were blocked at stage 0 or stage II did undergo the shift to solvent production and produced granulose (Jones et al. 1982; Long et al. 1984). Isolation of similar types of mutants have also been reported from several solvent-producing strains (Rogers and Palosaari 1987; Clark et al. 1989). The characteristics of this class of pleiotropic mutant suggest that they may represent mutations that affect some common regulatory mechanism. The relationship between spore-forming ability and solvent production has also been investigated in continuous culture (Gottshal and Morris 1982; Meinecke et al. 1984; Stephens et al. 1985). Under the appropriate conditions, the selection of asporogenous mutants which may or may not exhibit a loss of solvent production was found to occur during continuous fermentation. The degeneration of solvent-producing strains and loss of spore-forming ability have been reported to be associated with changes in colonial morphology (Adler and Crow 1987).

In *C. botulinum*, the onset of sporulation has been observed to be accompanied by a shift in the ratio of the acid end products that are produced. Sporulating cells undergo a shift from predominantly acetate production to predominantly butyrate production. Using asporogenic mutants, Emeruwa et al. (1974) were able to show that mutants blocked at the early forespore stage failed to undergo the shift in metabolism. A similar link between the induction of endospore formation and other physiological and morphological changes in *C. thermosaccharolyticum* was investigated by Hsu and Ordal (1970). Actively growing vegetative cells typically produce acetate, butyrate, and lactate. However, under appropriate culture conditions the fermentation goes through an intermediate morphological stage associated with a shift to ethanol production. An asporogenous mutant isolated by Landuyt and Hsu (1985) remained in the intermediate stage and continued to produce ethanol instead of sporulating. Asporogenous mutants of *C. thermocellum* have also been isolated by Tailliez et al. (1989a, 1989b). A mutant selected for asporogeny alone, exhibited a similar fermentation pattern to the wild-type strain but produced enhanced levels of ethanol. A second mutant isolated by selection for resistance to fluoropyruvate, was pleiotropic and also exhibited enhanced ethanol tolerance, improved cellulose degradation and ethanol hyperproduction.

Several species of saccharolytic clostridia synthesize an amylopectin-like polyglucosan called *granulose*. When cells are grown in carbohydrate-rich medium, granulose is deposited in large granules within the cell. Granulose normally accumulates at the inception of the stationary phase and serves as an endogenous reserve material. It has also been postulated that granulose synthesis could serve as an overflow mechanism to overcome a block in sporulation (Mackey and Morris 1974). Mutants of *C. pasteurianum* defective in granulose synthesis (Robson et al. 1974) and granulose degradation (Mackey and Morris 1974) were used to investigate the physiology of granulose synthesis and utilization. Mutants deficient in granulose synthesis were also isolated from *C. acetobutylicum* (Jones et al. 1982; Reysenbach et al. 1986). Two classes of mutants could be distinguished. The one class consisted of mutants blocked in the biosynthetic pathway, which did not affect the ability of cells to form solvents or endospores. The second class of mutants were those blocked in all later events including solvent production and the formation of clostridial forms, capsules, and endospores. These appeared to be regulatory mutants. Under the appropriate growth conditions most strains of *C. acetobutylicum* are able to produce substantial amounts of extracellular capsular material during the later stages of the fermentation. Mutants that were defective in the biosynthesis of capsular material were isolated by Long et al. (1984), and employed in immobilized cell studies (Largier et al. 1985).

Many saccharolytic clostridia are characterized by the ability to alter both carbon and electron flow through their metabolic pathways during the course of the fermentation. Typically, the ratio of the end products shifts towards the production of more highly reduced acids or alcohols under adverse growth conditions, which prevail during the later stages of the fermentation. The isolation of mutants with defects in the pathways leading to the production of acids or solvents have made a significant contribution to the understanding of the biochemistry and physiology of end-product formation in several different species of clostridia. The most effective method of selecting these various types of pathway mutants has been through the use of suicide substrates. In the majority of cases, the suicide substrates are toxic analogues of the normal end products produced during the fermentation. The parental strain possessing the full metabolic capability is able to metabolize the toxic substrates that are lethal to the cell. Mutants, defective in the particular pathway, are unable to metabolize these substrates and survive.

Allyl alcohol acts as a suicide substrate being oxidized to the toxic aldehyde acrolein by the action of alcohol dehydrogenases (Rando 1974). This compound has been used to isolate mutants from several different microorganisms that are defective in alcohol dehydrogenase activity. Resistance to allyl alcohol was used to select spontaneous mutants of *C. acetobutylicum*, which failed to produce butanol while, acetone and ethanol production remained unaffected (Dürre et al. 1986). Surprisingly, these mutants were found to be defective in butyraldehyde-dehydrogenase activity rather than ethanol or butanol-dehydrogenase activity. Rogers and Palosaari (1987) made use of a similar strategy and were able to isolate allyl alcohol-resistant mutants from *C. acetobutylicum* that were defective in butanol and ethanol production. These mutants produced significant amounts of butyraldehyde and contained normal levels of CoA-dependent butyraldehyde dehydrogenase but showed a marked reduction in activity of butanol and ethanol dehydrogenase activity.

This treatment and selection procedure also generated several mutants that were defective in both solvent production and the ability to produce endospores. These were similar to the solvent-defective mutants reported previously (Jones et al. 1982). A further class of mutants that produced low levels of acids and enhanced levels of solvents were isolated and appeared to be prematurely induced for solvent production and had enhanced ability to recycle acids.

A similar type of strategy using suicide substrates consisting of halogen analogues of acetate and butyrate was employed to select for mutants of *C. acetobutylicum* that were unable to produce acetone but in which butanol and ethanol production remained largely unaffected (Janati-Idrissi et al. 1987; Junelles et al. 1987). Mutants selected using 2-bromobutyrate were unable to re-metabolize acids following the shift to solventogenesis. Butanol production in these mutants was reduced but ethanol production was unaffected. In contrast, mutants selected for resistance to 4-chloroacetate and other acetate analogues produced mutants unable to shift to the solventogenic phase. Clark et al. (1989) also reported the use of 2-bromobutyrate to isolate a mutant of *C. acetobutylicum* exhibiting reduced acetone production and showed that the only solvent pathway enzyme present in reduced quantity was CoA-transferase. These workers also isolated a second class of mutant that showed reduced activity of all solvent pathway enzymes similar to other regulatory mutants previously reported. In addition, rifampicin resistance was used to isolate a variety of mutants exhibiting reduced levels of solvent production, with differing patterns of solvent enzyme activity, which also appeared to be regulatory.

Yet another type of suicide substrate was used by El Kanouni et al. (1989) to select for mutants of *C. acetobutylicum* resistant to toxic halogen analogues of pyruvate. Some of these mutants accumulated more lactate and acetoin and underwent a quicker transition from the acidogenic to the solventogenic phase. Enzymic assays revealed that acetate kinase, butyrate kinase, and acetoacetate decarboxylase activities decreased sooner in the mutants than in the parental strain and acids were more rapidly re-metabolized.

Winkelman and Clark (1984) reported the use of a proton suicide method for isolation of fermentation mutants of *E. coli* based on lethal effects caused by the synthesis of acids. In this procedure, bromate and bromide sodium salts incorporated into the culture medium react to release toxic bromine in an acid environment. This proton suicide technique was used by Cueto and Mendez (1990) to isolate mutants of *C. acetobutylicum*, which showed alteration in the pattern of acid production. The first class of mutants exhibited lower acid production and higher butanol production and showed greater acid-recycling activity. Solvent production was induced by lower levels of inducer in these mutants. The second class of mutants that were defective acetate producers, did not achieve full growth but produced normal levels of butyric acid. A third class of mutants produced higher levels of acids and lower levels of solvents and appeared to be defective in acid recycling and acid re-utilization and appeared to be regulatory mutants. A double mutant exhibiting both defective acetone and acetate production was isolated following two cycles of mutagenic and selection treatment with toxic halogen analogues (Jones et al. 1990).

Similar approaches for the isolation and characterization of mutants showing defects in metabolic pathway enzymes have been applied to several species of ethanol-producing clostridia. Mutants of *C. saccharolyticum* which were unable to use pyruvate as a carbon source, showed enhanced substrate utilization and produced more cell biomass and ethanol but less acetic acid (Murray et al. 1983). Fluoroacetate resistance was used to select for mutants of *C. thermosaccharolyticum* deficient in acetate kinase or phosphotransacetylase activity (Rothstein 1986). Mutants deficient in acetate production produced more ethanol than the parental strain. Low acid-producing strains of *C. thermocellum* were isolated using bromo-cresol purple dye selection (Duong et al. 1983). A mutant strain was found to accumulate ethanol and acetate in a ratio of 8:1 during fermentation of cellobiose compared to the 1:1 ratio obtained with the parental strain.

Another area in which specific types of clostridial mutants have been used to investigate physiological processes has been in carbohydrate uptake and utilization. As part of a study on carbohydrate transport, *C. pasteurianum* mutants which were defective in the transport of specific carbohydrates were isolated and characterized (Booth and Morris 1982). Spontaneous changes that occurred in cultures of *C. thermocellum*, which gave the adapted strains the ability to grow on sugar substrates that could not be used by the parental strain, were investigated by Nochur et al. (1990). It appears that the selection of mutations which arose during the lag phase occurred in response to the selection pressure exerted when glucose or fructose were used as the sole carbon source. Recently, Mori (1990) described the isolation of mutants from *C. thermocellum* that exhibited enhanced cellulase activity compared with the parental strain. A further area in which the isolation of mutants could play a potentially useful role, is in elucidating the basis of oxygen toxicity. The isolation of aerotolerant mutants of *C. perfringens* was reported (Zavadova et al. 1974) but little further work appears to have been done in this field.

4.1.3 Use of Mutants for Practical Applications

In addition to their role as genetic markers and their use in elucidating gene function and regulation, several types of mutants have been isolated primarily for their potential value in practical applications. In the majority of cases, these have been mutants that confer some type of resistance which does not occur in the parental strain.

One of the major limitations associated with the use of clostridial fermentations, for the production of acids or alcohols, is the low tolerance exhibited to these end products. Various laboratories have attempted to enhance end-product tolerance by isolating acid and alcohol-resistant mutants from several different clostridia. The most widely used procedure has been to expose mutagenized cultures to high concentrations of acids or alcohols. It has been assumed that increased tolerance to acids or alcohols would enable resistant strains to produce higher concentrations of these end products. Although this has proved to be the case in some instances, in general, this type of approach has only proved to be moderately effective.

One of the major limitations associated with the acetone-butanol (AB) fermentation has been the low final concentration of solvents produced. It is, therefore, not surprising that attempts have been made by several laboratories to isolate mutant strains of *C. acetobutylicum* which exhibit enhanced tolerance to butanol (Lin and Blascheck 1982; Hermann et al. 1985; Lemmel 1985; Baer et al. 1987). The primary action of butanol, which is the most toxic of the solvents produced, appears to be due to its chaotrophic effect, which results in increased membrane fluidity and permeability, causing the disruption of membrane function. Although butanol-tolerant strains have been isolated, which produce higher concentrations of solvents than parental strains, the highest solvent concentrations which have been reported have not been significantly higher than those obtained using the best industrial strains, under optimal fermentation conditions. Only trace amounts of butanol were produced by the butanol-resistant mutant reported by Baer et al. (1987), demonstrating that increased solvent tolerance need not necessarily be linked to higher solvent production.

A similar approach has been applied to the isolation of mutants exhibiting increased tolerance to ethanol. However, the mechanism by which ethanol toxicity occurs in thermophilic species like *C. thermohydrosulfuricum* and *C. thermocellum* appears to be different. The ethanol-tolerant mutants of *C. thermohydrosulfuricum* isolated by Lovitt et al. (1984) showed pleiotropic characteristics which included enhanced growth, alterations in the substrate range and alteration in the ratio of fermentation end products, in addition to enhanced tolerance to ethanol and other solvents. Ethanol-tolerant mutants from *C. thermocellum* were also reported by Wang et al. (1983) and Herrero et al. (1985) and two different types of ethanol-tolerant mutants have been reported recently by Tailliez et al. (1989a, 1989b). One mutant was selected though its ability to grow in 4% ethanol. It produced mainly ethanol during early growth and exhibited an improved ethanol yield. The other mutant that was resistant to fluoropyruvate exhibited a variety of pleiotropic characteristics including ethanol-tolerance, asporogeny, decreased hydrogen production and altered end-product ratio. It has been postulated that the low ethanol tolerance of the thermophilic clostridia occurs through end-product inhibition of enzymes of the electron flow pathways rather than through disruption of membrane structure and function (Lovitt et al. 1984; Herrero et al. 1985). A pyruvate negative mutant of *C. saccharolyticum* that exhibited increased ethanol tolerance also displayed other pleiotropic effects including an alteration in the ratio of end product produced (Murray et al. 1983).

The production of acetic acid by homoacetogenic clostridia has also been the focus of several studies. An acid-tolerant mutant of *C. thermoaceticum* that was isolated after selection in pH-controlled continuous culture has been investigated by Schwartz and Keller (1982) and Brumm (1987). The mutant strain, which exhibited high acetic acid productivity under conditions of low pH and high concentrations of acetate (Schwartz and Keller 1982), was able to grow on lactate as a sole carbon source (Brumm 1987). This appeared to be due to the ability of the mutant strain to grow in media of higher redox potential.

An autolysin-deficient mutant, which was isolated from *C. acetobutylicum* by Allcock et al. (1981) as part of a study investigating protoplast formation and regeneration, was also shown to confer increased cell stability and butanol tolerance to the mutant strain (Van der Westhuizen et al. 1982). Sporulation mutants of *C. perfringens* were used to investigate the relationship between autolysin production and endospore formation (Labbe and Tang 1983). In addition to autolysins, several species of clostridia also produce bacteriocins. Mutants of *C. pasteurianum* exhibiting decreased sensitivity to a bacteriocin were isolated by Clarke et al. (1982). These mutants were used to investigate the mode of action of the bacteriocin which was shown to inhibit the activity of the membrane ATPase in vegetative cells.

Phage infections proved to be a serious problem in the AB fermentation resulting in substantial economic losses. Various attempts were made to combat the problem by isolating phage-resistant mutants of *C. acetobutylicum* and related species but with limited success (see review by Ogata and Hongo 1979).

A further class of mutants, which has potential application in genetic procedures, has been those showing reduced levels of deoxyribonuclease (DNase) activity. DNase activity has been reported to occur in several clostridial species and has been implicated in the interference of DNA transformation procedures (Lin and Blaschek 1984; Burchhardt and Dürre 1990). In an attempt to overcome this problem, mutants defective in DNase activity were isolated from *C. perfringens* (Blaschek and Klacik 1984) and *C. acetobutylicum* (Burchhardt and Dürre 1990).

4.2 POTENTIAL FOR APPLICATIONS IN BIOTECHNOLOGY

Close to 100 recognized species of clostridia, together, make up a very large and diverse assemblage of bacteria that includes several distinct phylogenetic lines (Gottschalk et al. 1981; Cato and Stackebrandt 1989). The great physiological diversity, and the wide variety of habitats occupied by the clostridia, provides considerable scope for applications in biotechnology (Andreesen et al. 1989; Morris 1989). The majority of clostridial species are heterotrophs and have the capability of degrading complex macromolecules, but there are also a few species that are autotrophic with the ability to grow on simple carbon substrates. Although the majority of species are mesophiles, the genus also includes a substantial number of thermophilic species.

Well over half of the reported and characterized species can be assigned to the group which exhibit the saccharolytic type of nutrition (Gottschalk et al. 1981). Most of these saccharolytic species are nonpathogenic and are characterized by having simple growth requirements. Most are able to grow heterotrophically on a wide range of simple hexose and pentose sugars and sugar alcohols and the majority of the species are characterized by an ability to degrade complex polysaccharides by means of a wide array of extracellular enzymes. Although many are able to use complex organic nitrogen sources, the majority are able to meet their nitrogen requirements by utilizing ammonium or other simple nitrogen sources. Several species can fix nitrogen and are able to grow in nitrogen-limited environments. Both mesophilic and thermophilic saccharolytic clostridia are ubiquitous in nature and play an important role in the breakdown of plant material under anaerobic conditions. They are commonly associated with decaying plant material such as compost and silage and are frequently involved in the spoilage of foodstuffs. Polysaccharides derived from plant biomass such as starch, pectin, hemicellulose, and cellulose provide the major sources of substrates. Species that are able to degrade starch and pectin are common and many species are also able to degrade xylans and other hemicelluloses. There are several species, including both thermophiles and mesophiles, which are capable of degrading cellulose by means of extracellular cellulolytic enzymes that, in some cases, have been demonstrated to be associated in multi-enzyme cellulosome complexes. In addition to carbohydrates, some species of clostridia are able to ferment organic acids and alcohols.

The major potential use of the saccharolytic clostridia in biotechnological processes relates to their ability to metabolize a broad diversity of carbohydrate substrates which may be converted to a limited range of end products including carbon dioxide (CO_2), hydrogen, acids, and alcohols (Jones and Woods 1989a; Lungdahl et al. 1989). The metabolism of these species is characterized by the ability to dispose of substantial amounts of excess reducing power in the form of molecular hydrogen under optimal conditions. This enables carbon flow to be diverted away from the production of reduced acids and alcohols to the production of acetate which generates additional ATP for the cell. Among the acid-producing clostridia the most commonly formed acids are acetate and butyrate but many species have the ability to switch to lactate under suboptimal growth conditions. In addition, propionate, formate, lactate, succinate, and caprolate may also be produced by some species while valerate, isovalerate, and isobutyrate may be produced by clostridial species that ferment amino acids. Several species of butyric acid bacteria are able to undergo a shift from acid production to solvent production during fermentation. These species usually produce a mixture of solvents including butanol, ethanol, and acetone or isopropanol.

Among the thermophiles, most species produce a mixture of acids and solvents during growth. These normally consist of varying proportions of acetate and ethanol, however, butyrate has been reported to be formed in two species. The mixture of end products that are produced have potential for the production of bulk chemical feedstocks. However, the branched pathways leading to a mixture of reduced end products, coupled with low final concentrations of the end products, is a distinct disadvantage and imposes serious limitations on the use of fermentation processes for the production of these products. The formation of a single product during fermentation would give higher yields and be advantageous for recovery. The homoacetogenic clostridia constitute a unique group of acid producers that are able to grow autotrophically on hydrogen and CO_2 and several other single carbon compounds to form acetate alone.

Several species of clostridia are characterized by their ability to ferment nitrogen-containing compounds including amino acids, purines, pyridimines, and amide compounds. A wide variety of species are also able to degrade complex nitrogen-containing polymers such as proteins and nucleic acids (Andreesen et al. 1989). Many of the medically important and pathogenic species are characterized by being proteolytic and these species often produce a wide range of exoproteins, many of which may function as virulence factors. Several of the medically important species are also characterized by their ability to produce potent toxins and disease results primarily from contamination of wounds or foodstuffs.

In addition to being able to metabolize the wide range of organic compounds already mentioned, some species can catabolize aromatic, homocyclic, and heterocyclic compounds in the absence of oxygen (Morris 1989). This makes these highly reducing organisms particularly well suited and potentially useful for reductive bioconversions. One of the most notable features of the clostridia is the large range of complex macromolecules that can be degraded by means of a wide variety of extracellular enzymes. The group as a whole appears to have evolved a wide range of efficient catabolic enzymes, which gives many species an interesting potential as a source of enzymes with novel characteristics. Several of these enzymes have been identified as having potential application in the food, chemical, and pharmaceutical industry as well as in health care and research (Saha et al. 1989).

4.3 STRATEGIES FOR APPLICATIONS IN BIOTECHNOLOGY

Although the metabolic diversity found among the clostridia provides many potentially useful characteristics that might find application in biotechnology, few successful processes involving clostridial species have been established. With the exception of small scale use for enzyme and toxin production and in bioconversion, the only industrial scale process making use of pure cultures of single species of clostridia to emerge has been the AB fermentation. The intrinsic constraints imposed by the metabolic capabilities of naturally occurring species of clostridia mitigate against their effective utilization for practical applications. The need to find ways of overcoming some of these limitations presents an exciting but daunting challenge for biotechnologists.

The manipulation of environmental factors to provide physiological control can be used to optimize growth and product formation and achieve overproduction. In addition, improvements in the process technology can make a significant difference to economic viability of the particular fermentation. Improvements can be achieved by providing more efficient ways of pretreatment and processing of substrates and improved fermentation control. Improved productivity of the fermentation process may also be achieved through the use of continuous culture and biomass retention systems. More efficient and cost-effective ways of product recovery, effluent disposal, and downstream processing can also have a major influence on the economic viability of the process.

There are, however, limits to how far these approaches can be taken to overcome the constraints imposed by the physiology of a particular species. In many cases, the only feasible way of achieving the level of improvement required is by radically altering the basic metabolic processes of the particular strain or species. However, lack of fundamental knowledge relating to the genetics and regulation of the individual species coupled with the lack of functional laboratory procedures for genetic manipulation for most species of clostridia has hampered the application of this approach to clostridial fermentations. In many cases, mutagenesis and selection still provide the only practical way of attempting any type of genetic manipulation for most clostridial species. The use of mutagenesis and selection alone tend to be limited and the application of recombinant DNA technology and related approaches will be essential if real progress is to be made.

The nature of the limitations associated with the application of clostridial fermentations differ depending on the species being used and the type of process envisaged. Mixed culture systems employing various species of clostridia play a fundamental role in initial degradative processes in anaerobic digestion and waste treatment that have important applications in biotechnology. The involvement of mixed populations of clostridia in retting, composting, and silage making has been well documented, and more recently mixed culture systems have been identified as having scope in the treatment of contaminated soils. The potential for improving these mixed culture systems using genetically modified strains of clostridia would, however, appear to be rather limited due to the complexity of these processes. The likelihood of displacement of introduced genetically modified strains by the native wild-type stains, and the general lack of control over such systems, make use of genetically modified strains difficult to envisage.

The use of genetically modified strains in fermentation systems employing just one or two species maintained in pure culture presents less problems and offers more promise. Currently, the use of proteolytic species of clostridia for application in biotechnology appear to have more limited potential than the saccharolytic group.

Potential uses of the proteolytic group have mostly been focused on production of enzymes, toxins, and other cell metabolites. The main use of mutation and selection procedures would be in obtaining overproduction of cell metabolites and enzymes or enhancement of bioconversion efficiency. However, a fundamental limitation associated with the production of enzymes by anaerobic fermentation is the low yield of cell biomass compared with the high yield of end product, which occurs as a consequence of energy limitation, associated with anaerobic metabolism. One possibility of overcoming this problem is the transfer of genes coding for a particular enzyme or product to an aerobic host cell system.

A variety of mesophilic and thermophilic species, belonging to the saccharolytic group, have potential for use in the industrial production of chemical feedstocks. However, the only successful large-scale industrial process has been the AB fermentation utilizing *C. acetobutylicum* and related strains. This fermentation process has been in use for more than 75 years and is still operating in the People's Republic of China, although, by the early 1960s, it had become economically uncompetitive in most western countries. However, in spite of its long history as an industrial microorganism, only limited attempts were made to improve strains by selection or mutation. Since 1980, the amount of research carried out on this fermentation has undergone a substantial increase (Jones and Woods 1986; Rogers 1986). The use of cellulolytic clostridia for the production of ethanol has been the focus of a second major area of research. Both thermophilic and mesophilic species have attracted interest, but inherent limitations associated with these fermentations have hindered the development of functional fermentation processes.

All of the fermentations using saccharolytic clostridia for the conversion of biomass into alcohols or acids suffer similar intrinsic limitations. These include the high proportionate cost of the fermentation substrate, the low yield and final concentration of the required end product and the low productivity and complexity of the fermentation process. Together these limitations make the fermentation processes ineffective and uncompetitive.

Improvements in the fermentation and process technology can only go part way to overcoming these limitations. The development of industrial strains using genetic manipulation, to give improved characteristics and performance is clearly required if progress is to be made (Jones and Woods 1989b). The basic strategies required to make these fermentations economically viable are similar, and are applicable to all processes, irrespective of the species of saccharolytic clostridia involved, the type of substrate used and end products produced.

In these fermentations, the cost of the fermentation substrate accounts for a substantial portion of the cost of producing the product. Most common sugar- and starch-based fermentation substrates can be effectively used by a wide range of saccharolytic clostridial species. However, the relatively high cost of agriculturally derived substrates tends to preclude their use for the production of low cost chemical feed stocks. Various types of agricultural and industrial wastes have been investigated as alternatives, but the majority are not suitable because of low nutrient concentration, unsuitable nutrient composition, toxicity, or availability in sufficient volume. Lignocellulosic wastes, which are produced in huge volumes by forestry, agricultural, and industrial processing, have great potential as substrates for anaerobic fermentation processes, but the use of these recalcitrant substrates still awaits efficient and economical processes for the conversion to fermentable sugars. Many of the potentially useful species such as *C. acetobutylicum* are able to use pentose sugars such as xylose and arabinose which comprise the bulk of the pentose component of hemicellulose. However, lower yields are obtained, and hexoses are used in preference to pentoses. Although several species of thermophilic and mesophilic clostridia are able to metabolize cellulose directly to potentially useful end products, at present the yield, productivity, and concentration of end products achieved in these fermentations fall far short of what would be required for an industrial process.

There has been limited use of mutagenesis and selection to improve the efficiency and range of substrate utilization. A more promising approach involves the transfer of genes encoding for the enzyme complexes required for the degradation of cellulose to species such as *C. acetobutylicum* (see Chapter 12, this volume). The development of more compatible strains for species which can be used in co-culture systems could also provide a further target for genetic manipulation.

One of the major limitations associated with all saccharolytic clostridial fermentations is the relatively low concentration of end products that are produced during the course of the fermentation. The acids and alcohols that are produced are highly reduced, and are extremely toxic to biological systems. This results in these products becoming inhibitory when the concentration reaches a few percent in the culture medium. The mechanism of toxicity of both the types of end products has been identified as being primarily due to detrimental effects on cell membranes that affect integrity, stability, and function. In some instances, metabolic pathways and enzyme activity may also be disrupted but the final limitation appears to be due to a decrease in the efficiency of the membrane as a barrier and the associated disruption in function in energy generation and transport.

A straightforward approach using mutagenesis and selection to increase tolerance by selecting cells that can survive in high concentrations has been tried with several species, but has produced only limited improvement. Membrane fluidity is known to depend largely on the characteristics of the membrane phospholipids, such as chain length and degree of saturation. Currently, the limited amount that is known about these complex systems has hampered attempts to manipulate membrane composition and structure either physiologically or genetically. A second approach could encompass the development of more efficient proton pumping mechanisms as a way of overcoming proton re-entry. The application of these more sophisticated approaches is likely to be challenging and difficult and will have to wait for a better understanding to emerge.

A further limitation associated with saccharolytic clostridial fermentations is the low yield of the individual end products that are achieved in these fermentations. Due to the proportionate high cost of the substrate in these fermentations this can represent a considerable economic loss. The main reason for the low yields is the large degree of flexibility provided by the branched catabolic pathways resulting in the generation of diversity of end products. The clostridia are able to dispose of excess reducing power in the form of molecular hydrogen. This provides the most efficient route for the disposal of electrons and protons, as it allows the electron flow to be diverted away from the production of other highly reduced end products, enabling the cell to divert much of the carbon flow to acetate production, which generates additional ATP. Maximum hydrogen production only appears to be achievable under optimal growth conditions. As conditions in the fermentation change, and end products accumulate, the diversion of electrons away from hydrogen production to fatty acid or alcohol production increases. Alterations in electron flow away from hydrogen production has been achieved physiologically by several methods (see reviews by Jones and Woods 1986; Rogers 1986; Jones and Woods 1989a). This type of regulation would, however, be difficult or impossible to achieve economically in an industrial fermentation process. An attractive alternative would be to obtain the required metabolic shift by genetic manipulation.

In the AB fermentation, the most desirable product is butanol, and in the case of thermophilic fermentations, the most desirable product is ethanol. Ideally, if a complete diversion of electrons away from hydrogen production into a single reduced end product such as ethanol or butanol could be achieved the only other end product would be CO_2. In the ethanol fermentation, this would entail inactivating the pathways leading to hydrogen, acetate, and lactate production. In the butanol fermentation, the situation would be even more complex as in addition to these three pathways, the pathways leading to butyrate production and acid recycle, coupled to acetone production, would also have to be inactivated. Some progress in this direction has been achieved through the isolation of mutant strains using suicide substrates that exhibit defects in individual pathways. It is not clear what the consequences of inactivating energy generating pathways leading to acetate and butyrate would be. It is conceivable that growth rate and the amount of cell biomass produced could be substantially reduced. It is possible that the use of biomass retention systems such as cell immobilization or cell recycle could be used to overcome such limitations.

The relatively low productivity associated with fermentation processes using saccharolytic clostridia also affects the viability of these processes. This problem is greatest when the traditional batch type of fermentation process is employed. In the case of the AB fermentation there have been several approaches to try and overcome this limitation by employing novel laboratory scale and industrial scale fermentation processes (see reviews by Jones and Woods 1986; Rogers 1986; Jones and Woods 1989a). Single stage continuous fermentation has not been successful for industrial application due to the complexity of the fermentation and the oscillation that occurs between acid and solvent production. Two-stage and multi-stage continuous systems have, however, been used with some success. In addition, systems employing biomass retention by cell immobilization or cell recycle show considerable promise, particularly when coupled with continuous solvent extraction. Asporogenous mutants, granulose defective mutants, and noncapsulated mutants have shown promise for application in these systems (Largier et al. 1985). Genetically modified strains able to maintain high levels of alcohol production throughout the fermentation would be of great advantage in improving productivity. Genetically modified strains with improved capabilities for substrate binding, floc formation, biofilm formation, cell harvesting, and filtration could also play a role in increasing the effectiveness of this type of process technology.

The loss of solvent-producing ability and strain degeneration have been reported to occur after prolonged batch and continuous culture (see reviews by Jones and Woods 1986, 1989a). The development of strains with enhanced stability or resistance to autolysis might prove to be a feasible way of overcoming problems of culture degeneration.

4.4 CONCLUSION

In spite of the development of new technologies for genetic manipulation, mutagenesis and selection are likely to retain their importance in clostridial research for some time to come, particularly for those species where genetic studies are rudimentary or nonexistent. The main advantage of these techniques is that they are relatively straightforward, simple, and easy to adapt. The disadvantages remain the random, uncontrolled nature of the mutants that are generated and the often laborious and time-consuming methods required for screening and isolation of mutants without guarantee of success.

REFERENCES

Adler, H.I., and Crow, W. (1987) *Appl. Environ. Microbiol.* 53, 2496–2499.

Aharnowitz, Y., and Cohen, G. (1981) *Sci. Am.* 245, 140–152.

Allcock, E.R., Reid, S.J., Jones, D.T., and Woods, D.R. (1981) *Appl. Environ. Microbiol.* 42, 929–935.

Andreesen, J.R., Bahl, H., and Gottschalk, G. (1989) in *Clostridia* (Minton, N.P., and Clarke, D.J., ed.), pp. 27–53, Plenum Press, New York.

Baer, S.H., Blaschek, H.P., and Smith, T.L. (1987) *Appl. Environ. Microbiol.* 53, 2854–2861.
Beesch, S.C. (1952) *Eng. Proc. Dev.* 44, 1677–1682.
Bertram, J., and Dürre, P. (1989) *Arch. Microbiol.* 151, 551–557.
Bertram, J., Kuhn, A., and Dürre, P. (1990) *Arch. Microbiol.* 153, 373–377.
Blaschek, H.P., and Klacik, M.A. (1984) *Appl. Environ. Microbiol.* 48, 178–181.
Booth, I.R., and Morris, J.G. (1982) *Biosci. Rep.* 2, 47–53.
Bowring, S.N., and Morris, J.G. (1985) *J. Appl. Bacteriol.* 58, 577–584.
Brumm, P.J. (1987) *Biotechnol. Bioeng.* 32, 444–450.
Burchhardt, G., and Dürre, P. (1990) *Curr. Microbiol.* 21, 307–311.
Canard, B., and Cole, S.T. (1989) *Proc. Natl. Acad. Sci.* 6, 676–680.
Carlton, B.C., and Brown, B.J. (1981) in *Manual of Methods for General Bacteriology* (Gerhardt, P., ed.), pp. 222–242, American Society for Microbiology, Washington, DC.
Carrasco, A., and Soto, C. (1987) *J. Appl. Bacteriol.* 63, 539–543.
Cato, E.P., and Stackebrandt, E. (1989) in *Clostridia* (Minton, N.P., and Clarke, D.J., eds.), pp. 1–26, Plenum Press, New York.
Clarke, D.J., Kell, D.B., Morley, C.D., and Morris, J.G. (1982) *Arch. Microbiol.* 131, 81–86.
Clark, S.W., Bennett, G.N., and Rudolph, F.B. (1989) *Appl. Environ. Microbiol.* 55, 970–976.
Cueto, P.H., and Mendez, B.S. (1990) *Appl. Environ. Microbiol.* 56, 578–580.
Daldal, F. (1985) *Arch. Microbiol.* 142, 93–96.
Davies, A., Oultram, J.D., Pennock, A., et al. (1988) in *Genetics and Biotechnology of Bacilli*, vol. 2 (Ganesan, A.T., and Hoch, J.A., eds.), pp. 391–395, Academic Press, London.
Drake, J.W., and Baltz, R.H. (1976) *Ann. Rev. Biochem.* 45, 11–37.
Duncan, C.L., Strong, D.H., and Sebald, M. (1972) *J. Bacteriol.* 110, 378–391.
Duong, T.V.C., Johnson, E.A., and Demain, A.L. (1983) *Topic. Enzymol. Ferm. Biotechnol.* 7, 156–195.
Dürre, P., Kuhn, A., and Gottschalk, G. (1986) *FEMS Microbiol. Lett.* 36, 77–81.
El Kanouni, A., Junelles, A-M., Janati-Idrissi, R., Petitdemange, H., and Gay, R. (1989) *Curr. Microbiol.* 18, 139–144.
Emeruwa, A.C., Hawriko, R.Z., Halvorson, H., and Suzuki, I. (1974) *J. Bact.* 120, 74–80.
Gottschal, J.C., and Morris, J.G. (1982) *Biotechnol. Lett.* 4, 477–482.
Gottschalk, G., Andreesen, J.R., and Hippe, H. (1981) in *The Prokaryotes* (Starr, M.P., Stolp, H., Truper, H.G., Balows, A., and Schlegel, H.G., eds.), pp. 1767–1803, Springer-Verlag KG, Berlin.
Hermann, M., Fayolle, F., Marchal, R. et al. (1985) *Appl. Environ. Microbiol.* 50, 1238–1243.
Herrero, A.A., Gomez, R.F., and Roberts, M.F. (1985) *J. Biol. Chem.* 260, 7442–7451.
Hopwood, D. (1981) *Sci. Am.* 245, 154–178.
Hotchkiss, R.D., and Gabour, M.H. (1985) in *The Molecular Biology of the Bacilli, vol. 2* (Dubnau, D.A. ed.), pp. 109–149, Academic Press, Orlando, FL.
Hsu, E.J., and Ordal, Z.J. (1970) *J. Bacteriol.* 102, 369–376.
Janati-Idrissi, R., Junelles, A-M., El Kanouni, A., Petitdemange, H., and Gay, R. (1987) *Ann. Inst. Pasteur/Microbiol. (Paris)* 138, 313–323.

Jones, D.T., Chauveau, M.T.N., Youngleson, J.S., and Woods, D.R. (1990) in *Fermentation Technologies: Industrial Applications* (Yu, P., ed.), pp. 22–27. Elsevier Applied Science, New York.
Jones, D.T., Jones, W.A., and Woods, D.R. (1985) *General Microbiol.* 131, 1213–1216.
Jones, D.T., Van der Westhuizen, A., Long, S. et al. (1982) *Appl. Environ. Microbiol.* 43, 1434–1439.
Jones, D.T., and Woods, D.R. (1986) *Microbiol. Rev.* 50, 484–524.
Jones, D.T., and Woods, D.R. (1989a) in *Clostridia* (Minton, N.P., and Clarke, D.J., eds.), pp. 105–144, Plenum Press, New York.
Jones, D.T., and Woods, D.R. (1989b) in *Recombinant DNA and Bacterial Fermentations* (Thomson, J.A., ed.), pp. 255–276, CRC Press, Boca Raton, FL.
Junelles, A.-M., Janati-Idrissi, R., El Kanouni, A., Petitdemange, H., and Gay, R. (1987) *Biotechnol. Lett.* 9, 175–178.
Labbe, R.G., and Tang, S.S. (1983) *Can. J. Microbiol.* 37, 829–832.
Landuyt, S.L., and Hsu, E.J. (1985) in *Fundamental and Applied Aspects of Bacterial Spores* (Ding, G.J., Gould, G.W., and Ellar, D.J., eds.), pp. 477–493, Academic Press, New York.
Largier, S.T., Long, S., Santangelo, J.D., Jones, D.T., and Woods, D.R. (1985) *Appl. Environ. Microbiol.* 50, 477–481.
Lemmel, S.A. (1985) *Biotechnol. Lett.* 7, 711–716.
Lin, Y.L., and Blascheck, H.P. (1982) *Appl. Environ. Microbiol.* 45, 966–973.
Lin, Y.L., and Blascheck, H.P. (1984) *Appl. Environ. Microbiol.* 48, 737–742.
Long, S., Jones, D.T., and Woods, D.R. (1984) *Biotechnol. Lett.* 6, 529–534.
Lovitt, R.W., Longin, R., and Zeikus, J.G. (1984) *Appl. Environ. Microbiol.* 48, 171–177.
Lungdahl, L.G., Hugenholtz, J., and Wiegel, J. (1989) in *Clostridia* (Minton, N.P., and Clarke, D.J., eds.), pp. 145–192, Plenum Press, New York.
Mackey, B.M., and Morris, J.G. (1974) *FEBS Lett.* 48, 64–67.
Meinecke, B., Bahl, H., and Gottschalk, G. (1984) *Appl. Environ. Microbiol.* 48, 1064–1065.
Mendez, B.S., and Gomez, R.F. (1982) *Appl. Environ. Microbiol.* 43, 495–496.
Mori, Y. (1990) *Agric. Biol. Chem.* 54, 825–826.
Morris, J.G. (1989) in *Clostridia* (Minton, N.P., and Clarke, D.J., eds.), pp. 193–226, Plenum Press, New York.
Murray, W.D., Wemyss, K.B., and Kahn, A.W. (1983) *Eur. J. Appl. Microbiol. Biotechnol.* 18, 71–74.
Nochur, S.V., Roberts, M.F., and Demain, A.L. (1990) *FEMS Microbiol. Lett.* 71, 199–204.
Ogata, S., and Hongo, M. (1979) *Adv. Appl. Microbiol.* 25, 241–273.
Oultram, J.D., Davis, A., and Young, M. (1987) *FEMS Microbiol. Lett.* 42, 113–119.
Rando, R.R. (1974) *Biochem. Pharmacol.* 23, 2328–2331.
Reysenbach, A.L., Ravenscroft, N., Long, S., Jones, D.T., and Woods, D.R. (1986) *Appl. Environ. Microbiol.* 52, 275–281.
Robson, R.L., Robson, R.M., and Morris, J.G. (1974) *Biochem. J.* 144, 503–511.
Rogers, P. (1986) *Adv. Appl. Microbiol.* 31, 1–60.
Rogers, P., and Palosaari, N. (1987) *Appl. Environ. Microbiol.* 53, 2761–2766.
Rothstein, D.M. (1986) *J. Bacteriol.* 165, 319–320.

Saha, B.C., Lamed, R., and Zeikus, J.G. (1989) in *Clostridia* (Minton, N.P., and Clarke, D.J., eds.), pp. 227–264, Plenum Press, New York.
Schwartz, R.D., and Keller, F.A. (1982) *Appl. Environ. Microbiol.* 43, 117–123.
Sebald, M. (1968) *Ann. Inst. Pasteur (Paris)* 114, 265–267.
Sebald, M., and Cassier, M. (1969) in *Spores IV* (Campbell, L.L., ed.), pp. 306–316, American Society for Microbiology, Washington, DC.
Sebald, M., and Costilow, R.N. (1975) *Appl. Microbiol.* 29, 1–6.
Sebald, M., and Schaeffer, P. (1965) *Comptes. Rendus.* 260, 5398.
Stephens, G.M., Holt, R.A., Gottschal, J.C., and Morris, J.G. (1985) *J. Appl. Bacteriol.* 59, 597–605.
Tailliez, P., Girard, H., Longin, R., Beguin, P., and Millet, J. (1989a) *Appl. Environ. Microbiol.* 55, 203–206.
Tailliez, P., Girard, H., Millet, J., and Beguin, P., (1989b) *Appl. Environ. Microbiol.* 55, 207–211.
Tatsuki, T., Imagawa, T., Kitajma, H., et al. (1981) *Bikens J.* 24, 1–11.
Van der Westhuizen, A., Jones, D.T., and Woods, D.R. (1982) *Appl. Environ. Microbiol.* 44, 1277–1282.
Volk, W.A., Bizzini, B., Jones, K.R., and Macrina, F.L. (1988) *Plasmid* 19, 255–259.
Walker, G.C., (1984) *Microbiol. Rev.* 48, 60–93.
Wang, D.I.C., Avgerinos, G.C., Biocic, I., Wang, S.D., and Fang, H.Y. (1983) *Phil. Trans. R. Soc. London B* 300, 323–333.
Winkelman, J.W., and Clark, D.P. (1984) *J. Bacteriol.* 160, 687–690.
Woods, D.R., and Jones, D.T. (1984) *Adv. Microb. Physiol.* 27, 1–64.
Woolley, R.C., Pennock, A., Ashton, R.J., Davies, A., and Young, M. (1989) *Plasmid* 22, 169–174.
Young, M., Minton, N.P., and Staudenbauer, W.L. (1989) *FEMS Microbiol. Rev.* 63, 301–326.
Yu, P.L., and Pearce, L.E. (1986) *Biotechnol. Lett.* 8, 469–474.
Zavadova, M., Mikulik, K., Sebald, M. (1974) *Cc. Epidemiol. Mikrobiol. Imunol.* 23, 249–256.

CHAPTER
5

Development and Exploitation of Conjugative Gene Transfer in Clostridia

Michael Young

Genetic exchange between bacteria was discovered almost 50 years ago, when the occurrence of gene transfer from one strain of *Escherichia coli* K-12 to another by a process called *conjugation*, requiring close cell-to-cell contact, was documented by Lederberg and Tatum (1946). Strains able to act as donors contain a large plasmid, called the fertility or F factor (Hayes 1953). It was 30 years later that genetic exchange in *Clostridium perfringens* was first detected by Sebald and Bréfort (1975). As with *E. coli*, gene transfer required close cell-to-cell contact, and a large plasmid called pIP401 was implicated in the process. Another decade elapsed before conjugative gene transfer was documented in the nonpathogenic, saccharolytic species, *C. acetobutylicum* (Oultram and Young 1985; Reysset and Sebald 1985; Yu and Pearce 1986).

Within 15 years of its discovery, much had been learned about the role of the F factor and the mechanism of gene transfer during conjugation in *E. coli* (reviewed by Hayes 1968). Moreover, conjugation was routinely being employed

Work carried out in the author's laboratory was supported by the SERC Biotechnology Directorate and the Biotechnology Action Programme of the EEC.

in many laboratories as a convenient tool for assigning mutations to loci on the *E. coli* chromosome. A similar interval has now elapsed since conjugative plasmids were first detected in *C. perfringens*. The contrast with *E. coli* could hardly be more striking. Comparatively little has been learned about these plasmids in the interim period. It is curious that apart from acting as a useful source of homologous antibiotic resistance genes for vector development (reviewed by Young et al. 1989a and by Rood and Cole 1991), they have so far played only a minor role in the emergent field of clostridial genetics.

At the time when Sebald and Bréfort (1975) were undertaking their pioneering work on the transfer of plasmid pIP401 between strains of *C. perfringens*, little was known about conjugation in Gram-positive bacteria in general. Plasmids with fertility properties had been described in streptomycetes (Vivian 1971) and conjugative gene transfer between strains of *Enterococcus faecalis* had also been detected (Raycroft and Zimmermann 1964; Tomura et al. 1973; Jacob and Hobbs 1974). Since 1980, it has become clear that conjugative plasmids, as well as a novel class of genetic elements called *conjugative transposons* (Clewell and Gawron-Burke 1986), play a seminal role in mediating gene transfer between Gram-positive bacteria. Much of what is currently known about conjugative gene transfer in Gram-positive bacteria derives from the pioneering studies of D.B. Clewell and his colleagues, who have worked extensively on the promiscuous conjugative genetic elements found in enterococci and streptococci (reviewed by Clewell 1981, 1990). This chapter draws extensively on their findings. Since many of the conjugative plasmids and transposons found in enterococci and streptococci have an extremely broad host range, they have proved particularly useful for genetic analysis in a wide variety of Gram-positive bacteria.

Genetic strategies initially developed for use in other Gram-positive bacteria have recently been adapted for use in clostridia (reviewed by Young et al. 1989a and Rood and Cole 1991). A major driving force behind these developments has been a growing awareness of the biotechnological potential of the saccharolytic and toxigenic clostridia. Strategies that rely on conjugation for transferring exotic genetic elements from a variety of donor organisms have proved particularly useful in these early days of clostridial genetics. Their development and exploitation will form the main thrust of this chapter. The rather limited information currently available concerning indigenous conjugative plasmids and transposons will also be summarized. These indigenous elements are likely to assume greater importance in the future, as they are studied for their own intrinsic scientific interest and developed as genetic tools that might be particularly well adapted to functioning in their natural hosts.

5.1 CONJUGATIVE PLASMIDS

5.1.1 General Characteristics

Conjugative plasmids are large (usually circular) DNA elements that contain *tra* genes encoding functions required for their *tra*nsfer from one bacterium to another. They are frequently maintained at a low copy number per cell and often encode additional functions, such as resistance to antibiotics, by which their transfer from one cell to another can be followed. Several of the conjugative plasmids found in Gram-negative bacteria, particularly those belonging to incompatibility groups F, I, and P, have been extensively characterized (Willetts and Wilkins 1984). It seems likely that a common mechanism governs their transfer from cell to cell. The F-mediated conjugation is initiated when contact is made between appendages (pili) on the surface of plasmid-containing donor cells and recipient cells lacking the plasmid. Pilus retraction into the donor cell brings the mating bacteria (pairs or aggregates) into intimate contact. Plasmid transfer is initiated as a result of the activity of an endonuclease that introduces a nick in the phosphodiester backbone of one DNA strand at a specific site called the origin of transfer (*oriT*). As rolling circle replication proceeds in the donor cell, the displaced single strand is transferred to the recipient, starting with the 5′ extremity, to which the site-specific endonuclease remains attached. Complementary strand synthesis occurs in the recipient cell as conjugative transfer proceeds. Finally, after completion of DNA transfer, the endonuclease recognizes the 3′ terminus and ligates it to the 5′ extremity to which it has remained attached. These events are described in more detail in the review by Willetts and Wilkins (1984).

In addition to mediating their own transfer, many conjugative plasmids can also mobilize other nonconjugative DNA elements including, in the case of the F factor, the bacterial chromosome. Indeed, it is this property, rather than transfer of the plasmids themselves, that has made them such useful tools for undertaking gene transfer and genetic mapping in bacteria. In some cases, mobilization occurs via the formation of co-integrate DNA molecules. These can arise as intermediates in the transposition of certain mobile genetic elements from one replicon to another. Alternatively, co-integrate molecules may be generated as a result of homologous (RecA-dependent) recombination between identical sequences—often IS elements— present on both replicons. In other cases, mobilization may occur following the productive interaction of the origin-nicking protein complex, synthesized by a conjugative plasmid, with an *oriT* site on an otherwise unrelated, and physically discrete, co-resident replicon. As will become apparent, both mechanisms of conjugative mobilization have proved useful for transferring genes to clostridia.

5.1.2 Indigenous Conjugative Plasmids

Large conjugative R plasmids have been found in many strains of *C. perfringens* independently isolated from different environments (Bréfort et al. 1977; Rood et al. 1978; Rood 1983; Heefner et al. 1984; Abraham and Rood 1985a; Abraham et al. 1985) but, apart from one report of a conjugative plasmid associated with methyl mercury-resistance in *C. cochlearium* (Pan-Hou et al. 1980), they have not been detected in other species of *Clostridium*. In contrast to the considerable diversity that exists amongst the conjugative plasmids found in Gram-negative bacteria, those found in *C. perfringens* are all closely related; indeed, several independently isolated plasmids were indistinguishable from each other (Magot 1984; Abraham and Rood 1985a, 1985b) and they all confer resistance to tetracycline (*tetP*) via an efflux mechanism (Abraham et al. 1988). They are transferred by a process requiring close cell-to-cell contact between different strains of *C. perfringens* (Sebald and Bréfort 1975; Bréfort et al. 1977; Rood et al. 1978), but nothing is currently known about the conjugation mechanism. With the exception of *C. difficile*, in which pIP401 appears not to be stably inherited (Ionesco 1980), transfer of pIP401 to other clostridia has not been reported.

Some of the conjugative plasmids found in *C. perfringens* also encode an acetyl transferase that confers resistance to chloramphenicol (*catP*) (Abraham et al. 1985; Steffen and Matzura 1989). During conjugative transfer, loss of the chloramphenicol resistance determinant often occurs, associated with loss of a 6.2 kilobase pair (kbp) DNA segment (Bréfort et al. 1977; Abraham et al. 1985). When cloned in *E. coli* on temperature-sensitive plasmids, the chloramphenicol-resistance determinants of plasmids pIP401 and pJIR27 (Abraham et al. 1985) undergo RecA-independent transposition to the bacterial chromosome, confirming that they lie on transposons, denoted Tn*4451* and Tn*4452*, respectively (Abraham and Rood 1987). Deletion derivatives of the 54 kbp plasmid pIP401, which no longer confer resistance to chloramphenicol, are indistinguishable from the 47 kbp plasmid, pCW3 (Rood et al. 1978; Abraham et al. 1985).

Physical maps of several of these conjugative plasmids have been constructed (Magot 1984; Abraham and Rood 1985a; Abraham et al. 1985), but the only genetic markers so far assigned to them are *tetP* and *catP*. Methods for high frequency electro-transformation of *C. perfringens* have now been developed (Allen and Blaschek 1988, 1990; Kim and Blaschek 1989; Scott and Rood 1989) and over the next few years, as details of the genetic organization of these clostridial R-plasmids begin to emerge, it will become apparent whether the mechanism by which conjugation proceeds is similar to that outlined above for the conjugative plasmids found in Gram-negative bacteria.

5.1.3 Exotic Conjugative Plasmids in Clostridia

The saccharolytic clostridia seem to lack indigenous conjugative plasmids. But several investigators have reported the conjugative transfer of broad host range MLS^R plasmids originating in enterococci and streptococci to *C. acetobutylicum* (Oultram and Young 1985; Reysset and Sebald 1985; Yu and Pearce 1986). Among the plasmids transferred were pAMβ1, pIP501 and its derivative pVA797, and pJH4. Plasmid pAMβ1 has also been introduced into *C. butyricum* and *C. pasteurianum* by conjugation (Young et al. 1987) and *C. perfringens* by electrotransformation (Allen and Blaschek 1988). A physical map of the 26 kbp plasmid, pAMβ1, has been constructed and regions encoding MLS^R, replication and transfer functions identified (LeBlanc and Lee

1984). The MLS^R gene (*ermAM*/*ermB*) and the replication region have been sequenced (Brehm et al. 1987; Swinfield et al. 1990) and the precise position of the replication origin determined (Bruand et al. 1991). Both the MLS^R gene and the replication functions have been employed for developing cloning vectors suitable for use in clostridia and other Gram-positive bacteria (e.g., Trieu-Cuot et al. 1987a; Chambers et al. 1988; Williams et al. 1990a, 1990b). A physical map of the 35-kbp plasmid, pIP501 (Horodniceanu et al. 1976), has also been constructed (Evans and Macrina 1983) and its replication, conjugation, and host range functions are currently being characterized (Hershfield 1979; Behnke and Gilmore 1981; Krah and Macrina 1989, 1991; Brantl et al. 1990). No details have yet emerged concerning the mechanism(s) by which these broad host range streptococcal and enterococcal plasmids are transferred during conjugation.

Plasmid transfer efficiency varies considerably with the donor organism. For example, *Lactococcus lactis* is a particularly efficient donor of plasmid pAMβ1 in filter matings with *C. acetobutylicum*, whereas *Bacillus subtilis* is not (Oultram and Young 1985; Oultram et al. 1987). This is correlated with the tendency for pAMβ1 to suffer a specific large deletion affecting genes required for conjugative plasmid transfer during propagation in *B. subtilis* (van der Lelie and Venema 1987). Transfer frequencies are also rather variable from one experiment to another, depending on the physiological state of the mating bacteria and the relative numbers of donor and recipient organisms.

5.1.4 Mobilization of Nonconjugative Plasmids

Some of the conjugative plasmids found in *E. coli*, notably F, behave as episomes; i.e., they can integrate into the bacterial chromosome. When in this state they can mobilize chromosomal genes during their transfer. It was this extremely useful property that made these plasmids such important analytical tools in the early development of *E. coli* genetics. Unfortunately, there is little evidence to suggest that any of the conjugative plasmids established in clostridia, whether indigenous or exotic, can mobilize chromosomal genes from one strain to another (but see Bréfort et al. 1977). However, several instances of mobilization of co-resident nonconjugative plasmids have been documented.

Strain CPN50 of *C. perfringens* contains a 10.2-kbp bacteriocinogenic plasmid, pIP404 (Ionesco and Bouanchaud 1973; Ionesco et al. 1976), which can be transferred to other strains of *C. perfringens* by conjugation (Bréfort et al. 1977). Plasmid pIP404 is comparatively small and it seems likely that it is mobilized by a large conjugative plasmid co-resident in strain CPN50 (Ionesco et al. 1976). Although plasmid pIP404 has been completely sequenced (Garnier and Cole 1988a) and is probably the best studied of all clostridial plasmids (Garnier and Cole 1986, 1988b, 1988c; Garnier et al. 1987), nothing is currently known about the mechanism of its conjugative mobilization. Nor is it known whether the comparatively well-studied conjugative tetracycline resistance plasmids found in *C. perfringens* (see Section 5.1.2) can mobilize pIP404.

Plasmid pAMβ1 can mobilize nonconjugative plasmids to several different organisms including *Staphylococcus aureus* (Schaberg et al. 1982), *S. faecalis* (LeBlanc and Lee 1984; Smith and Clewell 1984; Smith 1985), *C. acetobutylicum* (Yu and Pearce 1986; Oultram et al. 1987), and probably also *Bacillus thuringiensis* (Lereclus et al. 1983). The mechanism of conjugative mobilization was not established in many of these reports, but in two cases it involved the transfer of plasmid co-integrates. These are considered in more detail in the next section.

5.1.4.1 Mobilization by the Formation of Co-Integrate Plasmids. Smith and Clewell (1984) showed that two incompatible plasmids sharing the same replication functions can become established in *Streptococcus sanguis* as a plasmid co-integrate. One of these plasmids was the pIP501 derivative, pVA797, which provided the functions necessary for conjugative transfer of the co-integrate molecule to *S. faecalis*. Resolution of the co-integrate occurred in *S. faecalis* and after growth in the absence of selective pressure, strains were readily recovered in which one or other of the incompatible constituent plasmids had been lost.

A similar approach has been adopted to mobilize a small nonconjugative plasmid from *B. subtilis* to *C. acetobutylicum* (Oultram et al. 1987). In this case, the *ermAM* gene from pAMβ1 was inserted into a small, nonconjugative, integrational plasmid lacking replication functions active in Gram-positive bacteria. When transformed into a *B. subtilis* strain harboring pAMβ1, it became established as a co-integrate molecule in which form it was transferred to and maintained in *C. acetobutylicum*. In view of the structural instability of pAMβ1 in *B. subtilis* (van der Lelie and Venema 1987) and the somewhat cumbersome nature of the experimental approach, conjugative plasmid mobilization from *B. subtilis* to *C. acetobutylicum* using pAMβ1 has not been widely employed for gene-transfer experiments. Nevertheless, this strategy was used to introduce a derivative of the extremely versatile integrational plasmid, pMTL21EC, containing the *leuBC* genes of *C. pasteurianum*, into *C. acetobutylicum* (Oultram et al. 1988a). Two independently isolated leucine auxotrophs were used as recipients. In both cases, the plasmid co-integrate was maintained intact and functional complementation of one of the auxotrophs was obtained. This approach has also been employed to introduce the *Pseudomonas putida xylE* gene into *C. acetobutylicum* (Minton et al. 1988). Although described only a few years ago, this gene transfer procedure has essentially been superseded by electrotransformation (Oultram et al. 1988b) and an alternative mobilization strategy that is described in the next section.

5.1.4.2 Mobilization by *Trans*-Acting Proteins at an *oriT* Site. Plasmids belonging to incompatibility groups P and Q that are found in Gram-negative bacteria have an extremely broad host range (Thomas and Smith 1987; Krishnapillai 1988). Their conjugation machinery shows a remarkable lack of specificity and they can mobilize nonconjugative plasmids from Gram-negative bacteria to many other bacteria, both Gram-negative (Simon et al. 1983) and Gram-positive (Trieu-Cuot et al. 1987a; Mazodier et al. 1989; Lazraq et al. 1990). Moreover, plasmid transfer to both yeast (Heinemann and Sprague 1989) and plants (Buchanan-Wollaston et al. 1987) has also been documented.

Trieu-Cuot et al. (1987a) exploited these remarkable properties of IncP plasmids to mobilize a small nonconjugative plasmid, pAT187, from *E. coli* to strains of *E. faecalis*, *L. lactis*, *Streptococcus agalactiae*, *B. thuringiensis*, *Listeria monocytogenes*, *S. aureus*, and *Bacillus sphaericus*, many of which had previously proved refractory to genetic exchange. Plasmid pAT187 has also been transferred to *C. acetobutylicum* (Williams et al. 1990a, 1990b) and suitably modified derivatives have been established in *Streptomyces* spp. (Mazodier et al. 1989) and *Mycobacterium smegmatis* (Lazraq et al. 1990) following conjugative mobilization from *E. coli.* Plasmid pAT187 contains the following three elements in the backbone of pBR322: (1) the pAMβ1 replication functions; (2) the *aphA-3* gene (KmR) from the *Campylobacter coli* plasmid pIP1433; (3) the *oriT* site from the IncP plasmid RK2. Suitable *E. coli* donor strains harbor an IncP "helper" plasmid which may be either autonomous or integrated in the bacterial chromosome (Simon et al. 1983). The IncP plasmid provides in *trans* all the proteins required for production of the origin-nicking "relaxosome" complex (Guiney et al. 1988; Pansegrau et al. 1988, 1990; Fürste et al. 1989; Ziegelin et al. 1989) and any other plasmid-encoded functions necessary for conjugative DNA transfer.

Several vectors containing the *oriT* region of RK2 are now available for effecting gene transfer to *C. acetobutylicum* by conjugative mobilization from *E. coli* (Williams et al. 1990a, 1990b). Most are based on the pMTL20 backbone (Chambers et al. 1988) into which has been incorporated the RK2 *oriT* site. The *aphA-3* (Km^R) gene originally used by Trieu-Cuot et al. (1987a), which is only poorly selective under anaerobic conditions, has been replaced by the MLS^R gene from pAMβ1 (Brehm et al. 1987). Derivatives containing a foreshortened segment of the pAMβ1 replication region (Swinfield et al. 1990) or the replication functions of the *C. butyricum* plasmid, pCB101 (Minton and Morris 1981; Collins et al. 1985; Young et al. 1989b), have been successfully introduced into *C. acetobutylicum*, where they replicate at comparatively high copy number. A low copy number vector containing *oriT* and the replication functions of the *Streptococcus cremoris* plasmid, pWV01 (Vosman and Venema 1983; Kok et al. 1984), has also been constructed (Williams et al. 1990b).

In spite of the advantages presented by this versatile strategy for the direct mobilization of plasmids from *E. coli* to other organisms, it is not yet clear whether it will find widespread application in clostridial genetics. Certain plasmid constructions appear to be structurally unstable, especially in the presence of the *trans*-acting proteins provided by the IncP helper plasmid co-resident in donor strains (D.I. Young, D.R. Williams, and M. Young, unpublished observations). In another case, a plasmid containing replication functions from pAMβ1 had a hot spot for insertion of IS*1* when propagated in *E. coli* donor strains, whereas a similar plasmid containing the same replication functions in the opposite orientation did not attract insertions of IS*1* (Williams et al. 1990b). Perhaps, for many applications, electrotransformation will become the preferred method for transferring genes to clostridia. Electrotransformation methods have already been adapted for use with certain strains of *C. perfringens* (Allen and Blaschek 1988, 1990; Kim and Blaschek 1989; Scott and Rood 1989) and *C. acetobutylicum* (Oultram et al. 1988b).

Conjugative transfer of IncP plasmids between *E. coli* strains, as well as conjugative mobilization mediated by IncP plasmids, is an extremely efficient process and in filter matings carried out at high donor to recipient ratios it is not unusual for the majority of the recipient bacteria to inherit the plasmid from the donor strain. The frequency of plasmid mobilization from *E. coli* to Gram-positive bacteria is several orders of magnitude lower, rarely exceeding a value of 10^{-6} per recipient. Plasmid-encoded pili are involved in conjugation between *E. coli* strains, but it is not known whether they are also involved in plasmid transfer to Gram-positive bacteria. Moreover, it has been assumed that a single DNA strand is transferred during Gram-negative/Gram-positive matings. Experimental support for this assumption has recently been obtained (D.I. Young, D.R. Williams, and M. Young, unpublished observations).

5.2 CONJUGATIVE TRANSPOSONS

5.2.1 General Characteristics

Conjugative transposons are an interesting class of mobile genetic elements found mostly, though not exclusively (see, e.g., Hecht and Malamy 1989), in Gram-positive bacteria, where they are usually, though not invariably (Christie et al. 1987) resident in the bacterial chromosome. They were discovered about 10 years ago in *Enterococcus faecalis* (Franke and Clewell 1981), and because of their extremely wide host range (Clewell and Gawron-Burke 1986) they have become important tools for genetic analysis in a variety of Gram-positive bacteria. There have been several recent reviews dealing with conjugative transposons (Clewell et al. 1985; Clewell and Gawron-Burke 1986; Clewell 1990; Trieu-Cuot et al. 1990) and they are also considered in Chapter 10 in this volume. Accordingly, only a very brief summary of their essential characteristics will be given below.

The best characterized examples are Tn*916* (16 kbp, *tetM*) and Tn*1545* (25 kbp, *tetM*, *ermAM*, *aphA-3*) originally discovered in *Ec. faecalis* strain DS16 and *Streptococcus pneumoniae* strain BM4200, respectively (Franke and Clewell 1981; Courvalin and Carlier 1986, 1987; Caillaud and Courvalin 1987). Most conjugative transposons reported to date contain *tetM* (Burdett et al.

1982; Clewell and Gawron-Burke 1986), the product of which appears to be an analogue of elongation factor G (Burdett 1991). They differ from other transposons in that they are able to move from a site in one cell to a different site in another cell by a process requiring close cell-to-cell contact (Franke and Clewell 1981; Gawron-Burke and Clewell 1982). The relatively large size of these elements, when compared with that of most classical transposons, can be attributed to the additional conjugation functions they encode. As more information becomes available, it will be of interest to compare their conjugation apparatus with that of the broad host-range plasmids also found in enterococci and streptococci.

The conjugation machinery associated with the conjugative transposons is remarkably nonspecific. Unlike the conjugative plasmids discussed previously, they are passively replicated after transfer, usually as part of the bacterial chromosome; hence, their host range is even wider than that of the conjugative plasmids. (Although transferable to *E. coli* from *Ec. faecalis*, plasmid pAMβ1 is unable to replicate in this organism—Trieu-Cuot et al. 1988.) Transfer of conjugative transposons between many different Gram-positive genera is well established (reviewed by Clewell and Gawron-Burke 1986) and it has recently been shown, using Tn*916*, that conjugative transfer to and from Gram-negative bacteria can also occur (Sen and Oriel 1990; Bertram et al. 1991). There can be little doubt that the conjugative transposons have played a seminal role in the spread of antibiotic-resistance genes between different groups of bacteria (Trieu-Cuot et al. 1985, 1987a, 1987b; Clewell and Gawron-Burke 1986; Brisson-Noël et al. 1988; Le Bouguénec et al. 1990).

Functional and molecular analysis of Tn*916* and Tn*1545* has provided some insights into the mechanism of their conjugative transposition. Insertional mutagenesis of Tn*916* in *E. coli* using Tn*5* served to identify those regions of the element implicated in transposon excision, transposition and conjugation (Jones et al. 1987; Yamamoto et al. 1987; Senghas et al. 1988). Mutant forms of the transposon unable to excise in *E. coli* were also unable to undergo transposition when re-introduced by transformation into *Ec. faecalis* protoplasts (Senghas et al. 1988). This supported the idea, proposed some years previously (Gawron-Burke and Clewell 1982, 1984), that the first step in transposition is transposon excision. A supercoiled circular transposition intermediate has been detected after excision of Tn*916* in *E. coli* (Scott et al. 1988) and genes concerned with transposon integration and excision have been identified near one extremity of Tn*1545* (Poyart-Salmeron et al. 1989, 1990).

These elements do not generate a duplication of the target sequence upon insertion (Caillaud and Courvalin 1987; Caillaud et al. 1987). Excision in *E. coli* can occur perfectly, in which case normal function is restored to an insertionally inactivated gene (Gawron-Burke and Clewell 1982, 1984; Nida and Cleary 1983; Courvalin and Carlier 1986; Gaillard et al. 1986; Kathariou et al. 1987). However, careful analysis of target sequences has shown that base-substitution mutations are frequently left after transposon excision in *E. coli* (Caillaud and Courvalin 1987; Caillaud et al. 1987; Clewell et al. 1988; Caparon and Scott 1989; Poyart-Salmeron et al. 1990). This is nicely accounted for by the proposition that transposon integration and excision is initiated by staggered cleavage of the target sequences adjacent to each extremity of the element. After excision of the circular transposition intermediate, the transposon ends are joined by a 6 or 7 base pair (bp) heteroduplex comprising two single strands from the target sequences lying on either side of the original insertion site (Caparon and Scott 1989). A similar heteroduplex would be left in the target after transposon excision. This may be repaired by the mismatch repair system of the host, unless replicated beforehand; hence the occurrence of base-substitution mutations after transposon excision in *E. coli.*

Transposition of Tn*916* and Tn*1545* is far from a random process; both elements have similar preferred d(A + T)-rich target sites for insertion (Poyart-Salmeron et al. 1990; Trieu-Cuot et al. 1990). It is not, therefore, surprising that they are proving useful tools for insertional mutagenesis and mutational cloning in organisms such as the mesophilic clostridia (see Chapter 10, this volume), whose DNA is particularly d(A + T)-rich. According to Trieu-Cuot et al. (1990) up to three mismatches are tolerated in the 6–7 bp heteroduplex formed when the circular transposition intermediate is generated by excision. Mismatch repair will generate one of two possible products, depending on which strand is corrected. Therefore, as these transposons move from one site to another the composition of the variable 6–7 bp core region formed after ligation of the transposon ends will change. This may well modify the insertion site specificity of the element (Trieu-Cuot et al. 1990). Several groups have reported that certain organisms appear to contain a hot spot for insertion of a particular conjugative transposon (Hill et al. 1985; Woolley et al. 1989; Mullany et al. 1990, 1991; Sen and Oriel 1990). If the sequence of the variable 6–7 bp core region is indeed a partial determinant of insertion site specificity (Trieu-Cuot et al. 1990), the tendency for an element to insert into a particular hot spot may be diminished if different donor organisms are employed. Also, for undertaking insertional mutagenesis it may be helpful to employ several different donors or, alternatively, one that harbors multiple copies (see below, and Sections 5.2.3 and 5.2.5) of the element.

Because of their extremely broad host range, the conjugative transposons are proving to be useful tools for working with organisms lacking well-developed genetic systems. As was originally proposed by Gawron-Burke and Clewell (1984), they have been used for mutating and cloning genes from many such organisms including *Streptococcus pyogenes* (Nida and Cleary 1983), *Listeria monocytogenes* (Gaillard et al. 1986; Kathariou et al. 1987), and *C. acetobutylicum* (Bertram and Dürre 1989; Bertram et al. 1990). A small (3.4 kbp) nonconjugative derivative of Tn*1545* called Tn*1545*-Δ*3* (Poyart-Salmeron et al. 1989, 1990) may prove particularly useful in this respect (Nassif et al. 1991). Several investigators, working with different elements in different bacteria, have noted that transcipients frequently inherit multiple copies of conjugative transposons (e.g., Gawron-Burke and Clewell 1984; Volk et al. 1988; Bertram and Dürre 1989; Woolley et al. 1989), which may be an unwanted complication in connection with their use for mutagenesis and mutational cloning. This feature also complicates their use for physical mapping of bacterial genomes (see Section 5.2.5).

Using a modified derivative of Tn*916*, denoted Tn$916\Delta E$, in which *tetM* was inactivated by insertion of an *erm* gene (Rubens and Heggen 1988), it has been shown that the presence of a resident copy of Tn*916* in a *S. pyogenes* recipient does not impede the transfer and establishment of a second copy of the element from a *B. subtilis* donor (Norgren and Scott 1991). The second copy was usually inserted elsewhere in the bacterial chromosome, but insertion by homologous recombination into the resident transposon was also detected, indicating that these elements may prove generally useful for affecting allelic replacement of genes cloned into them (see, e.g., Norgren et al. 1989).

Another potential application of the conjugative transposons concerns their use to mobilize nonconjugative elements (plasmids and perhaps, even, the bacterial chromosome) from one organism to another. This possibility was investigated by Naglich and Andrews (1988), who showed that the conjugation machinery of Tn*916* can mobilize two small nonconjugative plasmids, pC194 (Horinouchi and Weisblum 1982; Gros et al. 1987) and pUB110 (Lacey and Chopra 1974; McKenzie et al. 1986), from *B. subtilis* to *B. thuringiensis*. More recently, Torres et al. (1991) have obtained evidence that Tn*925* mobilizes chromosomal genes between strains of both *B. subtilis* and *Ec. faecalis*, at frequencies comparable to those observed for transfer of Tn*925* itself. Their data were consistent with the occurrence of a kind of cell fusion, permitting extensive recombination between the chromosomes of mating bacteria. These authors also observed an increase in transfer frequency if *Ec. faecalis* donor strains were grown in the presence of tetracycline (cf. Doucet-Populaire et al. 1991). However, no such effect was noted with *B. subtilis* donors.

The small body of information that exists concerning the behavior of conjugative transposons in clostridia is considered below.

5.2.2 Indigenous Conjugative Transposons

Several elements that resemble conjugative transposons have been found in the pathogenic, amino acid-fermenting clostridia. At least two have been detected in the chromosome of strains of *C. difficile*. One confers resistance to tetracycline (*tetM*) and has been transferred to *B. subtilis* as well as to other strains of *C. difficile* (Ionesco 1980; Smith et al. 1981; Wüst and Hardegger 1983; 1988; Hächler et al. 1987a; Mullany et al. 1990). This conjugative transposon appears to be rather similar to Tn*916*, with which it shares substantial homology, as revealed by DNA-DNA hybridization (Hächler et al. 1987a; Mullany et al. 1990). It seems to insert randomly into the *B. subtilis* chromosome, but has a preferred site for insertion into the *C. difficile* chromosome (Mullany et al. 1990).

Another element, which behaves like a conjugative transposon but is unusual in that it confers MLS^R and not Tc^R, is present in certain strains of *C. difficile*, (Wüst and Hardegger 1983) from which it can be transferred to *S. aureus* at low frequency (Hächler et al. 1987b). It may be related to an element found in *C. innocuum*, that also confers MLS^R and was transferable by a conjugation-like process from this organism to *C. perfringens* (Magot 1983). The MLS^R gene carried by the *C. difficile* element, denoted *ermZ*, shares homology with *ermB* found on the *S. aureus* transposon, Tn*551* (Hächler et al. 1987b) and is, therefore, similar to the *ermAM* genes found on pAMβ1 (Brehm et al. 1987) and Tn*1545* (Courvalin and Carlier 1987) and the *ermP* gene found on the *C. perfringens* plasmid pIP402 (Berryman and Rood 1989). However, only in the case of the chromosomally located *erm* genes found in *C. difficile* and *C. innocuum* is there evidence suggesting that they form part of a conjugative transposon.

5.2.3 Exotic Conjugative Transposons in Clostridia

There have been several reports of the transfer of enterococcal and streptococcal conjugative transposons to clostridia. The element that has been most widely used is Tn*916*, which has been transferred to *C. tetani* (Volk et al. 1988), *C. difficile* (Mullany et al. 1991), *C. acetobutylicum* (Davies et al. 1988; Bertram and Dürre 1989; Woolley et al. 1989; Mattsson and Rogers 1989), and *C. perfringens* (reviewed by Rood and Cole 1991). Other similar elements, including Tn*925* (Bertram and Dürre 1989) and Tn*1545* (Woolley et al. 1989), have also been established in *C. acetobutylicum*. In many cases, transfer was by filter mating, using a donor strain carrying one or more copies of the conjugative transposon in its chromosome. Recently (S. Allen and H.P. Blaschek, personal communication), electrotransformation has been used to transfer Tn*916* to *C. perfringens* on the nonreplicative "suicide" plasmid, pAM120 (Gawron-Burke and Clewell 1984). In most cases, insertions apppeared to occur at many different sites in the bacterial chromosome, but there was a preferred site for Tn*916* insertion in the chromosome of *C. difficile* and circumstantial evidence also suggested the presence of a preferred site for Tn*916* insertion in the chromosome of the NCIMB 8052 strain of *C. acetobutylicum* (Woolley et al. 1989). In those cases where there was no evidence of a hot spot for transposon insertion, multiple insertions in individual transconjugants were commonly encountered (see Section 5.2.1).

5.2.4 Mutagenesis and Mutational Cloning Using Conjugative Transposons

In our own experience with Tn*1545* in the NCIMB 8052 strain of *C. acetobutylicum* (Woolley et al. 1989; Wilkinson and Young 1993), most transposon insertions are "silent"; i.e., they are not associated with a readily recognizable phenotypic change in the transcipient, apart, that is, from acquisition of the Tn-encoded antibiotic-resistance phenotype. This is consistent with the known preference these transposons display for d(A + T)-rich target sequences (see Section 5.2.1). Although the DNA of the mesophilic clostridia is remarkably d(A + T)-rich in general, there is a noticeable difference in base composition between coding and noncoding sequences, the latter tending to be more d(A + T)-rich than the former (Young et al. 1989a). Hence, one might predict that these transposons would preferentially insert into noncoding regions. Nevertheless, insertions conferring various auxotrophies and others affecting solvent production, granulose production, and spore formation in *C. acetobutylicum* have been obtained (Mattsson and Rogers 1989; Bertram et al. 1990; D.I. Young, S.R. Wilkinson, and M. Young, unpublished observations) and are currently being characterized. This topic is considered in some detail in Chapter 10, this volume.)

5.2.5 Conjugative Transposons as Tools for Physical Analysis of Clostridial Genomes

Owing to their large size, conjugative transposons are potentially extremely useful for the construction of physical maps of bacterial genomes. Tn*1545*, which is a 25 kbp element, is being employed for the construction of a physical map of the chromosome of the NCIMB 8052 strain of *C. acetobutylicum* (Wilkinson and Young 1993). This organism has a large genome size (about 6.5 mbp). We have been constructing a physical map using the enzymes *Sma*I and *Apa*I, both of which generate about 30 fragments, ranging in size from <50 to >800 kbp. Tn*1545* insertions in fragments up to 600 kbp were readily detected by a change in mobility of the corresponding band during pulsed field-gel electrophoresis. Transposon insertions, most of which appear to be genetically silent (see Section 5.2.4), can be used in much the same way as conventional gene probes for assigning overlapping regions to restriction fragments generated by different enzymes. Several of the fragments generated by both *Sma*I and *Apa*I co-migrate as pairs during electrophoresis. This was confirmed by "splitting" of the associated bands into two in the DNA extracted from certain strains harboring Tn*1545* insertions. The analysis confirmed an earlier report (Woolley et al. 1989) that some strains harbor multiple copies (up to four have been detected) of Tn*1545*.

The physical map of the *C. acetobutylicum* NCIMB 8052 genome is nearly complete. Using a combination of cloned genes as hybridization probes and strains harboring transposon insertions, seven genome segments were initially assembled, varying in size from 200–1300 kbp. Results obtained with partially digested DNA samples suggest a probable order for some of these segments. However, a few of the smallest fragments (<50 kbp) containing rRNA genes are still unassigned. To complete and confirm the provisional map, a derivative of Tn*1545* has been constructed (G. Price and M. Young, unpublished observations), in which the *aphA-3* gene has been inactivated by introducing the *cat* gene from pC194 (Horinouchi and Weisblum 1982; Gros et al. 1987) together with a unique *Not*I restriction site. There is no *Not*I site in the genome of *C. acetobutylicum* NCIMB 8052. The modified transposon will be used to generate strains containing a unique *Not*I restriction site at different positions in the chromosome. Partially digested DNA samples, in conjunction with probes specific for Tn*1545* sequences on either side of the *Not*I site, will be used to verify the provisional map.

During this investigation, genomic DNA samples from several other *C. acetobutylicum* strains were compared after fragmentation with *Apa*I and *Sma*I (Wilkinson and Young 1993). The strains fell into three groups. One contains strains NCIMB 8052 and NI-4081 (Institut Pasteur strain formerly known as *C. saccharoperbutylacetonicum*), whose genomes gave identical *Apa*I and *Sma*I restriction profiles. Another contains the ATCC 824 and DSM 1731 strains, whose genomes gave similar *Apa*I and *Sma*I restriction profiles and appear to be considerably smaller (about 4.2 mbp) than those of strains in the former group. The South African strain, P262, lies in a third group and has the smallest genome (about 2.9 mbp). The *Apa*I and *Sma*I restriction profiles of strains from different groups were completely dissimilar. These findings add weight to a growing body of evidence (Woolley et al. 1989; Woolley and Morris 1990; Wilkinson and Young 1993) suggesting that these three groups of organisms, all known as *C. acetobutylicum*, are in fact quite unrelated.

5.3 CONCLUSION

It has become possible genetically to manipulate certain species of *Clostridium* within the last 5 years. Two organisms, the solventogenic species, *C. acetobutylicum*, and the pathogenic species, *C. perfringens*, have already assumed prominence in this regard (Young et al. 1989a; Rood and Cole 1991). This chapter has dealt with the development of gene transfer procedures based on conjugation; additional equally important gene transfer procedures are considered in some of the other chapters in this volume. It seems apparent that genetically speaking, the clostridia have almost "come of age." Until it becomes possible to exploit homologous recombination between incoming DNA molecules and the bacterial chromosome, many questions of fundamental scientific interest, as well as problems of a more applied nature, will remain experimentally inaccessible and these organisms will not assume ". . . their full and deserving place in biotechnology" (Dixon 1990). The development of gene transfer involving homologous recombination with the bacterial chromosome remains an important objective for future research on genetic manipulation of the clostridia.

REFERENCES

Abraham, L.J., and Rood, J.I. (1985a) *J. Bacteriol.* 161, 636–640.
Abraham, L.J., and Rood, J.I. (1985b) *Plasmid* 13, 155–162.
Abraham, L.J., and Rood, J.I. (1987) *J. Bacteriol.* 169, 1579–1584.
Abraham, L.J., Wales, A.J., and Rood, J.I. (1985) *Plasmid* 14, 37–46.
Abraham, L.J., Berryman, D.I., and Rood, J.I. (1988) *Plasmid* 19, 113–120.
Allen, S.P., and Blaschek, H.P. (1988) *Appl. Environ. Microbiol.* 54, 2322–2324.
Allen, S.P., and Blaschek, H.P. (1990) *FEMS Microbiol. Lett.* 70, 217–220.
Behnke, D., and Gilmore, M.S. (1981) *Mol. Gen. Genet.* 184, 115–120.
Berryman, D.I., and Rood, J.I. (1989) *Antimicrob. Agents Chemother.* 33, 1346–1353.
Bertram, J., and Dürre, P. (1989) *Arch. Microbiol.* 151, 551–557.
Bertram, J., Kuhn, A., and Dürre, P. (1990) *Arch. Microbiol.* 153, 373–377.
Bertram, J., Strätz, M., and Dürre, P. (1991) *J. Bacteriol.* 173, 443–448.
Brantl, S., Behnke, D., and Alonso, J.C. (1990) *Nucleic Acids Res.* 18, 4783–4790.
Bréfort, G., Magot, M., Ionesco, H., and Sebald, M. (1977) *Plasmid* 1, 52–66.
Brehm, J., Salmond, G., and Minton, N.P. (1987) *Nucleic Acids Res.* 15, 3177.
Brisson-Noël, A., Arthur, M., and Courvalin, P. (1988) *J. Bacteriol.* 170, 1739–1745.
Bruand, C., Ehrlich, S.D., and Jannière, L. (1991) *EMBO J.* 10, 2171–2177.
Buchanan-Wollaston, V., Passiatore, J.E., and Cannon, F. (1987) *Nature (London)* 328, 172–175.
Burdett, V., Inamine, J., and Rajagopalan, S. (1982) *J. Bacteriol.* 149, 995–1004.
Burdett, V. (1991) *J. Biol. Chem.* 266, 2872–2877.
Caillaud, F., Carlier, C., and Courvalin, P. (1987) *Plasmid* 17, 58–60.
Caillaud, F., and Courvalin, P. (1987) *Mol. Gen. Genet.* 209, 110–115.
Caparon, M.G., and Scott, J.R. (1989) *Cell* 59, 1027–1034.
Chambers, S.P., Prior, S.E., Barstow, D.A., and Minton, N.P. (1988) *Gene* 68, 139–149.
Christie, P.J., Korman, R.Z., Zahler, S.A., Adsit, J.C., and Dunny, G.M. (1987) *J. Bacteriol.* 169, 2529–2536.
Clewell, D.B. (1981) *Microbiol. Rev.* 45, 409–436.
Clewell, D.B. (1990) *Eur. J. Clin. Microbiol. Infect. Dis.* 9, 90–102.
Clewell, D.B., Fitzgerald, G.F., Dempsey, L., et al. (1985) in *Molecular Basis of Oral Microbial Adhesion* (Mergenhagen, S.E., and Rosan, B., eds.), pp. 194–203, American Society for Microbiology, Washington, DC.
Clewell, D.B., Flannagan, S.E., Ike, Y., Jones, J.M., and Gawron-Burke, C. (1988) *J. Bacteriol.* 170, 3046–3052.
Clewell, D.B., and Gawron-Burke, C. (1986) *Annu. Rev. Microbiol.* 40, 635–659.
Collins, M.E., Oultram, J.D., and Young, M. (1985) *J. Gen. Microbiol.* 131, 2097–2105.
Courvalin, P., and Carlier, C. (1986) *Mol. Gen. Genet.* 205, 291–297.
Courvalin, P., and Carlier, C. (1987) *Mol. Gen. Genet.* 206, 259–264.
Davies, A., Oultram, J.D., Pennock, A., Williams, D.R., Richards, D.F., Minton, N.P., and Young, M. (1988) in *Genetics and Biotechnology of Bacilli, vol. 2*, (Ganesan, A.T., and Hoch, J.A., eds.), pp. 391–395, Academic Press, Orlando, FL.
Dixon, B. (1990) *Bio/Technology* 8, 591.

Doucet-Populaire, F., Trieu-Cuot, P., Dosbaa, I., Andremont, A., and Courvalin, P. (1991) *Antimicrob. Agents Chemother.* 35, 185–187.
Evans, Jr., R.P., and Macrina, F.L. (1983) *J. Bacteriol.* 154, 1347–1355.
Franke, A.E., and Clewell, D.B. (1981) *J. Bacteriol.* 145, 494–502.
Fürste, J.P., Pansegrau, W., Ziegelin, G., Kroger, M., and Lanka, E. (1989) *Proc. Natl. Acad. Sci. USA* 86, 1771–1775.
Gaillard, J.L., Berche, P., and Sansonetti, P.J. (1986) *Infect. Immun.* 52, 50–55.
Garnier, T., and Cole, S.T. (1986) *J. Bacteriol.* 168, 1189–1196.
Garnier, T., and Cole, S.T. (1988a) *Plasmid* 19, 134–150.
Garnier, T., and Cole, S.T. (1988b) *Plasmid* 19, 151–160.
Garnier, T., and Cole, S.T. (1988c) *Molec. Microbiol.* 2, 607–614.
Garnier, T., Saurin, W., and Cole, S.T. (1987) *Molec. Microbiol.* 1, 371–376.
Gawron-Burke, C., and Clewell, D.B. (1982) *Nature (London)* 300, 281–284.
Gawron-Burke, C., and Clewell, D.B. (1984) *J. Bacteriol.* 159, 214–221.
Gros, M.F., te Riele, H., and Ehrlich, S.D. (1987) *EMBO J.* 6, 3863–3869.
Guiney, D.G., Deiss, C., and Simnad, V. (1988) *Plasmid* 20, 259–265.
Hächler, H., Kayser, F.H., and Berger-Bächi, B. (1987a) *Antimicrob. Agents Chemother.* 31, 1033–1038.
Hächler, H., Berger-Bächi, B., and Kayser, F.H. (1987b) *Antimicrob. Agents Chemother.* 31, 1039–1045.
Hayes, W. (1953) *J. Gen. Microbiol.* 8, 72–88.
Hayes, W. (1968) *The Genetics of Bacteria and their Viruses*, 2nd ed., Blackwell, Oxford.
Hecht, D., and Malamy, M.H. (1989) *J. Bacteriol.* 171, 3603–3608.
Heefner, D.L., Squires, C.H., Evans, R.J., Kopp, B.J., and Yarus, M.J. (1984) *J. Bacteriol.* 159, 460–464.
Heinemann, J.A., and Sprague Jr., J.F. (1989) *Nature (London)* 340, 205–209.
Hershfield, V. (1979) *Plasmid* 2, 137–149.
Hill, C., Daly, C., and Fitzgerald, G.F. (1985) *FEMS Microbiol. Lett.* 30, 115–119.
Horinouchi, S., and Weisblum, B. (1982) *J. Bacteriol.* 150, 815–825.
Horodniceanu, T., Bouanchaud, D., Biet, G., and Chabbert, Y. (1976) *Antimicrob. Agents Chemother.* 10, 795–801.
Ionesco, H. (1980) *Ann. Inst. Pasteur Microbiol.* 131, 171–179.
Ionesco, H., and Bouanchaud, D.H. (1973) *C.R. Acad. Sci.* Ser. D 276, 2855–2857.
Ionesco, H., Bieth, G., Dauguet, C., and Bouanchaud, D.H. (1976) *Ann. Inst. Pasteur Microbiol.* 127B, 283–294.
Jacob, A., and Hobbs, S.J. (1974) *J. Bacteriol.* 117, 360–372.
Jones, J.M., Gawron-Burke, C., Flannagan, S.E., et al. (1987) in *Staphylococcal Genetics* (Ferretti, J.J., and Curtiss III, R., eds.), pp. 54–60, American Society for Microbiology, Washington, DC.
Kathariou, S., Metz, P., Hof, H., and Goebel, W. (1987) *J. Bacteriol.* 169, 1291–1297.
Kim, A.Y., and Blaschek, H.P. (1989) *Appl. Environ. Microbiol.* 55, 360–365.
Kok, J., Van der Vossen, J.M.B.M., and Venema, G. (1984) *Appl. Environ. Microbiol.* 48, 726–731.
Krah, E.R., and Macrina, F.L. (1989) *J. Bacteriol.* 171, 6005–6012.
Krah, E.R., and Macrina, F.L. (1991) *Plasmid* 25, 64–69.
Krishnapillai, V. (1988) *FEMS Microbiol. Rev.* 54, 223–238.
Lacey, R., and Chopra, I. (1974) *J. Med. Microbiol.* 7, 285–297.

Lazraq, R., Clavel-Sérès, S., David, H.L., and Roulland-Dussoix, D. (1990) *FEMS Microbiol. Lett.* 69, 135–138.
Le Bouguénec, C., de Cespédès, G., and Horaud, T. (1990) *J. Bacteriol.* 172, 727–734.
LeBlanc, D.J., and Lee, L.N. (1984) *J. Bacteriol.* 157, 445–453.
Lederberg, J., and Tatum, E.L. (1946) *Nature (London)* 158, 558.
Lereclus, D., Menou, G., and Lecadet, M.-M. (1983) *Mol. Gen. Genet.* 191, 307–313.
McKenzie, T., Hoshino, T., Tanaka, T., and Sueoka, N. (1986) *Plasmid* 15, 93–103.
Magot, M. (1983) *FEMS Microbiol. Lett.* 18, 149–151.
Magot, M. (1984) *Ann. Inst. Pasteur Microbiol.* 135, 269–282.
Mattsson, D.M., and Rogers, P. (1989) *ASM Abstr.*, p. 153.
Mazodier, Ph., Petter, R., and Thompson, C. (1989) *J. Bacteriol.* 171, 3583–3585.
Minton, N.P., Brehm, J.K., Oultram, J.D., Swinfield, T.J., and Thompson, D.E. (1988) in *Anaerobes Today* (Hardie, J.M., and Borriello, S.P., eds.), pp. 125–134, John Wiley & Sons, Chichester.
Minton, N.P., and Morris, J.G. (1981) *J. Gen. Microbiol.* 127, 325–331.
Mullany, P., Wilks, M., Lamb, I., Clayton, C., Wren, B., and Tabaqchali, S. (1990) *J. Gen. Microbiol.* 136, 1343–1349.
Mullany, P., Wilks, M., and Tabaqchali, S. (1991) *FEMS Microbiol. Lett.* 79, 191–194.
Naglich, J.G., and Andrews, R.E. (1988) *Plasmid* 20, 113–126.
Nassif, X., Puaoi, D., and So, M. (1991) *J. Bacteriol.* 173, 2147–2154.
Nida, K., and Cleary, P.P. (1983) *J. Bacteriol.* 155, 1156–1161.
Norgren, M., Caparon, M.G., and Scott, J.R. (1989) *Infect. Immun.* 57, 3846–3850.
Norgren, M., and Scott, J.R. (1991) *J. Bacteriol.* 173, 319–324.
Oultram, J.D., Davies, A., and Young, M. (1987) *FEMS Microbiol. Lett.* 42, 113–119.
Oultram, J.D., Peck, H., Brehm, J.K., Thompson, D., Swinfield, T.J., and Minton, N.P. (1988a) *Mol. Gen. Genet.* 214, 177–179.
Oultram, J.D., Loughlin, M., Swinfield, T.J., Brehm, J.K., Thompson, D.E., and Minton, N.P. (1988b) *FEMS Microbiol. Lett.* 56, 83–88.
Oultram, J.D., and Young, M. (1985) *FEMS Microbiol. Lett.* 27, 129–134.
Pan-Hou, H.S.K., Hosono, M., and Imura, N. (1980) *Appl. Environ. Microbiol.* 40, 1007–1011.
Pansegrau, W., Balzer, D., Kruft, V., Lurz, R., and Lanka, E. 1990. *Proc. Natl. Acad. Sci. USA* 87, 6555–6559.
Pansegrau, W., Ziegelin, G., and Lanka, E. (1988) *Biochim. Biophys. Acta* 951, 365–374.
Poyart-Salmeron, C., Trieu-Cuot, P., Carlier, C., and Courvalin, P. (1989) *EMBO J.* 8, 2425–2433.
Poyart-Salmeron, C., Trieu-Cuot, P., Carlier, C., and Courvalin, P. (1990) *Molec. Microbiol.* 4, 1513–1521.
Raycroft, R.E., and Zimmermann, L.N. (1964) *J. Bacteriol.* 87, 799–801.
Reysset, G., and Sebald, M. (1985) *Ann. Inst. Pasteur Microbiol.* 136, 275–282.
Rood, J.I. (1983) *Can. J. Microbiol.* 29, 1241–1246.
Rood, J.I., and Cole, S.T. (1991) *Microbiol. Rev.* 55, 621–648.
Rood, J.I., Scott, V.N., and Duncan, C.L. (1978) *Plasmid* 1, 563–570.
Rubens, C.E., and Heggen, L.M. (1988) *Plasmid* 20, 137–142.

Schaberg, D.R., Clewell, D.B., and Glatzer, L. (1982) *Antimicrob. Agents Chemother.* 22, 204–207.

Scott, J.R., Kirchman, P.A., and Caparon, M.G. (1988) *Proc. Natl. Acad. Sci. USA* 85, 4809–4813.

Scott, P.T., and Rood, J.I. (1989) *Gene* 82, 327–333.

Sebald, M., and Bréfort, G. (1975) *C.R. Acad. Sci.* Ser. D 281, 317–319.

Sen, S., and Oriel, P. (1990) *FEMS Microbiol. Lett.* 67, 131–134.

Senghas, E., Jones, J.M., Yamamoto, M., Gawron-Burke, C., and Clewell, D.B. (1988) *J. Bacteriol.* 170, 245–259.

Simon, R., Priefer, U., and Pühler, A. (1983) *Bio/Technology* 2, 784–791.

Smith, C.J., Markowitz, S.M., and Macrina, F.L. (1981) *Antimicrob. Agents Chemother.* 19, 997–1003.

Smith, M.D. (1985) *J. Bacteriol.* 162, 92–97.

Smith, M.D., and Clewell, D.B. (1984) *J. Bacteriol.* 160, 1109–1114.

Steffen, C., and Matzura, H. (1989) *Gene* 75, 349–354.

Swinfield, T.J., Oultram, J.D., Thompson, D.E., Brehm, J.K., and Minton, N.P. (1990) *Gene* 87, 79–90.

Thomas, C.M., and Smith, C.A. (1987) *Annu. Rev. Microbiol.* 41, 77–101.

Tomura, T., Hirano, T., Ito, T., and Yoshioka, M. (1973) *Jpn. J. Microbiol.* 17, 445–452.

Torres, O.R., Korman, R.Z., Zahler, S.A., and Dunny, G.M. (1991) *Mol. Gen. Genet.* 225, 395–400.

Trieu-Cuot, P., Gerbaud, G., Lambert, T., and Courvalin, P. (1985) *EMBO J.* 4, 3583–3587.

Trieu-Cuot, P., Carlier, C., Martin, P., and Courvalin, P. (1987a) *FEMS Microbiol. Lett.* 48, 289–294.

Trieu-Cuot, P., Arthur, M., and Courvalin, P. (1987b) *Microbiol. Sci.* 4, 263–266.

Trieu-Cuot, P., Carlier, C., and Courvalin, P. (1988) *J. Bacteriol.* 170, 4388–4391.

Trieu-Cuot, P., Poyart-Salmeron, C., Carlier, C., and Courvalin, P. (1990) in *Proceedings of the Sixth International Symposium on Genetics of Industrial Microorganisms* (Heslot, H., Davies, J., and Florent, J., et al., eds.), Aug. 12–18, 1990, Strasbourg, France, pp. 195–205, Société Française de Microbiologie, Strasbourg, France.

van der Lelie, D., and Venema, G. (1987) *Appl. Environ. Microbiol.* 53, 2458–2463.

Vivian, A. (1971) *J. Gen. Microbiol.* 69, 353–364.

Volk, W.A., Bizzini, B., Jones, K.R., and Macrina, F.L. (1988) *Plasmid* 19, 255–259.

Vosman, B., and Venema, G. (1983) *J. Bacteriol.* 156, 920–921.

Wilkinson, S.R., and Young, M. (1993) *J. Gen. Microbiol.* 139, in press.

Willetts, N., and Wilkins, B. (1984) *Microbiol. Rev.* 48, 24–41.

Williams, D.R., Young, D.I., Oultram, J.D., Minton, N.P., and Young, M. (1990a) in *Clinical and Molecular Aspects of Anaerobes* (Borriello, S.P., ed.), pp. 242–245, Wrightson Biomedical, Petersfield, UK.

Williams, D.R., Young, D.I., and Young, M. (1990b) *J. Gen. Microbiol.* 136, 819–826.

Woolley, R.C., and Morris, J.G. (1990) *J. Appl. Bacteriol* 69, 718–728.

Woolley, R.C., Pennock, A., Ashton, R.J., Davies, A., and Young, M. (1989) *Plasmid* 22, 169–174.

Wüst, J., and Hardegger, U. (1983) *Antimicrob. Agents Chemother.* 23, 784–786.

Wüst, J., and Hardegger, U. (1988) *Zentralbl. Bakteriol. Mikrobiol. Hyg.* 267, 383–394.

Yamamoto, M., Jones, J.M., Senghas, E., Gawron-Burke, C., and Clewell, D.B. (1987) *Appl. Environ. Microbiol.* 53, 1069–1072.

Young, M., Minton, N.P., and Staudenbauer, W.L. (1989a) *FEMS Microbiol. Rev.* 63, 301–325.

Young, M., Oultram, J.D., Pennock, A., and Richards, D.F. (1987) in *Genetics of Industrial Microorganisms* (Alacevic, M., Hranueli, D., and Toman, Z., eds.), Part A, pp. 403–413, Pliva, Zagreb.

Young, M., Staudenbauer, W.L., and Minton, N.P. (1989b) in *Clostridia, Biotechnology Handbooks*, vol. 3 (Minton, N.P., and Clarke, D.J., eds.), pp. 63–103, Plenum Press, New York.

Yu, P.-L., and Pearce, L.E. (1986) *Biotechnol. Lett.* 8, 469–474.

Ziegelin, G., Fürste, J.P., and Lanka, E. (1989) *J. Biol. Chem.* 264, 11989–11994.

CHAPTER
6

Clostridial Cloning Vectors

Nigel P. Minton
John K. Brehm
Tracy-Jane Swinfield
Sarah M. Whelan
Margaret L. Mauchline
Nicola Bodsworth
John D. Oultram

The genus *Clostridium* represents a heterologous assemblage of microorganisms whose principal unifying characteristic is that they are Gram-positive, anaerobic, endospore-forming bacilli (Cato and Stackebrandt 1989). As a group they have a wide distribution in nature and exhibit extremely diverse fermentative abilities. This latter capability has fostered interest in harnessing certain *Clostridium* spp. in the biocatalytic production of chemical fuels and high-added value stereospecific chemicals, e.g., acetone and butanol from *C. acetobutylicum* (Jones and Woods 1989). Other clostridia elaborate proteinaceous materials that cause diseases in man and his animals, ranging from the severe (e.g., botulism, tetanus, and gas gangrene) to the debilitating (food poisoning and antibiotic-associated diarrhea and colitis). Given these factors,

Research was carried out within the framework of the Biotechnology Action Programme of the Commission of the European Communities (BAP-0046-UK).

it is perhaps not surprising that considerable efforts have been made since 1980 to further our understanding of the respective industrial and medical attributes of particular *Clostridium* sp. To facilitate these studies many laboratories around the world have turned to recombinant DNA methodology (Young et al. 1989a, 1989b).

The primary prerequisite for applying recombinant techniques to clostridia is the availability of gene transfer systems. Their development has been a major preoccupation of several laboratories worldwide. In this chapter, we will review the current status of clostridial host:vector systems. Currently, the only two organisms in which any real progress has been made are the industrial species *C. acetobutylicum*, and the medically important *C. perfringens*.

6.1 VECTOR TRANSFER

In both *C. acetobutylicum* and *C. perfringens* the route taken for developing gene transfer systems has, in the main, followed traditional procedures established in more genetically amenable Gram-positive bacteria, such as *Bacillus subtilis*. This has involved the derivation of an autonomously replicating element perceived as likely to replicate in the chosen clostridial host, and then using it to formulate a transformation procedure. These vectors have been, with notable exceptions, plasmids. Although plasmids have proven to be widespread in both the saccharolytic and proteolytic clostridia, the vast majority of the former are cryptic in function (Minton and Thompson 1991). In contrast, many of the plasmids found in *C. perfringens* have been shown to encode antibiotic resistant determinants (Bréfort et al. 1977; Abraham et al. 1985). The major consequence of this early finding was that the derivation of plasmids for use in *C. perfringens* relied on endogenous plasmids, while many of the vectors employed in studies with *C. acetobutylicum* were comprised of heterologous plasmid components.

6.1.1 Transformation of Protoplasts

After many years of fruitless, unreported attempts to develop transformation procedures by inducing some form of natural competence in clostridial cells, the breakthrough for would-be clostridial geneticists came with the publication of polyethylene glycol (PEG)-induced protoplast transformation procedures for *B. subtilis* and *Streptomyces* (Chang and Cohen 1979; Bibb et al. 1978). Thereafter, numerous reports appeared in the literature documenting the conversion of clostridial cells to stable, wall-less variants (protoplasts, autoplasts, or L forms) and their subsequent regeneration back to bacillary rods (reviewed in Minton and Thompson 1991). However, only in the case of *C. perfringens* and *C. acetobutylicum* did this necessary enabling technology culminate in the establishment of a successful transformation protocol.

6.1.1.1 *Clostridium perfringens.* Heefner et al. (1984) were the first authors to report on the successful transformation of *C. perfringens.* In their initial procedure, *C. perfringens* cells converted to stable L forms, by growth in the presence of penicillin (10 μg/ml) and 0.4 M sucrose, were shown to be transformable by pCW3 plasmid DNA in the presence of PEG. This large endogenous conjugal R-factor encodes for tetracycline (Tet) resistance, and had initially been cured from the *C. perfringens* host employed. The usefulness of this system was severely restricted by the finding that the L forms derived could not regenerate back to normal walled bacterial cells. A second procedure was, however, formulated which relied on the predilection of growing cells to undergo autolytic conversion to naked cells in the presence of an osmotic stabilizer, such as sucrose. The so-called *autoplasts* formed could both be transformed, and be induced to revert back to walled bacteria, albeit inefficiently. Crucially, autoplasts could not be transformed with DNA prepared from *Escherichia coli*, only from *C. perfringens* L forms. The authors then went on to make several rudimentary shuttle vectors (Squires et al. 1984) which could be introduced into L forms or autoplasts by the same basic procedure. This methodology was, however, of limited utility both with regard to transformation frequency (10^2 per μg DNA) and the cumbersome nature of the regeneration procedure employed, and the fact that shuttle vector DNA had first to be passaged through L forms. Furthermore, although other workers demonstrated that other strains of *C. perfringens* could also be converted to L forms capable of being transformed (Mahoney et al. 1988; Roberts et al. 1988), these same strains did not produce transformable autoplasts.

6.1.1.2 *Clostridium acetobutylicum.* Transformation of *C. acetobutylicum* protoplasts was first obtained by Reid et al. (1983) using phage CA1 DNA. Successful transfection was not reliant on the addition of PEG. Plaques did not form on the regeneration medium employed, but rather were assayed indirectly by plating regenerated growth on normal agar medium. Consequently, no quantitative estimate of the frequency of transfection could be obtained. In the following year, it was shown that the staphylococcal plasmid pUB110 could transform PEG-treated protoplasts of *C. acetobutylicum* (Lin and Blaschek 1984). This report remains unsubstantiated.

The most efficient and "workable" procedure for transforming *C. acetobutylicum* is that reported by Reysset et al. (1988). The success of this system appears to have been the beneficial measures taken to limit autolysin activity during protoplast regeneration. The pivotal steps were the isolation of an autolysin defective mutant, N1–4081, of the *C. acetobutylicum* strain N1-4 (formerly *C. saccharoperbutyl-acetonicum*) and the inclusion of autolysin inhibitors (choline or sodium polyanethole sulphonate) in the medium (Reysset et al. 1987). Using this improved regeneration procedure they were able to show PEG-dependent transfection of N1-4081 with phage HM3 DNA at frequencies of between 10^4–10^5 transformants per μg (Reysset et al. 1988).

Because no plasmid has been isolated from *C. acetobutylicum* to which a function has been attributed, plasmid transformation was subsequently demonstrated initially using derivatives of the broad-host-range enterococcal plasmid pAMβ1 (Reysset et al. 1988), and thereafter with various plasmids encoding antibiotic resistance from other Gram-positive organisms (Truffaut et al. 1989). Transformation of plasmid DNA occurred at equivalent frequencies to those obtained with phage HM3 DNA.

6.1.2 Transformation of Whole Cells

In 1983, the publication of a procedure for transforming bacterial cells using high-voltage electric fields (Shivarova et al. 1983), a procedure more normally associated with higher eucaryotic cells, signaled a change in the direction of subsequent attempts to derive transformation procedures for genetically intransigent bacterial species. As soon as the necessary apparatus became commercially available, a plethora of publications appeared describing the transformation of whole bacterial cells by electroporation (see Chassy et al. 1988; Luchansky et al. 1988).

6.1.2.1 *Clostridium acetobutylicum.* The first clostridial species to be transformed by electroporation was *C. acetobutylicum* NCIB 8052 (Oultram et al. 1988a). The procedure devised used a BioRad Gene Pulser and was typical of those reported for other Gram-positive bacteria, with maximum transformants occurring at voltages of 5.0 kV/cm. The plasmid vector employed was essentially a pAMβ1 derivative, pMTL500E (Table 6–1), and the frequencies obtained varied between 10^2–10^3 transformants per μg of DNA. The reproducibility of efficient transformation was, and remains, inconsistent. Factors contributing to this inconsistency have not been determined. The current procedure in use in this laboratory utilizes melibiose in place of sucrose as the osmotic stabilizer, and a smaller volume of cells (0.4 ml) by employing 0.2-cm cuvettes (field strength 1.25 kV, 25 μF, 100 ohms). Additionally, whenever possible the plasmid DNA is prepared from *B. subtilis*, rather than *E. coli*, as transformation frequencies are reproducibly higher.

There have been no other published reports of the successful transformation of *C. acetobutylicum* by electroporation. Indeed, the applicability of this method to other strains, in particular strain ATCC 824, has proven problematical. However, this strain has now been successfully electrotransformed with plasmid DNA (G. Bennett, E. Papoutsakis, and H. Blaschek, personal communication).

It is perhaps worth mentioning that the development of procedures for the transformation of whole cells of *C. acetobutylicum* have not just been limited to those reliant on electroporation. Yoshino et al. (1990) have recently reported that strain 86 can be induced to take up DNA by prior treatment with alkaline-Tris buffer. A similar procedure was previously shown to effect transformation of *C. thermohydrosulfuricum* (Soutschek-Bauer et al. 1985).

TABLE 6–1 Clostridial Cloning Vectors

Vector	Size (kb)	Replication Origins		Selectable Marker[1]		Restriction Enzyme Cloning Sites		References
		E. coli	Clostridium	E. coli	Clostridium	*Selectable*	*Nonselective*	
Vectors for *C. perfringens*:								
pJU12	11.6	ColE1	pJU121	*bla (tetP)*	*tetP*	NG	NG	Squires et al. 1984
pJU13	12.2	ColE1	pJU122	*bla (tetP)*	*tetP*	NG	*Hinc*II	Squires et al. 1984
pJU16	12.2	ColE1	pJU122	*bla (tetP)*	*tetP*	NG	*Hinc*II	Squires et al. 1984
pHR106	7.9	ColE1	pJU122	*bla (cat)*	*cat*	Ap^S - *Pst*I	*Hind*III, *Sal*I, *Bam*HI, *Cla*I, *Aat*II	Robert et al. 1988
pSB92A2	7.9	ColE1	pCP1	*bla (cat)*	*cat*	Ap^S - *Pst*I	*Bal*I, *Sal*I, *Bam*HI	Phillips-Jones 1990
pAK201	8.0	ColE1	pHB101	*cat*	*cat*	NG	*Eco*RI, *Cla*I, *Bam*HI, *Pvu*II, *Ava*I	Kim and Blaschek 1988
Vectors for *C. acetobutylicum*:								
pKNT11	6.5	ColE1	pIM13	*bla*	*erm*C	Ap^S - *Pst*I, *Pvu*I	*Bam*HI, *Eco*RI, *Pvu*II, *Ava*I	Truffaut et al. 1989
pKNT14	4.5	[none]	pIM13	NAP	*erm*C	Tc^S - *Kpn*I, Em^S - *Bcl*	*Eco*RI, *Ava*I *Bam*HI, *Pvu*II	Truffaut et al. 1989
pKNT18/19	4.9	ColE1	pIM13	*bla, lacZ'*	*erm*C	Lac^- - *Eco*RI, *Kpn*I, *Pst*I - *Sal*I, *Sma*I, *Xba*I	NG	Reysset, unpublished observations

(continued)

TABLE 6–1 (continued)

		Replication Origins		*Selectable Marker*[1]		*Restriction Enzyme Cloning Sites*		
Vector	*Size (kb)*	E. coli	Clostridium	E. coli	Clostridium	*Selectable*	*Nonselective*	*References*
pMTL500E	6.43	ColE1	pAMβ1	*bla*, *lacZ'*	*erm*C	Lac⁻ - *Aat*II, *Bgl*II, *Kpn*I, *Mlu*I, - *Pst*I, *Sal*I, *Sma*I, *Sph*I, - *Sst*I,*Stu*I, *Xba*I, *Xho*I		Oultram et al. 1988a
pMTL502E	7.52	ColE1	pAMβ1	*bla*, *lacZ'*	*erm*C	as pMTL500E, except *Kpn*I		Swinfield et al. 1990
pMTL500F	6.69	ColE1	pAMβ1	*bla*, *lacZ'*	*erm*C	as pMTL500E, + *Nde*I, *Nru*I	*Eco*RV	This laboratory
pMTL710	7.38	ColE1	pAMβ1	*bla*, *xyl*E	*erm*C	XylE⁺ - *Bgl*II, *Mlu*I, *Nco*I, *Pst*I, *Sph*I, *Stu*I, *Xho*I,	*Kpn*I, *Sma*I, *Sst*I	This laboratory

[1] The *C. perfringens* derived *tetP* and *cat* genes both express in *E. coli*. Where original descriptions of cloning vectors have not given data on the presence of unique cloning sites, this is indicated by NG.

6.1.2.2 *Clostridium perfringens.* Several different procedures have been described for the transformation of *C. perfringens* by electroporation. The initial study of Allen and Blaschek (1988) demonstrated that *C. perfringens* ATCC 3624A could be transformed with plasmid DNA at frequencies 100-fold higher than PEG-induced procedures for L-phase variants. Plasmids used were pAMβ1 (3.8×10^{-5} transformants per viable cell) and a vector derived from an endogenous *C. perfringens* plasmid, pHR106 (4.2×10^{-4} transformants per viable cell). In subsequent studies, attempts were made to determine the factors that influence electroporation induced transformation frequencies (Allen and Blaschek 1990). Increases in efficiency were effected by several factors, including the use of higher cell densities in a smaller cell sample volume (3×10^8 colony forming units per ml), the use of late exponential cells, and a reduction in the pH of the expression and selective media to 6.4. These, and other modifications improved the frequency of pAMβ1 transformation by 42-fold. Similar findings were reported by Phillips-Jones (1990) and Scott and Rood (1989), using *C. perfringens* strains P90.2.2 and 13, respectively. The latter study also demonstrated that maximal frequencies were only obtained if suspended cells were pretreated with lysostaphin to weaken the cell walls. Such treatment proved unnecessary for the same strain under the conditions employed by Allen and Blaschek (1990), who also showed that type B strains of *C. perfringens* may be recalcitrant to current electroporation procedures.

6.1.3 Conjugative Transfer

Parallel to efforts to devise transformation procedures for clostridia, investigations into conjugative methods have been undertaken. Surprisingly, although conjugative plasmids have been well documented in *C. perfringens* (Bréfort et al. 1977; Abraham et al. 1985), their potential as genetic tools has not been examined. Rather, it is the ability of *C. acetobutylicum* to act as a recipient in intergeneric plasmid transfer that has been exploited. Two separate studies demonstrated that broad-host-range enterococcal/streptococcal plasmids, such as pAMβ1, could be transferred from aerobic Gram-positive donors (e.g., *B. subtilis* and *Streptococcus lactis*) to *C. acetobutylicum* (Oultram and Young 1985; Yu and Pearce 1986). Indeed it was this early demonstration that pAMβ1 could replicate in this clostridial species that led to its frequent use as the vector DNA in the development of transformation protocols (see Sections 6.1.2.1 and 6.1.2.2). Thereafter, pAMβ1 transfer was exploited to mobilize small cloning vectors from *B. subtilis* to *C. acetobutylicum* in the form of cointegrate molecules (Oultram et al. 1987). More recently, it has proven possible to mobilize cloning vectors from *E. coli* to *C. acetobutylicum* (Williams et al. 1990a). This type of transfer is made possible by endowing the cloning vector with the origin of transfer (*oriT*) of plasmid RK2 and providing the donor with the necessary transfer functions. Conjugative plasmid transfer is more fully discussed in Chapter 5 in this volume.

6.2 CLOSTRIDIAL CLONING VECTOR COMPONENTS

Although not always the case, the initial demonstration of transformation in particular clostridial strains has subsequently led to the construction of specialized cloning vectors. The proviso is included, because, in certain instances, the shuttle vector was specifically constructed as the plasmid to be used in the pioneering transformation experiments, e.g., pMTL500E for *C. acetobutylicum* NCIB 8052 (Oultram et al. 1988a) and pSB92A2 for *C. perfringens* P90.2.2 (Phillips-Jones 1990). However, whatever the order of events, the usual route taken has been to create an *E. coli*/*Clostridium* shuttle vector. Thus, the majority of vectors are composed of an *E. coli* moiety (replicon and selective marker) and "clostridial" replication and antibiotic resistance functions. In every instance to date the *E. coli* replicon chosen has been that of plasmid ColE1, and the selective marker most often employed the *bla* gene of pBR322 or derivatives thereof. The clostridial moiety of such shuttle vectors has been founded on a multitude of different clostridial replication regions and antibiotic resistance genes. In certain instances, vectors have been designed to function in a host other than *E. coli*, e.g., *B. subtilis*.

The majority of cloning vectors constructed to date carry minimal numbers of unique restriction sites into which heterologous DNA may be cloned (see Table 6–1). In most instances, insertion of DNA fragments into those sites that are present does not cause any phenotypic change by which recombinants may be identified. In a few cases, the attributes of the pUC *lacZ'* alpha complementation system has been incorporated, such that recombinants may be selected, in an appropriate *E. coli* host, as achromogenic colonies in the presence of 5-bromo-4-chloro-3-indoyl-β-D-galactosidase (X-Gal). Currently, the only vectors that use this system are the pIM13-based vectors constructed at the Institut Pasteur (using the pUC18/19 polylinker, Yanisch-Perron et al. 1985), and the pAMβ1-based vectors constructed in this laboratory (using the extended pMTL20 polylinker sequence, Chambers et al. 1988). A comprehensive list of vectors either constructed for use, or found to function, in clostridia is illustrated in Table 6–1.

6.2.1 Selective Markers

The wide distribution of conjugal plasmids encoding antibiotic resistance determinants in different strains of *C. perfringens* (Bréfort et al. 1977; Abraham et al. 1985), has provided a range of endogenous selectable marker genes for plasmid constructions in this species. In *C. acetobutylicum* the complete absence, to date, of plasmids specifying antibiotic resistance has led to the selection of such genes from heterologous sources.

6.2.1.1 *Clostridium perfringens* Vectors. The two genes most commonly employed as selectable markers in *C. perfringens* vectors are *cat* of plasmid pIP401 (Bréfort et al. 1977; Abraham and Rood 1987) and *tetP* from plasmid pCW3 and related plasmids (Abraham et al. 1988). Plasmids using these selectable markers are listed in Table 6–1. In general, the concentrations of antibiotic used in the selection of cells carrying these plasmid-borne genes is 20–30 μg/ml of chloramphenicol (Cm) and 2 μg/ml of Tet. Both genes also confer the appropriate resistance on the *E. coli* host. Although the *erm* gene of pAMβ1 is known to function in *C. perfringens* (Allen and Blaschek 1988), this finding has not been further exploited in the construction of shuttle vectors for this species.

6.2.1.2 *Clostridium acetobutylicum* Vectors. In contrast to the situation with *C. perfringens*, genes specifying resistance to erythromycin (Em) represent the most reliable selectable markers for use in *C. acetobutylicum*. The *erm* gene of pAMβ1 has proven particularly useful. The availability of its entire nucleotide sequence (Brehm et al. 1987) facilitates more informed vector design, and the constitutive nature of its expression can allow growth of the organism at levels of antibiotic supplementation as high as 2 mg/ml. The level of antibiotic routinely used in our laboratory is 10 μg/ml. An *ermC* gene also forms the basis of selection for the *B. subtilis*/*Clostridium* vector pIM13 (Truffaut et al. 1989).

As with *C. perfringens*, use has also been made of *cat* and *tet* genes. In the case of pC194-derived *cat*, however, primary selection for Cm^R has not proved possible. Rather plasmids were first selected on the basis of Em^R, by virtue of an encoding *erm* gene, and then shown to have also become Cm^R (Oultram et al. 1987, 1988b). This finding is no doubt related to the observation that clostridia are frequently tolerant to Cm (O'Brien and Morris 1971). In *C. thermohydrosulfuricum*, this problem was circumvented by employing thiamphenicol (Soutschek-Bauer et al. 1985). However, we have had difficulty in applying such a strategy to *C. acetobutylicum* NCIB 8052.

Although the *tetC* gene of the staphylococcal plasmid pT127 was successfully employed in the construction of plasmids for use in *C. acetobutylicum* N1-4081 (Truffaut et al. 1989), we have experienced some difficulty in using the *tetP* gene of the *C. perfringens* plasmid pIP401 in *C. acetobutylicum* NCIB 8052. Insertion of this gene into the polylinker site of the cloning vector pMTL500E (Table 6–1), results in a vector, pMTL500ET, which can be selected for either on the basis of Em^R or Tet^R in both *E. coli* and *B. subtilis*. However, transformants of *C. acetobutylicum* NCIB 8052 may only be selected on the basis of Em^R. These transformants, however, are subsequently found to be resistant (at levels of 2 μg/ml) to Tet (M. Mauchline and S. Whelan, unpublished observations). Part of the problem appears to reside in the high background of spurious Tet^R cells that are present in a given population of this strain.

6.2.2 Clostridial Replication Functions

Replication functions of clostridial plasmids have been identified by the insertion of plasmid-derived restriction fragments into Gram-positive replicon probe vectors (see Section 6.4.1), and then testing the ability of the resultant recombinant plasmid to transform the clostridial strain under investigation. Replicon probe vectors in this context may be defined as an *E. coli* cloning vector encoding a Gram-positive resistance gene capable of expression in clostridia, which is unable to replicate in a Gram-positive bacterium. Some replicons have not been derived from clostridial plasmids (e.g., pAMβ1 and pIM13), but for the purpose of this review have been termed clostridial because they function in this host.

Before describing the characteristics of clostridial vector replicons, it is worthwhile reviewing the current status of replication analysis in other Gram-positive bacteria. It is now clear that small plasmids indigenous to many different Gram-positive bacteria form a highly interrelated family, termed "ss DNA plasmids" (Gruss and Ehrlich 1989). They all replicate via a single stranded (ss) DNA intermediate, using a rolling circle mechanism of replication, typical of *E. coli* filamentous phages (Baas 1985). A plasmid encoded protein (Rep) initiates DNA replication by introducing a nick in the positive DNA strand at a fixed position, termed the "plus" origin. Replication proceeds by synthesis of a new positive strand by 3′-OH extension from the nick, concomitant with displacement of the "old" parental positive strand. Synthesis of a new negative strand, using the displaced positive strand as template, is mediated by host factors acting at the "minus" origin (reviewed by Gruss and Ehrlich 1989). Although ss DNA plasmids have adopted an identical replication strategy, they appear, on the basis of the homology between their Rep proteins and plus origins of replication, to fall into at least three subfamilies (Gruss and Ehrlich 1989; Minton et al. 1988). Representatives of the first group are the staphylococcal plasmids, pC221 and pT181 (Brenner and Shaw 1985), those of the second group, the staphylococcal plasmid, pE194 (Villafane et al. 1987), and the streptococcal plasmid, pLS1 (Lacks et al. 1986), and those of the third group the *Staphylococcus aureus* plasmids, pC194 (Dagert et al. 1984) and pUB110 (McKenzie et al. 1986).

6.2.2.1 The pCB101 Replicon. The best characterized replication region of a saccharolytic clostridial plasmid is that of pCB101. This 6.05 kilobase (kb) cryptic plasmid was originally isolated from a bacteriocinogenic strain of *C. butyricum*, NCIB 7423 (Minton and Morris 1981). Preliminary analysis, using a replicon probe vector and *B. subtilis* as the host, localized replication functions to a 3.3-kb *Sau*3 A subfragment (Collins et al. 1985). Subsequent deletion and nucleotide sequence analysis has revealed the following facts. The minimum fragment required for replication (a 2.4-kb fragment composed of two contiguous *Taq*I fragments) contains three open reading frames (ORFs), designated B, C′, and C (Figure 6–1). The polypeptides encoded by the largest two ORFs are both essential for replication, and have been designated RepB (43,039 daltons, Da) and RepC (27,134 Da). The former protein exhibits significant homology (see Minton et al. 1990a) with known

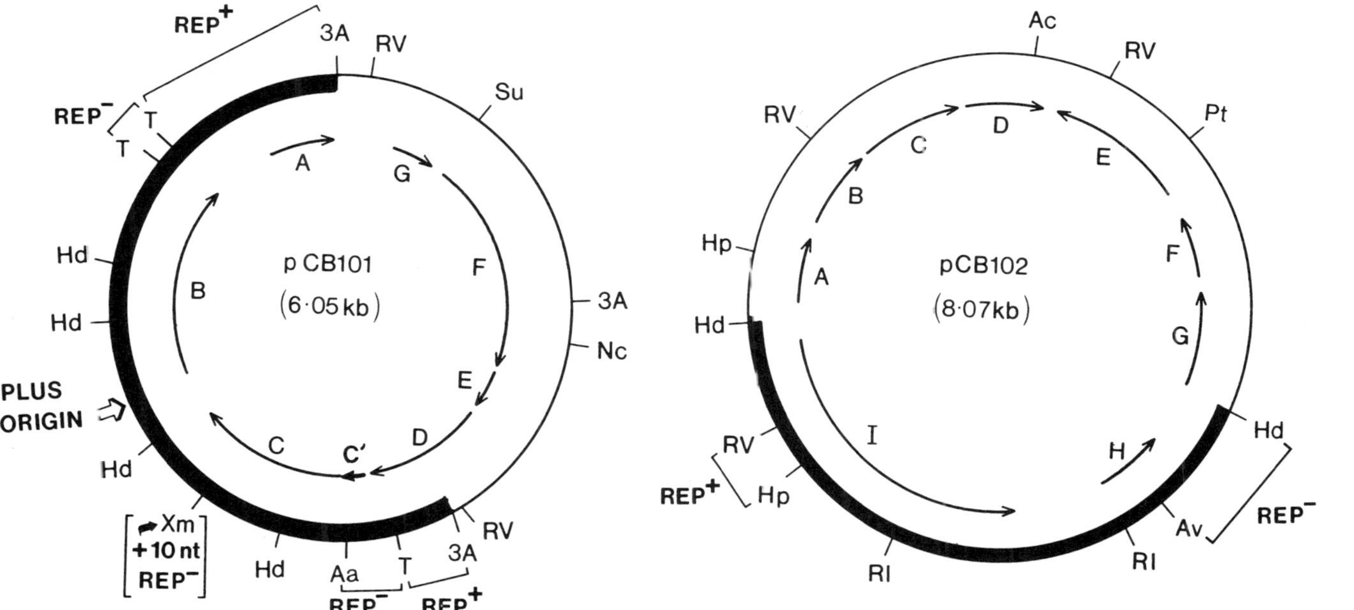

FIGURE 6–1 Plasmids pCB101 and pCB102. The extent of the above plasmids utilized in the construction of the plasmids listed in Table 6–2, pCB3 (pCB101) and pCB4 (pCB102), is indicated by a bold line. The major ORFs encoded by each plasmid, identified by nucleotide sequencing, are indicated by arrowed lines within the respective maps. Restriction enzyme sites are: Aa, *Aat*II; Ac, *Acc*I; Av, *Ava*III; Bg, *Bgl*II; Hd, *Hin*dIII; Hp, *Hpa*I; Nc, *Nco*I; Pt, *Pst*I; R1, *Eco*RI; RV, *Eco*RV; Su, *Stu*I; T, *Taq*I; Xm, *Xmn*I; 3A, *Sau*3A. The regions between bracketed sites represent specifically deleted segments of DNA, which in some cases destroyed the replicative ability of derivative plasmids (REP$^-$). The introduction of a frameshift in ORF C of pCB101 (by inserting 10 nucleotides at the indicated Xm site) also conferred a REP$^-$ phenotype on the resultant plasmid.

Gram-positive replication proteins, notably RepH of pC194 (23.2% identity) and RepB of pUB110 (18.2% identity). RepC exhibits no homology to any currently characterized replication protein. In common with replication proteins of this type (Murai et al. 1987; Josson et al. 1990), the C terminus of RepB plays an indispensable role in plasmid replication. Thus plasmid derivatives in which only nine codons have been removed from the 3′ end of *repB* are unable to replicate. Preceding *repB* is a sequence motif exhibiting substantial homology to the plus origins of pUB110 and pC194 (see Minton et al. 1990a), which has been shown to act as a "hot spot" for deletion formation in certain plasmid chimeras.

These data suggest that pCB101 belongs to the pUB110/pC194 subfamily of ss DNA plasmids, and must, therefore, replicate via a rolling circle mechanism. Indeed, experiments have demonstrated that ss DNA replication intermediates accumulate in *B. subtilis* cells carrying a pCB101-derived plasmid. The plasmid is, however, atypical in that it requires a second protein, RepC, for autonomous replication. Furthermore, recently undertaken comparative alignments have shown that the ORF C′ polypeptide exhibits distant homology to the RepA protein of pLS1, and more significant homology to the putative RepB protein of the *B. thuringiensis* plasmid pHD2 (McDowell and Mann 1991). Interestingly, the 3′-end of ORF C′ overlaps with the 5′ end of *repC*, and tentative evidence has been obtained suggesting that an ORF C′::RepC translational fusion protein can be produced under certain conditions. Plasmid pCB101 therefore appears to share features of at least two different classes of ss DNA plasmids, while possessing several characteristics currently unique in plasmidology.

6.2.2.2 The pCB102 Replicon. Shuttle vectors have also been constructed using the replicon of pCB102, a cryptic plasmid co-resident with pCB101 in *C. butyricum* NCIB 7423 (Minton and Morris 1981). The entire nucleotide sequence of pCB102 has been determined (D.E. Thompson, J.K. Brehm, and N.P. Minton, unpublished observations), and a total of nine putative ORFs are evident (see Figure 6–1). Early studies demonstrated that all the functions necessary for autonomous replication of pCB102 derivatives in *C. acetobutylicum* resided on a 3.57-kb *Hin*dIII fragment, which encodes two of the identified ORFs. The larger of the two, ORF I, encodes for a protein with a predicted M_r 68,800, while the second smaller ORF (H) encodes for a polypeptide of M_r 11,800. Neither protein exhibits homology to any known replication protein. Two pieces of evidence suggest that ORF I is not required for replication. Firstly, insertion of a 2.79-kb *Hpa*I-*Hin*dIII subfragment (which effectively deletes 36% of ORF I) into pMTL20E results in a plasmid that is still able to replicate in *C. acetobutylicum*. Secondly, deletion of the ORF I encoding DNA from between the *Eco*RV and *Hpa*I sites has no effect on the replicative ability of a pMTL20E recombinant plasmid. Other experiments suggest that ORF H is required for replication. Thus, a pMTL20E plasmid recombinant derived by inserting a 2.93-kb *Hin*dIII-*Ava*III pCB102-derived fragment was unable to transform *C. acetobutylicum*.

The analysis of pCB102 replication functions is still at an early stage, and the lack of homology of any of its components to other plasmid replicons means that no conclusions may be drawn as to its mode of replication. It is noteworthy that in contrast to the co-resident ss DNA plasmid pCB101, the ORFs of pCB102 are not encoded by the same DNA strand (see Figure 6–1). The confinement of ORFs to the one DNA strand is a general feature of ss DNA plasmids. None of the remaining seven ORFs that lie outside of the replication region show homology with any known protein. The polypeptide encoded by ORF C does, however, exhibit several features reminiscent of bacteriocins, including a high polarity index and a high content of glycine residues (7%). Furthermore, an extracellular location for its polypeptide is implied by its possession of a 27 amino acid motif, at its extreme N terminus, typical of Gram-positive signal peptides. The predicted M_r of the protein following removal of this putative signal peptide corresponds almost exactly to that estimated (32,500 Da) for the bacteriocin purified from *C. butyricum* NCIB 7423, butyricin 7423 (Clarke et al. 1975). If ORF C does correspond to the gene specifying this butyricin, then in comparison to other bacteriocins, the method adopted for secretion would be highly unusual.

6.2.2.3 The pIP404 Replicon. The 10.2-kb bacteriocinogenic, *C. perfringens* plasmid, pIP404 (Bréfort et al. 1977), represents the most thoroughly characterized of all clostridial plasmids. Its entire nucleotide sequence has been determined (Garnier and Cole 1988a), and functions have been assigned to 6 of the 10 ORFs identified. Three of the ORFs (*bcn*, *uviA*, and *uviB*) are involved in bacteriocin production/immunity (Garnier and Cole 1986, 1988b), one (*res*) is suggested to play a role in segregational stability (Garnier et al. 1987), while the remaining two are involved in replication (Garnier and Cole 1988c).

The minimal replication region was determined by cloning various pIP404-derived restriction fragments into a replicon probe vector, encoding the pC194 *cat* gene, and testing the ability of resultant recombinant plasmids to transform *B. subtilis* to Cm^R (Garnier and Cole 1988c). The smallest fragment identified, a 2.4-kb *Eco*RI-*Eco*RV fragment, carried two ORFs (4 and 5) and a d(A+T)-rich region comprising a family of repeated sequences. Further analysis showed that a functional ORF5 encoded polypeptide was essential for replication, and was, therefore, designated *rep*. While disruption of ORF4 did not abolish replication, the copy number of plasmids was significantly increased. ORF4 was therefore designated *cop*. Because the repeated region was also found to be essential for replication, and in view of its resemblance to replication origins in other plasmids, such as RK2 and pSC101, it is likely that these sequences represent the plasmid origin. A particularly novel aspect of the pIP404 replication functions was the discovery of a strong promoter 3′

to *rep*, which directed the transcription of a 150 nucleotide RNA molecule, termed RNA1. It followed that the 3′-ends of RNA1 and the 3′-end of the *rep* mRNA overlap by some 130 nucleotides. Although regulation of gene expression through the agency of antisense messengers is not uncommon, overlap is more usually seen at the 5′-end of the target mRNA transcript.

Paradoxically, although the pIP404 replicon has been well characterized, its use in *C. perfringens* vectors has not been reported, as functional characterization was undertaken in *B. subtilis*. However, early difficulties in transforming derivative vectors into *C. perfringens* have now been largely overcome (S.T. Cole, personal communication), and one can anticipate a major role for pIP404-derived vectors in the future manipulation of this clostridial species.

6.2.2.4 The pAMβ1 Replicon. The broad-host range conjugal plasmid pAMβ1 (Leblanc and Lee 1984) has played a prominent role in the development of clostridial transformation systems (see Sections 6.1.1.2 and 6.2.2.1). Therefore to facilitate future vector constructions, we determined the entire nucleotide sequence of its replication region (Swinfield et al. 1990). Of the 5.1-kb region sequenced, the minimum replication region resided on a 2.588–base pair (bp) *Hpa*I-*Acc*I fragment that carried two ORFs, ORF D (encoded polypeptide 11,762 Da) and ORF E (encoded polypeptide 57,380 Da). As the integrity of ORF E was shown to be essential for replication, it was designated *repE*. As with the homologous protein, RepA, of the related plasmids pSM19035 and pIP501 (Brantl et al. 1989, 1990; Sorokin and Khazak 1989), RepE was indirectly shown to positively regulate plasmid copy number. Because pSM19035 lacks an equivalent ORF, ORF D is probably not essential for replication. However, in pAMβ1 ORF D resides 5′ to *repE*, and evidence was obtained that suggests the transcription of the two genes is linked.

Outside of the replication region, one of pAMβ1's most striking features is a 459-bp region composed of two classes of directly repeated 9-bp sequences. The first 14 nonamer units conformed to consensus GTYGATCCA, while the latter 37 repeats could be represented by the sequence ACRGARCCA. Surprisingly, the encompassing ORF (C) is encoding. The polypeptide is, therefore, composed of reiterated tripeptide amino acid motifs, VDP (GTYGATCCA) and TEP (ACRGARCCA). The possession of an N terminus typical of procaryotic signal peptides (von Heijne 1983), and an amino acid sequence motif at its C terminus resembling membrane-anchor sequences (see Vos et al. 1989), prompted the suggestion that the ORF C polypeptide has a cell surface location and is involved in the conjugation process (Swinfield et al. 1990). This suggestion has been substantiated by the recent demonstration that the major pheromone-inducible protein PD78 of plasmid pPD1, believed to contribute to bacterial conjugation (Nakayama et al. 1990), also contains a tandem repeated sequence of XXP.

Since the publication of the sequence of the pAMβ1 replication region, Bruand et al. (1991) have gone on to investigate the mechanism of replication. The detection in *B. subtilis* cells of θ-shaped replication intermediates indicated that pAMβ1 undergoes unidirectional theta replication. By insertion of the replication terminus (*terC*) of the *B. subtilis* chromosome, they were able to map the site of initiation of DNA replication, to a region immediately 3′ to *repE* between nucleotides 4610 and 4660 (Swinfield et al. 1990). This alternative mode of replication to the more common strategy of rolling circle replication has important consequences with regard to vector structural stability (see Section 6.3).

6.2.2.5 The pIM13 Replicon. Plasmid pIM13 is a 2.246-kb *B. subtilis* plasmid that confers Em^R resistance on the host cell. Its physical characterization has shown that a single 18-kDa protein (RepL) is required for replication and that it encodes a constitutively expressed adenine methylase, highly homologous to that encoded by the *ermC* gene of the staphylococcal plasmid pE194 (Monod et al. 1986). Although sequenced, its replication functions have not been rigorously characterized. However, the presence of *palA* and *palB* lagging-strand conversion signals, and the fact that host cells accumulate ss DNA (Projan et al. 1987), clearly indicate that it belongs to the ss DNA class of plasmids. On the basis of similarities at the nucleotide sequence level and functional organization, Novick (1989) subdivides staphylococcal ss DNA plasmids into four distinct families, represented by the prototype plasmids pT181, pC194, pSN2, and pE194. Plasmid pIM13 falls into the pSN2 family. Plasmids from all four families are able to replicate in *B. subtilis*. The results of Truffaut et al. (1989) and Reysset (personal communication) suggest that only pIM13-like vectors may be efficiently maintained in *C. acetobutylicum*. Thus pBC16Δ1 (pC194 family), pT127 (pT181 family), and pE194 (family prototype) all replicate inefficiently in this clostridial species. The reasons for this are not clear, but it should be noted that the plasmid family to which pIM13 belongs are the only ss DNA plasmids in which the direction of replication is opposite to that of the direction of transcription of the *rep* gene (Novick 1989).

6.2.2.6 Miscellaneous Replicons. Although several other clostridial plasmid replicons have been used in the construction of shuttle vectors, they remain uncharacterized. These include the *C. perfringens* plasmids pJU121, pJU122, pCP1, and pHB101 (see Table 6–1). Similarly, in our own laboratory, vectors have been derived from the replicons of the *C. butyricum* plasmid pCB103 and the *C. paraputrificum* plasmid pCPA1 (formerly called pCP1, see Young et al. 1989a, 1989b), for use in *C. acetobutylicum*. One interesting factor to emerge from the general analysis of clostridial replicons is the ob-

TABLE 6–2 Segregational Stability of Clostridial Vectors

Vector	Size (kb)	*Source of Replicon*		*Replication in*		Segregational Instability[1]	Reference
		Plasmid	*Organism*	Bacillus	Clostridium		
pTYD101	7.1	pCS86	*C. acetobutylicum*	ND	+	6.9×10^{-2}	Yoshino et al. 1990
pTYD104	3.5	pCS86	*C. acetobutylicum*	ND	+	4.3×10^{-1}	Yoshino et al. 1990
pAK201	8.0	pHB101	*C. perfringens*	ND	+	4.3×10^{-3}	Kim and Blaschek 1988
pJU12	11.6	pJU121	*C. perfringens*	ND	+	8.0×10^{-4}	Squires et al. 1984
pSB92A2	7.9	pCP1	*C. perfringens*	–	+	ND	Phillips-Jones 1990
pCB4	7.10	pCB102	*C. butyricum*	–	+	ND	This laboratory
pCB5	9.50	pCB103	*C. butyricum*	–	+	ND	This laboratory
pCP3	10.40	pCPA1	*C. paraputrificum*	–	+	6.5×10^{-2}	This laboratory
pCB3	7.03	pCB101	*C. butyricum*	+	+	1.4×10^{-2}	This laboratory
pTG67	6.6	pIP404	*C. perfringens*	+	+	2.0×10^{-1}	Garnier and Cole 1988c
pTG36	6.0	pIP404	*C. perfringens*	+	+	1.2×10^{-1}	S.T. Cole, personal communication
pMTL500E	6.43	pAMβ1	*E. faecalis*	+	+	4.1×10^{-2}	Oultram et al. 1988a
pMTL531E	8.60	pAMβ1	*E. faecalis*	+	+	5.0×10^{-3}	This laboratory
pMTL500*Eres*$^+$	7.10	pAMβ1	*E. faecalis*	+	+	4×10^{-3}	This laboratory
pMTL500*Eres*$^-$	7.10	pAMβ1	*E. faecalis*	+	+	2×10^{-1}	This laboratory
pKNT11	6.5	pIM13	*B. subtilis*	+	+	5.0×10^{-2}	Truffaut et al. 1989
pKNT14	4.3	pIM13	*B. subtilis*	+	+	1.2×10^{-2}	Truffaut et al. 1989

[1] Expressed as plasmid loss per cell per generation. The data for segregational instabilities of plasmids pTG67, pTG36, pMTL500*Eres*$^+$ and pMTL500*Eres*$^-$ have only been derived in *B. subtilis*. Where no data are available, this is indicated by ND. The data for pAK201 were derived by assuming that in the experiment undertaken (Kim and Blaschek 1989), growth was for 60 generations.

servation that, where investigated, only two out of six clostridial replication regions function in *B. subtilis* (Table 6–2). The two exceptions are plasmids pIP404 and pCB101.

6.3 VECTOR STABILITY

Recombinant manipulation of a microorganism requires that the integrity of introduced DNA is maintained during plasmid replication, and that it is partitioned with high fidelity to each successive generation. It follows that any vector system developed must exhibit both structural and segregational stability. It is now well established that the problems of structural and segregational instability which have plagued the use of *B. subtilis* as a recombinant host were largely due to the type of plasmid employed in vector construction. The plasmids most frequently used were the small, staphylococcal antibiotic resistance plasmids that replicate by a rolling circle mechanism. A characteristic feature of this mode of replication is that it is easily perturbed. It follows that alterations to plasmid integrity (i.e., during vector construction or insertion of cloned DNA), or the use of plasmids in an alternative host, frequently leads to structural or segregational instability (see Gruss and Ehrlich 1989). These findings suggest that care should be taken in the choice of plasmid replicon in shuttle vector design. If the replicon used is from a ss DNA plasmid (e.g., pCB101) then careful consideration should be given to: (1) the position at which cloning sites are created and (2) the use of a plasmid indigenous to the clostridial species under investigation to avoid problems resulting from the inefficient use of lagging strand conversion signals.

Alternatively, use should be made of a plasmid that does not replicate by a rolling circle mechanism, e.g., pIP404 and pAMβ1. The case for using pAMβ1 is particularly convincing. Its mode of replication, a unidirectional theta mechanism, has been experimentally shown to result in 1000-fold greater structural stability in *B. subtilis* over a ss DNA plasmid (pC194) control (Jannière et al. 1990). Furthermore, the efficiency of cloning and maintenance of large DNA fragments is greatly improved (Jannière et al. 1990). We have constructed several different bifunctional *E. coli*/*Clostridium* shuttle vectors (Figure 6–2), exhibiting both low and high copy number (Oultram et al. 1988a; Swinfield et al. 1990). As the pAMβ1-derived replication fragment was inserted into the unique *Nhe*I site immediately external to the *lac*Z′ gene of pMTL20E/C, these vectors retain all the cloning advantages of the replicon probe vectors (see Section 6.4.1).

6.3.1 Structural Stability

The only reported example of structural instability in a clostridial plasmid is that exhibited by a vector (pTY10) based on the cryptic *C. acetobutylicum* plasmid pCS86. Although apparently stably maintained in *E. coli*, pTY10 was

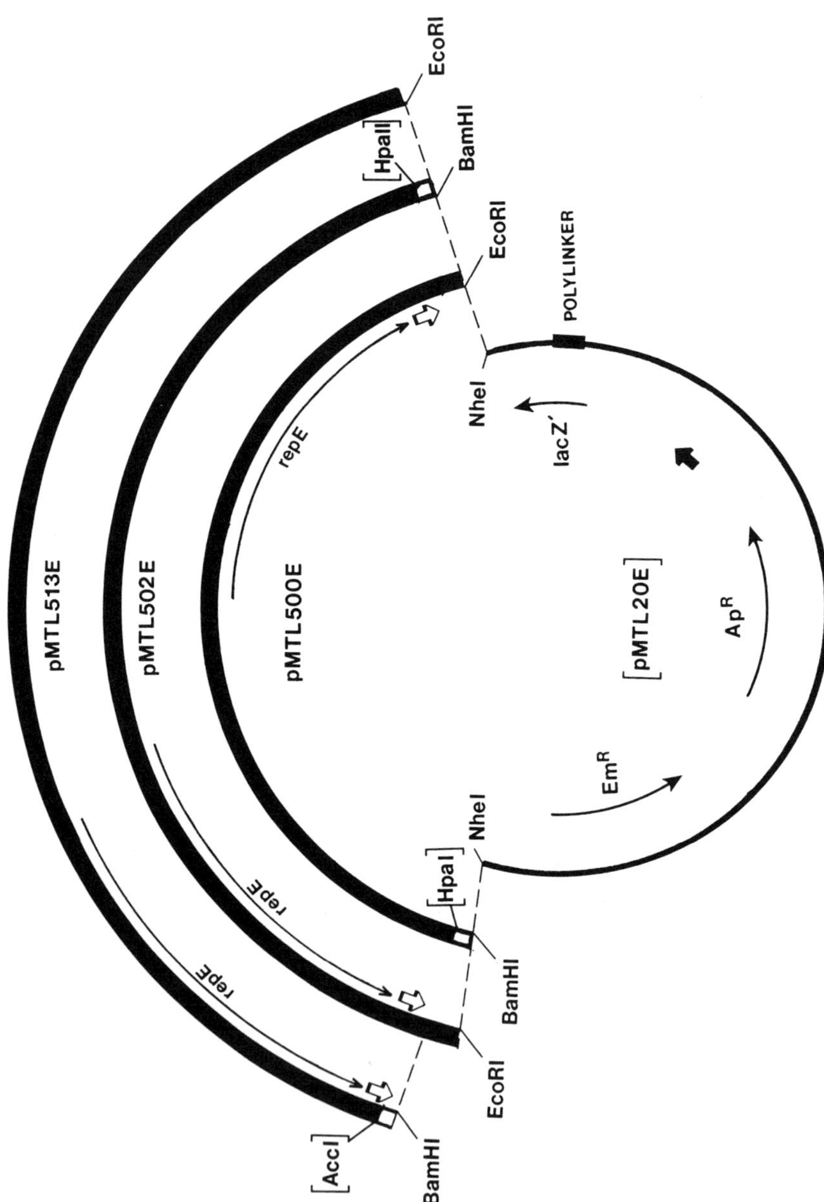

pMTL513E
pMTL502E
pMTL500E
[pMTL20E]
EcoRI
BamHI
[HpaII]
EcoRI
POLYLINKER
NheI
lacZ′
repE
repE
repE
ApR
EmR
NheI
[HpaI]
BamHI
EcoRI
[AccI]
BamHI

extremely unstable when introduced into *C. acetobutylicum* (Yoshino et al. 1990), generating a range of different sized deletion variants (from 4.0–7.6 kb). The scarcity of reported examples of structural instability is probably not because the problem does not exist, but rather reflects the fact that no systematic studies have been undertaken. Thus only gross abnormalities will have been detected. Nevertheless, we have observed no gross structural rearrangements in any of the vectors constructed in our laboratory, even in pCB101-based vectors. Similarly, vectors based on the ss DNA plasmid pIM13 have also not been observed to exhibit any measurable structural instability (G. Reysset, personal communication).

6.3.2 Segregational Stability

The experiments necessary to estimate the segregational stability of vectors in *Clostridium* have only been carried out with a handful of plasmids. When they have been undertaken, the different regimes and calculations used to estimate instability make comparisons difficult. It is, however, apparent that there are no reported examples of a vector that exhibits 100% segregational stability in the absence of antibiotic selection (see Table 6–2). The most stable element in the *C. perfringens* plasmid pJU12, described by Squires et al. (1984). The probability of plasmid loss per generation was reported to be 8×10^{-4} (i.e., after 20 generations, 98.4% of the cells would still contain the plasmid). The same plasmid was substantially more unstable when maintained in L-phase variants (probability of plasmid loss per generation, 1×10^{-3}). In the study of Kim and Blaschek (1989), 92% of cells were deemed to have lost plasmid pHR106 over a 3-day period of growth in the absence of antibiotic. The shuttle vector pAK201 was significantly more stable, with only 23% loss over the same time. Plasmids based on pIP404 proved to be dramatically unstable, although to date experiments have only been undertaken in *B. subtilis* (Garnier and Cole 1988c). In *C. acetobutylicum*, analogous problems of segregational instability have been observed with plasmids based on pIM13 (Truffaut et al. 1989; E. Papoutsakis, personal communication),

---

FIGURE 6–2 (shown on page 136) Cloning vectors based on the pAMβ1 replicon. All three illustrated plasmids were generated by insertion of the indicated pAMβ1-derived DNA (see Swinfield et al. 1990) fragment (bold line) into the *Nhe*I site of pMTL20E (thin line). The *lacZ'* is, therefore, functional (blue colonies in the presence of X-Gal) unless inactivated by subsequent insertion of heterologous DNA into one of the restriction sites of the pMTL20-derived (Chambers et al. 1988) polylinker region. Plasmid pMTL500E is a high copy number plasmid, while pMTL502E and pMTL513E have a low copy number. The open arrow downstream of *repE* is the pAMβ1 replication origin, the filled in arrow within the pMTL20E moiety is the ColE1 origin.

pSC86 (Yoshino et al. 1990), pCB101, pCPA1, and pAMβ1 (see Table 6–2). Interestingly, it has recently proved possible to select a mutant strain of *C. acetobutylicum* N1-4081, in which pIM13, and derivatives, are 100% segregationally stable (G. Reysset, personal communication). The nature of the defect in the host genome is presently unknown, but the generation time of this mutant was markedly longer.

6.3.3 Stability Determinants

One strategy adopted in other systems to attain segregational stability is to endow the cloning vector with a stability determinant. Although all plasmids must specify elements that ensure equipartition at cell division, no such elements have been experimentally identified on a clostridial plasmid. Nucleotide sequence analysis has demonstrated that pIP404 encodes a resolvase-like protein. In certain cases, segregational instability of plasmids can be caused by the predominant existence of the plasmid population as multimers, thereby reducing the overall number of plasmid units available for partition at cell division (Austin 1988). Plasmid encoded resolvases have been suggested to mediate plasmid stability by maintaining the plasmid population in the monomeric state through the agency of multimer resolution. No experimental evidence to support this contention for the pIP404 *res* gene has been described.

We have recently shown that the broad-host-range plasmid pAMβ1 also carries a *res* gene (Swinfield et al. 1991). Janniére et al. (1990) had observed that pAMβ1-based vectors which carried an additional 2 kb of pAMβ1 DNA to that sequenced by Swinfield et al. (1990) were segregationally more stable in *B. subtilis* than vectors such as pMTL500E. Insertion of this extra segment of DNA into the polylinker region of pMTL500E resulted in a plasmid, pMTL531E, which exhibited significantly improved segregational stability, both in *B. subtilis* and *C. acetobutylicum* (Table 6–2). Nucleotide sequence analysis of this fragment has shown it carries two ORFs (designated H and I), the expression of which may be transcriptionally/translationally coupled. The ORFH polypeptide (M_r of 23,930) exhibits substantial identity with bacterial site specific recombinases, including the resolvases of the Gram-positive transposon Tn*917* (30.3%), and the Gram-positive plasmids pI524 (31.6%) and pIP404 (27.1%) (Shaw and Clewell 1985; Rowland and Dyke 1988; Garnier et al. 1987). The second ORF (I) is incomplete. Its polypeptide (M_r of at least 55,555) shares 26% identity with *E. coli* DNA topoisomerase I (Tse-Dinh and Wang 1986). Additional sequence data positions both its translational stop codon and a C terminal "zinc finger" DNA-binding motif, homologous to those present in the *E. coli* topoisomerase enzyme (Tse-Dinh and Beran 1988), immediately 5′ to the previously sequenced (Brehm et al. 1987) pAMβ1 Em^R gene (J.D. Oultram, unpublished observations). The truncated ORF I polypeptide may therefore be inactive and play no role in stability.

Examination of *B. subtilis* cell lysates indicated that the pMTL500E (res^-) DNA was present in a highly polymerized form. As pMTL531E (res^+) DNA existed principally as the monomeric species, it seems likely that stabilization is mediated by multimer resolution. Confirmation that the *res* gene is the active element was obtained by constructing a pMTL500E derivative plasmid, pMTL500*Eres*$^+$, which carried the entire *res* gene but only 20 codons from the 5′-end of ORF I. Plasmid pMTL500*res*$^+$ exhibited a similar level of segregational stability to pMTL531E (Table 6–2). Furthermore, this increase in segregational stability was completely negated by the introduction of a frame shift into the coding region of *res* at a unique *Sty*I site (pMTL500*Eres*$^-$, Table 6–2). Utilization of this element could prove generally useful to the stabilization of clostridial vectors.

6.4 SPECIALIZED CLOSTRIDIAL VECTORS

Having devised transformation systems and the basic cloning vectors, attention will undoubtedly focus on the provision of clostridial vectors with specialist functions. In the following sections we will document some of the vectors of a more specialized nature constructed, or under construction, in this laboratory.

6.4.1 Replicon-Probe Vectors

Replicon probe vectors have proved particularly useful in both the construction of clostridial cloning vectors, and, once a clostridial replication origin has been identified, in its subsequent physical characterization. Such vectors essentially comprise an *E. coli* plasmid, possessing unique cloning sites and encoding an antibiotic resistance gene capable of being expressed in a clostridial cell. As *E. coli* plasmid origins such as ColE1 cannot replicate in *Clostridium* spp., clostridial plasmid derived restriction fragments may be inserted and the ability of the resultant chimerae to replicate in a clostridial host tested. The pMTL series of replicon probe vectors represent a particularly versatile collection of plasmids of this type (Swinfield et al. 1990; Minton et al. 1990a). Based on the cloning vectors pMTL20/21 (Chambers et al. 1988), these vectors offer a choice of three different Gram-positive resistance genes known to express in clostridia, namely, the pAMβ1 *erm*C gene (Brehm et al. 1987), pMTL20E/21E; the pC194 *cat* gene (Minton et al. 1990c), pMTL20C/21C, and; the *tetP* gene of the *C. perfringens* plasmid pIP401 (Abraham et al. 1988), pMTL20T/21T. All members of this plasmid series carry a functional copy of the *E. coli lac*Z′ gene, which encompasses a multiple cloning site region containing six additional cloning sites (*Aat*II, *Xho*I, *Nco*I, *Bgl*II, *Mlu*I, and *Stu*I) to that of the pUC18/19 polylinker region. Thus, *E. coli* clones harboring recombinant plasmids, derived by insertion of heterologous DNA, may be easily distinguished from transformants carrying parental plasmids by supplementation of the selective agar medium with X-Gal. Whereas parental plasmids confer a blue coloration on colonies, the inactivation of the *lacZ′* gene in recombinant plasmids results in colorless colonies.

The pMTL20E/21E vector pair have proved particularly useful in the construction of vectors in this laboratory, for the reasons discussed in Section 6.2.1. They have also formed the basis of a different type of replicon-probe vector, which has the facility to be introduced into *Clostridium* spp. by conjugative means. These vectors, pMTL30/31, were constructed (Williams et al. 1990a) by inserting a segment of DNA carrying the *oriT* of plasmid RK2, in the two possible orientations, into the *Nhe*I site of plasmid pMTL20E. The provision of *oriT* means that either plasmid may be mobilized from an appropriate *E. coli* donor to suitable recipient cells, providing the necessary transfer functions are supplied in trans. They were constructed (Williams et al. 1990a) following the demonstration that pAMβ1-derived plasmids encoding *oriT* could be conjugatively transferred from appropriate *E. coli* cells to *C. acetobutylicum* NCIB 8052, and have proved of great utility in assessing the ability of several Gram-positive replicons to function in this *Clostridium* sp., e.g., the broad host-range streptococcal plasmid pWV01 (Kok et al. 1984) was shown to replicate in strain NCIB 8052 (Williams et al. 1990b; see Chapter 5, this volume).

6.4.2 A Clostridial Expression Vector

To facilitate the expression of cloned genes in *C. acetobutylicum* we have constructed a pMTL500E derivative carrying the transcriptional/translational signals of the *C. pasteurianum* ferredoxin (Fd) promoter. Physiological studies had previously established that the Fd gene is relatively highly expressed in *C. pasteurianum* cells, grown on iron-sufficient media, and the Fd protein can represent up to 2% of the cells' soluble protein (Rabinowitz 1972). Using an oligonucleotide probe derived from the published nucleotide sequence (Graves et al. 1985), the Fd gene was cloned as a 650-bp *Sau*3A fragment, converted to a portable *Eco*RI fragment, and site-directed mutagenesis (SDM) employed to substitute the coding region of the structural gene with the polylinker cloning region of pMTL20. The resultant expression cartridge (see Minton et al. 1990a) therefore comprised multiple cloning sites flanked at the 5′-end by the transcriptional initiation signals of the Fd gene, and at the 3′-end by the Fd transcriptional termination signals (Graves and Rabinowitz 1986). Following insertion of a promoter-less copy of the pC194 *cat* gene, Fd::*cat* was cloned into pMTL500E and the plasmid obtained electroporated into *C. acetobutylicum* NCIB 8052. Measurement of chloramphenicol transacetylase (CAT) levels in the resultant transformants indicated that the *cat* gene was expressed at reasonably high levels, with recombinant CAT representing around 7% of the cells' soluble protein. Furthermore, the Fd promoter directed the efficient expression of *cat* in both *E. coli* and *B. subtilis* (see Minton et al. 1990b), in addition to the clostridial host.

The above results encouraged us to undertake the construction of a more refined expression vector, pMTL500F (Figure 6–3). This vector essentially represents pMTL500E in which the promoter signals directing the expression of *lac*Z (the *E. coli* derived *lac po* region) have been replaced with those of

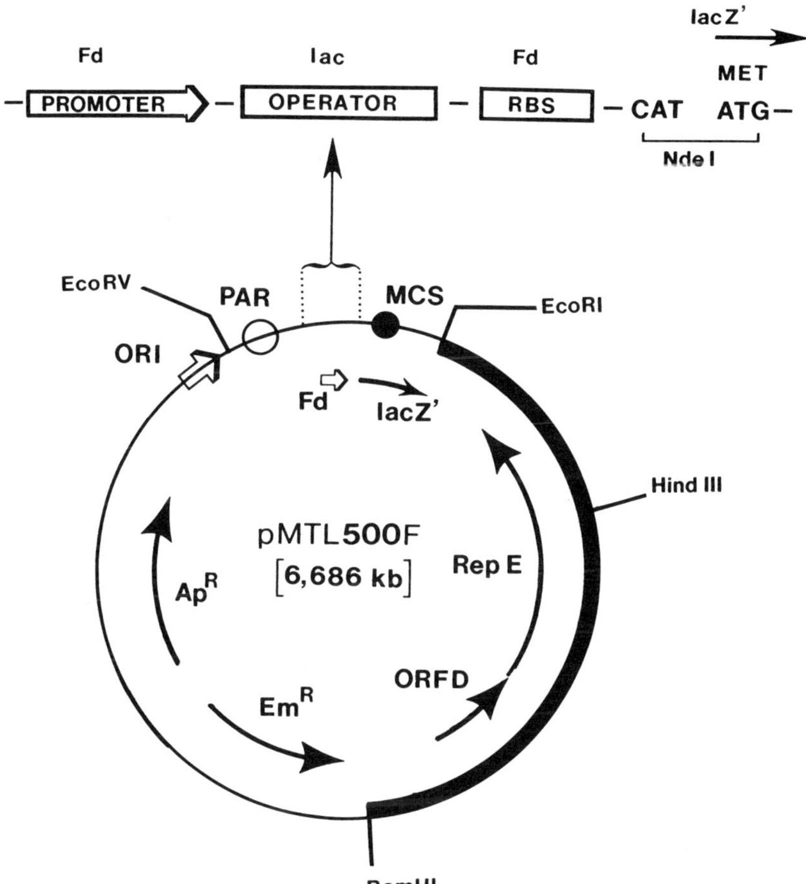

FIGURE 6–3 The *C. acetobutylicum* expression vector, pMTL500F. Plasmid pMTL500F was constructed by replacing the *lac po* region of PMTL500E with the indicated modified (see Minton et al. 1990a) ferrodoxin (Fd) promoter. During its derivation, plasmid pMTL500F also acquired the pSC101 stability function, *par* (PAR). The ATG trinucleotide of the indicated *Nde*I restriction recognition site corresponds to the AUG translational start codon of *lacZ'*, and is immediately preceded by the Fd ribosome-binding site (RBS). The multiple cloning sites (MCS) are those of pMTL20 (Chambers et al. 1988).

Fd. This substitution was achieved by an indirect cloning route (J.K. Brehm, S.M. Whelan, and N.P. Minton, unpublished observations), but the principal steps involved were: (1) the creation, by SDM, of a *Hpa*I site at the Fd +1; (2) the insertion of a synthetic *lac* operator sequence into this created site; (3) the creation of *Nde*I restriction sites (CATATG) in pMTL500E and the Fd cartridge DNA sequence, such that the ATG of this palindrome represented the AUG start codon of the *lacZ′* and Fd gene, respectively, and; (4) substitution of the *lac*Z′ promoter of pMTL500E with that of the modified Fd gene, making use of the created *Nde*I sites. It follows that in pMTL500F, transcription of *lac*Z′ DNA is under the control of the Fd promoter, and translation of transcribed mRNA is reliant on the Fd ribosome-binding site (RBS). The salient features of pMTL500F are therefore the provision of multiple-cloning sites 3′ to transcriptional signals capable of directing the efficient expression of genes in both Gram-negative and Gram-positive hosts. As Fd is active in *E. coli*, pMTL500F retains the cloning advantages of its progenitor plasmid, pMTL500E, with regard to blue to white selection on agar medium supplemented with X-Gal. Heterologous genes inserted into the polylinker cloning sites may rely on their endogenous ribosome binding site for translation of Fd-directed transcripts, or that of the Fd gene. Translational coupling to the Fd RBS could be achieved in two ways: (1) by gene fusion to 5′ codons of *lacZ′*, following insertion into the polylinker region, or (2) by creating an *Nde*I site over the AUG start codon of the heterologous gene, and using this site to position it precisely adjacent to the Fd RBS.

The insertion of a synthetic copy of the *lac* operator into the Fd promoter region was undertaken with a view to placing transcription under regulatory control. It is assumed that when pMTL500F is introduced into a cell that overproduces *lac*I repressor protein, transcription from the Fd promoter will be blocked. To test this supposition, the promoterless *cat* gene was inserted into both pMTL500E and pMTL500F and the resultant plasmids transformed independently into *E. coli* JM83 containing the ColE1 compatible plasmid pNM52, which encodes *lac*I^q (Gilbert et al. 1986). Exponentially growing cells carrying both sets of plasmids were then subjected to IPTG induction and the production of CAT monitored. The results obtained are illustrated in Figure 6–4. It can be seen that Fd-directed expression of *cat* is clearly induced by the addition of IPTG, and furthermore, the degree of regulatory control is equivalent to the natural *lac* promoter of pMTL500E.

To achieve similar control in *C. acetobutylicum* it will be necessary to elicit the expression of *lac*I. This will require the coupling of the *lac*I gene to a clostridial promoter, as exemplified by the analogous *B. subtilis* system of Peschke et al. (1985). Our intention had been to use the promoter region of a previously cloned *leu*B gene from *C. pasteurianum* (Oultram et al. 1988a). However, nucleotide sequence analysis of the cloned DNA fragment encoding this gene has shown it may form part of an operon, and therefore may not possess transcriptional signals (J.D. Oultram, unpublished observations). Alternative promoter signals could be obtained using the promoter probe vector

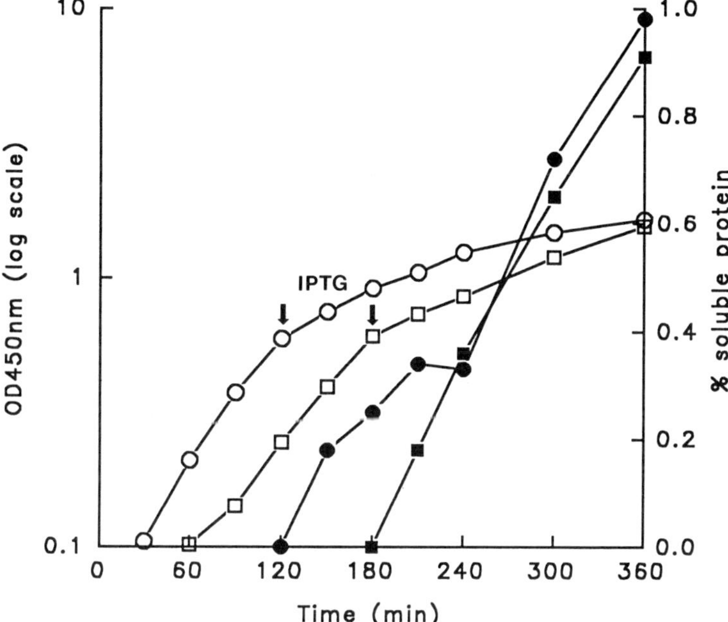

FIGURE 6–4 Inducible expression of the pC194 *cat* gene cloned in pMTL500E and pMTL500F. A promoterless copy of the pC194 *cat* gene, excised from pMTL20C (Swinfield et al. 1990) as a 0.8-kb *Mnl*I fragment, was inserted into the *Sma*I site of pMTL500E and pMTL500F, such that transcription depended on the *lac* or ferrodoxin (Fd) promoter, respectively. The two recombinant plasmids were independently introduced into *E. coli* TG1 containing the *lac*I^q-encoding plasmid pNM52 (Gilbert et al. 1986), and the two clones grown in 2XYT broth to an OD_{450} of 0.6. At this point, expression was induced by addition of IPTG (indicated by an arrow) to a final concentration of 1 mM. CAT activity of cells carrying pMTL500E (●) or pMTL500F (■) is expressed as % cell soluble protein. The culture OD_{450} of cells carrying pMTL500E and pMTL500F is indicated by (○) and (□), respectively.

pMTL710 (see Section 6.4.3). Once obtained, a clostridial promoter::*lac*I fusion could be located either, on a co-resident, pMTL500F compatible plasmid, on pMTL500F itself or integrated into the host chromosome. All three possibilities are being investigated. A potential compatible plasmid, pMTL520, has been constructed, composed of the origin of replication of the *E. coli* plasmid p15A, the *tet* gene (Abraham et al. 1988) of the *C. perfringens* plasmid pJIR71 and the pCB101 replicon. As the pJIR71 *tetP* gene has subsequently proved inappropriate for use in *C. acetobutylicum* NCIB 8052 (see Section 6.2.1.2), pMTL520 will have to be modified to encode an alternative antibiotic resistance gene before it can be used in the proposed two plasmid system.

6.4.3 A Clostridial Promoter-Probe Vector

To facilitate the isolation of clostridial promoters, we have constructed a promoter-probe vector, based on pMTL500E, using a promoterless copy of the pseudomonad *xylE* gene as the reporter (Figure 6–5). This particular reporter gene has proved of great value in other bacterial systems as the encoding enzyme, catechol 2,3-oxygenase, is easily assayed both in cell lysates and in situ. Thus, cells expressing this enzyme activity catalyze the conversion of catechol to 2-hydroxymuconic semialdehyde and turn yellow when sprayed with an aqueous catechol (2%, wt/vol) solution. Although derived from a Gram-negative host, the *xyl*E gene possesses a RBS sequence which is efficiently utilized by Gram-positive bacteria (Zukoski et al. 1983), including *C. acetobutylicum* (see Minton et al. 1990a). Using an indirect route, several unique restriction enzyme cloning sites were positioned 5′ to the translational initiation codon of *xylE*, and the modified gene inserted into a derivative of pMTL500E in which the *lac*Z′ and *lac po* region had been deleted. Because *xylE* is located 3′ to the RNA II promoter of the ColE1 replication origin,

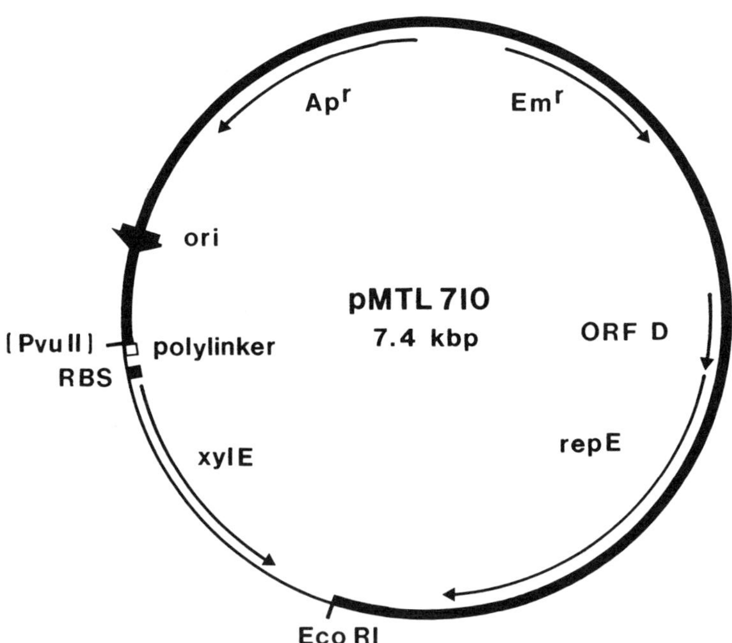

FIGURE 6–5 The promoter probe vector pMTL710. This plasmid was constructed essentially by replacing the *lac po*/*lacZ′* region of pMTL500E with a promoterless copy of the pseudomonad *xylE* gene. The thickened portion of the plasmid equates to the DNA derived from pMTL500E, the bracketed *Pvu*II site being destroyed during pMTL710's construction. Restriction sites present in the polylinker are listed in Table 6–1, RBS = ribosome binding site.

the gene is weakly expressed in *E. coli*, and the resultant colonies exhibit a slight yellow coloration when sprayed with catechol. As this ColE1 promoter does not function in Gram-positive bacteria, *B. subtilis* and *C. acetobutylicum* transformants are colorless.

Genomic fragments of *B. subtilis* and *C. acetobutylicum*, derived by cleavage with various restriction enzymes (i.e., *Mbo*I and *Taq*I), were ligated to the cut pMTL710 DNA and the ligation mixtures transformed in *E. coli*. All Ap^R transformants obtained (more than 2000) were pooled, and used to prepare a heterogenous bulk plasmid DNA preparation. This DNA was used to transform *B. subtilis* and *C. acetobutylicum*, and the Em^R transformants obtained sprayed with catechol to detect cloned DNA fragments with promoter activity. Of the numerous promoter elements identified, 12 have been characterized in more detail with regard to level of *xylE* expression in the two Gram-positive hosts, and nucleotide sequence. To date, it is apparent that the promoters found exhibit substantially weaker activity than the *C. pasteurianum* Fd promoter. They, therefore, may represent ideal candidates for use as the transcriptional signals for the *lac*I gene (see Section 6.4.2). Nucleotide sequence analysis of the promoter-bearing fragments cloned from *C. acetobutylicum* NCIB 8052 has revealed the presence of incomplete ORFs which, by comparative analysis of the encoded polypeptides, has given some insight into the identity of the genes from which the cloned promoters are derived. Different putative encoded polypeptides exhibit homology to; *E. coli* CTP synthetase protein (PyrG), 6-phospho-β-glucosidase, histidine permease, and O-acetylserine (thiol) lyase (CysK), and to the *Azotobacter vinelandii* NifA regulatory protein (I. Burr and J.D. Oultram, unpublished observations).

6.4.4 A Potential Stability Probe/Suicide Vector

Although the minimum pAMβ1-derived fragment capable of supporting the replication of plasmid chimera in *B. subtilis* was shown to be a 2588-bp *Hpa*I-*Acc*I fragment (Swinfield et al. 1990) subsequent studies have indicated that the 229-bp *Acc*I-*Eco*RI fragment adjacent to this region may be important in replication. Thus, a low copy number plasmid carrying the entire 5.1-kb *Eco*RI fragment in which this small *Acc*I-*Eca*RI fragment was deleted, pMTL20Cβ13, had an apparent copy number too low to estimate by the method employed (densitometric scanning of total cell lysates). Furthermore, when *B. subtilis* cells carrying pMTL20Cβ13 were grown for 10 generations, in the absence of antibiotic selection, and plated to give single colonies on antibiotic-free agar medium, of the 100 colonies tested in any one experiment only one or two (sometimes zero) colonies still retained antibiotic resistance. This degree of instability could be capitalized upon to detect plasmid encoded determinants of segregational stability. An analogous vector to pMTL20Cβ13 has, therefore, been constructed by inserting the 4.9-bp *Eco*RI-*Acc*I pAMβ1 replicon fragment into the *Nhe*I site of pMTL20E. The resultant vector, pMTL513E (see Figure 6–2), therefore retains all the cloning advantages of pMTL20E with regards to ease of recombinant selection and exhibits comparable segregational instability to pMTL20Cβ13. The vector has two potential uses. It could be used to: (1) screen random, or specific, restriction fragments from clostridial plasmids for the ability to confer improved segregational stability on pMTL513E; or (2) it could form the basis of a suicide delivery system to facilitate the integration of DNA into the *C. acetobutylicum* genome by homologous recombination. Both uses are currently being investigated.

6.5 APPLICATION OF DEVELOPED VECTORS

Although clostridial vectors for use in *C. acetobutylicum* and *C. perfringens* have been available for some time, little use has been made of them with regard to gene manipulation in vivo. In *C. perfringens*, Scott and Rood (1990) demonstrated that the shuttle vector pHR106 could be used to directly clone the *tetP* gene into strain L13. In our own studies, we have shown that the cloned *C. pasteurianum leuB* gene could be introduced into a *C. acetobutylicum* NCIB 8052 Leu$^-$ auxotroph, where complementation occurred (Oultram et al. 1988a, 1988b). Other laboratories have started reintroducing genes encoding key enzymes of primary metabolism. Thus, a cloned copy of the gene encoding acetoacetyl-CoA decarboxylase has been transferred into *C. acetobutylicum* ACTC 824 inserted in a pIM13-based vector, pFNK3 (E. Papoutsakis and G. Bennett, personal communication). Preliminary data indicate that enzyme activity is 12- and 34-fold higher than the plasmid-free host during early exponential and stationary phase, respectively. Furthermore, the level of enzyme produced was more than 20-fold greater in stationary phase cells than during exponential growth, consistent with induction of the encoding gene with the onset of solventogenesis.

The beneficial potential of recombinant manipulation has been highlighted by recent work in this laboratory, where we were able to generate a strain of *C. acetobutylicum* capable of growth on the soluble cellulose-like substrate, lichenan (see Minton et al. 1990b). Although *C. acetobutylicum* NCIB 8052 possesses β-glucosidase activity, and is therefore capable of growth on cellobiose, it does not produce the necessary endoglucanases to degrade lichenan down to cellobiose. The endoglucanase encoded by the *C. thermocellum celA* gene (Schwarz et al. 1986) efficiently depolymerizes long chain β-glucans, but its activity against chain lengths below about five glucose residues is poor. In contrast, the CelC endoglucanase (Schwarz et al. 1988a) is less active against long chains, but efficiently reduces short chains down to the level of cellobiose (two glucose units in length; Schwarz et al. 1988b). Thus, in theory, the coordinate action of both endoglucanases should efficiently break down soluble β-glucan to cellobiose and hence permit NCIB 8052 to grow on β-glucans as sole sources of energy. An artificial operon, composed of the *C. thermocellum celC* and *celA* genes (under the transcriptional control of the *celC* promoter), was therefore constructed and transferred into strain NCIB 8052 inserted in the shuttle vector pMTL500E. The clone produced was subsequently shown to grow on lichenan as the sole carbon source, confirming the feasibility of altering the fermentation range of *C. acetobutylicum* by genetic means.

6.6 CONCLUSION

The number and sophistication of vectors developed for use in *Clostridium* spp. has now reached a level where in vivo manipulation of individual species can be considered. Indeed, in the case of *C. acetobutylicum*, the first tentative steps have been taken. Although much remains to be done, particularly with regard to segregational stability, in the coming years we can expect a proliferation of experiments describing the recombinant manipulation of clostridial cells. This approach should prove particularly informative in unravelling the molecular complexities of bacterial pathogenesis, the genetic and biochemical basis of the regulation of primary metabolism, and biopolymer degradation. From a biotechnologists' point of view, the information gained should ultimately lead to the generation of strains with improved fermentation characteristics.

REFERENCES

Abraham, L.J., Berryman, D.I., and Rood, J.I. (1988) *Plasmid* 19, 113–120.

Abraham, L.J., and Rood, J.I. (1987) *J. Bacteriol.* 161, 636–640.

Abraham, L.J., Wales, A.J., and Rood, J.I. (1985) *Plasmid* 14, 37–46.

Allen, S.P., and Blaschek, H.P. (1988) *Appl. Environ. Microbiol.* 54, 2322–2324.

Allen, S.P., and Blaschek, H.P. (1990) *FEMS Microbiol. Lett.* 70, 217–220.

Austin, S.J. (1988) *Plasmid* 20, 1–9.

Baas, P.D. (1985) *Biochim. Biophys. Acta* 825, 111–139.

Bibb, M.J., Ward, J.M., and Hopwood, D.A. (1978) *Nature (London)* 274, 398–400.

Brantl, S., Nowak, A., Behnke, D., and Alonso, J.C. (1989) *Nucleic Acids Res.* 17, 10110.

Brantl, S., Behnke, D., and Alonso, J.C. (1990) *Nucleic Acids Res.* 18, 4783–4790.

Bréfort, G., Magot, M., Ionesco, H., and Sebald, M. (1977) *Plasmid* 1, 52–66.

Brehm, J.K., Salmond, G.P.C., and Minton, N.P. (1987) *Nucleic Acids Res.* 15, 3177.

Brenner, D.G., and Shaw, W.V. (1985) *EMBO J.* 4, 561–568.

Bruand, C., Ehrlich, S.D., and Jannière, L. (1991) *EMBO J.* 10, 2171–2177.

Cato, E.P., and Stackebrandt, E. (1989) in *Clostridia* (Minton, N.P., and Clarke, D.J., eds.), pp. 1–26, Plenum Press, New York.

Chambers, S.P., Prior, S.E., Barstow, D.A., and Minton, N.P. (1988) *Gene* 68, 139–149.

Chang, S., and Cohen, S.N. (1979) *Mol. Gen. Genet.* 168, 111–115.

Chassy, B.M., Mercenier, A., and Flickinger, J. (1988) *TIBTECH* 6, 303–309.

Clarke, D.J., Robson, R.M., and Morris, J.G. (1975) *Antimicrob. Agents Chemother.* 7, 256–264.

Collins, M.E., Oultram, J.D., and Young, M. (1985) *J. Gen. Microbiol.* 131, 2097–2105.

Dagert, M., Jones, I., Goze, A., et al. (1984) *EMBO J.* 3, 81–86.

Garnier, T., and Cole, S.T. (1986) *J. Bacteriol.* 168, 1189–1196.

Garnier, T., and Cole, S.T. (1988a) *Plasmid* 19, 134–150.

Garnier, T., and Cole, S.T. (1988b) *Mol. Microbiol.* 2, 607–614.

Garnier, T., and Cole, S.T. (1988c) *Plasmid* 19, 151–160.

Garnier, T., Saurin, W., and Cole, S.T. (1987) *Mol. Microbiol.* 1, 371–376.

Gilbert, H.J., Blazek, R., Bullman, H.M.S., and Minton, N.P. (1986) *J. Gen. Microbiol.* 132, 151–160.

Graves, M.C., Mullenbach, G.T., and Rabinowitz, J.C. (1985) *Proc. Natl. Acad. Sci. USA* 82, 1653–1657.

Graves, M.C., and Rabinowitz, J.C. (1986) *J. Biol.Chem.* 261, 11409–11415.

Gruss, A., and Ehrlich, S.D. (1989) *Microbiol. Rev.* 53, 231–241.

Heefner, D.L., Squires, C.H., Evans, R.J., Kopp, D.J., and Yarus, M.J. (1984) *J. Bacteriol.* 159, 460–464.

Janniére, L., Bruand, C., and Ehrlich, S.D. (1990) *Gene* 87, 53–61.

Jones, D.T., and Woods, D.R. (1989) in *Clostridia* (Minton, N.P., and Clarke, D.J., eds.), pp. 105–144, Plenum Press, New York.

Josson, K., Soetaert, P., Michiels, F., Joos, H., and Mahillon, J. (1990) *J. Bacteriol.* 172, 3089–3099.

Kim, A.Y., and Blaschek, H.P. (1989) *Appl. Environ. Microbiol.* 55, 360–365.

Kok, J., Van der Vossen, J.M.B.M., and Venema, G. (1984) *Appl. Environ. Microbiol.* 48, 726–731.

Lacks, S.A., Lopez, P., Greenberg, B., and Espinosa, M. (1986) *J. Mol. Biol.* 192, 753–765.

Leblanc, D.J., and Lee, L.N. (1984) *J. Bacteriol.* 157, 445–453.

Lin, Y.-L., and Blaschek, H.P. (1984) *Appl. Environ. Microbiol.* 48, 737–742.

Luchansky, J.B., Muriana, P.M., and Klaenhammer, T.R. (1988) *Mol. Microbiol.* 2, 637–646.

Mahony, D.E., Mader, J.A., and Dubel, J.R. (1988) *Appl. Environ. Microbiol.* 54, 264–267.

McDowell, D.G., and Mann, N.H. (1991) *Plasmid* 25, 113–120.

McKenzie, T., Hoshino, T., Tanaka, T., and Sueoka, N. (1986) *Plasmid* 15, 93–103.

Minton, N.P., Brehm, J.K., Oultram, J.D., et al. (1990a) in *Clinical and Molecular Aspects of Anaerobes* (Borriello, S.P., ed.), pp. 187–201, Wrightson Biomedical Publishing Ltd., Petersfield, UK.

Minton, N.P., Brehm, J.K., Oultram, J.D., et al. (1990b) in *Proceedings of the Sixth International Symposium on the Genetics of Industrial Microorganisms* (Heslot, H., Davies, J., Florent, J., et al., eds.), Aug. 12–18, 1990, Strasbourg, France, pp. 759–770, Société Francaise de Microbiologie, Strasbourg, France.

Minton, N.P., and Morris, J.G. (1981) *J. Gen. Microbiol.* 127, 325–331.

Minton, N.P., Oultram, J.D., Brehm, J.K., and Atkinson, T. (1988) *Nucleic Acids Res.* 16, 3101.

Minton, N.P., and Thompson, D.E. (1991) in *Anaerobes in Human Disease* (Duerden, B.I., and Drassar, D.I., eds.), pp. 38–61, Edward Arnold, London.

Minton, N.P., Swinfield, T.-J., Brehm, J.K., and Oultram, J.D. (1990c) *Nucleic Acids Res.* 18, 1651.

Monod, M., Denoya, C., and Dubnau, D. (1986) *J. Bacteriol.* 167, 138–147.
Murai, M., Miyashita, H., Araki, H., Seki, T., and Oshima, Y. (1987) *Mol. Gen. Genet.* 210, 92–100.
Nakayama, J., Nagasawa, H., Isogai, A., Clewell, D.B., and Akinori, S. (1990) *FEBS Lett.* 267, 81–84.
Novick, R.P. (1989) *Annu. Rev. Microbiol.* 43, 537–565.
O'Brien, R.W., and Morris, J.G. (1971) *J. Gen. Microbiol.* 68, 307–318.
Oultram, J.D., Davies, A., and Young, M. (1987) *FEMS Microbiol. Lett.* 42, 113–119.
Oultram, J.D., Loughlin, M., Swinfield, T.-J., et al. (1988a) *FEMS Microbiol. Lett.* 56, 83–88.
Oultram, J.D., Peck, H., Brehm, J.K., et al. (1988b) *Mol. Gen. Genet.* 214, 177–179.
Oultram, J.D., and Young, M. (1985) *FEMS Microbiol. Lett.* 27, 129–134.
Peschke, U., Beuck, V., Bujard, H., Gentz, R., and Le Grice, S. (1985) *J. Mol. Biol.* 186, 547–555.
Phillips-Jones, M.K. (1990) *FEMS Microbiol. Lett.* 66, 221–226.
Projan, S.J., Monod, M., Narayanan, C.S., and Dubnau, D. (1987) *J. Bacteriol.* 169 5131–5139.
Rabinowitz, J. (1972) *Methods Enzymol.* 24, 431–446.
Reid, S.J., Allcock, E.R., Jones, D.T., and Woods, D.R. (1983) *Appl. Environ. Microbiol.* 45, 305–307.
Reysset, G., Hubert, J., Podvin, L., and Sebald, M. (1987) *J. Gen. Microbiol.* 133, 2595–2600.
Reysset, G., Hubert, J., Podvin, L., and Sebald, M. (1988) *Biotechnol. Technol.* 2, 199–204.
Roberts, I., Holmes, W.M., and Hylemon, P.B. (1988) *Appl. Environ. Microbiol.* 54, 268–270.
Rowland, S.J., and Dyke, K.G.H. (1988) *FEMS Microbiol. Lett.* 50, 253–258.
Schwarz, W.H., Grabnitz, F., and Staudenbauer, W.L. (1986) *Appl. Environ. Microbiol.* 51, 1293–1299.
Schwarz, W.H., Schimming, S., Rucknagel, K.P., et al. (1988a) *Gene* 63, 23–30.
Schwarz, W.H., Schimming, S., and Staudenbauer, W.L. (1988b) *Appl. Microbiol. Biotechnol.* 29, 25–31.
Scott, P.T., and Rood, J.I. (1989) *Gene* 82, 327–333.
Shaw, J.H., and Clewell, D.B. (1985) *J. Bacteriol.* 164, 782–796.
Shivarova, N., Förster, W., Jacob, H.-E., and Grigorova, R. (1983) *Z. Allg. Mikrobiol.* 23, 595–599.
Sorokin, A.V., and Khazak, V.E. (1989) in *Genetic Transformation and Expression* (Butler, L. O., Harwood, C., and Moseley, B.E.B., eds.), pp. 269–281, Intercept Ltd., Wimborne, UK.
Soutschek-Bauer, E., Hartl, L., and Staudenbauer, W.L. (1985) *Biotechnol. Lett.* 7, 705–710.
Squires, C.H., Heefner, D.L., Evans, R.J., Kopp, B.J., and Yarus, M.J. (1984) *J. Bacteriol.* 159, 465–471.
Swinfield, T.-J., Janniére, L., Ehrlich, S.D., and Minton, N.P. (1991) *Plasmid* 26, 209–221.
Swinfield, T.-J., Oultram, J.D., Thompson, D.E., Brehm, J.K., and Minton, N.P. (199) *Gene* 87, 79–90.
Truffaut, N., Hubert, J., and Reysset, G. (1989) *FEMS Microbiol. Lett.* 58, 15–20.

Tse-Dinh, Y-C., and Beran, R.K. (1988) *Adv. Gene Technol.* 85, 85.

Tse-Dinh, Y-C. and Wang, J.C. (1986) *J. Mol. Biol.* 191, 321–331.

Villafane, R., Bechhofer, D.H., Narayanan, C.S., and Dubnau, D. (1987) *J. Bacteriol.* 169, 4822–4829.

von Heijne, G. (1983) *Eur. J. Biochem.* 133, 17–21.

Vos, P., Simons, G. Siezen, R.J., and de Vos, W.M. (1989) *J. Biol. Chem.* 264, 13579–13585.

Williams, D.R., Young, D.I., Oultram, J.D., Minton, N.P., and Young, M. (1990a) in *Clinical and Molecular Aspects of Anaerobes* (Borriello, S.P., ed.), pp. 239–246, Wrightson Biomedical Publishing Ltd., Petersfield, UK.

Williams, D.R., Young, D.I., and Young, M. (1990b) *J. Gen. Microbiol.* 136, 819–826.

Yanisch-Perron, C., Vieira, J., and Messing, J. (1985) *Gene* 19, 103–119.

Yoshino, S., Yoshino, T., Hara, S., Ogata, S., and Hayashida, S. (1990) *Agric. Biol. Chem.* 54, 437–441.

Young, M., Minton, N.P., and Staudenbauer, W.L. (1989a) *FEMS Microbiol. Lett.* 63, 301–326.

Young, M., Staudenbauer, W.L., and Minton, N.P. (1989b) in *Clostridia* (Minton, N.P., and Clarke, D.J., eds.), pp. 63–103, Plenum Press, New York.

Yu, P.-L., and Pearce, L.E. (1986) *Biotechnol. Lett.* 8, 469–474.

Zukoski, M.M., Gaffney, D.F., Speck, D., et al. (1983). *Proc. Natl. Acad. Sci. USA* 80, 1101–1105.

CHAPTER

7

Transformation/ Electrotransformation of Clostridia

Gilles Reysset
Madeleine Sebald

Development of biotechnology in clostridia involves cloning and studying the structure and expression of clostridial genes in *Escherichia coli* or alternative hosts easy to manipulate genetically such as *Bacillus subtilis* or yeast. But this approach may not be sufficient because of the possible lack of heterologous gene expression. Even in case of successful results, it is important to be capable to reintroduce clostridial genes from the host in which they were cloned into the original species, for a theoretical or applied biotechnological purpose. This may be achieved by conjugative gene transfer as discussed in Chapter 5 by M. Young in this volume. Nevertheless, this sort of technique is still subject to many constraints, and transformation is a basic alternative technique for introducing DNA in a strain of interest. Data have been accumulated on transformation in many bacterial species since the development of both protoplast regeneration in *Streptomyces* (Bibb et al. 1978; Hopwood 1981) and *B. subtilis* (Chang and Cohen 1979) and electroporation (Shigekawa and Dower 1988).

Those techniques make any bacterial strain potentially transformable, even in the absence of natural transformation. Both techniques have been developed successfully for DNA transformation of a few *Clostridium* spp. and will be reviewed.

7.1 PROTOPLAST TRANSFORMATION

This technique is based upon the general easiness of introducing DNA into cell wall-less protoplasts and involves the ability of protoplasts to regenerate into walled vegetative cells. A pioneer work on *C. acetobutylicum* was published as early as 1982 (Allcock et al. 1982) and partially or successful data were later published on *C. pasteurianum* ATCC 6013 (Minton and Morris 1983), *C. saccharoperbutylacetonicum* N1–4 (later referred to as *C. acetobutylicum*) (Yoshino et al. 1984; Reysset et al. 1987), *C. perfringens* (Heefner et al. 1984; Stal and Blaschek 1985; Roberts et al. 1988; Mahony et al. 1988), *C. thermohydrosulfuricum* (Soutschek-Bauer et al. 1985), thermophilic clostridia (Pettinari et al. 1989; Kurose et al. 1989). We will not analyze in detail each of these publications, this has been done in recent reviews (Young et al. 1989; Reysset 1993), but try to sort out the general features and present limitations of protoplast transformation and regeneration. Protoplasts are generally made in complex media and stabilized with appropriate concentrations of divalent salts and osmostabilizer. Sucrose is generally used at about 0.3 M, but alternative osmostabilizers have been used, either a nonfermented sugar, e.g., lactose at 15% (wt/vol) for *C. pasteurianum* (Minton and Morris 1983) or a fermented sugar, e.g., xylose at 0.25 M for *C. acetobutylicum* (Reysset et al. 1987).

In developing a regeneration technique, the difficulty is obtaining a protoplast formation frequency high enough not to mistake protoplast regeneration for the residual vegetative growth. This necessitates careful controls for assessing the residual vegetative cells by diluting them in a hypotonic solution and plating in an isotonic medium. Any residual growth obtained in these conditions has to be substracted from the apparent protoplast regenerants obtained by dilution and plating on isotonic media. It is generally difficult to obtain a high regeneration frequency, and techniques have to be developed leading to virtually 100% of protoplast formation. Protoplasts are usually obtained by using lysozyme or a β-lactam antibiotic inhibiting murein biosynthesis, and preculture of the cells in the presence of glycine may sensitize the cell walls to those agents (Allcock et al. 1982; Stal and Blaschek 1985).

The critical step in protoplast regeneration is to inhibit autolytic activity which antagonizes the protoplast regeneration process. This was demonstrated in *C. perfringens* (Heefner et al. 1984) and *C. tertium* (Knowlton et al. 1984) and the authors succeeded in reducing the autolytic activity by adding gelatin as suggested for *B. subtilis* (Landman and Forman 1969). In the case of *C. perfringens*, Heefner et al. (1984) succeeded in obtaining L-form cells incapable of regenerating. A similar result was obtained with *C. perfringens* strain L13 (Mahony et al. 1988; Roberts et al. 1988). The regeneration of autoplasts of *C. perfringens* and their transformation by indigenous or recombinant plasmids has also been reported by Heefner et al. (1984) and Squires et al. (1984), but needs to be confirmed.

In our laboratory, when trying to develop protoplast regeneration of *C. acetobutylicum* strains, we first eliminated strains with uncharacterized high autolytic activities and decided to use strain N1-4 (Hongo 1960) as a model. Although high, its autolytic activity was well documented (Yoshino et al. 1962). The strain had the additional potential advantage of forming autoplasts in response to sucrose addition at 0.3 M (Ogata et al. 1975). These autoplasts turned out to be capable of reversion to L-form cells with a very poor regeneration frequency. Finally, we succeeded in obtaining an efficient (40%) protoplast regeneration system by isolating an autolytic-deficient mutant (strain N1-4081). The strain required a mixture of lysozyme (100 $\mu g\ ml^{-1}$) and penicillin G (20 $\mu g\ ml^{-1}$) for protoplast formation and the addition of autolysin inhibitors, sodium polyanethol sulfonate (1 $mg\ ml^{-1}$) or choline (4 $mg\ ml^{-1}$) (Podvin et al. 1988; Reysset et al. 1988) for regeneration. Incidentally, autolysin inhibition by choline was at that time documented in *Streptococcus pneumoniae* only, but we were lucky enough to encounter in strain N1-4 an autolytic system similar to that of *S. pneumoniae*. As in *S. pneumoniae*, choline was shown to be a component of N1-4 techoic acid and we suggest that exogenous choline interferes with choline residues of N1-4 techoic acid as ligands for *C. acetobutylicum* autolysin (Podvin et al. 1988). We further suggest that autolysin inhibition by choline may contribute to protoplast regeneration in other *C. acetobutylicum* strains because, in several strains tested, choline addition to exponential cells inhibits the cell division as in strain N1–4 and N1-4081 (L. Podvin, unpublished observations).

The next step in protoplast transformation consists of introducing DNA from phage or plasmid origin into the protoplasts. There is no particular difficulty in developing the required conditions, except the usual barriers due to nonspecific nuclease activity or host-controlled restriction/modification systems, particularly in the case of heterologous DNA. A method of overcoming this difficulty consists of using isogenic DNA, either phage or plasmid DNA (Reid et al. 1983; Reysset et al. 1988). The heating of protoplasts at 60°C before transformation has been recommended by Lin and Blaschek (1984) as a method to inactivate the nuclease activity of *C. acetobutylicum* protoplasts.

7.2 TRANSFORMATION OF VEGETATIVE CELLS

Alternative techniques consist of introducing DNA into cells without cell wall removal. This may be achieved by cell envelope permeabilization through either chemical or physical procedures.

7.2.1 Chemical Permeabilization

Bacterial cell competence may be induced by chemical treatment. The procedure was initially developed in Gram-positive bacteria by Takahashi et al. (1983) for a non-naturally competent species of *Bacillus*, *Bacillus brevis*. The *B. brevis* procedure was based on a Tris-HCl treatment of the cells (50 mM Tris-HCl buffer at an alkaline pH), followed by induction of DNA uptake with PEG 6000 [28% (wt/vol)]. This technique was successfully used for transforming *C. thermohydrosulfuricum* (Soutschek-Bauer et al. 1985) and *C. acetobutylicum* (Yoshino et al. 1990). For *C. thermohydrosulfuricum* DSM 568, the authors used Tris-HCl 50 mM at pH 8.3, and PEG at 35% (wt/vol). Kanamycin-resistant transformants were obtained with pUB110 DNA at a frequency of 4.10^{-6} (transformants per viable cell); transformants were also obtained with pGS13, a derivative of pUB110 carrying the chloramphenicol acetyltransferase of pC194. For *C. acetobutylicum*, Yoshino et al. (1990) treated cells with 50 mM Tris-HCl at pH 8.5 and used PEG at a final concentration of 22% (wt/vol).

In the case of *B. brevis*, washing with alkaline Tris-buffer was shown to remove protein surface layer from the cell wall (Takahashi et al. 1983).

7.2.2 Electroporation/Electrotransformation

DNA transformation was also recently achieved in bacteria and eukaryotic cells by treating them with high-voltage pulses of electricity, a procedure referred to as *electroporation*. The overall procedure is designated as electrotransformation. The precise mechanism by which the cell envelope is thus made permeable to DNA uptake is still unknown, and the more popular but still unproved hypothesis is the transient creation of pores in the cell outer membranes allowing passive influx of the DNA molecules (Calvin and Hanawalt 1988). Electrotransformation was developed in *C. perfringens* and *C. acetobutylicum* and data obtained with *C. perfringens* strain L13 are well documented (Scott and Rood 1989). These authors recommended using bacteria harvested early in the logarithmic phase of growth and treating them with lysostaphin. With the plasmid vector pHR106 as the DNA source and in the conditions used (capacitance 25 μF, voltage 6.25 kV cm^{-1}, time constant about 7.5 ms), there was a linear relationship between the transformation frequency and the DNA concentration from 0.1–1 μg DNA per cuvette, with a maximum of 3×10^5 transformants per μg DNA. In the reported experiments the transformation frequency was estimated from colony-forming units on chloramphenicol agar, after an overnight incubation in broth without antibiotic; cell division during phenotypic expression cannot be excluded, and therefore an overestimation of the transformation frequency is obtained.

Other procedures have been applied to other strains of *C. perfringens* using mid-log phase cells (Allen and Blaschek 1988) or late stationary phase cells (Kim and Blaschek 1989; Allen and Blaschek 1990; Phillips-Jones 1990). The antibiotics used by Allen and Blaschek (1990) for selection were chloramphenicol for plasmids pHR106 and pAK201, tetracycline for pIP401, erythromycin for pAMβ1 and derivatives. Phenotypic expression was achieved in 3 h and a 3-h incubation time was also used by Phillips-Jones (1990) to allow expression of a chloramphenicol resistance gene from plasmid pSB92A2, the resistance marker originated from pIP401. In *C. acetobutylicum*, electrotransformation procedures have been developed for strain ATCC 8052 (Oultram et al. 1988), N1-4 derivatives (Reysset 1993) and E.T. Papoutsakis (personal communication). Using mid-log cells, frequencies of 4–5 $\times$ 10^4 transformants per μg of DNA were obtained with recombinant plasmids pMTL500E [6.4 kilobase (kb)] (Oultram et al. 1988) and pKNT11 (6.5 kb) (Reysset 1993).

7.3 CONCLUSION

Electrotransformation is an alternative technique to protoplast transformation/regeneration for introducing DNA into clostridia, but the transformation frequency has still to be improved. The technique will benefit from the use of new types of electroporation apparatus leading to a lower cell lethality and permitting higher electric discharges. It will also benefit from a better understanding of the electroporation process such as killing by poration and control of cell division following poration. A potentially useful technique involves cell-to-cell DNA electrotransfer of nonconjugable plasmids (Pfau and Yonderian 1990; Summers and Withers 1990).

In addition to DNA transformation, protoplast formation and regeneration has permitted protoplast fusion in *C. acetobutylicum* P 262 (Jones et al. 1985) and plasmid curing in *C. acetobutylicum* N1-4 (G. Reysset, unpublished observations) according to the technique of Novick et al. (1980). Electroporation is also a potential method of plasmid extraction (Kim et al. 1990) or plasmid curing as the DNA transfer across the cell envelope is not oriented. The technique has already been applied to plasmid curing in *E. coli* (Heery et al. 1989).

REFERENCES

Allcock, E.R., Reid, S.J., Jones, D.T., and Woods, D.R. (1982) *Appl. Environ. Microbiol.* 43, 719–721.

Allen, S.P., and Blaschek, H.P. (1988) *Appl. Environ. Microbiol.* 54, 2322–2324.

Allen, S.P., and Blaschek, H.P. (1990) *FEMS Microbiol. Lett.* 70, 217–220.

Bibb, M.J., Ward, J.M., and Hopwood, D.A. (1978) *Nature (London)* 274, 398–400.

Calvin, N.M., and Hanawalt, P.C. (1988) *J. Bacteriol.* 170, 2796–2801.

Chang, S., and Cohen, S.N. (1979) *Mol. Gen. Genet.* 168, 111–115.

Heefner, D.L., Squires, C.H., Evans, R.J., Kopp, B.J., and Yarus, M.J. (1984) *J. Bacteriol.* 159, 460–464.

Heery, D.M., Powell, R., Gannon, F., and Dunican, L.K. (1989) *Nucleic Acids Res.* 17, 10131.

Hopwood, D.A. (1981) *Annu. Rev. Microbiol.* 35, 237–272.

Hongo, M. (1960) U.S. patent no. 2,945,786.

Jones, D.T., Jones, W.A., and Woods, D.R. (1985) *J. Gen. Microbiol.* 131, 1213–1216.
Kim, A.Y., and Blaschek, H.P. (1989) *Appl. Environ. Microbiol.* 55, 360–365.
Kim, A.Y., Vertes, A.A., and Blaschek, H.P. (1990) *Appl. Environ. Microbiol.* 56, 1725–1728.
Knowlton, S., Ferchak, J.D., and Alexander, J.K. (1984) *Appl. Environ. Microbiol.* 48, 1246–1247.
Kurose, N., Miyazaki, T., Kakimoto, T., et al. *J. Ferment. Bioeng.* 68, 371–374.
Landman, O.E., and Forman, A. (1969) *J. Bacteriol.* 99, 576–589.
Lin, Y., and Blaschek, H.P. (1984) *Appl. Environ. Microbiol.* 48, 737–742.
Mahony, D.E., Mader, J.A., and Dubel, J.R. (1988) *Appl. Environ. Microbiol.* 54, 264–267.
Minton, N.P., and Morris, J.G. (1983) *J. Bacteriol.* 155, 432–434.
Novick, R., Sanchez-Rivas, C., Gruss, A., and Edelman, I. (1980) *Plasmid* 3, 348–358.
Ogata, S., Choi, K.H., and Hongo, M. (1975) *Agric. Biol. Chem.* 39, 1247–1254.
Oultram, J.D., Loughlin, M., and Swinfield, T.J., et al. *FEMS Microbiol. Lett.* 56, 83–88.
Pettinari, M.J., Ivanier, S.E., and Méndez, B.S. (1989) *FEMS Microbiol. Lett.* 58, 255–258.
Pfau, J., and Youderian, P. (1990) *Nucleic Acids Res.* 18, 6125.
Phillips-Jones, M.K. (1990) *FEMS Microbiol. Lett.* 66, 221–226.
Podvin, L., Reysset, G., Hubert, J., and Sebald, M. (1988) in *Anaerobes Today*, (Hardie, J.M., and Borriello, S.P., ed.), pp. 135–140, John Wiley & Sons, Chichester.
Reid, S.J., Allcock, E.R., Jones, D.T., and Woods, D.R. (1983) *Appl. Environ. Microbiol.* 45, 305–307.
Reysset, G. (1993) *Genetics and Molecular Biology of Anaerobes* (Sebald, M., ed.), pp. 111–119, Springer-Verlag, New York.
Reysset, G., Hubert, J., Podvin, L., and Sebald, M. (1987) *J. Gen. Microbiol.* 133, 2595–2600.
Reysset, G., Hubert, J., Podvin, L., and Sebald, M. (1988) *Biotechnol. Tech.* 2, 199–204.
Roberts, I., Holmes, W.M., and Hylemon, P.B. (1988) *Appl. Environ. Microbiol.* 54, 268–270.
Scott, P.T., and Rood, J.I. (1989) *Gene* 82, 327–333.
Shigekawa, K., and Dower, W.J. (1988) *Biotechniques* 6, 742–751.
Soutschek-Bauer, E., Hartl, L., and Staudenbauer, W.L. (1985) *Biotechnol. Lett.* 7, 705–710.
Squires, C.H., Heefner, D.L., Evans, R.J., Kopp, B.J., and Yarus, M.J. (1984) *J. Bacteriol.* 159, 465–471.
Stal, M.H., and Blaschek, H.P. (1985) *Appl. Environ. Microbiol.* 50, 1097–1099.
Summers, D.K., and Withers, H.L. (1990) *Nucleic Acids Res.* 18, 2192.
Takahashi, W., Yamagata, H., Yamaguchi, K., Tsukagoshi, N., and Udaka, S. (1983) *J. Bacteriol.* 156, 1130–1134.
Yoshino, S., Ogata, S., and Hayashida, S. (1982) *Agric. Biol. Chem.* 46, 1243–1248.
Yoshino, S., Ogata, S., and Hayashida, S. (1984) *Agric. Biol. Chem.* 48, 249–250.
Yoshino, S., Yoshino, T., Hara, S., Ogata, S., and Hayashida, S. (1990) *Agric. Biol. Chem.* 54, 437–441.
Young, M., Minton, N.P., and Staudenbauer, W.L. (1989) *FEMS Microbiol. Rev.* 63, 301–326.

CHAPTER 8

Cloning, Structure, and Expression of Acid and Solvent Pathway Genes of *Clostridium acetobutylicum*

E.T. Papoutsakis
George N. Bennett

The use of *Clostridium acetobutylicum* for acetone and butanol production from starches has been discussed in the review of Jones and Woods (1986). The availability of petrochemicals has limited the commercial exploitation of the fermentation for bulk production of butanol and acetone; however, there is continued interest in the process as the organism exhibits growth on a wide variety of feedstocks including certain materials that would otherwise present disposal problems (e.g., apple pomace, cheese whey, and liquids from the paper industry). Other economic factors include the potential recovery of the CO_2 and H_2 produced, and the use and value of the bacterial biomass in animal feed supplements. The potential of genetic engineering to improve

We would like to thank H. Bahl, P. Dürre, and J.-S. Chen for copies of recent results cited here. Research on *C. acetobutylicum* at Rice University and Northwestern University has been sponsored by the National Science Foundation (under grants BCS-8912094 and BCS-8912209), and by the U.S. Department of Agriculture.

the product pattern and overall yield have renewed scientific interest in the solvent-forming clostridia.

8.1 SOLVENT METABOLISM IN *CLOSTRIDIUM ACETOBUTYLICUM*

8.1.1 Metabolic Pathways

The solvent-producing pathway of *C. acetobutylicum* has been thoroughly discussed in recent reviews (Jones and Woods 1986, 1989) and is only briefly summarized here. A diagram of the solventogenic pathway is shown in Figure 8–1. Acetyl-coenzyme A (CoA) is considered as the starting compound for the pathways (Figure 8–1). Side pathways exist that can compete with the conversion of pyruvate to acetyl-CoA. These pathways are responsible for the formation of acetoin and lactate. Acetyl-CoA serves as the precursor of ethanol via acetaldehyde as an intermediate during solvent production. The enzymes acetaldehyde dehydrogenase and alcohol (ethanol) dehydrogenase carry out the conversion. The NAD(P)H formed in the metabolic steps of glycolysis serves as the reducing cofactor. In the acid-producing phase acetyl-CoA is converted to acetate and butyrate. Acetate is synthesized by a process resulting in ATP formation. The acetyl-CoA is converted to acetyl-phosphate by phosphotransacetylase (PTA) followed by enzymatic hydrolysis to acetate by acetate kinase (AK). Butyrate production involves a thiolase (acetyl-CoA acetyltransferase) to condense acetyl-CoA to yield acetoacetyl-CoA. Acetoacetyl-CoA has three primary functions in the production of acids and solvents: (1) that leading to the acetone production; (2) the energetically neutral butyrate and acetate uptake by acetoacetyl-CoA:acetate/butyrate:CoA transferase (CoAT); and (3) as the precursor in the pathway of butyryl-CoA formation. Pathway (1) yields acetoacetate, produced as a part of taking up acids by the CoAT (pathway 2) as an intermediate. Acetoacetate decarboxylase (AADC) removes CO_2 from acetoacetate to yield acetone. Pathway (3) reduces the keto group of acetoacetyl-CoA in a fashion similar to that occurring in fatty acid metabolism by the sequential action of 3-hydroxybutyryl-CoA dehydrogenase, crotonase, and butyryl-CoA dehydrogenase, respectively. These reductions apparently use NADH as the reducing cofactor (Ljungdahl et al. 1989). At butyryl-CoA, the pathway branches into either that of solvent formation via the reduction to butyraldehyde, by butyraldehyde dehydrogenase (BAD) and then the further reduction to butanol by butanol dehydrogenase (BDH). The production of the acidic metabolite butyrate proceeds from butyryl-CoA analogous to that seen in the formation of acetate from acetyl-CoA by first invoking conversion to a phosphorylated form (butyryl phosphate) by the enzyme butyrate kinase (BK) followed by its hydrolysis by the enzyme phosphotransbutyrylase (PTB). This set of reactions also results in the production of one molecule of ATP.

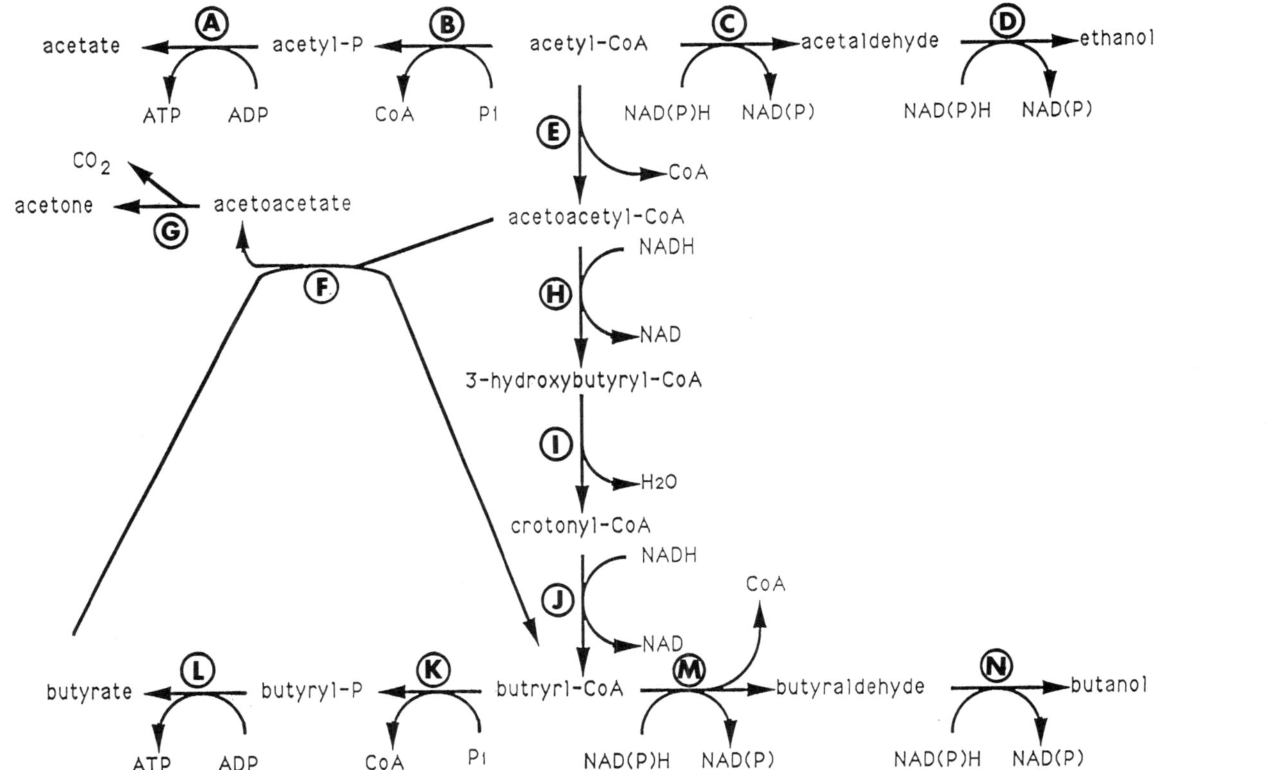

FIGURE 8–1 Solventogenic pathways in clostridia. Arrows indicate the direction of carbon flow and cofactor utilization. Enzymes are indicated by letters as follows: (A) acetate kinase; (B) phosphotransacetylase; (C) acetaldehyde dehydrogenase; (D) ethanol dehydrogenase; (E) acetyl-CoA acetyltransferase (thiolase); (F) acetoacetyl-CoA:acetate/butyrate:CoA transferase (CoA-transferase); (G) acetoacetate decarboxylase; (H) 3-hydroxybutyryl-CoA dehydrogenase; (I) crotonase; (J) butyryl-CoA dehydrogenase; (K) phosphotransbutyrylase; (L) butyrate kinase; (M) butyraldehyde dehydrogenase; (N) butanol dehydrogenase. The reduction of pyruvate to lactate by the lactate dehydrogenase, and the uptake of acetate by the CoA transferase are not specifically shown.

With this overview of the metabolic pathways, we now consider the role of the various enzymes in metabolic regulation of solvent formation and the genetic aspects of regulation of levels of those enzymes that are induced during solvent formation.

8.1.2 General Regulation of the Metabolic Pathway

This topic is considered in this volume (see Chapter 2) and in previous reviews (Rogers 1986; Bahl and Gottschalk 1988; Jones and Woods 1989) and will not be discussed here except in regard to features used in cloning, screening, or expression studies. A brief consideration is given for various enzymes as they are separately discussed. In this review, the enzymes induced during the solvent-producing phase are grouped and those that appear at all phases are grouped. From analysis of the specific activities of various enzymes during acidogenic and solventogenic phases, the enzymes that appear to be induced are: BAD (Dürre et al. 1987; Palosaari and Rogers 1988; Yan et al. 1988), AADC (Andersch et al. 1983; Yan et al. 1988; Hüsemann and Papoutsakis 1989a, 1989b), BDH (Dürre et al. 1987; Yan et al. 1988; Palosaari and Rogers 1988; Welch et al. 1992), and CoAT (Andersch et al. 1983; Hartmanis and Gatenbeck 1984; Hüsemann and Papoutsakis 1989a, 1989b; Welch et al. 1992). Enzymes that act primarily to form butyrate; BK and PTB are not greatly induced (Palosaari and Rogers 1988; Hüsemann and Papoutsakis 1989a, 1989b; Welch et al. 1992) although there have been reports of variation in levels upon the onset of solventogenesis (Andersch et al. 1983; Hartmanis and Gatenbeck 1984). The enzymes converting acetoacetyl-CoA to butyryl-CoA have not been studied extensively.

Inhibitors of protein synthesis by transcriptional inhibitors (rifampicin) or translational inhibitors (chloramphenicol) demonstrated that new synthesis of BDH, CoAT, and AADC were reduced (Ballongue et al. 1985; Palosaari and Rogers 1988; Welch et al. 1992). These experiments have further shown that BAD activity is relatively unstable in vivo after further protein synthesis is blocked (Palosaari and Rogers 1988; Welch et al. 1992). BAD activity declines rapidly after adding protein synthesis inhibitors in contrast to the several other enzymes of the acid or solvent pathway that appear to be stable in vivo.

The culture conditions needed to induce enzymes of solvent production have been a subject of considerable interest (George and Chen 1983; Ballongue et al. 1985; Ballongue et al. 1989; Hüsemann and Papoutsakis 1989b). The appearance of the induced enzymes requires low pH as well as the presence of butyrate, and more detailed analysis thus revealed a close correlation with the internal concentration of undissociated butyric acid (Monot et al. 1984; Gottwald and Gottschalk 1985; Terracciano and Kashket 1986; Hüsemann and Papoutsakis 1988; Hüsemann and Papoutsakis 1990). Thus, the conversion of acids to solvents seems to be a stress-related mechanism to detoxify the excess acid and allow limited further growth (Zerner et al. 1966; Huang et al. 1985, 1986; Ballongue et al. 1987).

8.1.3 *Clostridium acetobutylicum* Strains and Mutants

Early work used various natural isolates and industrial and academic work has yielded detailed inoculation preparation protocols for practical handling and producing stocks for long-term reproducibility (Gabriel 1928; Rodgers et al. 1946). Work with *C. acetobutylicum* has now focused on a small number of specific characterized stocks available from established culture collections. The strains most widely used in current studies include *C. acetobutylicum* P262, NCIB 8052, ATCC 824, DSM 792, DSM 1732, and *C. acetobutylicum* B643. Mutants of *C. acetobutylicum* deficient in solvent-related enzymes have been isolated and characterized by several groups in an effort to manipulate the pattern of solvents formed. Strains exhibiting lower acid production or enhanced formation of butanol have been identified (Rogers and Palosaari 1987). Mutants that produce quantities of butyraldehyde and contain only low levels of butanol dehydrogenase have been reported (Rogers and Palosaari 1987). A 2-bromobutyrate resistant mutant has lost the ability to produce acetone (Junelles et al. 1987). The lack of acetone production may be due to a loss of CoAT activity as was reported in a nonsolvent-producing mutant (Clark et al. 1989). Other mutants defective in producing solvents have been isolated, and in many of these strains more than one enzymatic activity has been lost (Clark et al. 1989; Petersen et al. 1989; Bertram et al. 1990). Among the most frequently lost enzyme activities are the CoAT, BAD, and AADC. Transposon mutagenesis has been recently used to create nonsolvent-producing mutants (Mattson and Rogers 1989; Bertram et al. 1990). This approach shows promise for future molecular analysis of solvent production regulation.

8.2 STRATEGIES AND METHODOLOGY FOR CLONING AND IDENTIFYING SOLVENT STAGE GENES

The presence of a strong nuclease and inefficient lysis hampered DNA isolation from *C. acetobutylicum* in early work. Technical advances in preparing protoplasts from cells grown in glycine supplemented media (Allcock et al. 1982) and the incorporation of anaerobic conditions early in the DNA preparation allowed long clonable DNA to be isolated (Zappe et al. 1986). The growth stage of the culture is critical. Using minor adaptations of standard lysis and DNA-extraction protocols, the isolation of large DNA (>40 kilobase, kb) suitable for making lambda phage and cosmid libraries has been accomplished.

Table 8–1 provides a summary of solvent-related genes from *C. acetobutylicum*, the properties of the enzymes, and the strategy used in cloning.

TABLE 8–1 Summary of Enzymes Related to Solvent Production in *C. acetobutylicum* and Cloning of their Genes

Enzyme	*Size and Number of Subunits in Native Enzyme*	*Reference for Purification*	*References for Cloning*	*Type of Cloning Method Used*	C. acetobutylicum *Strain Used as Source of DNA*
AK	Not reported	Not reported	Not reported		
PTA	Not reported	Not reported	Not reported		
BK	2 (39 kDa)	Hartmanis 1987	Cary et al. 1988	Genetic complementation	ATCC 824
PTB	8 (31 kDa)	Weisenborn et al. 1989	Cary et al. 1988	Genetic complementation	ATCC 824
Thiolase	4 (44 kDa)	Weisenborn et al. 1989	Petersen and Bennett 1990a	Antibody immunoscreening	ATCC 824
3-OH acyl CoADH	Not reported	Not reported	Youngleson et al. 1989	Genetic complementation associated	P 262
Crotonase	4 (43 kDa)	Waterson et al. 1972	Not reported		
CoA Transferase	2 (26 kDa) plus 2 (28 kDa)	Weisenborn et al. 1989	Cary et al. 1990a	Synthetic oligonucleotide hybridization	ATCC 824
AADC	8 (28 kDa)	Zerner et al. 1966	Petersen and Bennett 1990b; Gerischer and Dürre 1990	Synthetic oligonucleotide hybridization	ATCC 824 DSM 792
BDH-NADH dep	2 (42 kDa)	Welch et al. 1989	Petersen et al. 1991	Synthetic oligonucleotide hybridization	ATCC 824
ADH-NADPH dep	—[1]	Not reported	Youngleson et al. 1988	Genetic complementation	P 262
BAD	2 (56 kDa)	Palosaari and Rogers 1988	Not reported		
Lactate dehydrogenase	4 (40 kDa)	Freier and Gottschalk 1987	Contag et al. 1990	Genetic complementation	B 643

[1] Number of subunits in native enzyme not established; probably 2.

Reviews of clostridial genetics have summarized cloned clostridial genes (Young et al. 1989a, 1989b). Three general approaches have been used to clone a gene from another bacterium into *Escherichia coli* (Cary et al. 1990a). The most direct method is to complement a known *E. coli* mutant with the corresponding activity from *C. acetobutylicum* through introduction of the cloned gene into the appropriate *E. coli* strain. This method involves functional protein expression in the host and allows selection or identification of the desired recombinant clone through complementation of a suitable *E. coli* host strain. Specifically, tailored *E. coli* mutant strains have been used with a suitable selection medium or by assessment of growth on an indicator medium either directly or with replica plating. If a simple assay system is available, and large recombinant molecules are being screened, it is sometimes possible to identify a desired clone by directly assaying several random colonies. Suitable mutant strains of *E. coli* have been prepared by mutagenesis of a highly efficient transformation recipient strain. This was the route followed in the case of cloning the *adh* gene in which an *adh*$^-$ derivative of HB101 was used as a recipient strain (Youngleson et al. 1988). *E. coli* strains bearing the desired mutations along with other unnecessary markers can also be used. The much-reduced transformation efficiency of the desired mutant strain may require the introduction of an initial amplification step. Such a procedure includes transforming a high-efficiency strain with a DNA modification system like that of the specialized host and then preparing a plasmid pool from this transformed stock for use as the source of DNA to subsequently transform the specialized mutant strain. For example, in cloning BK and PTB genes the initial transformation was into the high-efficiency transforming strain, DH5, followed by isolation of total plasmid DNA and subsequent transformation into strain LJ32 that allowed selection on butyrate (Cary et al. 1988). In complementation procedures, the gene not only must be expressed but the gene product must function in the intracellular conditions of *E. coli* even under circumstances where the expression of the gene is not optimal, and where conditions of pH and substrate levels do not favor full enzyme activity. Factors of this sort can prevent detection of an initial recombinant by complementation even though the highly expressed gene can carry out the reaction in *E. coli*. This type of complication was observed in cloning of the PTB and BK genes. However, these two gene products together allow uptake of butyrate and growth of an *E. coli ato* mutant on this short chain fatty acid. However, in the cloning of the CoAT gene, which should be capable of the same activation and uptake of butyrate, we did not observe clones readily by complementation in the same strain, LJ32, apparently due to its low expression levels. After plasmids bearing the CoAT gene were available, plate tests showed that some of them, in which the enzyme was sufficiently expressed based on assay values, could indeed complement the *ato* mutant. These CoAT recombinants yielded smaller colonies than the colonies produced by complementation with plasmids incorporating the highly expressed BK and PTB genes.

The second technique, that of using immunological detection, requires protein expression that serves as an antigen but does not require functional enzymatic activity or even the production of a completely intact protein. This method has been productively used in cloning genes from many species where only an antibody to the desired protein is available, and little other information concerning the gene exists. To employ this technique efficiently, a pure preparation of the protein is used directly as an antigen for antibody preparation. If the antibody is a mixture of proteins or is uncharacterized, methods for subtracting out positive but not relevant clones must be worked out. This strategy has been effective in approaching complex situations. The isolated antibody fraction must be free of cross-reaction with *E. coli* proteins, vector encoded proteins, or other highly antigenic proteins from the original organism (i.e., *C. acetobutylicum*). When the above conditions are met, screening of plaques from a λ-phage library can yield positive recombinants (Petersen et al. 1989; Petersen and Bennett 1990a). Special protein fusion vectors and strains are available for stabilizing labile or poorly expressed protein (Yong and Davis 1984). Rabbit, sheep, or mouse antibodies to the purified protein are the most commonly used primary antibodies. Visualization of positive plaques is performed by using a second antibody conjugate [e.g., antibody prepared in goat that recognizes rabbit immunoglobulin G (IgG) and which is conjugated to alkaline phosphatase to allow colorimetric detection] or the positive primary antibody locations can be detected by incubation with labeled protein A.

The third approach is more technologically demanding than the others. It involves an oligonucleotide probe hybridization screening experiment to identify the specific recombinant λ-phage plaque that bears a sequence complementary to the probe. To carry out this method, the protein must be purified and the amino acid sequence of either the N-terminal portion or an internally derived peptide must be determined. Once the amino acid sequence is available, use of the genetic code in a "reverse translation" mode produces a nucleic acid sequence with all possible nucleotides at each position that could possibly encode that amino acid sequence. Since *C. acetobutylicum* is very A+T rich (72%; Cummins and Johnson 1971), sequence analysis of several genes has shown a high preference for A or T in the codon third position as well as certain other favored codons for particular amino acids (e.g., Arg; compiled in Young et al. 1989b and Youngleson et al. 1989b). The availability of information on codon preferences and third position preferences reduces the number of ambiguous positions one needs to consider in designing a probe with appropriate length with as few degeneracies as possible. This favorable feature of *C. acetobutylicum* codons is somewhat diminished by the requirement for a longer probe to yield high specificity since the oligonucleotide sequence is likely to be comprised primarily of A–T base pairs with a consequently lower T_m. To improve specific hybridization with the probe, alterations in hybridization conditions have been employed such as using quarternary ammonium salts (DiLella and Woo 1987). Even then, a longer probe is often required. This is common if the amino acid sequence available for the protein contains several amino acids that are encoded by four or six codons. To avoid great increases in probe degeneracy resulting from three- or fourfold ambiguities in individual positions, inosine can be placed as a nondestabilizing base to allow a longer probe to be designed and synthesized (Martin and Castro 1985). The approach of using a long oligonucleotide has been successfully employed in cloning the NADH-dependent BDH (Petersen et al. 1990, 1991). In that screening, the probe used was a 32-fold degenerate 62-mer containing eight inosine residues. In work with probes, concerns about specific hybridization signals must always be tested to avoid problems because of hybridization to the λ-vector sequences or limited hybridization with other segments of *C. acetobutylicum* DNA. After a suitable oligonucleotide mixture is designed and labelled, it is tested in Southern hybridizations to demonstrate that it gives a unique hybridization to a fragment in the chromosomal DNA. It then can be used to screen plaques from a λ library (Cary et al. 1990a). Recombinant lambda phage giving positive signals are rescreened and tested by Southern hybridization to see if they contain specific restriction fragments similar in size to that of the chromosome digested with the same enzyme. Preliminary mapping of the phage clone then allows smaller fragments to be subcloned into plasmid vectors for testing the presence of the gene by assay of the enzyme activity in growing host cells for positive confirmation of the presence of the gene on the plasmid. The cloning of smaller and overlapping fragments and assay of the strains bearing them for enzyme activity locates the gene to a small portion of the DNA segment in the initial lambda clone. Transposon mutagenesis can also be used to inactivate the gene and allow the coding region to be localized (Cary et al. 1988).

8.3 EXPRESSION OF CLONED *CLOSTRIDIUM ACETOBUTYLICUM* GENES IN *ESCHERICHIA COLI*

Several genes from *C. acetobutylicum* have been found to be expressed well in *E. coli* (Table 8–2). The enzymatic activities found in extracts is compared to those reported for *C. acetobutylicum*. In certain cases, this assessment may be limited by the possibility that not all of the protein produced in *E. coli* was enzymatically active and, in certain cases, different assays have been used at different times by various workers. The high-level expression may not have been expected considering the dramatic difference in A–T composition between the two bacteria and the effects of unusual codon usage. The increased usage of relatively rare *E. coli* codons in expressing *C. acetobutylicum* genes does not prevent the high-level production of the proteins in *E. coli* (Youngleson et al. 1989b). This reference also documents the preferred codon usage of several sequenced *C. acetobutylicum* genes. A compilation of ribosome-binding sites for *C. acetobutylicum* genes is presented in Figure 8–2. These

TABLE 8–2 Expression of Cloned *C. acetobutylicum* Genes in *E. coli*

Enzyme	*Specific Activity of Cloned Gene Product in* E. coli *Extracts*	*Reference*	*Fold Expression Compared to Typical Specific Activity of* C. acetobutylicum	*Expression System*[1]
PTB	23	Cary et al. 1988	3	CP
BK	1.4	Cary et al. 1988	3	CP
Thiolase	3.5	Petersen and Bennett 1990a	1	CP
BHBD	4300	Youngleson et al. 1989b	—[2]	VP
CoAT	3.2	Cary et al. 1990a	9	VP
AADC	3.1	Petersen and Bennett 1990b	90	CP
BDHI	3.1	Petersen et al. 1991	40	CP
BDHII	0.4	Petersen et al. 1991	10	CP
ADH	1.3	Youngleson et al. 1989a	—[2]	VP

[1] The gene was expressed either by sequences within the clostridial insert (CP) or from a promoter on the vector (VP).

[2] The activity values were reported assayed by significantly different methods, so no accurate comparison can be made.

ribosome-binding sites are rather similar to those compiled from various *Clostridium* sp. by Young et al. (1989a) and display extensive complementary sequences to the 3′-terminal region of *B. subtilis* 16S RNA. The greater length of the complementary region and the increased number of guanosine residues give the complex with the 16S RNA more stability than is usually found for the *E. coli* association (Hager and Rabinowitz 1985) and this would act to favor expression in the foreign host. This is an interesting point considering the exceptionally high A–T rich nature of *C. acetobutylicum* DNA. In contrast to the apparently minor difficulty in translating the coding sequence, transcription of the genes has been a limitation on expression levels in certain cases. This might not have been considered to be a difficulty since the A–T rich nature of *C. acetobutylicum* DNA should randomly provide a likely template for RNA polymerase recognition even if the normally used segments are not sufficient to be recognized by the *E. coli* RNA polymerase. The poor expression of certain solvent stage genes may be explained by the requirement of special factors for transcription not present in *E. coli*. In some cases, the natural promoter element may not be present within the cloned segment. In those cases, protein expression can be attained when the insert is properly oriented close to a vector promoter. Young et al. (1989b) have compared a

GENE	SEQUENCE	REFERENCE
adh	AUUUUUAGGAGGUAUAGAUUUAUG	Youngleson et al. 1989a
hbd	GAUUUUGAGGAGGAUUUAUCUAUG	Youngleson et al. 1989b
eng	UUUAUAAUAGGGGGUAUUAACUUG	Zappe et al. 1988
gln	AUGUAAAGGGGGAGUUGUAAAAUG	Janssen et al. 1988
adc	AAAUUUAGGAAGGUGACUUUUAUG	Gerischer and Dürre 1990
*act*B CoAT (28 kDa)	UAGUAAAGGAGCCUGCAUAAAAUG	D.J. Petersen, unpublished observations; Gerischer and Dürre 1990
*act*A CoAT (26 kDa)	AAUUUAAAAGGAGGGAUUAAAAUG	Petersen 1991
buk	GUUAAGUGGAGGAAUGUUAACAUG	Petersen 1991
ptb	GUAAAAGGGAGUGUACGACCAGUG	Petersen 1991
thl	AAAUUUAGGAGGUUAGUUAGAAUG	Petersen 1991
flox	UUAAUUAGGAGGAUUUUAUCAAUG	Santangelo et al. 1991
*bdh*A	UAUUACUAGGAGGUAAGAAGUAUG	Walter et al. 1992
*bdh*B	AUCUAAACAGGAGGGGUUAAAGUG	Walter et al.1992
*dna*J	AAUAUUUAGGUGGUGAAAGGGAUG	Narberhaus et al. 1992
*dna*K	AAUAUAUAGGAGGUUUUUAUAAUG	Narberhaus et al. 1992
*gro*EL	UAAUUCAGGGAGGGAUUCUAAAUG	Narberhaus and Bahl 1992
*gro*ES	UUUUUAAGGAGGGGUUUAAAAAUG	Narberhaus and Bahl 1992
*grp*E	AGUAGAGUAAGAGGUGAGCUUAUG	Narberhaus et al. 1992
lyc	AAUUAGAAAGGGAAUGAUAGAAUG	Croux and Garcia 1991
*xyn*B	AAACUAGGAGGGGAAUUAAAUAUG	Zappe et al. 1990

FIGURE 8–2 Ribosome-binding sites of *C. acetobutylicum* genes. Genes are as indicated. The region complementary to the 16S ribosomal RNA sequence and acting as the Shine-Dalgarno sequence is included. The initiation codon AUG, GUG, or UUG is also underlined and used for alignment.

few known clostridial promoter sequences to those of *E. coli* and other organisms. In general, normal promoters in the Gram-positive organisms are close to the consensus sequences found with *E. coli*.

8.4 PERSPECTIVE AND CLONING OF GENES FOR INDIVIDUAL ENZYMES: NON-INDUCED ENZYMES

In this section, brief descriptions of individual enzymes are presented along with information concerning the genes for that enzyme. The cloning approach used and the major findings from characterization of individual genes are also briefly covered.

8.4.1 Phosphotransbutyrylase and Butyrate Kinase

The PTB (EC 2.3.1.19) and BK (EC 2.7.2.7) play a major role in energy metabolism of *C. acetobutylicum*. The PTB catalyzes the direct transfer of the butyryl moiety of butyryl-CoA to inorganic phosphate, while BK carries out a transfer of phosphate between butyryl-phosphate and ADP resulting in ATP production and butyrate. The reactions are similar energetically and mechanistically to the acetate kinase pathway.

Gavard et al. (1957) first reported PTB activity in *C. acetobutylicum*. Subsequently, extracts of *C. butyricum* and *C. sporogenes* (Valentine and Wolfe 1960) were shown to contain the enzyme. PTB from *C. butyricum* was partially purified (Valentine and Wolfe 1960) and it was confirmed that it was distinct from PTA. The specificity for butyryl-CoA was established. The *C. acetobutylicum* enzyme was purified to homogeneity and showed the native PTB consisted of an octamer of identical 31,000 kilodalton (kDa) subunits (Wiesenborn et al. 1989a). Although PTA has not been characterized from *C. acetobutylicum*, the PTB does not resemble PTA from other clostridial species in size or number of subunits (Robinson and Sagers 1972; Drake et al. 1981). The specificity for butyryl-CoA that identified PTB is important in the energy metabolism of *C. acetobutylicum*. The enzyme exhibits $<2\%$ relative activity with acetyl-CoA as compared to butyryl-CoA, while isovaleryl-CoA and *n*-valeryl-CoA were 95 and 73% as effective as substrates, respectively.

The purification of PTB from *C. beijerinckii* B593 was reported (Thompson and Chen 1990). It resembled PTB from *C. acetobutylicum* ATCC 824 (Wiesenborn et al. 1989a). A high K_m for phosphate was noted and the authors suggested that low phosphate levels late in growth (during solvent phase) might limit PTB activity in the direction of butyric acid formation. This factor increased the availability of butyryl-CoA for butanol formation. This is possible as phosphate limitation favors solvent production (Bahl et al. 1982a, 1986). Also noted was the ability of this PTB to react with acetoacetyl-CoA

(K_m = 1.1 mM) in addition to its reaction with butyryl-CoA (K_m = 0.04 mM) which raises questions as to whether any acetoacetyl phosphate is formed in vivo.

As a branch point enzyme, the PTB might be expected to compete with BAD for butyryl-CoA, and to be in relatively higher amount during the acid production phase. The direction of metabolism is apparently due to more BAD production upon entering the solvent phase. High levels of PTB and BAD do not coincide; BAD is an inducible enzyme, whereas significant levels of PTB are present throughout the various phases. Although in some reports specific activities of PTB (and BK) decreased considerably at the end of the acidogenic phase from peak values (Andersch et al. 1983; Hartmanis and Gatenbeck 1984), other enzyme assays have found little change in PTB levels upon the shift to solvent-producing phase (Palosaari and Rogers 1988; Hüsemann and Papoutsakis 1989a). The enzyme was sensitive to pH changes within the range of typical fermentations, with an optimal pH of about 8.0, bounded by sharp declines in activity with both increasing and decreasing pH (Wiesenborn et al. 1989a). The internal pH of the cell can decrease from 7 to as low as 5.5 during a typical fermentation because of the accumulation of butyric and acetic acids. This decrease would also act to lower PTB activity and reduce the flow to acids. The decrease in PTB activity with decreasing pH also agrees with in vivo studies (Andersch et al. 1983; Hartmanis and Gatenbeck 1984). Since the PTB/BK pathway produces butyrate, the pH control of PTB can be viewed as feedback inhibition by butyric acid.

Butyrate kinase catalyzes the reversible formation of butyrate from butyryl phosphate while ATP is formed concomitantly. The enzyme has been found in many species of clostridia, including *C. acetobutylicum*, *C. pasteurianum*, *C. sporogenes*, and *C. butyricum*. Butyrate kinase has been purified from *C. butyricum* (Twarog and Wolfe 1962), *C. tetanomorphum* (Twarog and Wolfe 1963), and *C. acetobutylicum* (Hartmanis 1987). As the enzyme constitutes a site for ATP synthesis, the enzyme level or its activity might be expected to be tightly regulated (Ballongue et al. 1986b). Indeed, during the acid formation stage, both PTB and BK exhibit high specific activity. Phosphate-limited cultures might reduce the amounts of butyryl-P formed, thus limiting the BK substrate, or shifting the equilibrium toward butyrate uptake (Bahl et al. 1982a). Increased ATP levels appearing during solventogenesis (Meyer and Papoutsakis 1989) also would shift the equilibrium toward uptake of butyrate. PTB has lowered activity in the presence of excess ATP. This may limit use of the pathway in solvent-producing states.

Purified BK is a dimer of identical 39,000 Da subunits with a pI value of 5.6. It exhibits broad acid specificity, using C_2 to C_6 straight and branched-chain carboxylic acids, with butyrate and valerate giving the highest activity (Hartmanis 1987). The enzyme was less active on the shorter chain acids and could not use acetate. The apparent K_m value for butyrate was 14 mM and BK was inactive with dicarboxylic acids. The kinase from *C. tetanomorphum* phosphorylated butyrate, valerate, isobutyrate, and propionate. Propionate was phosphorylated 30% as efficiently as butyrate with the *C. tetanomorphum* enzyme while the *C. butyricum* kinase phosphorylated propionate and butyrate at equal rates. Amino acid composition of BK showed four Cys residues per subunit, and although 1 mM dithiothreitol was necessary for optimal activity, treatment with a variety of sulfhydryl-modifying reagents did not result in inhibition (Hartmanis 1987).

The pH optimum of BK was determined only in the direction of butyryl-phosphate formation exhibiting a maximum activity around pH 7.5. A sharp decline was observed at pH values below 7.0, decreasing to 10% at pH 5.5 (Hartmanis 1987). The pH dependency is quite similar to that of PTB. Unlike PTB, the specific activity of BK has been reported to increase upon entering solventogenesis (Hartmanis and Gatenbeck 1984); thus increased levels of the enzyme may partially compensate for its inactivation by low pH. Other reports have not shown a large change in BK activity upon the shift to solvent stage (Ballongue et al. 1986a; Hüsemann and Papoutsakis 1989a).

Under certain conditions, the PTB/BK pathway can be reversible (Valentine and Wolfe 1960; Meyer et al. 1986; Hüsemann and Papoutsakis 1989a). Taking advantage of the reverse reaction, Cary et al. (1988) succeeded in cloning the genes encoding PTB + BK by complementation of *E. coli ato fadR* mutants on butyrate as a sole carbon source. The *fadR* mutation allows growth of *ato*$^+$ cells on butyrate; however, the *ato* mutation eliminates the ability to activate the acid. Analysis of plasmids allowing growth on butyrate revealed the PTB gene was located in close proximity to the BK gene on small subcloned fragments. The coordinate function of both of these genes was required to allow the uptake and growth on butyrate of the *E. coli fadR ato* mutant LJ32, as elimination of either gene prevented growth on the selective medium. The genes may form an operon as they were adjacent on the *C. acetobutylicum* chromosome within a region of about 2.5 kb, separated by no more than 250 base pair (bp). Recent DNA sequencing analysis has confirmed the orientation of the two genes. The finding that they are both transcribed in the same direction suggests an operon arrangement may account for the overall similarity in production of these enzymes. A promoter within the cloned *C. acetobutylicum* DNA segment was highly active in the expression of these genes in *E. coli*. PTB and BK production yielded about 1% of total cellular protein and the proteins were easily detected in Comassie stained SDS-PAGE of *E. coli* cells harboring pBR322 derivatives bearing these genes.

The finding of an operon-type structure of the PTB-BK in *C. acetobutylicum* contrasts with the gene order of the operon recently found in *E. coli* for the acetate kinase-phosphotransacetylase genes (Yamamoto-Otake et al. 1990). In Figure 8–3, the overall arrangement of clustered *C. acetobutylicum* solvent-related genes is presented. It will be interesting to know what gene arrangement is found in the other organisms where acetate uptake for methane production uses this route (Aceti and Ferry 1988).

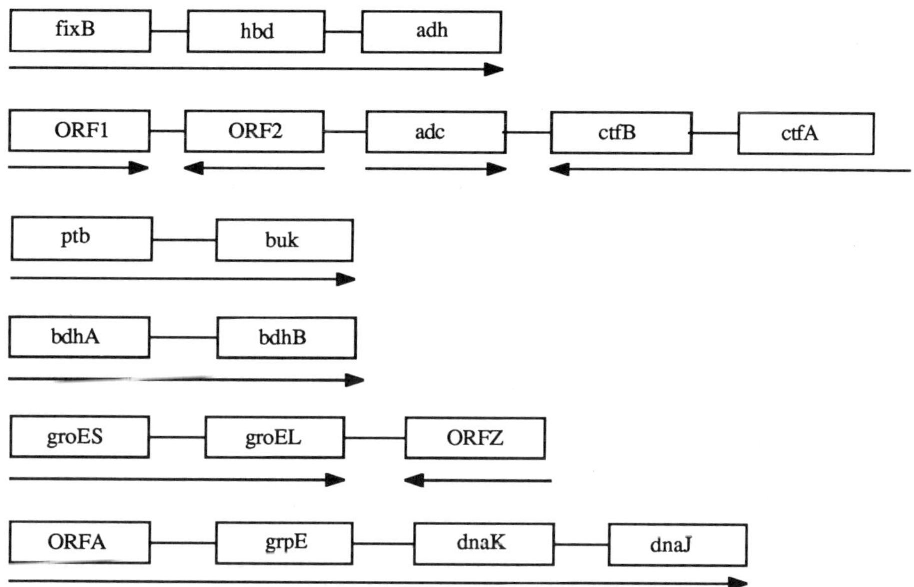

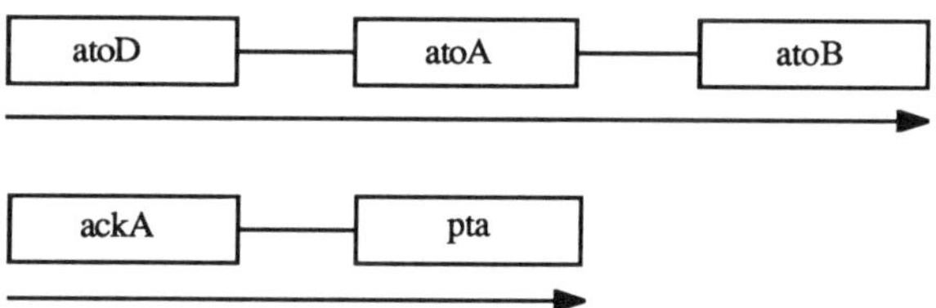

FIGURE 8–3 Arrangement of gene clusters. The arrows indicate the directions in which the genes are transcribed (they do not represent RNA transcripts). The relative size of the genes and intergenic regions are not to scale. References: *fix*B, *hbd*, *adh* (Lin 1992; Youngleson et al. 1989b); *adc*, *ctf*A, *ctf*B (Gerischer and Dürre 1990; Petersen 1991); *ptb*, *buk* (Cary et al. 1988; Petersen 1991); *bdh*A, *bdh*B (Walter et al., 1992); *gro*ES, *gro*EL (Narberhaus and Bahl 1992); *grp*E, *dna*K, *dna*J (Narberhaus e al. 1992); *ato*D, *ato*A, *ato*B (Jenkins and Nunn 1987); *ack*A, *pta* (Yamamoto-Otake et al. 1990).

8.4.2 Phosphotransacetylase and Acetate Kinase

Although the genes encoding PTA and AK have not been cloned from *C. acetobutylicum*, some measurements of enzyme activity and physiological studies have been reported. Molecular weights (kDa) of PTAs (EC 2.3.1.8) of different species vary from 88 kDA in *C. thermoaceticum* (Drake et al. 1981), 63 and 75 kDa in *C. acidiurici* (Robinson and Sagers 1972), to 60 kDa in *C. kluyveri* (Klotsch 1969). The native enzyme from *E. coli* B is very large with an estimated 450-kDa complex being formed (Shimizu et al. 1969). The gene encoding *pta* from *E. coli* K-12 has been cloned and encodes an 81-kDa subunit (Yamamoto-Otake et al. 1990).

Phosphotransacetylase has not been purified or characterized from *C. acetobutylicum*. Stadtman (1955) purified the enzyme from *C. kluyveri*. The K_m for acetyl-CoA was reported as 0.64 mM (Bergemeyer et al. 1963). The enzyme was moderately active (10–50%) with propionyl-CoA, but was unable to use butyryl-CoA (Stadtman 1955). The enzyme exhibited a broad pH optimum from 7.4 to 8.4 and activity was irreversibly lost upon dialysis and was not restored by known cofactors. Similar properties were noted for phosphotransacetylase purified from *E. coli* B.

Increases in PTA and PTB levels were observed upon CO-gassing of glucose–limited cultures of *C. acetobutylicum*. The highest specific activities were seen at pH 4.25 in pH-controlled batch fermentations while almost no activity was found in cultures maintained at pH 5.0 or 6.0 (Hüsemann and Papoutsakis 1989a). Cultures grown in phosphate-limited chemostats of PTA had a sevenfold lower specific activity at pH 6.0 than at pH 4.3 (Andersch et al. 1983). Levels of PTA in batch fermentations declined throughout the fermentation, with a sharp drop at the onset of solventogenesis (Hartmanis et al. 1984). Increased reducing power also affects the regulation of *C. acetobutylicum* PTB. Although increased levels of NADH occur with the onset of solventogenesis, a decrease in PTA activity with increasing NADH concentrations has not been shown. Partial inactivation of the *C. kluyveri* PTA at relatively high ATP concentrations (0.01 M) has been noted (Stadtman 1955).

Acetate kinase (EC 2.7.2.1) has been purified from several bacterial species, *Desulfovibrio vulgaris* (Mannens et al. 1988), *Methanosarcina thermophila* (Aceti and Ferry (1988), and *E. coli* (Rose 1955), but the enzyme has not been characterized in clostridia. This may be due to the low levels of the enzyme present: the specific activity of AK in crude extracts of *C. kluyveri* was about 80-fold less than found in *E. coli* (Rose 1955). Generally, a divalent metal ion (Mg^{2+} or Mn^{2+}) requirement has been reported for the enzyme (Rose 1955; Aceti and Ferry 1988; Mannens et al. 1988). The *E. coli* enzyme exhibited a very high K_m for acetate of 300 mM, and 470 mM for propionate. The V_{max} in the acetate-forming direction was five times greater than that for the acetyl-phosphate-forming direction, thus favoring the formation of ATP. The gene encoding the AK from *E. coli* K-12 has been cloned and encodes a protein of about 43 kDa subunit molecular weight (Matsuyama et al. 1989). The gene is located adjacent to *pta* and the two genes may form an operon transcribed from the *ackA* promoter (Yamamoto-Otake et al. 1990).

Studies on the levels of acetate kinase in *C. acetobutylicum* revealed a decline to about 15% maximal activity at the onset of the solvent-producing stage (Andersch et al. 1983; Hartmanis et al. 1984). Acetate kinase activity also decreased as the acetic acid concentration in the medium increased (Ballongue et al. 1986a). Although reversal of the AK/PTA pathway has been suggested as a route for uptake of acetate under certain fermentation conditions (Valentine and Wolfe 1960), the decline in both of these enzymatic activities upon solventogenesis suggests that the majority of the acetate uptake does not occur by a direct reversal of the acetate-forming pathway (Hartmanis et al. 1984). Phosphate-limited batch fermentations where the pH was controlled at either 6.0 or 4.3 showed a threefold reduction in specific activity of acetate kinase at the lower pH. Acetate kinase levels remained relatively constant in both butyrate pulsing and carbon monoxide (CO) gassing experiments (Hüsemann and Papoutsakis 1989a).

8.4.3 Thiolase

The *ato* thiolase described in bacteria (Thiolase II; acetyl-CoA:acetyltransferase, EC 2.3.1.9) catalyzes the acetylation of acetyl-CoA to form acetoacetyl-CoA. A separate *fad* thiolase (Thiolase I, EC 2.3.1.16) has been identified and distinguished from the *ato* thiolase by physical properties, subcellular localization, and substrate specificity. Thiolase I has the ability to cleave both long- and short-chain 3-ketoacetyl-CoA substrates. In *E. coli*, the thiolase I is constitutively expressed whereas thiolase II is inducible. Different classes of thiolase activity have been reported in yeasts and higher eukaryotes as well.

Thiolases have been isolated from three clostridial species: *C. pasteurianum* (Berndt and Schlegel 1975), *C. kluyveri* (Hartmanis and Stadtman 1982), and *C. acetobutylicum* (Wiesenborn et al. 1988). The three species are significantly different and the thiolase from each has unique specificities relating to its major metabolic role. For example, the major metabolic products of *C. pasteurianum* are acetate and butyrate, while *C. kluyveri* produces mostly butyrate, caproate, and succinate. In the acid-producing phase, the thiolase acts as a branch-point enzyme, with the two branch pathways leading to the formation of butyric and acetic acids. Thiolase condensation to acetoacetyl-CoA competes with the acetate-producing pathway, specifically phosphotransacetylase, for the available pool of acetyl-CoA. The metabolic and energetic consequences of this divergence include the ability of the acetate-forming pathway to yield 4 mol of ATP while 3 mol of ATP can be theoretically made by allowing acetyl-CoA to produce butyrate. However, butyrate formation is neutral in NADH consumption and formation whereas acetate formation leads to excess production of NADH. During solventogenesis, the thiolase also can act as a control point since it competes with acetaldehyde dehydrogenase for acetyl-CoA. The relative activities of these two enzymes could influence the ratio of butanol plus acetone to ethanol.

Two thiolases have been found in *C. pasteurianum* (Berndt and Schlegel 1975), but only one thiolase has been reported in *C. acetobutylicum* (Wiesenborn et al. 1988). The thiolase appears to be expressed coordinately with at least two of the three other enzymes that, together with the thiolase, convert acetyl-CoA to butyryl-CoA (Hartmanis and Gatenbeck 1984). In *C. acetobutylicum*, thiolase appears to play a role in acid uptake as it functions in the thermodynamically unfavorable condensation of two molecules of acetyl-CoA that yield the substrate (acetoacetyl-CoA) used by CoAT in its uptake of acetate and butyrate. The thiolase in *C. acetobutylicum* is not dramatically induced during solventogenesis (Yan et al. 1988).

The *C. acetobutylicum* thiolase is similar to other bacterial thiolases in size and subunit association. The native enzyme is composed of four 44-kDa monomers (Wiesenborn et al. 1988). The thiolase from *C. acetobutylicum* exhibits high activity throughout the usual physiological pH range of 5.5–7.0 (Huang et al. 1985), unlike the thiolase of *C. pasteurianum* which has a sharper pH optimal at pH 8.0 (Berndt and Schlegel 1975). It is interesting that changes in internal pH do not appear important in the regulation of thiolase and this may relate to the role of the thiolase in forming solvents acetone and butanol as well as butyrate. The condensation reaction is extremely sensitive to micromolar amounts of CoASH, so the relative amounts of acetyl-CoA and CoASH may be the key factors regulating the metabolic flow through this branch point, and hence influence the ratio of C_2 vs. C_3+C_4 end products (Wiesenborn et al. 1988).

Using antibodies raised to a sample of the purified *C. acetobutylicum* thiolase (Wiesenborn et al. 1988), Petersen and Bennett (1990a) have succeeded in cloning the thiolase from *C. acetobutylicum*. Plaques from a lambda-phage library were screened using the anti-thiolase antibody prepared in a rabbit.

The N-terminal amino acid sequence was obtained from protein sequencing of the purified protein; however, the sequence did not allow very unique probes to be designed due to the highly degenerate codons of common amino acids present in the analyzed segment of the protein. Oligonucleotide probes were nonetheless made and tested but did not afford highly specific hybridization in Southern hybridizations with chromosomal DNA and in library screening. They did, however, prove useful in analyzing the location of the gene subfragments of the phage giving positive tests with the anti-thiolase antibody. The hybridization experiments allowed the N-terminal region to be located and the orientation of the gene to be determined. The results were confirmed by limited DNA sequencing around the N-terminus of the thiolase gene. Hybridization of the cloned thiolase segment with chromosomal DNA indicated key sites and fragments were identical between the cloned segment and chromosomal DNA. Western blots demonstrated expression of the thiolase protein in *E. coli*. Enzyme assays indicated a high level of thiolase activity in *E. coli* DH5 bearing the *C. acetobutylicum* thiolase gene (pTeco11). The molecular weight of the cloned thiolase (42 kDa) corresponded to that of the purified enzyme. Hybridization studies using the cloned thiolase gene as a probe have identified only one thiolase gene present in the genome and further showed that the thiolase gene is not adjacent to the CoAT genes (Petersen 1991) as in the arrangement that exists in *E. coli* (Jenkins and Nunn 1987). Hence, if another thiolase gene is present, it would be expected to have a different nucleotide sequence.

8.4.4 Enzymes Involved in the Conversion of Acetoacetyl-CoA to Butyryl-CoA

The enzymes from *C. acetobutylicum* that carry out the conversion of acetoacetyl-CoA to butyryl-CoA have not been studied extensively. The overall similarity of these reactions with those of fatty acid metabolism suggest possible structural or mechanistic conservation with these enzymes. Activities of the enzymes of the pathway from acetyl-CoA to butyryl-CoA have been studied during the shift to solvent production. The thiolase, 3-hydroxybutyryl dehydrogenase, and crotonase were coordinately expressed and all exhibited maximal activity at the end of growth (Hartmanis and Gatenbeck 1984).

Recently, the gene encoding β-hydroxybutyryl-CoA dehydrogenase (BHBD) (EC 1.1.1.35) was found to be adjacent to the *adh*1 gene earlier reported (Youngleson et al. 1988). The gene encoding this enzyme was identified by computer-analyzed amino acid sequence homology derived from translation of an open reading frame (ORF) sequence upstream of the *adh* gene (Youngleson et al. 1989b). The 282 amino acids (aa) of the ORF indicated a M_r for the protein of 31,435 and the sequence was 46% identical in amino acids to the mitochondrial fatty acid β-oxidation enzyme from pig (Bitar et al. 1980) and 38% identical to that of the 3-hydroxyacyl-CoA dehydrogenase (HAD) portion of the bifunctional HAD-enoyl-CoA hydratase enzyme from rat peroxisomes (Osumi et al. 1985; Ishii et al. 1987). There was a segment of six amino acids (82–87) in the *C. acetobutylicum* enzyme conserved exactly in the three proteins. The function of this segment is unknown. A lower similarity (about 24%) was found to λ-crystallin of rabbit lens. An overall fit within the β-α-β structural arrangement of nucleotide-binding fold of other dehydrogenases (Rossmann et al. 1974) was located to a segment of the BHBD of *C. acetobutylicum*. Secondary structure prediction programs suggested the BHBD was an α-helix-rich protein and that it exhibited some relation to the known structure of the pig mitochondrial 3-hydroxyacyl-CoA dehydrogenase.

The BHBD gene was separated from the *adh* gene by a 354-bp intergenic region that did not have sequences typical of a classical transcription terminator. Regions that could form some stem-loop structures were earlier identified but any function of these is unclear (Youngleson et al. 1989a). Another

ORF upstream from the BHBD gene existed but it was not completely within the pCADH100 clone. The possible relation of this ORF to other *C. acetobutylicum* proteins has not been presented, but it is intriguing to speculate that it may encode part of a metabolically related enzyme. The BHBD and ADH are not expressed from a promoter of clostridial origin so the upstream *orf* and promoter region are apparently yet to be reported. The location of the hydroxybutyryl-CoA dehydrogenase (BHBD) gene near that for the NADPH-dependent alcohol dehydrogenase (Youngleson et al. 1989a) is an example of clustering of metabolic genes in *C. acetobutylicum* and hints that the metabolically related crotonase and the butyryl-CoA dehydrogenase genes might also be encoded in some cluster.

Crotonase (EC 4.2.1.17) from *C. acetobutylicum* has been isolated and characterized (Waterson et al. 1972) and is stable in a lyophilized form. The crotonase has a native molecular weight of 158 kDa and subunit size estimated at 40–43 kDa, as determined from sodium dodecyl sulfate (SDS) gel electrophoresis and sedimentation studies in denaturing conditions (6 M guanidine hydrochloride), suggesting the native protein exists in a tetrameric form. Crotonase exhibits an extremely high turnover rate, with a similar K_m and apparent equilibrium constant to the bovine liver enzyme. The *C. acetobutylicum* enzyme is specific for short chain fatty acyl-CoA compounds, with only crotonyl-CoA and hexenoyl-CoA being hydrated. This specificity is in contrast to the broad action of crotonase from bovine liver on C4–C16 acids (Waterson and Hill 1972). In crude extracts of *C. acetobutylicum*, activity toward the longer chain forms was found, indicating more than one hydratase is present in the organism. The other enzyme apparently is involved in the formation of membrane constituents. The isolated crotonase was sensitive to high concentrations of crotonyl-CoA, a property also unlike bovine liver crotonase. While sequences are unknown, the overall amino acid composition is rather similar between the bacterial and bovine enzymes. It will be interesting to know the relatedness of these enzymes and the gene structure and arrangement, especially since the hydroxybutyryl-CoA dehydrogenase has been sequenced and compared (Youngleson et al. 1989b).

The characterization of the butyryl-CoA dehydrogenase (EC 1.3.99.2) from *C. acetobutylicum* has not been reported, although Hartmanis and Gatenbeck (1984) reported a low activity in extracts. The corresponding enzyme from another microorganism, *M. elsdenii*, has been isolated and the flavoprotein dissociated (Van Berkel et al. 1988).

8.5 PERSPECTIVE AND CLONING OF GENES FOR INDIVIDUAL ENZYMES: INDUCED OR SOLVENT STAGE ENZYMES

8.5.1 CoA-Transferase

CoA-transferases (EC 2.8.3.9) are involved in the activation of carboxylic acids to the respective CoA thioester while releasing a carboxylic acid from another species of CoA thioester. This transfer of the CoA moiety conserves the energy of the thiol ester, and is reversible. During acetone and butanol production by *C. acetobutylicum*, the uptake of acetate and butyrate from the medium occurs via the CoA-transferase, and is coupled to acetone production by AADC that removes acetoacetate and serves to pull the CoAT reaction in the direction of acetate and butyrate uptake (Hartmanis et al. 1984). Different types of CoA-transferases have been identified in bacteria. They typically have relatively broad substrate specificities. In most bacteria, the enzyme functions in the uptake of substrates for energy production and in structural components, while in butanol-forming clostridia the uptake of acids by the CoA-transferase is primarily to detoxify the medium (Wiesenborn et al. 1989b). Thus, the metabolic role of CoAT in *C. acetobutylicum* is fundamentally different from that of other CoA-transferases. For example, the CoAT from *C. kluyveri* functions mainly in butyrate formation. This different role is reflected in the substrate specificity pattern of the enzyme. The preferred substrates for the *C. acetobutylicum* enzyme were shown to be acetate, propionate, and butyrate (4:2:1 ratio) of rates of utilization, and the K_m values reported were very high, ranging from 660–1200 mM (Wiesenborn et al. 1989b). In contrast, the CoAT in *Clostridium* sp. SB4 exhibited much lower K_m values ranging from 0.8–29 mM. The intracellular concentrations of acetate and butyrate in *C. acetobutylicum* during the solvent stage usually are not above 300 and 700 mM, respectively. At these subsaturating levels, the CoAT would be sensitive to changes in concentrations of acetate and butyrate within the cell. The CoAT has been shown to be inducible upon entering the solvent stage (Andersch et al. 1983; Yan et al. 1988; Hüsemann and Papoutsakis 1989b), although low levels of this enzyme can be detected in acid-producing cells (Andersch et al. 1983; Yan et al. 1988).

The CoAT of *C. acetobutylicum* has recently been purified (Wiesenborn et al. 1989b) and exhibits some similarities to other bacterial CoA-transferases. The protein is a heterotetramer, composed of two subunits of M_r 26,000 and 28,000 (Cary et al. 1990b) compared to the subunits of 23,000 and 26,000 for the *E. coli* enzyme (Sramek and Frerman 1975) and 23,000 and 25,000 for the CoAT from *Clostridium* sp. strain SB4 (Barber et al. 1978). Relative-activity measurements in the range of pH 5.4–7.8 indicate a broad pH optimum from pH 6.4–7.8, and declining to about half at pH 5.4 (Wiesenborn et al. 1989b). The decrease in enzymatic activity at pH 5.4 corresponds to the lower internal pH existing during the solvent stage but the induction of synthesis of the enzyme overcomes this effect. This factor could also prevent excessive acid uptake before the AADC is induced which appears to slightly trail the induction of the CoAT.

The gene encoding the CoAT from *C. acetobutylicum* has been cloned and the enzyme has been expressed in *E. coli* (Cary et al. 1990b). The cloning was achieved by hybridization of plaques from a genomic phage library with oligonucleotides designed to the N-terminal amino acid sequence obtained from the individual purified protein subunits. In this case, a fourfold degenerate 18-mer corresponding to amino acids 7–12 of the 26-kDa subunit and a 12-fold degenerate 20-mer corresponding to residues 1–7 of the 28-kDa subunit proved to be sufficiently specific for hybridization experiments. Oligonucleotides generated to each of the two subunits were used as a mixture to further identify DNA fragments from recombinant phage that had initially given positive hybridization signals. Fragments that hybridized with both subunit probes were subcloned and *E. coli* cells harboring plasmids containing these DNA inserts were analyzed for CoAT activity. The enzyme was not expressed well from a promoter of clostridial origin in *E. coli*, but was expressed when oriented appropriately with the *lac* promoter in the vector. In *E. coli*, the two subunits of the CoAT are encoded as part of an operon (*ato* operon) that is involved in the degradation of short chain fatty acids (Jenkins and Nunn 1987). The cloned genes encoding the two clostridial CoAT subunits were adjacent to one another, oriented in the same direction, and had an intervening region between the genes of only 27 bp (J.W. Cary, unpublished observations). Thus, they may be expressed in a coordinate fashion in an operon type of arrangement similar to that found in *E. coli*. Beyond the termination codon is a long stable hairpin structure characteristic of transcription termination sequences. The distal CoAT subunit gene has been sequenced (Gerischer and Dürre 1990) and its molecular weight of 23,622 determined from the amino acid sequence was less than that appearing on SDS gels (28 kDa). Hybridization studies have failed to find the *C. acetobutylicum* thiolase gene in close proximity to the cloned CoAT gene. Although the *C. acetobutylicum* CoAT can function in *E. coli* as evidenced by complementation of *E. coli ato* mutants, the genes encoding the CoAT and the enzyme itself appear to share little homology with their *E. coli* counterparts as analyzed by both Southern and Western blot experiments (Cary et al. 1990b).

8.5.2 Acetoacetate Decarboxylase

The enzyme acetoacetate decarboxylase (EC 4.1.1.4) has an important role in solvent production as it catalyzes the formation of acetone by decarboxylation of the acetoacetate produced by uptake of acids by the CoAT. The removal of acetoacetate effectively pulls the less-favorable CoAT reaction towards the uptake of acids. AADC was found to be highly inducible, with an increase in activity correlating with the onset of solvent production (Andersch et al. 1983; Ballongue et al. 1985; Hüsemann and Papoutsakis 1989a). Although the CoAT is also highly inducible at the onset of solventogenesis, its activity appears somewhat earlier in the fermentation (Andersch et al. 1983; Janati-Idrissi et al. 1988). As little solvent is normally produced before the AADC is induced, it appears that the presence and action of AADC energetically aids the uptake of acids from the medium.

The enzyme was first studied by Davies (1943) who investigated the metabolic precursors of acetone and butanol. Acetoacetate was the only metabolite whose decarboxylation was catalyzed by the AADC (Davies 1943). Chemical analysis has identified no prosthetic groups associated with the enzyme. The enzyme from *C. acetobutylicum* B-527 was crystallized (Zerner et al. 1966) and this material was used in more detailed enzymological studies aimed at elucidating the mechanism of the reaction. The reaction proceeds by Schiff-base formation at an active site lysine (Warren et al. 1966); however, no pyridoxal phosphate was necessary for the decarboxylation. Amino acid sequencing of an active site peptide revealed no homology with the active sites of other decarboxylases (Laursen and Westheimer 1966) and more recent comparisons also showed no homology (Gerischer and Dürre 1990).

The protein consists of an octamer of identical subunits of about 28 kDa. Information from inhibition studies has shown that either two subunits combine to form a single active site, or that one-half of the subunits in the complex are inactive (Davies 1943; Zerner et al. 1966). The enzyme is activated by heating at 70° for 1 h (an approximately twofold increase in activity) whereas an inactivation is observed at temperatures above 75°C (Pomeroy 1970). AADC has a pH optimum at about 5.0, well suited for the production of acetone at acidic pH values (Davies 1943). It is insensitive to oxygen, and withstands high concentrations of acetone. Indeed, this property has been used in early stages of purification procedures that employ extracting the enzyme from acetone powders (Fridovich 1963; Petersen and Bennett 1990b). Interestingly, a loss of activity is found in the usually prepared cell extracts made by sonication.

The gene encoding the AADC has been cloned by hybridization screening of a genomic phage library (Petersen and Bennett 1990b) or a plasmid pUC9 library (Gerischer and Dürre 1990) using oligonucleotide probes generated to the N-terminal amino acid sequence of the purified protein. The enzyme was found to be well expressed in *E. coli* from a promoter of clostridial origin in both of the previous studies. The gene encoding AADC, called *adc*, has been sequenced (Gerischer and Dürre 1990). The gene encoded a protein of 244 aa and the active site peptide sequence previously determined (Laursen and Westheimer 1966) was found. From the amino acid sequences of AADC as deduced from the nucleotide sequence, the secondary structure of the protein was predicted and it was noted that the C terminus was unusually hydrophobic. The gene was preceded by a ribosome-binding site (Figure 8–2) showing reasonable homology to consensus models (Shine and Dalgarno 1974) and the gene terminated with tandem UAA stop codons. Downstream of *adc* is a strong transcription terminator and beyond are the oppositely oriented CoAT genes (Gerischer and Dürre 1990; Cary et al. 1990b; Petersen and Bennett 1990b). The transcription terminator has a strong stem bounded by AT regions enabling it to serve as a transcription terminator in both directions, thus allowing termination of both the CoAT and AADC encoding transcripts. The role of the terminator in any regulation is unknown. The region upstream of the gene encoding AADC has also been sequenced (Gerischer and Dürre 1990), and two ORFs were found, one (ORF1) has homology to an α-amylase of *B. subtilis*. Sequencing of the 5′ upstream region has allowed the *adc* promoter to be defined and the transcription initiation site has been identified by RNA experiments (U. Gerischer and P. Dürre, unpublished observations). Such an unusual gene arrangement has not previously been found in clostridia (Figure 8–3). The independent transcription of each gene may account for their independent expression levels during cell growth. Comparison of the transcriptional promoter regions with those of acidogenic enzymes such as BK and PTB may provide some further insight into induction mechanisms.

The close proximity of the AADC and CoAT genes may account for the loss in activity of both enzymes in mutants and degenerate cultures (Clark et al. 1989; Petersen et al. 1989). Cloning of the *C. acetobutylicum* gene encoding AADC coupled with the ability to introduce AADC-gene bearing plasmids into *C. acetobutylicum* mutants (Clark et al. 1989) will enable genetic analysis of the mechanism of regulation in vivo. Studies to allow optimization of expression of this gene for metabolic engineering purposes can also be undertaken. The AADC gene has already been reintroduced into a *C. acetobutylicum* AADC mutant and shown to be functionally expressed in vivo as is discussed in Section 8.7.

8.5.3 Butyraldehyde Dehydrogenase

The activity of an enzyme (EC 1.2.1.10) catalyzing the formation of acetaldehyde or butyraldehyde from acetyl- or butyryl-CoA using NADH or NADPH as a cofactor was observed in *C. kluyveri* (Burton and Stadtman 1953; Lurz et al. 1979). Early purifications did not completely distinguish if the enzyme was a separate protein from the alcohol dehydrogenase activity as these two activities are often combined in other organisms [e.g., *E. coli* (Rudolph et al. 1968; Shone and Fromm 1981)]. Further characterization of the *C. kluyveri* enzyme has been reported (Smith and Kaplan 1980). The enzyme that converts acetyl- and butyryl-CoA to acetaldehyde and butyraldehyde, respectively, has been purified from *C. acetobutylicum* NRRL B643 (Palosaari and Rogers 1988) and from *C. beijerinckii* 592 (Yan and Chen 1990). The *C. acetobutylicum* NRRL B643 enzyme has a native M_r of 115,000 and subunit M_r of 56,000. The enzyme from *C. beijerinckii* B592 has a native M_r of around 100,000 and subunit M_r of approximately 55,000 and the acetaldehyde and butyraldehyde dehydrogenase activities copurified from the organism. *C. acetobutylicum* (strain 6A), a nonsporulating, nonsolvent-producing strain, contained no BAD activity (Palosaari and Rogers 1988). Other mutants lacking BAD activity have also been reported (Clark et al. 1989; Petersen and Bennett 1989). Although efforts to clone the gene encoding BAD by the techniques used previously with other enzymes are ongoing, the cloning of the gene encoding BAD has not yet been reported.

8.5.4 Butanol Dehydrogenase

Long-term attention directed toward the understanding of BDH regulation and enzymatic properties has been evident since early enzymological investigations of the fermentation. Only a few bacteria are known to produce butanol as a major fermentation product and these include the *Clostridium* spp. *C. aurantibutyricum*, *C. beijerinckii*, *C. tetanomorphum*, and *C. puniceum*. Results as to the nature of the enzyme functioning as an alcohol dehydrogenase were confusing and interpretation of findings depends on the method of preparation or the type of assay used. The BDH is an inducible enzyme, although not to the extent of the AADC and even after induction the specific activity attained was rather low (Andersch et al. 1983). Dürre et al. (1987) separated two enzyme activities by ultracentrifugation and clarified the situation considerably. Two distinct BDHs with different cofactor requirements and different pH ranges have been detected in *C. acetobutylicum* (Dürre et al. 1987; Welch et al. 1989; Welch 1990). The NADPH-dependent BDH exhibits a broad pH optimum between pH 7.8 and 8.5 and maintains stability in oxygen (Dürre et al. 1987). The NADH-dependent BDH, however, yielded similar activities at both pH 7.8 and 6.0

Once the existence of two BDHs was established, consideration of the physiological roles of each is to be considered. Since the NADH-dependent BDH has a lower pH optimum corresponding more closely to the physiological conditions in the cell during the solvent stage, it makes this the most likely candidate for the primary butanol-forming enzyme. However, the preference for butyraldehyde (Dürre et al. 1987) and induction studies (Palosaari and Rogers 1988) suggested a role of the NADPH-dependent BDH.

An NADH-dependent BDH was purified and existed as a dimer composed of identical 42-kDa subunits (Welch et al. 1989). This enzyme has a K_m for butyraldehyde of 16 mM and the activity was 50-fold greater in the butanol-forming direction. Since intracellular levels of butyraldehyde are only estimated to be about 1 mM (Hartmanis et al. 1984), the NADH-dependent BDH would be sensitive to the level of the aldehyde. The purified enzyme was reported to be about 46-fold greater in activity for the reduction of butyraldehyde as a substrate than for acetaldehyde (Welch et al. 1989). The overall substrate specificity and subunit composition of this BDH was closer to that of mammalian dehydrogenases than for other bacterial alcohol dehydrogenases (ADHs).

Cloning studies and further enzyme isolation and characterization have revealed more complexity in the BDH story. An allyl alcohol resistant (*adh*$^-$) mutant of *E. coli* HB101 was transformed with a plasmid library of *C. acetobutylicum* P262 DNA (Youngleson et al. 1988) and cells that exhibited enhanced sensitivity were analyzed for alcohol dehydrogenase activity. The alcohol dehydrogenase activity was encoded within a 3.2-kb DNA insert. However, the properties of the dehydrogenase activity were quite different from that expected for the major BDH; most importantly, the enzyme was active with ethanol as a substrate, showing just fourfold greater activity with butanol. On this basis, it was classified as an ADH. This ADH was found to be NADPH specific, and its cellular role remains to be identified. The nucleotide sequence of *adh*1 gene of *C. acetobutylicum* P262 was determined and a region of 1164 bp encoded a 388 aa residue protein (M_r 43,274; Youngleson et al. 1989a). This is near the size found for purified BDH enzyme (Welch et al. 1989). Comparison of the amino acid sequence of the ADH showed that it most closely resembled a type 3-ADH as defined by Jörnvall et al. (1987). The *C. acetobutylicum* ADH contained 39% identical residues compared to the iron containing *Z. mobilis* ADH and 37% when compared with ADH4 of *S. cerevisiae.* The sequence did not contain the conserved glycine-rich region typically found within the β-α-β fold in the zinc-containing ADHs (Jörnvall et al. 1987). Since the *C. acetobutylicum adh*1 gene encodes a NADPH-dependent enzyme, this is not surprising, as the fold is more prominently found in NADH-dependent enzymes. Polypeptide secondary structure prediction of the ADH indicates it to be unusually high in α-helix structure and low in β-sheets. This feature also correlates with the classification of the *C. acetobutylicum* P262 NADPH-dependent ADH as a type 3-ADH and clearly distinguishes it from the type 1- and 2-ADHs from horse liver or *Drosophila melanogaster* that have about equal proportions of α-helix and β-sheet structures. Downstream of the *adh* gene, a strong 25-bp stem loop structure ($\Delta G = -16$ kcal/mol) was found and it was followed by a sequence of five Us. This structure resembles a classical rho-independent transcription terminator (Rosenberg and Court 1979).

More recently, a gene encoding the NADH-dependent BDH was isolated (Petersen et al. 1991). N-terminal sequencing of the purified NADH-dependent BDH provided an amino acid sequence. In this case, a long but less specific oligonucleotide was designed. A 32-fold degenerate 62-mer oligonucleotide bearing eight inosine residues at positions of exceptional variability was synthesized and used to screen a genomic phage library. A DNA fragment showing hybridization with the probe was identified from phage exhibiting positive hybridization and was subcloned into pUC19. *E. coli* harboring this recombinant plasmid, pBDH51, was found to have increased NADH-dependent BDH activity. Expression of plasmid-borne proteins in Maxicells and Western blotting of whole cell extracts showed a single band of M_r about 43,000 that correlated with the BDH activity.

During this time of cloning of the BDH gene, a second NADH-dependent BDH was purified using a combination of dye resin columns (Welch 1990). This enzyme (BDH I) is more reactive towards acetaldehyde than the previously reported NADH-dependent butanol dehydrogenase (now called BDH II). The two BDHs differ in pH optimum, cofactor requirements, $K_{m_{butyraldehyde}}$, $K_{m_{NADH}}$ and pI (Welch 1990). The role of BDH I may be much greater in ethanol production. After purification of BDH I and BDH II to homogeneity, N-terminal sequence studies showed several differences between the two within the first 20 aa. Only limited amino acid sequence homology was found with the previously cloned ADH. Hybridization studies between the cloned *adh* and *bdh* genes showed no homology under the conditions used. This indicates that the ADH and BDHs are distinct proteins as expected from enzymological experiments. Studies were conducted to further examine the BDH I and BDH II found in *E. coli* bearing pBDH51. The enzymes were purified from *E. coli* using the same method used for preparation of these enzymes from *C. acetobutylicum*. Two separate protein fractions were isolated that closely resembled the BDHs isolated from *C. acetobutylicum*. The two enzymes are both encoded within the same cloned 8-kb clostridial DNA fragment (Petersen et al. 1991). Their close proximity may suggest coordinate regulation or some more elaborate arrangement could be present. Each isozyme is well expressed in *E. coli*, although one is found in approximately sevenfold higher activity. In addition to the NADH-linked BDH I and II activities, enzyme extracts of *E. coli* harboring the 8-kb clostridial DNA demonstrated an additional NADPH-linked BDH (Petersen et al. 1991). The exact role of each enzyme in the production of ethanol and butanol and their induction during the solventogenic switch remains to be explored. The use of antibodies or nucleotide segments specific for each enzyme/gene will be essential for elucidating their roles.

8.5.5 Production of Other Solvent-Related Metabolites

Although *C. acetobutylicum* normally produces only low amounts of lactate, certain species of clostridia (*C. thermolacticum* sp. nov.) produce lactate as a major metabolite. Growth conditions have been reported where lactate is a major product of metabolism in *C. acetobutylicum* (Simon 1974). The most favorable conditions for lactate formation are pH values above 5.0 with CO or potassium cyanide or when growth is limited by low concentrations of iron or sulfate (Kubowitz 1934; Hanson and Rodgers 1946; Katagiri et al. 1960). Upon growth in continuous culture under iron or sulfate limitation, butanol and acetone are produced at pH 4.5, but above pH 5.0 lactate is formed (Bahl et al. 1982b; Bahl and Gottschalk 1984; Bahl et al. 1986). Under these conditions, the increased concentration of fructose 1,6-bisphosphate, an activator of lactate dehydrogenase, favors production of lactate (Freier and Gottschalk 1987). Mutants of *C. acetobutylicum* resistant to the pyruvate halogen analogs 3-fluoropyruvate or 3-bromopyruvate were found to accumulate more acetoin and lactate in normal batch cultures regulated at pH 4.8 (El Kanouni et al. 1989).

Lactate dehydrogenase (EC 1.1.1.27) was purified from *C. acetobutylicum* DSM 1731 and the native enzyme was a tetramer of 159 kDa with a pH optimum of 5.8 (Freier and Gottschalk 1987). The enzyme catalyzed only the reduction of pyruvate and was activated by the presence of fructose-1,6-bisphosphate with calcium or magnesium ions as positive effectors as is typical for LDH from Gram-positive organisms (Garvie 1980). The *C. acetobutylicum* enzyme was unstable and had little activity below pH 5.0. The highest specific activity (0.81 U/mg) was observed in iron-limited cells where lactate was produced or in cells grown on complex media. Little lactate was found in solvent-producing cells (0.01 U/mg) (Freier and Gottschalk 1987). Two thermostable lactate dehydrogenase isoenzymes were purified from *C. thermo-*

hydrosulfuricum, which were activated differently by fructose-1,6-bisphosphate (Turunen et al. 1987).

The gene encoding LDH was recently cloned from *C. acetobutylicum* B643 (Contag et al. 1990). *E. coli* PRC436, an *alhc adh*$^+$ *acd* strain, was used (Cunningham and Clark 1986) that could not grow anaerobically on glucose. After transformation with a *C. acetobutylicum*, DNA plasmid library and selection for growth anaerobically on glucose, plasmids were isolated and metabolic studies were undertaken. *E. coli* strains harboring a *C. acetobutylicum ldh* containing plasmid (pPC37) had higher production of lactate and lower production of ethanol and acetic acid. Strains deficient in LDH could also be complemented for lactate production by pPC37. Enzyme assays also showed high levels of LDH expressed from the plasmid. Maxicell analysis revealed a protein band of about 38 kDa that corresponded rather closely to the 36 kDa proposed on the basis of SDS gels of purified *C. acetobutylicum* LDH (Freier and Gottschalk 1987). Southern hybridization results of genomic *C. acetobutylicum* DNA using the cloned *ldh* gene as a probe were consistent with the restriction map of the cloned segment.

8.6 CLUSTERS OF METABOLIC GENES IN *CLOSTRIDIUM ACETOBUTYLICUM*

The genes already cloned have been found to be clustered in a fashion similar to operons described in other species. Figure 8–3 shows the orientation of some of these genes. Even in the most analyzed cases of solventogenesis-related genes in *C. acetobutylicum*, a polycistronic messenger has not been shown clearly, nor have the ends of a polycistronic mRNA been defined. RNA experiments on the gene clusters will further define the precise transcription of these genes and lead to more detailed studies of their regulation. In the case of the genes encoding the AADC and CoAT, the unusual convergent arrangement was unexpected; however, there are examples of this rare situation in other organisms. The overall genetic map location of the various solvent-related genes is not known so it is not known how distant the individual clusters might be on the chromosome. Studies using the fractionation of DNA fragments made by rare restriction endonucleases coupled to the availability of gene specific probes promises to give a picture of this aspect of chromosome structure.

8.7 EXPRESSION OF CLONED GENES IN *CLOSTRIDIUM ACETOBUTYLICUM*

8.7.1 Gene Transfer Mechanisms

The ability to express heterologous and homologous genes in *C. acetobutylicum* to alter the cell metabolism is critical to the creation of overproducing strains for industrial applications. There are two mechanisms for introducing gene-carrying vectors to bacterial cells, conjugation or transformation. Both mechanisms have been explored in the past ten years or so for transforming *C. acetobutylicum* and related strains and have been recently reviewed (Minton and Oultram 1988; Young et al. 1989a). Conjugal plasmid transfer to *C. acetobutylicum* from streptococci, *E. coli*, *Bacillus subtilis*, and *C. acetobutylicum*, has been extensively reviewed (Davies et al. 1988; Minton and Oultram 1988; Young et al. 1989a). Several investigators have reported on the protoplast transformation of clostridia (including *C. acetobutylicum*), with the most recent ones reporting efficiencies of up to 6 $\times$ 10^6 transformants per μg DNA. Key difficulties that lead to low transformation efficiencies for *C. acetobutylicum* and related strains include the high autolysin activity of the cells, which prevents regeneration of the bacterial wall, and the strong general nuclease activity. Thus, the high transformation efficiency of 6 $\times$ 10^6 transformants per μg DNA was achieved with the autolysin-deficient mutant *C. acetobutylicum* N1-4081 using sodium polyanethole sulfonate to inhibit the residual N-acetyl muramidase activity (Reysset et al. 1988; Truffaut et al. 1989). The induction, isolation, and characterization of DNase-deficient mutants of *C. acetobutylicum* DSM 792 was recently reported (Burchhardt and Dürre 1990). Such nuclease deficient mutants would presumably be amenable to more efficient transformation, in addition to allowing easier DNA isolation; however, these mutants exhibit substantially reduced product-formation capabilities, and in particular, inability to form butanol, thus rendering them unsuitable for practical applications. Natural transformation of *C. thermohydrosulfuricum* was discussed earlier (Young et al. 1989a), and a low efficiency protocol for transformation of *C. acetobutylicum* using Tris-PEG was described recently (Yoshino et al. 1990). Transformation by electroporation (electrotransformation will be the term used in the remaining part of this chapter) of *C. acetobutylicum* NCIB 8052 was first reported by Oultram et al. (1988a), and is now becoming the method of choice for the transformation of other *C. acetobutylicum* strains in several laboratories, including ours.

Our early efforts to introduce foreign DNA into the *C. acetobutylicum* ATCC 824 cells (all subsequent discussion of work from our laboratory will refer to this strain) by conjugative transfer from streptococcal species did not produce satisfactory results, probably because of the combined strong detrimental effects of the restriction system, as discussed below, and the general nuclease and autolysin activities of our strain. Our efforts turned to the development of an electrotransformation protocol as soon as the new Bio-Rad electroporation device (with a Gene Pulser, Bio-Rad Laboratories, Richmond, CA) suitable for electrotransformation of bacteria (Dower et al. 1988; Calvin and Hanawalt 1988) became commercially available. The electrotransformation protocol that Oultram et al. (1988a) published, despite repeated attempts, did not yield any successful results with our *C. acetobutylicum* strain, but a protocol that is a modification of the published *E. coli* protocols (Dower et al. 1988; Calvin and Hanawalt 1988) eventually allowed satisfactory electrotransformation of our cells (Mermelstein et al. 1992). Briefly, our protocol is as follows. We grow the cells in reinforced clostridia medium (RCM, Difco, Detroit, MI, initial pH 5.2) until an OD_{600} of 1.2, which in this medium corresponds to the late exponential growth phase. Cells (a total volume of 60 ml) are harvested and centrifuged anaerobically at room temperature, and are then resuspended and washed in ice-cold electrotransformation buffer (ETB, 5 mM NaH_2PO_4, 272 mM sucrose, pH 7.4). After a second centrifugal harvest, the cells are resuspended in 2.1-ml ice-cold ETB, from which 0.7 ml are dispensed into three sterile electroporation cuvettes (0.4-cm electrode gap). Cells are chilled for 5 min in the cuvettes, and DNA dissolved in distilled water (up to 20 μl total) is then added. The mixture is chilled for another 2 min before exposing the cells to a 6.25 kV/cm current pulse from a 25 μF capacitor (τ = about 7.5 ms). The cell/DNA mixture is then poured immediately into a 10 ml screw-cap tube containing RCM (pH = 5.2) for 4 h of outgrowth. Cells are then centrifuged and resuspended in 900 μl of pH 5.2 RCM and plated on selective RCM plates (40 μg/ml erythromycin, pH 5.8). The results shown in Table 8–3 have been obtained with this protocol.

8.7.2 Vectors for Introducing Foreign DNA into Clostridia and the Restriction System of *Clostridium acetobutylicum* ATCC 824

Vectors used for gene transfer into clostridia, by conjugation or transformation, through early 1989 have been reviewed by Brehm et al. (1988), Minton and Oultram (1988), and Young et al. (1989a). These include several shuttle vectors for *C. acetobutylicum* from *E. coli* or *B. subtilis* as have been tabulated by Young et al. (1989a). The selectable markers of these shuttle vectors include Macrolide, Lincosamide, and Streptogramin B antibiotic resistance (MLS^R) from pAMβ1 or pIM13, Km^R from Tn *1545*, or Tc^R from pT127. These vectors employ replicons that originate from *S. faecalis* (pAMβ1), *B. subtilis* (pIM13), *C. butyricum* (pCB101), or *C. paraputrificum* (pCP1). No detailed copy number studies have been reported other than a characterization of high or low copy number based on the ease with which plasmid DNA can be visualized in cleared lysates. For example, plasmid pIM13 and its derivatives pKNT11 and pKNT14 are maintained in *C. acetobutylicum* N1-4081 with a copy number that "appears closer to its equivalent in *Streptococcus* (10 copies per genome) than in *B. subtilis* (200 copies per genome)" (Truffaut et al. 1989). No systematic plasmid stability studies have been reported either, and no detailed investigations of structural or segregational instability of these plasmids are known. To date, and on the basis of limited data, these plasmids show substantial segregational instability. Notably, in *C. acetobutylicum* N1-4081, after 40 generations in the absence of antibiotic (erythromycin) pressure, plasmids pIM13, pKNT11, and pKNT14 remain in 8, 27, and 52% of the total cell population, respectively (Truffaut et al. 1989).

More recent work not included in the aforementioned reviews include the use of a *C. acetobutylicum-E.coli* shuttle vector (pTY10) constructed from a high copy number plasmid (pCS86, 3.0 kb in size, 10–20 copies per cell) isolated from a laboratory strain of *C. acetobutylicum* and from the commercially available *E. coli* plasmid pKK232-8 (Yoshino et al. 1990). The plasmid was transferred into *C. acetobutylicum* cells by a Tris-PEG treatment. The CAT-gene (from pKK232-8, for chloramphenicol resistance) was used for selection in both *E. coli* and *C. acetobutylicum*. pTY10 was substantially deleted in *C. acetobutylicum*. One deletion derivative (pTYD101) was maintained at a high copy number (100 copies per cell) in *C. acetobutylicum* but could not transform *E. coli*. This plasmid was also segregationally unstable. Yoon et al. (1991) have more recently reported the construction of an *E.coli*/*B. subtilis*/*C. acetobutylicum* shuttle vector (pCAB32). The plasmid was derived from a cryptic plasmid (pCA134) isolated from a *Clostridium* sp. and the plasmid pHV32. The CAT gene was used to select the *C. acetobutylicum* ATCC 4259 electrotransformants on chloramphenicol plates. No transformation efficiencies, copy number or plasmid-stability data were reported.

In our laboratory we have followed two different approaches for constructing shuttle vectors for the introduction of product-formation genes into *C. acetobutylicum* ATCC 824. The first family of plasmids was generated by ligating a fragment from the *C. butyricum* cryptic plasmid pCBU2 (Luczak et al. 1985) to a derivative of pBR322 containing the erythromycin (Em)-resistance gene from pMU1328. One of the plasmids (pSYL2) was introduced into *C. acetobutylicum* ATCC 824 by electrotransformation with very low and variable efficiency (0–90 transformants per μg DNA) when the plasmid was isolated from *E. coli* (Table 8–3). Plasmid prepared from *C. acetobutylicum* gave much higher transformation efficiencies (8×10^4 transformants per μg DNA). As we discuss later, this was due to the strong restriction system in *C. acetobutylicum* ATCC 824, which cuts GC-rich DNA fragments (such as pBR322 and other *E. coli*-originating plasmids) with very high frequency

TABLE 8–3 Correlation between the Number of *Cac*824I Restriction Sites on Various Plasmids and the Efficiency of Electrotransformation of *C. acetobutylicum* 824 by These Plasmids

Plasmid	*No. of* Cac*824I Restriction Sites*	*Transformants per μg DNA*
pIM13	1	6×10^4
pFNK1	2	4×10^3
pSYL2	47	0–90
pBR322[1]	42	—[1]
pUC19[1]	19	—[1]

[1] Plasmids do not contain Em-resistance genes or Gram-positive replicons, and could not thus be tested for their ability to electrotransform *C. acetobutylicum* 824.

(Table 8–3). pSYL2 was maintained at a copy number of about 14 plasmids per cell (data not shown), and rather stably, in *C. acetobutylicum* in the absence of antibiotic pressure (Em) (more than 75% of cells retained the plasmid after 60 generations, Figure 8–4). The plasmid replicates in both *B. subtilis* and *Streptococcus faecalis*. It remains undeleted in *C. acetobutylicum* and plasmid prepared from *C. acetobutylicum* retransforms *E. coli* effectively.

In the second approach, we have attempted to construct shuttle *E. coli*/*C. acetobutylicum* vectors using Gram-positive replicons from *S. ferrus* (pMU1328) and *B. subtilis* (pIM13) plasmids (Truffaut et al. 1989; Young et al. 1989a). Various chimeras between these plasmids and either pBR322 or pUC19 (both are *E. coli* plasmids) could not transform *C. acetobutylicum*. Vector pMU1328 could transform *C. acetobutylicum* but was maintained in the strain as a specific deletion derivative lacking the ColE1 origin of replication. We therefore concluded that the transformation difficulties encountered with plasmids containing *E. coli*-plasmid portions were due either to the instability of the ColE1 origin in *C. acetobutylicum* or the existence of a restriction system that recognizes the ColE1 sequences. The latter was easier to elucidate, and indeed we have shown that *C. acetobutylicum* contains a strong restriction system that cuts ColE1 plasmids with very high frequency (Table 8–3). This restriction endonuclease (named *Cac*824I) recognizes the

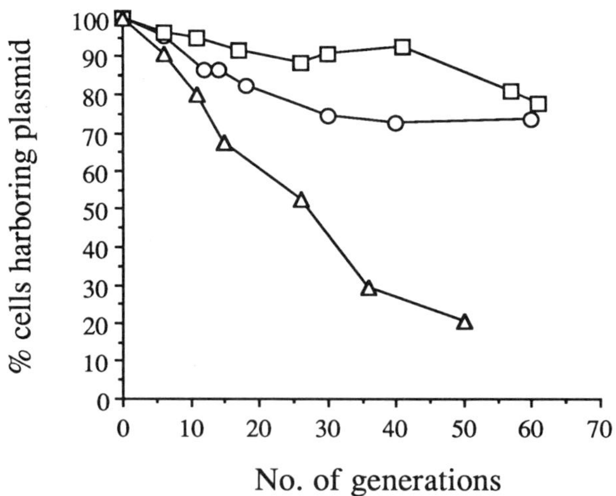

FIGURE 8–4 Stability of plasmid vectors: pFNK1 (□), pFNK5 (○), and pSYL2 (△). Cells harboring plasmids were serially subcultured by transferring culture (up to 10% vol/vol, depending on the stage of growth) into a fresh medium that does not contain antibiotic for about 60 generations. Stability is represented by percent cells harboring plasmid after a given number of generations, and was determined by plating an appropriate dilution onto antibiotic (erythromycin) versus nonantibiotic containing plates followed by comparison of the number of colonies appeared.

sequence 5′-GCNGC-3′, where N is any DNA base, requires high salt concentrations (an optimal of about 150 mM NaCl, see Figure 8–5), is unique among other known clostridia restriction endonucleases (Young et al. 1989a), and has two known isoschizomers, *Fnu*4HI (Figure 8–5) and *Fbr*I. The recognition sequence was determined by analyzing the restriction digests of high A+T DNA of known sequence. A key factor that allows the detection of this restriction activity in the *C. acetobutylicum* 824 cell extracts is the preparation of cell extracts from washed protoplasts. This reduces substantially the secreted and wall-associated general nuclease activity, thus permitting an unambiguous demonstration of the *Cac*824I activity. The data in Table 8–3 demonstrate that this restriction activity is a major barrier for the transformation of this strain. The presence of a restriction system may explain the low transformation efficiencies and the plasmid-deletion patterns reported with other strains as we discussed previously. We applied the same protocol we used for strain 824 to investigate if *C. acetobutylicum* NCIB 8052 is free from a restriction system, as had been reported earlier (Richards et al. 1988). We found that indeed it contains no detectable restriction endonuclease activity.

With the knowledge that the restriction activity of *Cac*824I would make any *E. coli*/*C. acetobutylicum* shuttle vector practically useless for the strain

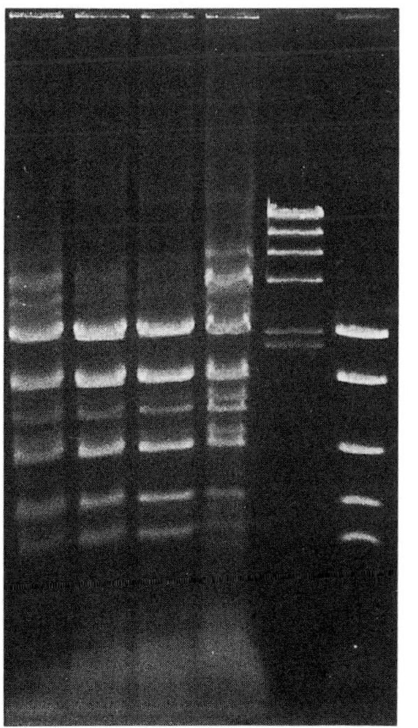

FIGURE 8–5 NaCl dependency of *Cac*824I activity as analyzed by ethidium bromide stained digests of DNA electrophoresed in 0.9% agarose. From left to right; *Cac*824I digests of pMU1328 (1 h at 37°C) in 10 mM Tris (pH 7.9), 10 mM $MgCl_2$, 2% crude extract and NaCl at concentrations indicated above lanes, *Hin*dIII digest of λ DNA, *Fnu*4HI digest of pMU1328.

C. acetobutylicum 824, we opted for the development of a *B. subtilis/C. acetobutylicum* 824 shuttle vector based on the *B. subtilis* plasmid pIM13, which had been shown earlier (Truffaut et al. 1989) to function well in *C. acetobutylicum*. This led to the development of the pFNK family of plasmids, which we used to express several *C. acetobutylicum* genes in *C. acetobutylicum* 824 (Figure 8–6). The basal plasmid pFNK1, which contains no *C. acetobutylicum* genes, contains the OrfII and MLS^R portions of pIM13 as well as a multiple cloning site from pUC9. We determined (results not shown) that pFNK1, and presumably its derivatives, is maintained in *C. acetobutylicum* 824 with a copy number of about 7, which is consistent with the reported observation (Truffaut et al. 1989) that pIM13, and its derivatives, is maintained in *C. acetobutylicum* N1-4081 with a copy number of 10. pFNK1 was found to be stable in *C. acetobutylicum* 824, but its derivatives (pFNK3 and pFNK5, see the next section) containing primary-metabolic genes from *C. acetobutylicum* were found to be rather unstable (Figure 8–4).

8.7.3 Expression of Primary-Metabolic Genes in *Clostridium acetobutylicum*

To date, apart from plasmid replication and antibiotic-selection genes, we are aware of the expression of only one cloned gene (the *C. pasteurianum* leucine biosynthesis gene) in *C. acetobutylicum* (Oultram et al. 1988b). We used the pFNK1 *B. subtilis/C. acetobutylicum* shuttle vector to express several cloned genes related to the primary metabolism of *C. acetobutylicum* in the 824 strain. These genes include (see Sections 8.4 and 8.5): the PTB, AADC, and the BDHs. The expression vector for the PTB gene, pFNK5, is shown in Figure 8–6. The other vectors are similar, and details have been published (Mermelstein et al. 1991a). All fermentations using recombinant 824 strains overexpressing proteins of the primary metabolism were carried out in the presence of 100 μg/ml erythromycin to promote plasmid stability. It is, however, known that erythromycin is chemically unstable at pH values below 5.5. In the fermentations that will be discussed below, the pH was mostly below 5.5, and therefore, substantial plasmid loss may underestimate the potential of these plasmids for overexpressing the corresponding proteins. Nevertheless, activity assays and protein gels show good expression levels of all proteins we have investigated. Our data also show that the solvent-producing capabilities of the recombinant strains remain at least as good as the unmodified 824 strain. Also, the recombinant strains were found to grow as fast as the unmodified 824 strain, even in the presence of Em (results not shown).

The gene for PTB was obtained from vector pJC7 (Cary et al. 1988) using a *Sma*I/*Pvu*II fragment. PTB is overexpressed in *C. acetobutylicum* 824 in a fashion similar to the single copy gene in the unmodified *C. acetobutylicum* (results not shown). pFNK5 plasmid was prepared from *B. subtilis* ATCC 33677 (results not shown), and was introduced into *C. acetobutylicum* 824 by electrotransformation as discussed previously. The PTB activity increases by

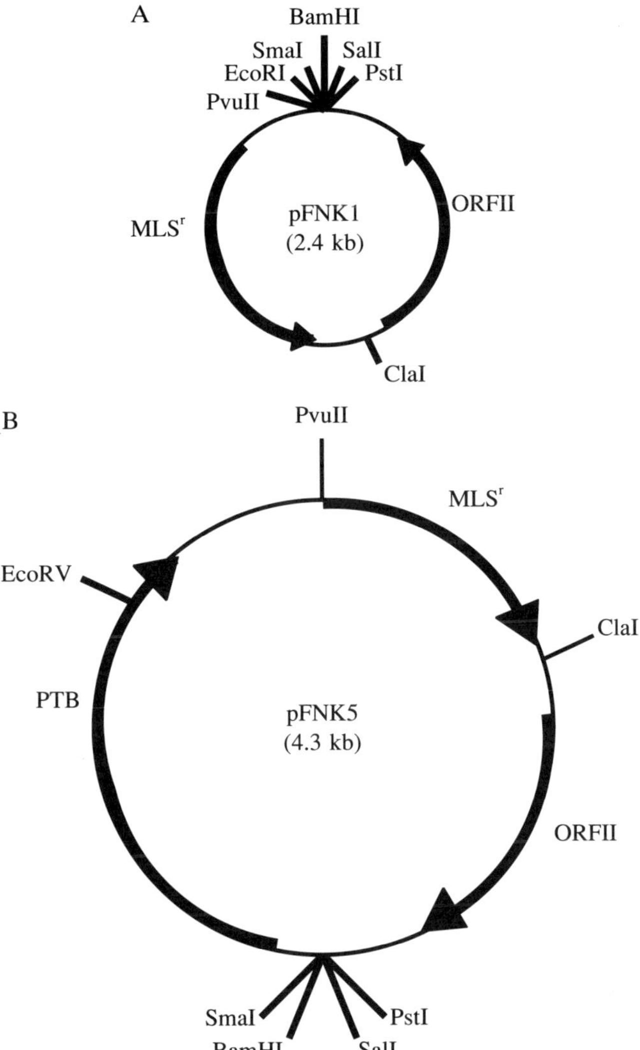

FIGURE 8–6 Physical map of vector pFNK1 (A) and its derivative, pFNK5 (B) containing the *C. acetobutylicum* ATCC 824 phosphotransbutyrylase (PTB) gene. The macrolide, lincosamide, and streptogramin B antibiotic resistance gene (MLS^R) and putative replication protein (ORFII) are derived from the *B. subtilis* plasmid pIM13. Bold arrows indicate structural genes and their transcriptional direction.

more than twofold during exponential growth in both unmodified and modified cells. The recombinant strain was found to contain a 3- to 20-fold higher PTB activity than the unmodified strain, depending on the stage of culture, pH and medium composition. This increased enzymatic activity is consistent with SDS-PAGE gels of cell extracts that show a major band around 31 kDa

(Figure 8–7), presumably corresponding to the subunits of the native PTB octamer (Cary et al. 1988; Wiesenborn et al. 1989a). These results demonstrate that the PTB gene contains a strong promoter, which might be useful for overexpressing other proteins in *C. acetobutylicum*. We are currently working on cloning this promoter. The overexpression of PTB in *C. acetobutylicum* results in a higher final ratio of [butyrate]/[butanol] (concentrations) in batch, predominantly acidogenic fermentations without pH control (results not shown). This is consistent with the expectation that higher PTB activities will compete with the butyraldehyde dehydrogenase for butyryl-CoA (Figure 8–1), thus resulting in increased butyrate/butanol ratios. We are currently investigating the effect of an overexpressed PTB on the product-formation patterns of strain 824 over a wide range of pH values.

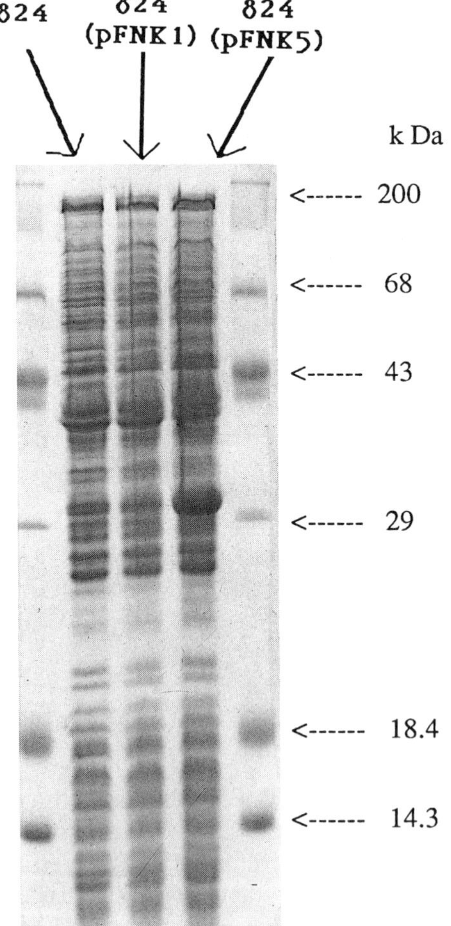

FIGURE 8–7 Analysis of uncontrolled pH batch fermentation whole-cell extracts by Coomassie blue stained 12.5% polyacrylamide-SDS gel electrophoresis. *C. acetobutylicum* ATCC 824 and the indicated plasmid containing derivatives were grown in CSM (Cary et al. 1988) and harvested during late exponential growth. The outside lanes contain standards of molecular masses indicated to the right. A band of increased intensity at about 31,000 Da corresponding to the phosphotransbutyrylase subunit molecular mass can be detected in 824(pFNK5).

Overexpression of AADC in strain 824 and its mutant M5 (which lacks a functional AADC) (Clark et al. 1989) did not affect their patterns of product formation (results not shown) but provided valuable information nevertheless. Plasmid pFNK3, which contains the AADC gene from plasmid pDP2 (Petersen and Bennett 1990b), was again prepared from *B. subtilis* ATCC 33677, and was used to electrotransform both strain 824 and M5. Batch fermentations of transformants were carried out without pH control, and formation of acetone by the recombinant strains was not different from the acetone formation by the unmodified strains. Strain M5(pFNK3) was not expected to form acetone since it also lacks CoA-transferase activities (Clark et al. 1989), which is the enzyme that precedes AADC in the acetone formation pathway (Figure 8–1). Cells collected from the exponential and stationary fermentation phases of strains 824, 824(pFNK3) and M5(pFNK3), were assayed for AADC activities. The results are shown in Table 8–4. AADC activities up to 45 times higher than the activities in the unmodified 824 strain were detected in the recombinant strains. These results show that AADC is expressed from pFNK3 in a regulated fashion similar to that of the chromosomal gene, and that the AADC-gene promoter is also a strong promoter, and thus potentially useful for overexpressing other enzymes in *C. acetobutylicum*. SDS-PAGE gels show a predominant band at 28 kDa (results not shown), which corresponds to the size of the AADC (Petersen and Bennett 1990b). The intensity of this band on the SDS-PAGE gels suggests that AADC is the major protein expressed in the recombinant strains. As was discussed in Section 8.1.3, the fact that in most reported cases the *C. acetobutylicum* mutants have lost multiple enzymatic activities (such as the most frequently lost CoAT, AADC, and

TABLE 8–4 Acetoacetate Decarboxylase Activities as Determined by Constant Pressure Differential Respirometry (Petersen and Bennett 1990b) Using Whole Cell Suspensions of the Indicated Strains Grown in CSM, pH 6.2, and Assayed at Exponential (E) or Stationary (S) Growth Phases. One Unit is Defined as the Amount of Enzyme Catalyzing the Evolution of 1 μmol of CO_2 Per Min at Standard Temperature and Pressure

	U/mg Protein	
Strain	*E*	*S*
824	0.8	4.8
824(pFNK1)	0.7	5.1
824(pFNK3)	9.9	171
M5	0.0	0.0
M5(pFNK1)	0.0	0.0
M5(pFNK3)	7.8	224

BAD activities, as is the case for mutant M5) could be logically interpreted to imply a mutation of a common regulatory protein. The restoration of the AADC activities in mutant M5 by the plasmid pFNK3 suggests that this may not be the case for mutant M5.

Finally, we have examined the overexpression in *C. acetobutylicum* of a cluster of genes (Petersen et al. 1991) containing at least two types of BDHs, one NADH dependent and the other NADPH dependent. We cloned a 3.5-kb *Eco*RI fragment, in both orientations, from the original plasmid pBDH51 containing the BDH-gene cluster (on an 8-kb clostridial DNA) into pUC9 to obtain two plasmids, pET89a and pET89b. When introduced into *E. coli* HB101, both of these plasmids resulted in expression of NADH- and NADPH-dependent activities that were indistinguishable from the activities obtained with cells containing pBDH51.The NADH-linked activities in the recombinant *E. coli* were typically seven times higher than the activities contained in the unmodified host, while the NADPH-linked activities were about 2 to 2.5 times higher (results not shown). A 3.5-kb DNA from pET89b was then cloned into the vector pFNK1 to produce the plasmid pFB7 which was prepared from *B. subtilis* and was used to electrotransform *C. acetobutylicum* 824. SDS-PAGE gels from 824(pFB7) cell extracts show a major band at 42 kDa (data not shown), which corresponds to the subunit size of dimeric BDH proteins (Petersen et al. 1991). All NADH- and NADPH-linked BDHs have indistinguishable molecular weights (Petersen et al. 1991), so the SDS-PAGE gels cannot identify which BDH type is overexpressed in *C. acetobutylicum*. Enzyme-activity assays have shown that while the NADH-linked BDH is identical in the recombinant and unmodified 824 strains, there is a 3.5- to 4-fold higher NADPH-linked BDH activity in the recombinant strain. These activities are lower than what is suggested from the SDS-PAGE gels that show that the 42-kDa proteins are unambiguously the predominant proteins in the cells. In pH 5 controlled fermentations, compared to strain 824, the recombinant strain grows equally fast but to higher cell densities, takes butyric acid up slower, produces butanol slower, produces more acetic acid, takes acetic acid up slower, and produces acetone slower. Currently, we are examining the reasons for these unexpected differences in the expression of the BDH genes from plasmid pFB7 in *E. coli* and *C. acetobutylicum*. Understanding the changes in the patterns of product formation brought about by the presence of this plasmid will require substantial additional work. Finally, the plasmid-borne BDH genes appear to be expressed in a regulated and similar fashion to the chromosomal genes (data not shown).

In summary, we have shown that genes for the primary-metabolic enzymes can be overexpressed without, in general, detrimentally affecting the solvent-producing capabilities of the cells, and that this overexpression can alter the product-formation patterns of the cells without apparently reducing the cell growth rate. What remains a challenge for future research is to study and determine the necessary genetic alterations that will bring about beneficial changes in product formation, such as overproduction of butanol.

8.8 CONCLUSION

Understanding the mechanism of the solventogenic switch is important to the commercial application of fermentations using *C. acetobutylicum*. This regulatory feature also represents an example of gene regulation in response to environmental stimuli and perhaps will reveal more about stress responses in *C. acetobutylicum* and the relation of solvent production genes to other stress genes and global control mechanisms. The cloning of genes involved in the production of acids and solvents has only positioned workers to conduct detailed studies on the mechanisms of gene regulation in this organism. DNA sequence analysis of these genes, coupled with transcriptional studies and the isolation and characterization of regulatory mutations via transposon mutagenesis will greatly increase our understanding of those regions within the DNA responsible for this regulation and the proteins involved. The ability to introduce genes or modified versions back into *C. acetobutylicum* is just beginning and it will surely lead to information on how specific sequences are involved in particular physiological responses during solvent induction. This ability is also of great potential importance in advancing recombinant technology for commercial applications with this organism.

REFERENCES

Aceti, D.J., and Ferry, J.G. (1988) *J. Biol. Chem.* 263, 15444–15448.

Allcock, E.R., Reid, S.J., Jones, D.T., and Woods, D.R. (1982) *Appl. Environ. Microbiol.* 43, 719–721.

Andersch, W., Bahl, H., and Gottschalk, G. (1983) *Eur. J. Appl. Microbiol. Biotechnol.* 18, 327–332.

Bahl, H., Andersch, W., and Gottschalk, G. (1982a) *Eur. J. Appl. Microbiol. Biotechnol.* 15, 201–205.

Bahl, H., Andersch, W., and Gottschalk, G. (1982b) *Eur. J. Appl. Microbiol. Biotechnol.* 14, 17–20.

Bahl, H., and Gottschalk, G. (1984) *Biotechnol. Bioeng. Symp.* 14, 215–223.

Bahl, H., and Gottschalk, G. (1988) in *Biotechnology* vol. 6b, (Rehm, H.-J., and Reed, G., ed.), pp. 1–30, VCH Verlagsgesellschaft, Weinheim, Germany.

Bahl, H., Gottwald, M., Kuhn, A., et al. (1986) *Appl. Environ. Microbiol.* 52, 169–172.

Ballongue, J., Amine, J., Masion, E., Petitdemange, H., and Gay, R. (1985) *FEMS Microbiol. Lett.* 29, 273–277.

Ballongue, J., Amine, J., Petitdemange, H., and Gay, R. (1986a) *Biomass* 121–129.

Ballongue, J., Amine, J., Petitdemange, H., and Gay, R.(1986b) *FEMS Microbiol. Lett.* 35, 295–301.

Ballongue, J., Janati-Idrissi, R., Petitdemange, H., and Gay, R.(1989) *J. Appl. Bacteriol.* 67, 611–617.

Ballongue, J., Masion, E., Amine, J., Petitdamange, H., and Gay, R. (1987) *Appl. Microbiol. Biotechnol.* 26, 568–573.

Barber, H.A., Jeng, I.-M., Neff, N., et al. (1978) *J. Biol. Chem.* 253, 1219–1225.

Bergemeyer, H.U., Holz, G., and Lang, G. (1963) *Biochem. Z.* 338, 114–121.
Berndt, H, and Schlegel, H.G. (1975) *Arch. Microbiol.* 103, 21–30.
Bertram, J., Kuhn, A., and Dürre, P. (1990) *Arch. Microbiol.* 153, 373–377.
Bitar, K.G., Perez-Aranda, A., and Bradshaw, R.A. (1980) *FEBS Lett.* 116, 196–198.
Brehm, J.K., Oultram, J.D., Thompson, D.E., Swinfield, T.J., Peck, H., Young, M., and Minton, N.P. (1988) In *Genetics and Biotechnology of Bacilli*, vol. 2 (A.T. Ganesan and J.A. Hoch, eds.), pp. 409–414, Academic Press, Orlando, FL.
Burchhardt, G., and Dürre, P. (1990) *Curr. Microbiol.* 21, 307–311.
Burton, R.M., and Stadtman, E.R. (1953) *J. Biol. Chem.* 20, 873–890.
Calvin, N.M., and Hanawalt, P.C. (1988) *J. Bacteriol.* 170, 2796–2801.
Cary, J.W., Petersen, D.J., Bennett, G.N., and Papoutsakis, E.T. (1990a) *Ann. NY Acad. Sci.* 589, 67–81.
Cary, J.W., Petersen, D.J., Papoutsakis, E.T., and Bennett, G.N. (1988) *J. Bacteriol.* 170, 4613–4618.
Cary, J.W., Petersen, D.J., Papoutsakis, E.T., and Bennett, G.N. (1990b) *Appl. Environ. Microbiol.* 56, 1576–1583.
Clark, S.W., Bennett, G.N., and Rudolph, F.B. (1989) *Appl. Environ. Microbiol.* 55, 970–976.
Contag, P.R., Williams, M.G., and Rogers, P. (1990) *Appl. Environ. Microbiol.* 56, 3760–3765.
Croux, C., and Garcia, J.L. (1991) *Gene* 104, 25–31.
Cummins, C.S., and Johnson, J.L. (1971) *J. Gen. Microbiol.* 67, 33–46.
Cunningham, P.R., and Clark, D.P. (1986) *Mol. Gen. Genet.* 205, 487–493.
Davies, A., Oultram, J.D., Pennock, D.R., Williams, D.F., Richards, N.P., Minton, and Young, M. (1988) in *Genetics and Biotechnology of Bacilli* (A.T. Ganesan and J.A. Hoch, eds.), vol. 2, pp. 391–395, Academic Press, Orlando, FL.
Davies, R. (1943) *Biochem. J.* 37, 230–238.
DiLella, A.G., and Woo, S.L.C. (1987) *Methods Enzymol.* 152, 447–451.
Dower, W.J., Miller, J.F., and Ragsdale, C.W. (1988) *Nucleic Acid. Res.* 16, 6127–6145.
Drake, H.L., Hu, S.-I., and Wood, H.G. (1981) *J. Biol. Chem.* 256, 11137–11144.
Dürre, P.., Kuhn, A., Gottwald, M., and Gottschalk, G. (1987) *Appl. Microbiol. Biotechnol.* 26, 268–272.
El Kanouni, A., Junelles, A.-M., Janati-Idrissi, R., Petitdemange, H., and Gay, R. (1989) *Curr. Microbiol.* 18, 139–144.
Freier, D., and Gottschalk, G. (1987) *FEMS Microbiol. Lett.* 43, 229–233.
Fridovich, I. (1963) *J. Biol. Chem.* 238, 592–598.
Gabriel, C.L. (1928) *Ind. Eng. Chem.* 20, 1063–1067.
Garvie, E.I.(1980) *Microbiol. Rev.* 44, 106–139.
Gavard, R., Hautecoeur, B., and Descourtieux, H. (1957) *C. R. Acad. Sci. (Paris)* 244, 2323–2326.
George, H.A., and Chen, J.S. (1983) *Appl. Environ. Microbiol.* 467, 321–327.
Gerischer, U., and Dürre, P. (1990) *J. Bacteriol.* 172, 6907–6918.
Gottwald, M., and Gottschalk, G. (1985) *Arch. Microbiol.* 143, 42–46.
Hager, P.W., and Rabinowitz, J.C. (1985) in *The Molecular Biology of the Bacilli*, vol. 2 (Dubnau, D., ed.), pp. 1–32, Academic Press, New York.
Hanson, A.M., and Rodgers, N.E. (1946) *J. Bacteriol.* 51, 568–569.
Hartmanis, M.G.N. (1987) *J. Biol. Chem.* 262, 617–621.

Hartmanis, M.G.N., and Gatenbeck, S. (1984) *Appl. Environ. Microbiol.* 47, 1277–1283.
Hartmanis, M.G.N., Klason, T., and Gatenbeck, S. (1984) *Appl. Microbiol. Biotechnol.* 20, 66–71.
Hartmanis, M.G.N., and Stadtman, E.R. (1982) *Proc. Natl. Acad Sci. USA* 79, 4912–4916.
Huang, L., Forsberg, C.W., and Gibbins, L.N. (1986) *Appl. Environ. Microbiol.* 51, 1230–1234.
Huang, L., Gibbins, L.N., and Forsberg, C.W. (1985) *Appl. Environ. Microbiol.* 50, 1043–1047.
Hüsemann, M.H.W., and Papoutsakis, E.T. (1988) *Biotech. Bioeng.* 32, 843–852.
Hüsemann, M.H.W., and Papoutsakis, E.T. (1989a) *Appl. Microbiol. Biotechnol.* 30, 585–595.
Hüsemann, M.H.W., and Papoutsakis, E.T. (1989b) *Appl. Microbiol. Biotechnol.* 31, 435–444.
Hüsemann, M.H.W., and Papoutsakis, E.T. (1990) *Appl. Environ. Microbiol.* 56, 1497–1500.
Ishii, N., Hijkata, M., Osumi, T., and Hashimoto, T. (1987) *J. Biol. Chem.* 262, 8144–8150.
Janati-Idrissi, R., Junelles, A.M., Petitdemange, H., and Gay, R. (1988) *Ann. Inst. Pasteur Microbiol.* 139, 683–688.
Janssen, P.S., Jones, W.A., Jones, D.T., and Woods, D.R. (1988) *J. Bacteriol.* 170, 400–408.
Jenkins, L.S., and Nunn, W.D. (1987) *J. Bacteriol.* 169, 42–52.
Jones, D.T., and Woods, D.R. (1986) *Microbiol. Rev.* 50, 484–524.
Jones, D.T., and Woods, D.R.(1989) in *Biotechnology Handbooks, vol. 3, Clostridia* (Minton, N.P., and Clarke, D.J., ed.), pp. 105–144, Plenum Press, New York.
Jörnvall, H., Persson, B., and Jeffery, J. (1987) *Eur. J. Biochem.* 167, 195–201.
Junelles, A.M., Janati-Idrissi, R., El Kanouni, A., Petitdemange, H., and Gay, R. (1987) *Biotechnol. Lett.* 9, 175–178.
Katagiri, H., Imai, K., and Sugimori, T. (1960) *Bull. Agric. Chem. Soc. Jpn.* 24, 163–172.
Klotsch, H.R. (1969) *Methods Enzymol.* 13, 381–386.
Kubowitz, F. (1934) *Biochem. Z.* 274, 285–298.
Laursen, R.A., and Westheimer, F.H. (1966) *J. Am. Chem. Soc.* 88, 3426–3430.
Lin, F.-P. (1992) Ph.D. thesis, University of Cape Town, Cape Town, South Africa.
Ljungdahl, L.G., Hugenholtz, J., and Wiegel, J. (1989) in *Biotechnology Handbooks, vol. 3, Clostridia* (Minton, N.P., and Clarke, J.J., ed.), pp. 145–191, Plenum Press, New York.
Luczak, H., Schwarzmoser, H., and Staudenbauer, W.L. (1985) *Appl. Microbiol. Biotechnol.* 23, 114–122.
Lurz, R., Mayer, F., and Gottschalk, G. (1979) *Arch.Microbiol.* 120, 255–262.
Mannens, G., Slegers, G., and Claeys, A. (1988) *Biotechnol. Lett.* 10, 563–568.
Martin, F.H., and Castro, M.M. (1985) *Nucleic Acids Res.* 13, 8927–8938.
Matsuyama, A., Yamamoto, H., and Nakano, E. (1989) *J. Bacteriol.* 171, 577–580.
Mattson, D.M., and Rogers, P. (1989) Abstr. 0–39, Annual Meeting, American Society for Microbiology, Washington, DC.
Mermelstein, L.D., Welker, N.E., Bennett, G.N., and Papoutsakis, E.T. (1992) *Bio/Technology* 10, 190–195.

Meyer, C.L., and Papoutsakis, E.T. (1989) *Appl. Microbiol. Biotechnol.* 30, 450–459.
Meyer, C.L., Roos, J.W., and Papoutsakis, E.T.(1986) *Appl. Microbiol. Biotechnol.* 24, 159–167.
Minton, N.P., and Oultram, J.D. (1988) Microbiol. Sci. 5, 310–316.
Monot, F., Engasser, J.M., and Petidemange, H. (1984) *Appl. Microbiol. Biotechnol.* 19, 422–426.
Narberhaus, F., and Bahl, H. (1992) *J. Bacteriol.* 174, 3282–3289.
Narberhaus, F., Giebeler, K., and Bahl, H. (1992) *J. Bacteriol.* 174, 3290–3299.
Osumi, T., Ishii, N., Hijikata, M., et al. (1985) *J. Biol. Chem.* 260, 8905–8910.
Oultram, J.D., Loughlin, M., Swinfield, T.J., Brehm, J.K., Thompson, D.E., and Minton, N.P. (1988a) *FEMS Microbiol. Lett.* 56, 83–88.
Oultram, J.D., Peck, H., Brehm, J.K., Thompson, D.E., Swinfield, T.J., and Minton, N.P. (1988b) *Mol. Gen. Genet.* 214, 177–179.
Palosaari, N.R., and Rogers, P. (1988) *J. Bacteriol.* 170, 2971–2976.
Petersen, D.J. (1991) Ph.D. thesis, Rice University, Houston, TX.
Petersen, D.J., and Bennett, G.N. (1990a) Abstr. 0-76, Annual Meetings, American Society for Microbiology, Washington, DC.
Petersen, D.J., and Bennett, G.N. (1990b) *Appl. Environ. Microbiol.* 56, 3491–3498.
Petersen, D.J., Cary, J.W., and Bennett, G.N. (1989) Abstr. I–124, Annual Meeting, American Society for Microbiology, Washington, DC.
Petersen, D.J., Welch, R.W., Rudolph, F.B., and Bennett, G.N. (1991) *J. Bacteriol.* 173, 1831–1834.
Petersen, D.J., Welch, R.W., Walter, K.A., et al. (1990) Abstr. p. 86, *Engineering Foundation Conferences: Progress in Recombinant DNA Technology*.
Pomeroy, A. (1970) Ph.D. thesis, Duke University, Durham, NC.
Reysset, G., Hubert, J., Podvin, L., and Sebald, M. (1988) *Biotechnol. Tech.* 2, 199–204.
Richards, D.F., Linnett, P.E., Oultram, J.D., and Young, M. (1988) *J. Gen. Microbiol.* 134, 3151–3157.
Robinson, J.R., and Sagers, R.D. (1972) *J. Bacteriol.* 112, 465–473.
Rodgers, N.E., Henika, R.H., and Hanson, A.M. (1946) *J. Bacteriol.* 51, 569–570.
Rogers, P. (1986) *Adv. Appl. Microbiol.* 31, 1–60.
Rogers, P., and Palosaari, N. (1987) *Appl. Environ. Microbiol.* 53, 2761–2766.
Rose, I.A. (1955) *Methods Enzymol.* 1, 591–595.
Rosenberg, M., and Court, D. (1979) *Annu. Rev. Genet.* 13, 319–353.
Rossmann, M.G., Moras, D., and Olsen, K.W. (1974) *Nature* (London) 250, 194–199.
Rudolph, F.B., Purich, D.L.,and Fromm, H.J. (1968) *J. Biol. Chem.* 243, 5539–5545.
Santangelo, J.D., Jones, D.T., and Woods, D.R. (1991) *J. Bacteriol.* 173, 1088–1095.
Shimizu, M., Suzuki, T., Kameda, K.-Y., and Abiko, Y. (1969) *Biochim. Biophys. Acta* 191, 550–558.
Shine, J., and Dalgarno, L. (1974) *Proc. Natl. Acad. Sci. USA* 71, 1342–1346.
Shone, C.C., and Fromm, H.J. (1981) *Biochemistry* 20, 7494–7501.
Simon, E. (1947) *Arch. Biochem.* 13, 237–243.
Smith, L.M., and Kaplan, N.D. (1980) *Arch. Biochem. Biophys.* 203, 663–675.
Sramek, S.J., and Frerman, F.E. (1975) *Arch. Biochem. Biophys.* 171, 14–26.
Stadtman, E.R. (1955) *Methods Enzymol.* 1, 596–599.
Terraiciano, J.S., and Kashket, E.R. (1986) *Appl. Environ. Microbiol.* 52, 86–91.
Thompson, D.K., and Chen, J.-S. (1990) *Appl. Environ. Microbiol.* 56, 607–613.
Truffaut, N., Hubert, J., and Reysset, G. (1989) *FEMS Microbiol. Lett.* 58, 15–20.

Turunen, M., Parkkinen, E., Londesborough, J., and Korhola, M. (1987) *J. Gen. Microbiol.* 133, 2865–2873.
Twarog, R., and Wolfe, R.S. (1962) *J. Biol. Chem.* 237, 2472–2477.
Twarog, R., and Wolfe, R.S. (1963) *J. Bacteriol.* 86, 112–117.
Valentine, R.C., and Wolfe, R.S. (1960) *J. Biol. Chem.* 235, 1948–1956.
Van Berkel, W.J.H. Van den Berg, W.A.M., and Mueller, F. (1988) *Eur. J. Biochem.* 178, 197–207.
Walter, K.A., Bennett, G.N., and Papoutsakis, E.T. (1992) *J. Bacteriol.* 174, 7149–7158.
Warren, S., Zerner, B., and Westheimer, F.H. (1966) *Biochemistry* 5, 817–823.
Waterson, R.M., and Hill, R.L. (1972) *J. Biol. Chem.* 247, 5258–5265.
Waterson, R.M., and Castellino, F.J., Hass, G.M., and Hill, R.L. (1972) *J. Biol. Chem.* 247, 5266–5271.
Welch, R.W., (1990) Ph.D. thesis, Rice University, Houston, TX.
Welch, R.W., Clark, S.W., Bennett, G.N., and Rudolph, F.B. (1992) *Enzyme Microbiol. Technol.* 14, 277–283.
Welch, R.W., Rudolph, F.B., and Papoutsakis, E.T. (1989) *Arch. Biochem. Biophys.* 273, 309–318.
Wiesenborn, D.P., Rudolph, F.B., and Papoutsakis, E.T. (1988) *Appl. Environ. Microbiol.* 54, 2717–2722.
Wiesenborn, D.P., Rudolph, F.B., and Papoutsakis, E.T. (1989a) *Appl. Environ. Microbiol.* 55, 317–322.
Wiesenborn, D.P., Rudolph, F.B., and Papoutsakis, E.T. (1989b) *Appl. Environ. Microbiol.* 55, 323–329.
Yamamoto-Otake, H., Matsuyama, A., and Nakano, E. (1990) *Appl. Microbiol. Biotechnol.* 33, 680–682.
Yan, R.-T., and Chen, J.-S. (1990) *Appl. Environ. Microbiol.* 56, 2591–2599.
Yan, R.-T., Zhu, C., Golemboski, C., and Chen, J. (1988) *Appl. Environ. Microbiol.* 54, 642–648.
Yong, R.A., and Davis, R.W. (1984) *Proc. Natl. Acad. Sci. USA* 84, 1194.
Yoon, K., Lee, J., and Kim, B. (1991) *Biotechnol. Lett.* 13, 1–6.
Yoshino, S., Yoshino, T., Hara, S., Ogata, S., and Hayashida, S. (1990) *Agric. Biol. Chem.* 54, 437–441.
Young, M., Minton, N.P., and Staudenbauer, W.L. (1989a) *FEMS Microbiol. Rev.* 63, 301–326.
Young, M., Staudenbauer, W.L., and Minton, N.P. (1989b) in *Biotechnology Handbooks, vol. 3, Clostridia* (Minton, N.P., and Clarke, D.J., ed.), pp. 63–103, Plenum Press, New York.
Youngleson, J.S., Jones, W.A., Jones, D.T., and Woods, D.R. (1989a) *Gene* 78, 355–364.
Youngleson, J.S., Jones, D.T., and Woods, D.R. (1989b) *J. Bacteriol.* 171, 6800–6807.
Youngleson, J.S., Santangelo, J.D., Jones, D.T., and Woods, D.R. (1988) *Appl. Environ. Microbiol.* 54, 676–682.
Zappe, H., Jones, W.A., Jones, D.T., and Woods, D.R. (1988) *Appl. Environ. Microbiol.* 54, 1289–1292.
Zappe, H., Jones, D.T., and Woods, D.R. (1986) *J. Gen. Microbiol.* 132, 1367–1372.
Zappe, H., Jones, W.A., and Woods, D.R. (1990) *Nucleic Acids Res.* 18, 2179.
Zerner, B., Coutts, S.M., Lederer, F., Waters, H.H., and Westheimer, F.H. (1966) *Biochemistry*, 5, 813–816.

CHAPTER

9

Molecular Analysis and Expression of Nitrogen Metabolism and Electron Transport Genes of *Clostridium*

D.R. Woods
J. Santangelo

In batch culture, solvent-producing *Clostridium* spp. produce H_2, CO_2, acetate, and butyrate during the initial growth phase (acidogenic phase), which results in a decrease in the pH of the culture medium. As the culture enters the stationary growth phase and the cells assume the characteristic clostridial morphology, the metabolism of the cells undergoes a shift to solvent production (solventogenic phase). During the solventogenic phase, the re-assimilation of acids occurs concomitantly with the continued consumption of carbohydrate that results in an increase of the pH of the culture medium. Initiation of a third phase involving endospore formation and the production of a forespore septum terminates the production of solvents.

The regulation and factors involved in triggering the transitions from the acidogenic phase to the solventogenic phase and to the sporulation phase are complex and involve internal and external pH, acid end products, nutrient, and electron flow (Jones and Woods 1986). Since nitrogen levels and electron flow are important in understanding the control of these transitions, we have been studying the molecular biology of nitrogen metabolism and electron flow in *C. acetobutylicum*. These subjects are reviewed in this chapter.

9.1 NITROGEN LIMITATION

In batch cultures, the shift from the acidogenic phase to the solventogenic phase requires an excess of the carbon source and nitrogen, and only acids are produced when the concentrations of sugar (Monot and Engasser 1983; Long et al. 1984) and nitrogen (Long et al. 1984) are limited. Furthermore, sporulation does not occur under conditions of sugar or nitrogen starvation. This is in contrast to *Bacillus* spp. where nutrient depletion has been identified as the trigger that induces sporulation (Young and Mandelstam 1979). In *C. acetobutylicum*, the initiation of sporulation only occurred under conditions in which growth was limited in the presence of excess glucose and ammonia (Long et al. 1984).

Long et al. (1984) investigated the effect of nitrogen limitation on the production of solvents and showed that when the concentration of ammonia in the medium was decreased to a level that resulted in a decrease of biomass, the rate of glucose consumption decreased and residual glucose remained at the end of the fermentation. Under these nitrogen-limited conditions, solvents were not produced and the motile rod-shaped cells did not differentiate into clostridial stage cells. The failure to produce solvents under these conditions appeared to be due to the inability to produce threshold concentrations of acid end products required for the induction of solventogenesis. Low solvent yields were also obtained in ammonium-limited chemostat cultures (Gottschal and Morris 1981; Andersch et al. 1982), but Monot and Engasser (1983) reported that solvent production was possible in a nitrogen-limited chemostat maintained at a low pH and run at a low dilution rate (0.038/1). The ammonia concentrations used in these studies were significantly higher than those observed to affect cell growth and glucose consumption in batch culture. Furthermore, acetate was fed at concentrations approaching the threshold level required for acetate accumulation and induction.

In the biosynthesis of nitrogenous compounds, the amino acids, glutamine, and glutamate have a pivotal role. They provide cellular nitrogen for the synthesis of purines, pyrimidines, amino sugars, most of the amino acids and metabolites such as NAD and *p*-aminobenzoate. Three enzymes, glutamine synthetase (GS, EC 6.3.1.2), glutamate synthetase (GOGAT, EC 1.4.1.13) and glutamate dehydrogenase (GDH, EC 1.4.1.4) catalyze the three main reactions generally used by bacteria for ammonia assimilation.

To determine whether the levels of GS varied during the different growth phases of *C. acetobutylicum*, S. Rosenrauch, S.J. Reid, and D.R. Woods (unpublished observations) investigated the activity of GS in *C. acetobutyli-*

cum grown in batch culture in a complete and minimal medium. In both media, there were two peaks of GS activity that were observed immediately prior to the initiation of solventogenesis and sporulation. These peaks in GS activity coincided with the transition stages in the growth cycle and suggested that GS may play a role in the triggering of solventogenesis and sporulation. These results and the importance of GS and the *glnA* gene region in the regulation of nitrogen metabolism prompted an investigation of the GS of *C. acetobutylicum*.

9.2 GLUTAMINE SYNTHETASE AND THE *glnA* GENE

GS catalyzes the reaction:

$$\text{L-glutamate} + NH_4^+ + \text{ATP} \longrightarrow \text{L-glutamine} + \text{ADP} + P_i$$

The structure and regulation of the GS structural gene, *glnA*, has been extensively studied in the enteric bacteria (Magasanik and Neidhardt 1987; Reitzer and Magasanik 1987; Magasanik 1988). The *glnA* gene of these Gram-negative aerobic bacteria is part of a complex operon (*glnALG*) (Pahel et al. 1982) the transcription of which is regulated by the products (NR_I, NR_{II}) of the *glnG* (*ntrC*) and *glnL* (*ntrB*) genes, respectively (Ninfa and Magasanik 1986). The NR_I and NR_{II} proteins also regulate the expression of other operons involved in nitrogen metabolism (global nitrogen regulatory system) (Kustu et al. 1979). A sigma factor (σ^{54}) encoded by the *glnF* (*ntrA*) gene is specifically required for the transcription of nitrogen-regulated (*ntr*) promoters (Hirschman et al. 1985; Hunt and Magasanik 1985). These promoters do not contain canonical −35 and −10 sequences, but instead have the consensus (−26) CTGGYAYR-N_4-TTGCA(−10) (Beynon et al. 1983; Ausubel 1984; Gussin et al. 1986). In *Escherichia coli*, σ^{54} has been implicated in the regulation of two anaerobically inducible enzymes, formate dehydrogenase and hydrogenase isoenzyme 3, which are part of the formate hydrogenlyase pathway (Birkmann et al. 1987).

The *glnALG* operon contains three promoters, *glnAp1* and *glnAp2*, located upstream of the *glnA* gene (Reitzer and Magasanik 1985) and *glnLp* situated within the *glnAL* intergenic region (Ueno-Nishio et al. 1983, 1984). *glnAp1* and *glnLp* are $E\sigma^{70}$ promoters repressed by NR_I and serve to maintain basal levels of GS, whereas the σ^{54}-dependent promoter *glnAp2* is activated by phosphorylated NR_I when nitrogen becomes limiting.

There is no evidence for the presence of global-regulatory *ntr*-genes in the Gram-positive endospore-forming aerobe, *Bacillus subtilis*. It appears that in *B. subtilis*, GS regulates its own expression (Schreier et al. 1985; Strauch et al. 1988). The structural *glnA* gene is preceded by an open reading frame (ORF) encoding a 136 amino acid (aa) polypeptide that contains a DNA binding helix-turn-helix motif and shows similarities to diverse regulatory proteins (Cro, TrpR). Although the precise mechanism is unknown, it has been suggested that the *glnR* protein acts in concert with GS to regulate expression of the *glnA* gene in *B. subtilis* (Strauch et al. 1988; Schreier et al. 1989; Zhang et al. 1989). It appears that the *B. cereus glnRA* operon has a similar regulatory system (Nakano et al. 1989). Although *B. subtilis* does not have an analogous global *ntr* regulatory system present in enteric bacteria, there is evidence for the involvement of the *glnA* protein in the regulation of urease and asparaginase as well as two genes in Tn*917-lacZ* fusion strains (*nrg*-21, *nrg*-29) where the expression of β-galactosidase is affected by the GS protein (Atkinson and Fisher 1991). The regulation of GS, urease, asparaginase, *nrg*-21 and *nrg*-29, appears to involve some common mechanisms.

GS enzymes of eukaryotic origin are octamers (Prusiner and Stadtman 1973) whereas the GS enzymes of eubacteria and archaebacteria are dodecamers of M_r of approximately 600,000 composed of a single type of subunit, the M_r of which falls in the range of 44,000 to 59,000 (Streicher and Tyler 1980; Bhatnagar et al. 1986).

Until recently, it appeared that prokaryotes had two forms of GS, termed GSI and GSII. The majority of bacteria investigated have GSI enzymes, but members of the *Rhizobiaceae* contain both GSI and GSII enzymes. GSI is the typical prokaryotic GS whereas GSII is similar to eukaryotic GS enzymes (Carlson and Chelm 1986). Prokaryotic GSI subunits investigated vary in length from 444 to 474 aa (Janssen et al. 1988), whereas the GSII subunits of eukaryotes investigated vary in length from 355 to 373 aa (Gebhardt et al. 1986; Hayward et al. 1986; Tischer et al. 1986). The eukaryotic GSII subunits lack the C-terminal portion of the prokaryotic GSI subunit including the adenylylation site. The GSII subunit from *Bradyrhizobium japonicum* is 329 aa long and also lacks the terminal portion of the GSI subunit and the adenylylation site (Carlson and Chelm 1986). In a comparison of prokaryotic and eukaryotic GSI and GSII enzymes, Rawlings et al. (1987) showed that although the amino acid similarity between the enzymes was only approximately 15%, the major part of this similarity was located in five regions corresponding to β-sheets in the *Salmonella typhimurium* enzyme (Almassy et al. 1986). These regions are strongly conserved in all GS enzymes analyzed to date (Janssen et al. 1988). A feature of the five conserved amino acid regions is that they are all associated with the proposed GS activity site (Almassy et al. 1986). Regions II to V are β-strands closely associated with two Mn^{2+} cations of one subunit, while region I contains the Trp residue which is thought to complete the active site formed between adjacent subunits.

Although it seemed that all GS from prokaryotes would belong to either GSI or GSII enzymes, recently two novel and different GS enzymes have been characterized from *Bacteroides fragilis* and *Butyrivibrio fibrisolvens*.

The *glnA* gene from *B. fragilis* was cloned and sequenced, and the biochemistry of the GS enzyme investigated (Southern et al. 1986; Hill et al. 1989). The GS subunit is larger than any of the other bacterial GS subunits and consists of 729 aa with a calculated M_r of 82,827. The M_r of the *B. fragilis* GS holoenzyme is approximately 490,000 and, in contrast to the GS of other bacteria, it is a hexamer. The *B. fibrisolvens* GS subunit is similar to the *B. fragilis* GS subunit in that it is large (701 aa) and shows 40% aa identity and antigenic homology (J.K. Goodman and D.R. Woods, unpublished observations).

9.3 GLUTAMINE SYNTHETASE AND THE *glnA* GENE OF *CLOSTRIDIUM ACETOBUTYLICUM*

The *C. acetobutylicum glnA* gene was cloned and complemented the *glnA* lesion in an *E. coli glnALG* deletion mutant (Usdin et al. 1986). The *C. acetobutylicum* GS belonged to the GSI class of enzymes and was a dodecamer with a M_r of approximately 600,000. The subunit contained 444 aa and had a calculated M_r of 49,630. The five conserved amino acid regions identified in GSI enzymes were present in the GS from *C. acetobutylicum* (Almassy et al. 1986; Janson et al. 1986; Janssen et al. 1988).

Studies on the crystallized GS from *S. typhimurium* (Almassy et al. 1986) indicated that the subunit contained a central loop between amino acids 156 and 173. This region is cleaved by several proteases with different specificities (Dautry-Varsat et al. 1979; Monroe et al. 1985). It is interesting that the cloned *C. acetobutylicum* GS appears to lack 26 aa residues involved in the formation of the central loop in *S. typhimurium* (Janson et al. 1986). The absence of a central loop may be a feature of GS from Gram-positive bacteria, since the *B. subtilis* GS also appears to lack a central loop (Strauch et al. 1988). The cloned *C. acetobutylicum* GS is not subject to adenylylation (Usdin et al. 1986), and not only lacks the tyrosine residue associated with GS adenylylation in *E. coli*, but also shows little similarity with the 18 aa adjacent to the tyrosine residue in the *E. coli* GS.

The *glnA* gene from *C. acetobutylicum* has a G-C content of 28% and the codon usage was strongly biased toward the use of codons in which A and U predominate (Janssen et al. 1988). The upstream regulatory region of the *glnA* gene contained two functional extended Gram-positive promoter consensus sequences (p_1 and p_2). The upstream regulatory region was also characterized by a complex palindromic sequence immediately upstream of p_1 between nucleotides 130 to 167. A 380 base pair (bp) region between nucleotides 160 to 540 had a very high DNA curvature score of 8.2 (calculated according to Plaskon and Wartell 1987) and contained five copies of a direct repeat, with the consensus sequence 5′-ATATTGTAA-3′. Similar 9-bp sequences (consensus 5′-ATATTGTTT 3′) occur nine times in the *B. subtilis glnA/gltC* regulatory region (Bohannon and Sonenshein 1989), five times in the *B. cereus glnRA* intergenic region (Nakano et al. 1989) and in the target DNAs of several LysR family proteins in *E. coli* (Kölling and Lother 1985; Henikoff et al. 1988). The significance of the complex palindromic sequence and the region with the high DNA curvature score is unknown at present.

Primer extension experiments with mRNA extracted from *C. acetobutylicum* and *E. coli* indicated that in *C. acetobutylicum* there were two start points adjacent to p_1 and p_2, respectively, but in *E. coli* there was an additional start point (Janssen et al. 1990). In *E. coli*, initiation of transcription was regulated by nitrogen and a downstream region was implicated in the regulation of transcript initiation by nitrogen.

Interesting features of the *C. acetobutylicum glnA* gene were the presence of a third extended Gram-positive promoter consensus sequence (p_3) and an extensive stretch of inverted repeat sequences, both of which were located immediately downstream of the structural *glnA* gene. Promoter p_3 was oriented towards the *glnA* gene and controlled the transcription of a putative antisense RNA, complementary to 43 bases of the 5′ end of the *glnA* mRNA.

Promoter p_3 is a functional promoter and, fused upstream of a *lac* promoter probe vector, produced β-galactosidase in *E. coli* cells (Janssen et al. 1988). Primer extension experiments with mRNA extracted from *C. acetobutylicum* and *E. coli* indicated the presence of a single transcript start point at adjacent nucleotides 4 and 5 bp from p_3 (Janssen et al. 1990). The role of antisense RNA in gene regulation is to act as an inhibitor of gene expression as a consequence of either transcriptional interference or specific RNA-RNA intermediates (Green et al. 1986). Increased production of antisense RNA would be expected to result in a decrease in the production of a specific gene product. To provide evidence for the involvement of antisense RNA in the regulation of GS, Janssen et al. (1990) constructed an up-promoter mutant in p_3 and showed that in a promoterless fusion vector, the mutant p_3 increased the expression of a tetracycline resistance gene. This mutant p_3 gene caused a decrease in GS activity in *E. coli* cells containing the *C. acetobutylicum glnA* mutant plasmid. This result was consistent with the prediction that p_3 controlled the production of antisense RNA. Recently, we subcloned the p_3 and antisense region onto a high copy number plasmid and determined the effect of the overproduction of antisense RNA from this high copy number plasmid in the same *E. coli* cells containing the *C. acetobutylicum glnA* region on a compatible low copy number plasmid. As predicted, the production of GS activity was repressed in these cells (I.P. Fierro-Monti, S.J. Reid, and D.R. Woods, unpublished observations). In a complementary experiment, deletion of p_3 from the *glnA* region resulted in an increase in GS production in *E. coli* cells. This result is consistent with the prediction that p_3 controls the production of antisense RNA. Although it appears that the antisense RNA has a role in down-regulating GS expression, it was not involved in regulation by nitrogen in *E. coli* (Janssen et al. 1990).

The importance of the downstream 158-bp inverted repeat (IR) sequences (located between p_3 and the end of the *glnA* gene) in the regulation of the *glnA* gene was determined by the construction of the deletion plasmids that lacked the IR region, but retained p_3 (Janssen et al. 1990). *E. coli* YMC11 *glnALG* cells containing these plasmids grown in nitrogen-excess and nitrogen-limiting media produced very low levels of GS activity. The removal of the IR sequences will affect the secondary structure(s) of the downstream mRNA and these changes could play a role in termination of transcription (Morgan et al. 1985) or in the stability of the mRNA (Belasco et al. 1985; Wong and Chang 1986). In these deletion plasmids, promoter p_3 is closer to the end of the *glnA* gene, and termination of transcription from p_3 may also be affected by the deletion of the IR sequences. This could result in transcription from p_3 into the *glnA* gene, which would affect normal transcription and production of *glnA* mRNA. Although the levels of GS activity in *E. coli* cells containing the deleted plasmids were low, they were still regulated by nitrogen.

The sophisticated genetic studies involving deleted and mutant plasmids of the *glnA* gene from *C. acetobutylicum* have been carried out in *E. coli glnA* deletion strains. Although these studies have been useful in the molecular characterization of the *C. acetobutylicum glnA* gene, *E. coli* is a heterologous host and it is important that similar studies are done in *C. acetobutylicum*. There is no doubt that the promoters p_1, p_2, and p_3 are expressed in *C. acetobutylicum*, but their role in the regulation of GS activity under different growth conditions is unknown. We have initiated experiments to study the regulation of WT and mutant *glnA* gene regions in *C. acetobutylicum*. An important result has been the detection of the predicted antisense RNA in *C. acetobutylicum* P_{262} cells (I.P. Fierro-Monti, S.J. Reid, and D.R. Woods, unpublished observations). A RNA protection assay and specific probe were used to detect the antisense RNA. Furthermore, we showed that the amount of antisense in the *C. acetobutylicum* cells varied depending on the concentration of nitrogen in the medium (I.P. Fierro-Monti, S.J. Reid, and D.R. Woods, unpublished observations), suggesting the involvement of the antisense RNA in regulation by nitrogen in *Clostridium*. The regulation of the *C. acetobutylicum glnA* gene is complex and involves mechanisms operating at the level of transcription and translation. The regulatory systems differ from those of other Gram-negative and Gram-positive bacteria studied to date. The analysis of the regulation of the *C. acetobutylicum glnA* gene will contribute to the overall understanding of the regulation of nitrogen metabolism in bacteria.

9.4 PROTEASES

Proteolytic activity is important in nitrogen metabolism for the release of peptides and amino acids. In the Gram-positive sporulating aerobe, *B. subtilis* serine metallo-proteases have been identified that are produced at a specific stage in the sporulation process (Sonenshein 1989). The appearance of these proteases has been used to identify the sporulation phase I. The production of proteases during sporulation of *C. acetobutylicum* P_{262} has not been observed (S. Long, D.T. Jones, and D.R. Woods, unpublished observations). The only report of proteolytic activity in *C. acetobutylicum* is in strain

ATCC824. An extracellular protease in culture supernatants of *C. acetobutylicum* ATCC824 was reported by Uchino et al. (1968). Croux et al. (1990) described the purification and characterization of this protease and showed that it was a heat-labile acidolysin (optimum pH 5.0), which was very sensitive to metal-chelating agents and phosphoramidon but was unaffected by sulfhydryl reagents. This protease appears to be the first phosphoramidon-sensitive, acidic zinc metalloprotease reported.

9.5 ENERGY AND ELECTRON DISTRIBUTION IN *CLOSTRIDIUM ACETOBUTYLICUM*

The growth of anaerobes is limited by the rate of energy-producing reactions (Thauer et al. 1977). In *C. acetobutylicum*, the glycolytic breakdown of 1 mol glucose to 2 mol pyruvate results in the net production of 2 mol ATP and 2 mol of NADH. During acidogenesis, 1 mol of ATP is produced per mole of acyl phosphate converted to acid end product. Therefore, the net energy yield obtained when 1 mol of glucose is converted to 2 mol of acetate is 4 mol of ATP. Since 2 mol of acetyl-coenzyme A (CoA) are required for the production of 1 mol of butyryl-CoA, the net energy yield obtained when 1 mol of glucose is converted to 1 mol of butyrate is 3 mol of ATP. In a normal batch fermentation, the acetate to butyrate ratio is approximately 2:3 (mol/mol) (Jones and Woods 1986). Accounting for the acetate to butyrate ratio, Rogers (1986) calculated the ATP yield for the acidogenic growth phase to be about 3.25 mol ATP per mol of glucose consumed, with a thermodynamic efficiency of approximately 62%. Apart from ATP produced during glycolysis, no further ATP production occurs during solventogenesis via the synthesis of acetone, butanol, and ethanol. Therefore, the ATP balance for the solventogenic growth phase drops to 2 mol ATP per mol of glucose consumed.

The energetically favorable production of acetate results in a net generation of 2 mol NADH. During both acidogenesis and solventogenesis 2 mol of reducing power, in the form of NADH, are consumed in the conversion of 2 mol of acetyl-CoA to 1 mol of butyryl-CoA. Therefore, even though butyrate production yields less ATP than does acetate production, it is redox neutral. Since the breakdown of glucose to acetyl-CoA yields 2 mol of NADH and considering that only a portion of the resulting acetyl-CoA is converted to butyrate, excess NADH must either accumulate, or be cycled through ferredoxin (Fd) via the NADH-Fd-oxidoreductase. The latter would result in reduced Fd and hence require increased hydrogen production for electron disposal and recycling of Fd to its oxidized state. The solvent-producing pathways provide additional routes for the disposal of NADH, where both aldehyde dehydrogenases, as well as the alcohol dehydrogenases, each consume 1 mol of NADH per mol of substrate converted to product. This implies that any accumulation of NADH and hence the activities of the NADH-Fd-oxidoreductase and hydrogenase enzymes, all play a role in the switching of the carbon flow from acid production to solvent production.

9.6 REGULATION OF ELECTRON FLOW

In *C. acetobutylicum*, the shift from acidogenesis to solventogenesis is accompanied by a decrease in hydrogen production and an increase in CO_2 production. This switch in carbon flow appears to be directly linked to the reduction of hydrogen production and Kim and Zeikus (1985) observed that the specific rate of hydrogen production decreased in stages during the course of a batch fermentation. The specific hydrogenase activity in whole cells from acid-producing cultures was shown to be approximately 2.2 times higher than that measured from solvent-producing cultures. These authors demonstrated that hydrogenase activity was not affected by either pH or acid concentration, and concluded that a decrease in hydrogen production during solventogenesis is a result of regulation of hydrogenase production as opposed to an inhibition of enzyme activity. In contrast to these findings, Andersch et al. (1983) reported that hydrogenase activity was similar in both acid- and solvent-producing cells, although hydrogenase activity in solvent-producing cells could only be detected in the assay after an initial lag period of 10 to 15 min. In this case, it was concluded that the hydrogenase from the solvent-producing cells was present in an inactive form and the conditions used for the assay activated the enzyme after a lag period.

Hydrogenase activity, and therefore solvent production, could be altered by changing the partial pressure of hydrogen in the culture vessel. When conditions are reached where the partial pressure of hydrogen is increased, there is a resulting decrease in hydrogen production and a stimulation of butanol and ethanol production, but not acetone production (Yerushalmi et al. 1985). Under these conditions, the H^+/H_2 redox potential is lowered and the flow of electrons from reduced Fd to molecular hydrogen is thermodynamically unfavorable (Jungermann et al. 1971). Because the hydrogenase was inhibited, the flow of electrons from reduced Fd was shifted to the generation of NADH and NADPH via the actions of the respective Fd-oxidoreductase. Stimulation of ethanol and butanol production was, therefore, necessary for the disposal of electrons carried by these reduced pyridine nucleotides. Analogous results were obtained with *C. butyricum* where an increase in the partial pressure of hydrogen resulted in an increase in the amount of butyrate produced with respect to acetate (Crabbendam et al. 1985).

The fate of reduced Fd determines the electron distribution in the cell in that it can either transfer electrons via hydrogenases to produce hydrogen, or electrons can be transferred via the appropriate Fd-oxidoreductase to yield reduced pyridine nucleotides. Therefore, the activities of NADH-Fd-oxidoreductase, NADPH-Fd oxidoreductase, and hydrogenase in effect control the direction of carbon flow by controlling the electron flow within the cell (Jungermann et al. 1973; Jones and Woods 1986). Furthermore, CoA and acetyl-CoA are known to act as allosteric inhibitors and activators of NADH-Fd-oxidoreductase, respectively (Jungermann et al. 1973). Since a high concentration of NADH inhibits the production of NADH via the NADH-Fd-oxidoreductase, and considering the role of CoA and acetyl-CoA in controlling this enzyme, the ratios of acetyl-CoA:CoA and NAD^+:NADH must play a principal role in the regulation of electron flow, and hence carbon flow, in *C. acetobutylicum* (Datta and Zeikus 1985; Jones and Woods 1986). In addition, it has been postulated that the ratios of these compounds can function as sensors for both ATP regeneration and hydrogen production (Datta and Zeikus 1985).

9.7 METRONIDAZOLE AND CLOSTRIDIAL ELECTRON TRANSPORT SYSTEMS

Metronidazole, 1-(β-hydroxyethyl)-2-methyl-5-nitroimidazole, is a nitroimidazole drug that must be reduced to elicit a bacteriocidal effect (Kucers and Bennett 1987). Investigations have shown that metronidazole is active against most obligately anaerobic bacteria (Tally et al. 1981; Noble and Tally 1984). It has been repeatedly suggested that reduction of the nitro group of metronidazole is necessary for drug activation. The actual mechanisms of metronidazole activation have not been directly elucidated, and possible theories include direct electron transfer via reduced electron carriers, as well as active reduction via enzymatic routes.

O'Brien and Morris (1972) showed that when sublethal concentrations of metronidazole were added to exponentially growing *C. acetobutylicum* cultures, the drug rapidly disappeared from the culture medium. When metronidazole was added to crude extracts of *C. acetobutylicum*, hydrogen production was halted until no metronidazole remained in the assay flask. The production of CO_2 was not altered, indicating that the phosphoroclastic breakdown of pyruvate to acetyl-CoA continued. The effect of metronidazole on the uptake of hydrogen by the purified *C. pasteurianum* hydrogenase was also examined. It was shown that metronidazole caused no inhibition of the hydrogenase, and, in fact, extended the phase of rapid consumption of hydrogen. For each mmol of metronidazole supplied, approximately 3 mmol of additional hydrogen was used. These authors concluded that, in *C. acetobutylicum*, metronidazole did not directly inhibit hydrogen production, but inhibited the action of the hydrogenase by serving as an alternative and preferred acceptor of electrons from reduced Fd.

Nonenzymatic transfer of electrons to metronidazole was implied to occur in a hydrogenase-linked assay system developed by Chen and Blanchard (1979). In the reverse reaction, a purified hydrogenase consumes hydrogen gas forming reduced Fd (or flavodoxin), which in turn, transfers electrons to metronidazole forming the reduced toxic intermediate(s). Therefore, the concentration of the electron carrier (Fd or flavodoxin) determines the amount of metronidazole that is reduced. After stopping the reaction, the amount of unreduced metronidazole was measured spectrophotometrically. Critical to the design and working of this assay are the facts that metronidazole reduction is irreversible and that once reduced, metronidazole loses its absorption peak at 320 nm.

Lockerby et al. (1984) demonstrated the requirement of Fd for the activation of metronidazole. Cell-free extracts of *C. pasteurianum* were capable of reducing metronidazole as measured by the uptake of hydrogen gas. Hydrogen uptake only occurred in the presence of metronidazole. However, removal of Fd from the cell-free extracts by diethylaminoethyl (DEAE) cellulose treatment eliminated hydrogen uptake in the presence of metronidazole. These authors concluded that metronidazole preferentially scavenges electrons from reduced Fd, thereby depriving other Fd-linked systems of essential reducing equivalents. The existence of a "metronidazole reductase" system that enzymatically transfers electrons from reduced Fd to metronidazole was suggested, however no evidence or speculation was provided concerning the components of this system.

The role of enzymes in metronidazole reduction has been postulated for several years. Unfortunately, many reports simply cite the responsible enzyme or enzymes as being "nitroreductases" (Edwards et al. 1973; Rosenkranz and Speck 1975; Wardman and Clarke 1976; Tally et al. 1978; Chrystal et al. 1980; McLafferty et al. 1982). This term most likely arose from the fact that the nitro group of metronidazole needs to be reduced in order for activation, and may not refer to specific nitroreductase systems found in many microorganisms.

Narikawa (1986) implicated pyruvate-Fd-oxidoreductase activity with metronidazole susceptibility. Metronidazole uptake, *p*-nitro benzoic acid reduction, pyruvate-Fd-oxidoreductase activity and metronidazole sensitivity was determined for 41 different microorganism species covering 19 different genera. All of the obligate anaerobes tested and four of the facultative anaerobes were capable of reducing *p*-nitro benzoic acid. The ability to reduce *p*-nitro benzoic acid correlated directly with an ability to decrease the concentration of metronidazole in the culture medium. However, only the obligate anaerobes were identified as being susceptible to metronidazole, and this susceptibility correlated with pyruvate-Fd-oxidoreductase activity.

Lockerby et al. (1985) studied the effects of metronidazole reduction on the phosphoroclastic breakdown of pyruvate. It was shown that the addition of metronidazole to cell-free extracts of *C. pasteurianum* caused an increase in acetyl-phosphate production. It was also shown that Fd removal from the cell-free extracts resulted in the complete loss of acetyl phosphate production. It was concluded that the role of the phosphoroclastic reaction with respect to metronidazole activation was to provide electrons through reduced Fd that would normally be evolved as hydrogen gas. Stimulation of phosphoroclastic activity was presumed to be due to the fact that the reduction of metronidazole recycles Fd to its oxidized state more rapidly than the hydrogenase, indicating that hydrogenase activity is the rate-limiting step in the phosphoroclastic breakdown of pyruvate. Again, the presence of a "metronidazole reductase" system was assumed to carry out the transfer of electrons from reduced Fd to metronidazole, however, no evidence was provided as the components of this system.

In an elegant study, Church et al. (1988) demonstrate that the bidirectional hydrogenase-1 of *C. pasteurianum* is the metronidazole reductase that these authors (Lockerby et al. 1985) mentioned previously. Strong stoichiometric evidence was presented indicating that metronidazole replaces protons as the substrate for the hydrogenase enzyme. Together with reduced Fd, the hydrogenase transfers electrons to metronidazole yielding oxidized Fd and reduced metronidazole. In this case, no hydrogen gas would be produced, which is in agreement with previous reports that showed hydrogen production was inhibited in the presence of metronidazole (Edwards and Mathison 1970; O'Brien and Morris 1972). A similar role for this hydrogenase enzyme has been shown in reducing 2-, 4-, and other 5-nitroimidazole drugs (Church et al. 1990).

9.8 USE OF METRONIDAZOLE TO ISOLATE ELECTRON TRANSPORT GENES

Although the spectrum of activity of metronidazole transcends many taxonomic boundaries, initial reports suggested that its toxicity was limited to strict anaerobes and microaerophiles, and that it was inactive against aerobes and facultative anaerobes except in concentrations that far exceeded those therapeutically attainable (Prince et al. 1969; Ralph and Clarke 1978; Onderdonk et al. 1979). Prince et al. (1969) tested metronidazole on a group of facultative anaerobes, and concluded that it had little clinical use for treatment of infections caused by these organisms. These authors did report that *E. coli* was slightly more sensitive than the other organisms tested, but the antibiotic concentration necessary for bacteriocidal action was still well above the obtainable serum concentrations. Since one of our objectives was to clone *C. acetobutylicum* electron transport genes using *E. coli* as the host, it was important to understand the interactions between this organism and metronidazole.

It has been established that the bacteriocidal toxic derivatives of metronidazole cause DNA damage that is responsible for the cytotoxic effects of this drug (Knight et al. 1978). Chrystal et al. (1980) proposed that reduced metronidazole can interact with DNA to exert the bacteriocidal effect, or with water to form acetamide. Measuring metronidazole sensitivities of various *E. coli* strains that carried one or more genetic lesions in their DNA repair systems, Jackson et al. (1984) confirmed that DNA was the target molecule of this drug. The wild-type strains were the least susceptible to metronidazole, whereas lesions in DNA repair genes conferred greater sensitivity to metronidazole. In all *E.*

coli strains tested, metronidazole disappeared from the culture medium indicating that *E. coli* has the capacity to activate metronidazole. In a similar study, Yeung et al. (1984) also tested the bacteriocidal activity of metronidazole against DNA repair mutant strains of *E. coli*. These DNA repair mutants were more sensitive to metronidazole than their respective parental strains. However, if these DNA repair mutants also lacked the ability to reduce nitrate and chlorate, they were no more susceptible to metronidazole than their wild-type parents. It was shown that metronidazole disappeared from the culture medium more slowly in these nitroreductase-deficient strains. Since acetamide concentration is proportional to the concentration of the active derivatives of metronidazole, these authors expressed metronidazole sensitivity in terms of acetamide concentration. This evaluation showed that all of the strains with lesions in DNA repair genes were more susceptible to the reduced active derivatives of metronidazole than the wild-type strains, whether or not they had diminished nitroreductase activity. This indicates that in *E. coli*, nitroreductase activity is obligatory for the activation of metronidazole, but not for the DNA damaging activity of the reduced derivatives.

Chrystal et al. (1980) presented a model relating metronidazole metabolism to bacteriocidal action. A similar model is shown in Figure 9–1. Met-

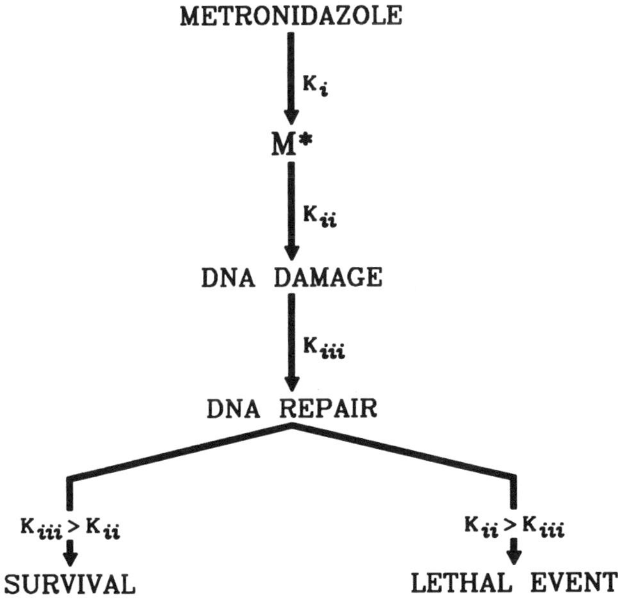

FIGURE 9–1 A model relating metabolism of metronidazole to cell death. Metronidazole is reduced to its active derivative M* by the cell. M* reacts with DNA causing damage, which can be repaired by the host cell's DNA repair systems. K_i is the rate of reduction of metronidazole, K_{ii} is the rate of DNA damage, and K_{iii} is the rate of DNA repair. If the rate of DNA damage (K_{ii}) is greater than the rate of DNA repair (K_{iii}), cell death occurs.

ronidazole is reduced to its active form M*, which can react with DNA resulting in DNA damage. If the rate of DNA damage (K_{ii}) is greater than the rate of repair (K_{iii}), cell death occurs. Therefore, a cell with an efficient DNA repair mechanism is able to tolerate higher levels of M* than a cell with a less-efficient DNA repair system. Since the rate of formation of M* (K_i) is proportional to the rate of DNA damage, two mechanisms of resistance to metronidazole exist. One mechanism involves an efficient DNA repair system and the other involves the lack of the ability to reduce metronidazole to its active intermediate M*.

A nitroreductase deficient, *recA*$^-$ *E. coli* mutant (F19) was isolated after transposon mutagenesis, and was shown to be less sensitive to metronidazole than its *recA*$^-$ parent, *E. coli* CC118 (Santangelo et al. 1991). This mutant has the desired traits of a diminished ability to reduce metronidazole, presumably due to the nitroreductase deficiency, while remaining highly sensitive to the reduced active derivatives of metronidazole, due to the *recA* mutation. *E. coli* F19 was, therefore, chosen as the cloning host to screen for recombinant plasmids containing *C. acetobutylicum* DNA coding for electron transport genes that activate metronidazole.

9.9 CHARACTERIZATION OF *CLOSTRIDIUM ACETOBUTYLICUM* ELECTRON TRANSPORT GENES

The *E. coli* F19 *recA*, nitroreductase-deficient mutant proved to be a most suitable host for the isolation of recombinant plasmids. Screening experiments resulted in the isolation of 25 different recombinant plasmids that contain *C. acetobutylicum* DNA coding for genes involved in the activation of metronidazole (Santangelo et al. 1991). These 25 isolated metronidazole sensitive clones were classified into five groups (I–V) based on their susceptibility to metronidazole, where class V was the most sensitive group and class I was the least sensitive group. All clones were more sensitive to metronidazole than the *E. coli* F19 mutant.

Since it has been reported that clostridial enzymes, pyruvate-Fd-oxidoreductase (Narikawa 1986) and hydrogenase (Church et al. 1988, 1990) may be responsible for the reduction of metronidazole, cell extracts of all 25 clones were tested for these enzymes. Since the enzyme activity responsible for metronidazole reduction in *E. coli* involves a nitroreductase (Yeung et al. 1984), the clones were also tested for nitroreductase activity. No significant nitroreductase activity or pyruvate-Fd-oxidoreductase activity was detected in any of the clones.

The theory that Narikawa (1986) proposed, that the enzyme pyruvate-Fd-oxidoreductase is responsible for metronidazole reduction to its active intermediates, assumes that metronidazole replaces oxidized Fd in the phosphoroclastic breakdown of pyruvate to acetyl-CoA (Figure 9–2). It is, therefore, surprising that none of the clones isolated in this study demonstrated nitro-

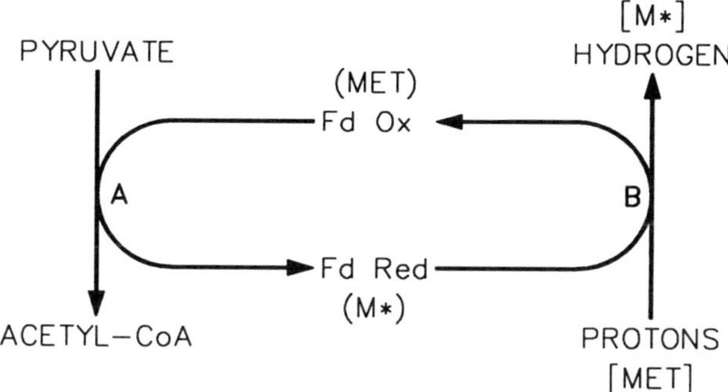

FIGURE 9–2 A simplified schematic of the *C. acetobutylicum* electron transport system. The relationship between oxidized ferredoxin (Fd_{ox}), reduced ferredoxin (Fd_{red}), pyruvate-Fd-oxidoreductase (A) and hydrogenase (B) is shown. In the absence of metronidazole, electrons are transferred to Fd_{ox} by pyruvate-Fd-oxidoreductase. Hydrogenase transfers electrons from Fd_{red} to protons to form hydrogen gas. In the presence of metronidazole, electrons may be transferred to metronidazole (MET) by pyruvate-Fd-oxidoreductase to form the reduced toxic derivative of metronidazole (M*), which causes DNA damage (Narikawa 1986). Alternatively, MET may replace protons and be reduced to [M*] by the action of the hydrogenase enzyme (Church et al. 1988, 1990).

reductase or pyruvate-Fd-oxidoreductase activity, but this does not rule out the possibility that *C. acetobutylicum* nitroreductase or pyruvate-Fd-oxidoreductase can reduce metronidazole to its active form. The theory of Church et al. (1988, 1990), that the hydrogenase enzyme is responsible for the activation of metronidazole, is based on the assumption that metronidazole replaces protons as the final electron acceptor (Figure 9–2). Only one clone, pMET15B2, consistently tested positive for hydrogenase activity, while all other clones showed hydrogenase levels comparable to the negative control. Since only one of the 25 different recombinant plasmids encoded an enzyme that was likely to be involved in metronidazole reduction suggests that either these enzyme activities are unstable and difficult to assay in *E. coli* or there are some interesting genes encoding proteins not before considered to be involved in metronidazole reduction.

Because of the lack of identifiable physiological characteristics, molecular biological techniques were employed to further classify these clones. DNA hybridization studies revealed that four clones, pMET13A, pMET15B2, pMET030, and pMET190, were linked on an 11.1-kilobase (kb) DNA fragment from the *C. acetobutylicum* chromosome that contained at least two genes involved with electron transport systems capable of activating metronidazole (Santangelo et al. 1991). With the exception of the hydrogenase positive clone, pMET15B2, all of these isolates demonstrated a class V sensitivity to metronidazole. This indicated that clones pMET13A, pMET030,

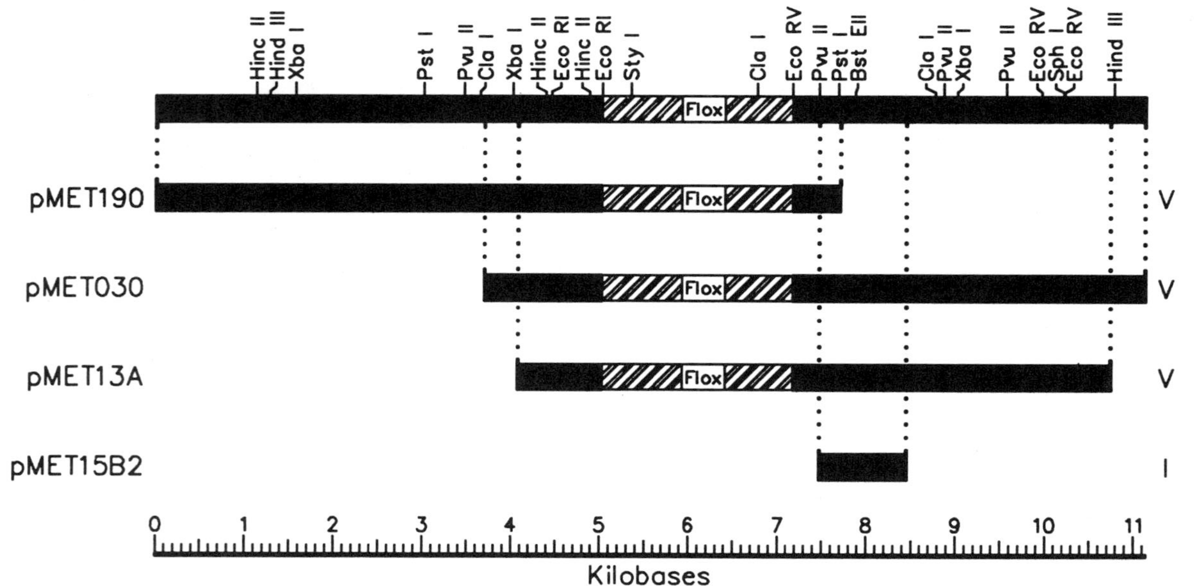

FIGURE 9–3 Restriction endonuclease map showing the relationship between clones pMET13A, pMET15B2, pMET030, and pMET190. The arbitrary metronidazole sensitivity class is shown on the right of each clone map. The striped region represents the area necessary to exhibit the class V metronidazole sensitivity phenotype in *E. coli* F19. Reproduced with permission from Santangelo et al. (1991).

and pMET190 all contained the same active region from the *C. acetobutylicum* chromosome, and this hypothesis was verified using restriction endonuclease mapping and deletion techniques. A restriction endonuclease map showing the relationship between these four clones is presented in Figure 9–3. In conjunction with the DNA hybridization studies, restriction endonuclease deletion studies verified that the region responsible for class V sensitivity to metronidazole was the same for clones pMET13A, pMET030, and pMET190. This region was localized to a 2-kb *Eco*RI-*Eco*RV fragment (see Figure 9–3). The insert DNA of pMET190 and pMET13A were in opposite orientations with respect to vector DNA indicating that in *E. coli*, the gene responsible for metronidazole activation is most likely transcribed from a promoter present on the insert.

9.10 MOLECULAR BIOLOGICAL ANALYSIS OF THE *CLOSTRIDIUM ACETOBUTYLICUM* FLAVODOXIN GENE

Since no enzyme activity could be detected in any of the class V metronidazole-sensitive isolates, and considering that the same active region was isolated as three separate cloning events (pMET13A, pMET030, and pMET190), obtaining the DNA sequence of the identified *Eco*RI-*Eco*RV active fragment was deemed necessary for identification of the gene or genes responsible for metronidazole activation.

The nucleotide sequence of the *Eco*RI-*Eco*RV active region of pMET13A was determined to be 1902 bp in length, and contained two partial ORFs and one complete ORF. Deletion of any of the coding region of the complete ORF resulted in the loss of the ability to activate metronidazole. At least 46 bp of the intergenic region upstream of this complete ORF were also required for metronidazole activation, indicating the presence of a promoter that was functioning in *E. coli*.

The deduced amino acid sequence from this complete ORF was used to search DNA and protein databases. The best similarity to this deduced amino acid sequence was obtained with the flavodoxin sequences from *Azotobacter vinelandii* (55%) (Bennett et al. 1988), *Clostridium* MP (51%) (Tanaka et al. 1974), *Anabaena variabilis* (55%) (Leonhardt and Straus 1989), and *Klebsiella pneumoniae* (56%) (Drummond 1985). Based on the alignments to these and other flavodoxins, the fact that this ORF was identified as the gene responsible for class V sensitivity to metronidazole, and the fact that Chen and Blanchard (1979) showed that reduced flavodoxin can transfer electrons to metronidazole, the sequence corresponding to this ORF was identified as coding for a flavodoxin from *C. acetobutylicum* P_{262} (Santangelo et al. 1991).

Wakabayashi et al. (1989) aligned the amino acid sequences of seven flavodoxins, and this alignment was used as the basis for the comparison shown in Figure 9–4, which includes the *C. acetobutylicum* flavodoxin. The

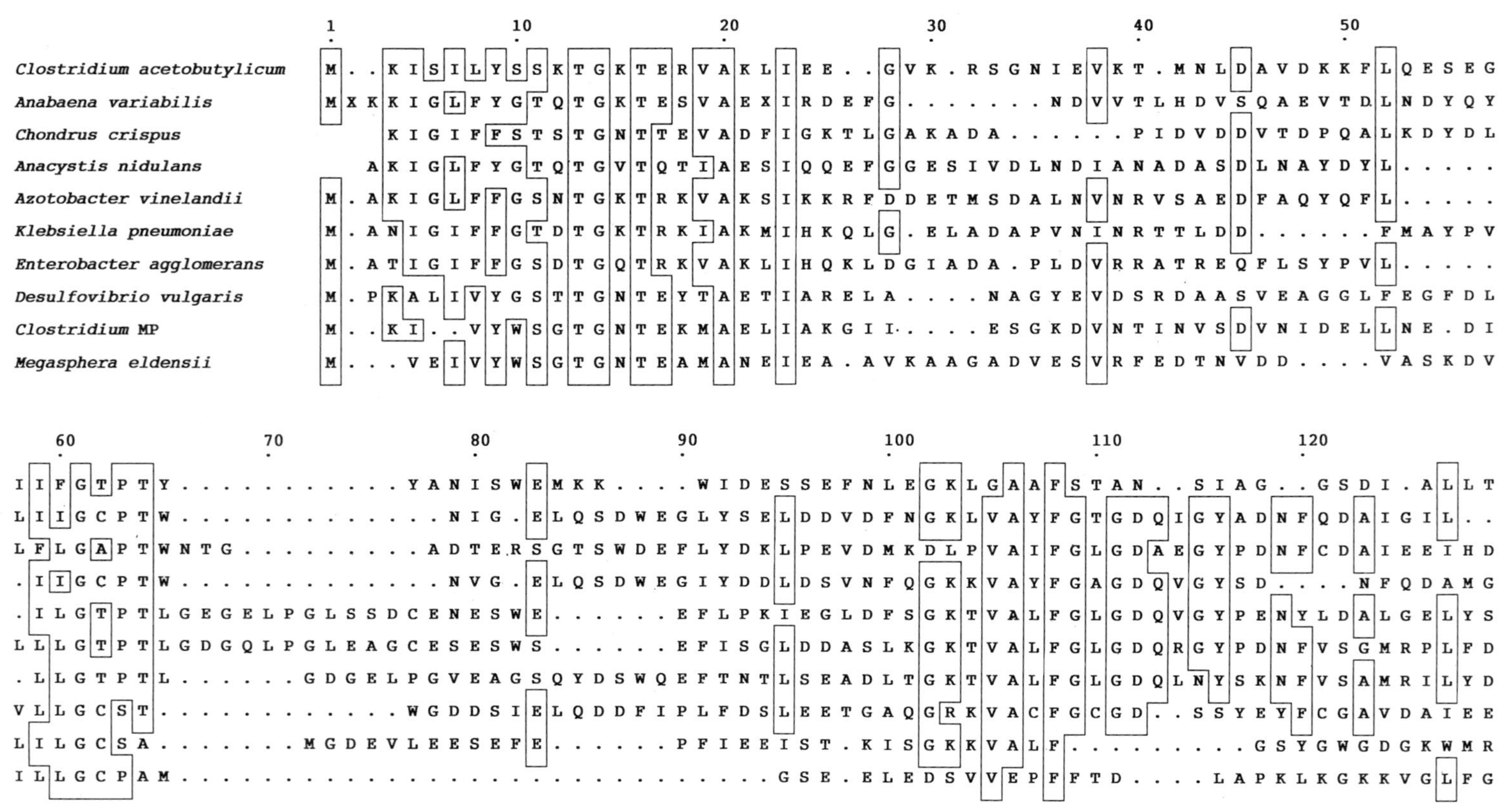
1 10 20 30 40 50
Clostridium acetobutylicum M..KISILYSSKTGKTERVAKLIEE..GVK.RSGNIEVKT.MNLDAVDKKFLQESEG
Anabaena variabilis MXKKIGLFYGTQTGKTESVAEXIRDEFG.......NDVVTLHDVSQAEVTDLNDYQY
Chondrus crispus KIGIFFSTSTGNTTEVADFIGKTLGAKADA......PIDVDDVTDPQALKDYDL
Anacystis nidulans AKIGLFYGTQTGVTQTIAESIQQEFGGESIVDLNDIANADASDLNAYDYL.....
Azotobacter vinelandii M.AKIGLFFGSNTGKTRKVAKSIKKRFDDETMSDALNVNRVSAEDFAQYQFL.....
Klebsiella pneumoniae M.ANIGIFFGTDTGKTRKIAKMIHKQLG.ELADAPVNINRTTLDD......FMAYPV
Enterobacter agglomerans M.ATIGIFFGSDTGQTRKVAKLIHQKLDGIADA.PLDVRRATREQFLSYPVL.....
Desulfovibrio vulgaris M.PKALIVYGSTTGNTEYTAETIARELA....NAGYEVDSRDAASVEAGGLFEGFDL
Clostridium MP M..KI..VYWSGTGNTEKMAELIAKGII....ESGKDVNTINVSDVNIDELLNE.DI
Megasphera eldensii M...VEIVYWSGTGNTEAMANEIEA.AVKAAGADVESVRFEDTNVDD....VASKDV
60 70 80 90 100 110 120
IIFGTPTY............YANISWEMKK....WIDESSEFNLEGKLGAAFSTAN..SIAG..GSDI.ALLT
LIIGCPTW..............NIG.ELQSDWEGLYSELDDVDFNGKLVAYFGTGDQIGYADNFQDAIGIL..
LFLGAPTWNTG..........ADTERSGTSWDEFLYDKLPEVDMKDLPVAIFGLGDAEGYPDNFCDAIEEIHD
.IIGCPTW..............NVG.ELQSDWEGIYDDLDSVNFQGKKVAYFGAGDQVGYSD....NFQDAMG
.ILGTPTLGEGELPGLSSDCENESWE......EFLPKIEGLDFSGKTVALFGLGDQVGYPENYLDALGELYS
LLLGTPTLGDGQLPGLEAGCESESWS......EFISGLDDASLKGKTVALFGLGDQRGYPDNFVSGMRPLFD
.LLGTPTL......GDGELPGVEAGSQYDSWQEFTNTLSEADLTGKTVALFGLGDQLNYSKNFVSAMRILYD
VLLGCST..............WGDDSIELQDDFIPLFDSLEETGAQGRKVACFGCGD..SSYEYFCGAVDAIEE
LILGCSA.......MGDEVLEESEFE......PFIEEIST.KISGKKVALF..........GSYGWGDGKWMR
ILLGCPAM.....................GSE.ELEDSVVEPFFTD....LAPKLKGKKVGLFG

```
130                 140                 150                 160                 170                 180                 190                 200
 .                   .                   .                   .                   .                   .                   .                   .
I L N H L M . . . . V K G M L V Y S . . . . G G V A F G K P K T H L G Y V H I N E I Q E N E D E N A R I F G E R I A N K V K Q I F . . . . . . .
. . . . . E E E I S Q R G G K T V G Y W S T D G Y D F N D S K A . L R N G K F V G L A L D E D N Q S D L T D D R I K S W V A Q L K S E F G L . .
C F A K . . . . . . . Q G A K P V G F S N P D D Y D Y E E S K S . V R D G K F L G L P L D M V N D Q I P M E K R V A G W V E A V V S E T G V . .
I L . . . E E K I S S L G S Q T V G Y W P I E G Y D F N E S K A . V R N N Q F V G L A I D E D N Q P D L T K N R I K T W V S Q L K S E F G L . .
F F K D . . . . . . . R G A K I V G S W S T D G Y E F E S S E A V V D G K F V G L A L D L D N Q S G K T . D E R V A A W L A Q I A P E F G L S L
A L S A R G A Q M I G S W P N E G Y E F S A S S A L E G D R F V G L V L D Q D N Q F D Q T E A R . . . . . . . . L A S W L E E I K R T V L . . .
L V I A R G A C . . V V G N W P R E G Y K F S F S A A L L E N N E F V G L P L D Q E N Q Y D L T . . . . . E E R I D S W L E K L K P A V L . . .
K L K N L G A E I V Q D G L R I . . . . . . D G D P R A A R D D I V G W A H . . D V R G A I . . . . . . . . . . . . . . . . . . . . . . . . .
D F E E R . . . . . . . . . . . . . . M N G Y G C V V V E T P . . . . . L I V Q N E P D E A E Q D C I E F G K K I A N I . . . . . . . . . . . .
S Y . . . . . . . . G W G S G E W M D A W K Q R T E D T G A T V I G T A I V N E M P D N A P E . . C K E L G E A A A K A . . . . . . . . . . . .
```

FIGURE 9–4 Comparison of the amino acid sequence of the *C. acetobutylicum* flavodoxin to nine other flavodoxin amino acid sequences. The flavodoxin sequences are from *A. variabilis* (Leonhardt and Straus 1989), *Chondrus crispus* (Wakabayashi et al. 1989), *Anacystis nidulans* (Laudenbach et al. 1988), *A. vinelandii* (Bennett et al. 1988), *K. pneumoniae* (Drummond 1985), *Enterobacter agglomerans* (W. Klingmüller, personal communication), *Desulfovibrio vulgaris* (Dubourdieu and Fox 1977; Krey et al. 1988), *Clostridium* MP (Tanaka et al. 1974) and *Megasphera eldensii* (Tanaka et al. 1973). The boxed regions indicate sequences with identical amino acid residues. Chemically related amino acids are not indicated. Reproduced with permission from Santangelo et al. (1991).

C. acetobutylicum flavodoxin shows some degree of similarity to the other flavodoxins. The invariant regions are located primarily at the flavin mononucleotide (FMN)-binding domains (the redox-active prosthetic group), which are included in positions 11–23, 58–64, and 101–114 (Wakabayashi et al. 1989).

While iron-containing Fd is one of the most important components of biological electron transport chains, it can be degraded as a source of iron under iron-limiting conditions (Schönheit et al. 1979; Ragsdale and Ljungdahl 1984). Certain organisms, such as members of the genus *Clostridium*, induce the synthesis of the non-iron-containing electron transfer protein flavodoxin under conditions of iron stress (Pardo et al. 1990). Ragsdale and Ljungdahl (1984) showed that when *C. formicoaceticum* was grown in a medium with a low iron content, the concentration of flavodoxin reached at least 2% of the total soluble protein. Flavodoxins mediate electron transfer at a low redox potential between prosthetic groups of microbial proteins in an analogous fashion to Fd. When the production of flavodoxin is induced (as in the clostridia) it generally replaces Fd in many oxidation-reduction reactions, but at a slower rate (Ragsdale and Ljungdahl 1984). In other organisms such as *A. vinelandii*, flavodoxin is constitutively produced and hence has a unique biological function (Tollin and Edmondson 1980).

All known flavodoxins have low molecular weights (14,500–23,000 g mol^{-1}) and are usually anionic (Tollin and Edmondson 1980). In general, flavodoxins from nonphotosynthetic bacteria have a molecular weight of approximately 15,000 (short flavodoxins), while flavodoxins from cyanobacteria and algae have a molecular weight of approximately 20,000 (long flavodoxins) (Wakabayashi et al. 1989). Flavodoxins consist of a single polypeptide chain lacking disulfide bonds (cysteine content varies from one to five residues). Flavodoxins noncovalently bind 1 mol of FMN. Binding studies indicate that all flavodoxins have structurally analogous coenzyme-binding sites (D'Anna and Tollin 1972). Amino acid alignments support this theory, as the ribityl phosphate-binding regions of different flavodoxins have a similar primary structure. With reference to Figure 9–4, Ser-11, Thr-13, and Thr-16 hydrogen bond to the FMN phosphate, whereas Thr-64 and Asp-176 hydrogen bond to the 4′-hydroxy group of the ribityl side chain (Wakabayashi et al. 1989).

E. coli is known to produce small amounts of Fd and flavodoxin, and while the biological role of Fd in this organism has not been completely elucidated, the role of flavodoxin has been established. Reduced flavodoxin is involved in the conversion of pyruvate formate lyase from an inactive to an active form (Knappe et al. 1984; Conradt et al. 1984). The mechanism of the reduction of flavodoxin in *E. coli* was elucidated by Blaschkowski et al. (1982), who showed that two enzymes exist that are capable of carrying out this reduction. The first enzyme is the CoA-acetylating pyruvate flavodoxin oxidoreductase, which is analogous to the pyruvate-Fd-oxidoreductase more commonly associated with the phosphoroclastic breakdown of pyruvate in anaerobes. The second enzyme is an NADPH-dependent oxidoreductase which these workers identified as the flavoprotein "component R" previously described by Fujii and Huennekens (1974).

If the cloned *C. acetobutylicum* flavodoxin is responsible for activating metronidazole in *E. coli* F19 by direct electron transfer, reasonable amounts of this flavodoxin must exist in a reduced form. The pyruvate flavodoxin oxidoreductase and the NADPH-dependent oxidoreductase are both minor proteins in *E. coli* and hence may not be present in large enough amounts to reduce significant quantities of the cloned flavodoxin. This suggests that either the cloned flavodoxin is reduced by other mechanisms, or that it is not entirely responsible for activating metronidazole. Since reduced flavodoxin is capable of converting the pyruvate formate lyase to its active form, perhaps the cloned flavodoxin merely enhances this effect. This could result in a higher conversion rate of pyruvate to acetyl-CoA and formate, which in turn could lead to induction of the formate dehydrogenase-H (Stewart 1988). This would reduce the overall oxidation-reduction potential of the cell, which is known to cause an increase in metronidazole sensitivity in *E. coli* (Onderdonk et al. 1979).

9.11 CONCLUSION

While the structure-function relationships of Fds have been studied extensively, these studies are not as advanced for the flavodoxins. Flavodoxins are among the smallest flavoproteins, and therefore represent the simplest models for studying the chemistry and biochemistry of this group of proteins. Since the production of flavodoxin in the clostridia occurs under conditions of iron limitation, the cloned *C. acetobutylicum* flavodoxin can provide a platform for studying the genetics of induction in this industrially important microorganism. Furthermore, since iron limitation is a form of stress, this was the first account of the cloning of a stress-induced gene from *C. acetobutylicum*. Since acid and solvent toxicity also represent a form of stress to this organism, an understanding of the genetic switches affected by stress is key to the understanding of the acetone-butanol-ethanol fermentation.

REFERENCES

Almassy, R.J., Janson, C.A., Hamlin, R., Xuong, N.H., and Eisenberg, D. (1986) *Nature (London)* 323, 304–309.

Andersch, W., Bahl, H., and Gottschalk, G. (1982) *Biotechnol. Lett.* 4(1), 29–32.

Andersch, W., Bahl, H., and Gottschalk, G. (1983) *Appl. Microbiol. Biotechnol.* 18, 327–332.

Atkinson, M.R., and Fisher, S.H. (1991) *J. Bacteriol.* 173(1), 23–27.

Ausubel, F.M. (1984) *Cell* 37, 5–6.

Belasco, J.G., Beatty, J.T., Adams, C.W., vonGabain, A., and Cohen, S.N. (1985) *Cell* 40, 171–181.

Bennett, L., Jacobson, M., and Dean, D.R. (1988) *J. Biol. Chem.* 264, 1364–1369.
Beynon, J., Cannon, M., Buchanan-Woolaston, V., and Cannon, F. (1983) *Cell* 34, 665–671.
Bhatnagar, L., Zeikus, J.G., and Aubert, J.-P. (1986) *J. Bacteriol.* 165, 638–643.
Birkmann, A., Sawers, R.G., and Bock, A. (1987) *Mol. Gen. Genet.* 210, 535–542.
Blaschkowski, H.P., Neuer, G., Ludwig-Festl, M., and Knappe, J. (1982) *Eur. J. Biochem.* 123, 563–569.
Bohannon, D.E.,and Sonenshein, A.L. (1989) *J. Bacteriol* 171(4), 4718–4727.
Carlson, T.A., and Chelm, B.K. (1986) *Nature* (*London*) 322, 568–570.
Chen, J.S., and Blanchard, D.K. (1979) *Anal. Biochem.* 93, 216–222.
Chrystal, E.J.T., Koch, R.L., McLafferety, M.A., and Goldman, P. (1980) *Antimicrob. Agents Chemother.* 18, 566–573.
Church, D.L., Rabin, H.R., and Laishley, E.J. (1988) *Biochem. Pharmacol.* 27, 1525–1534.
Church, D.L., Rabin, H.R., and Laishley, E.J. (1990) *J. Antimicrob. Chemother.* 25, 15–23.
Conradt, H., Hohmann-Berger, M., Hohmann, H.-P., Blaschkowski, H.P., and Knappe, J. (1984) *Arch. Biochem. Biophys.* 228, 133–142.
Crabbendam, P.M., Neijssel, O.M., and Tempest, D.W. (1985) *Arch. Microbiol.* 142, 375–382.
Croux, C., Paquet, V., Goma, G., and Soucaille, P. (1990) *Appl. Environ. Microbiol.* 56(12), 3634–3642.
D'Anna, J.A., and Tollin, G. (1972) *Biochemistry* 11, 1073–1080.
Datta, R., and Zeikus, J.G. (1985) *Appl. Environ. Microbiol.* 49, 522–529.
Dautry-Varsat, A., Cohen, G.N., and Stadtman, E.R.(1979) *J. Biol. Chem.* 254, 3124–3128.
Drummond, M.H. (1985) *Biochem. J.* 232, 891–896.
Dubourdieu, M., and Fox, J.L. (1977) *J. Biol. Chem.* 252, 1453–1463.
Edwards, D.I., and Mathison, G.E. (1970) *J. Gen. Microbiol.* 63, 297–302.
Edwards, D.I., Dye, M., and Carne, H. (1973) *J. Gen. Microbiol.* 76, 135–145.
Fujii, K., and Huennekens, F.M. (1974) *J. Biol. Chem.* 249, 6745–6753.
Gebhardt, C., Oliver, J.E., Forde, B.G., Saarelainen, R., and Miflin, B. (1986) *EMBO J.* 5, 1429–1435.
Gottschal, J.C., and Morris, J.G. (1981) *FEMS Microbiol. Lett.* 12, 385–389.
Green, P.J., Pines, O., and Ihouye, M. (1986) *Ann. Rev. Biochem.* 55, 569–597.
Gussin, G.N., Ronson, C.W., and Ausubel, F.M. (1986) *Annu. Rev. Genet.* 20, 567–591.
Hayward, B.E., Hussain, A., Wilson, R.H., Lyons, A., Woodcock, V., McIntosh, B., and Harris, T.J.R. (1986) *Nucleic Acids Res.* 14, 999–1008.
Henikoff, S., Haughn, G.W., Calvo, J.M., and Wallace, J.C. (1988) *Proc. Natl. Acad. Sci. USA* 85, 6602–6606.
Hill, R.T., Parker, J.R., Goodman, H.J.K., Jones, D.T., and Woods, D.R. (1989) *J. Gen. Microbiol.* 135, 3271–3279.
Hirschman, J., Wong, P.-K., Keener, K. S., and Kustu, S. (1985) *Proc. Natl. Acad. Sci. USA* 82, 7525–7529.
Hunt, T.P., and Magasanik, B. (1985) *Proc. Natl. Acad. Sci. USA* 82, 8453–8457.
Jackson, D., Salem, A., and Coombs, G.H. (1984) *J. Antimicrob. Chemother.* 13, 227–236.

Janson, C.A., Kayne, P. S., Almassy, R. J., Grunstein M., and Eisenberg, D. (1986) *Gene* 46, 297–300.

Janssen, P.J., Jones, D.T., and Woods, D.R. (1990) *Mol. Microbiol.* 4, 1575–1583.

Janssen, P.J., Jones, W.A., Jones, D.T, and Woods, D.R. (1988) *J. Bacteriol.* 170, 400–408.

Jones, D.T., and Woods, D.R. (1986) *Microbiol. Rev.* 50, 484–524.

Jungermann, K., Rupprecht, E., Ohrloff, R., Thauer, R.K., and Decker, K. (1971) *J. Biol. Chem.* 246, 960–963.

Jungermann, K., Thauer, R.K., Leimenstoll, G., and Decker, K. (1973) *Biochim. Biophys. Acta* 305, 268–280.

Kim, B.H., and Zeikus, J.G. (1985) *Dev. Ind. Microbiol.* 26, 1–14.

Knappe, J., Neugebauer, F.A., Blaschkowski, H.P., and Gänzler, M. (1984) *Proc. Natl. Acad. Sci. USA* 81, 1332–1335.

Knight, R.C., Skolimowski, I.M., and Edwards, D.I. (1978) *Biochem. Pharmacol.* 27, 2089–2093.

Kölling, R., and Lothger, H. (1985) *J. Bacteriol.* 164(1), 310–315.

Krey, G.D., Vanin, E.F., and Swenson, R.P. (1988) *J. Biol. Chem.* 263, 15436–15443.

Kucers, A., and Bennett, N.M. (1987) in *The Use of Antibiotics: A Comprehensive Review with Clinical Emphasis*, pp. 1290–1329, Heinemann Medical Books, London.

Kustu, S.G., McFarland, N.C., Hui, S.P., Esmon, B., and Ames, G.F. (1979) *J. Bacteriol.* 138(1), 218–234.

Laudenbach, D.E., Reith, M.E., and Straus, N.A. (1988) *J. Bacteriol.* 170, 258–265.

Leonhardt, K.G., and Straus, N.A. (1989) *Nucleic Acids Res.* 17, 4384.

Lockerby, D.L., Rabin, H.R., and Laishley, E.J. (1985) *Antimicrob. Agents Chemother.* 27, 863–867.

Lockerby, D.L., Rabin, H.R., Bryan, L.E., and Laishley, E.J. (1984) *Antimicrob. Agents Chemother.* 26, 665–669.

Long, S., Jones, D.T., and Woods, D.R. (1984) *Appl. Microbiol. Biotechnol.* 20, 256–261.

Magasanik, B. (1988) *TIBS* 13, 475–479.

Magasanik, B., and Neidhardt, F.C. (1987) in Escherichia coli *and* Salmonella typhimurium; *Cellular and Molecular Biology* (Neidhardt, F.C., Ingraham, J.L., Low, K.B., et al., eds.), pp. 1318–1325, American Society for Microbiology, Washington, DC.

McLafferty, M.A., Koch, R.L., and Goldman, P. (1982) *Antimicrob. Agents Chemother.* 21, 131–134.

Monot, F., and Engasser, J.M. (1983) *Biotechnol. Lett.* 5(4), 213–218.

Monroe, D.M., Noyes, C.M., Griffith, M.J., Lundblad, R.L., and Kingdon, H.S. (1985) *Curr. Top. Cell. Regul.* 27, 361–372.

Morgan, W.D., Bear, D.G., Litchman, B.L., and von Hippel, P.H. (1985) *Nucleic Acids Res.* 13, 3739–3754.

Nakano, Y., Kao, C., Tanaka, E., Kimura, K., and Horikoshi, K. (1989) *J. Biochem.* 106, 209–215.

Narikawa, S. (1986) *J. Antimicrob. Chemother.* 18, 565–574.

Ninfa, A.J., and Magasanik, B. (1986) *Proc. Natl. Acad. Sci. USA* 85, 5909–5913.

Noble, J.I., and Tally, F.P. (1984) in *Antimicrobial Therapy* (Ristuccia, A.M., and Cunha, B.A., eds.), pp. 255–263, Raven Press, New York.

O'Brien, R.W., and Morris, J.G. (1972) *Arch. Microbiol.* 84, 225–233.

Onderdonk, A.B., Louie, T.J., Tally, F.P., and Bartlett, J.G. (1979) *J. Antimicrob. Chemother.* 5, 201–210.

Pahel, G., Rothstein, D.M., and Magasanik, B. (1982) *J. Bacteriol.* 150, 202–213.

Pardo, M.B., Gómez-Moreno, C., and Peleato, M.L. (1990) *Arch. Microbiol.* 153, 528–530.

Plaskon, R.R., and Wartell, R.M. (1987) *Nucleic Acid Res.* 15, 785–796.

Prince, H.N., Grunberg, E., Titsworth, E., and DeLorenzo, W.F. (1969) *Appl. Microbiol.* 18, 728–730.

Prusiner, S., and Stadtman, E.R. (1973) *The Enzymes of Glutamine Metabolism,* Academic Press, New York.

Ragsdale, S.W. and Ljungdahl, L.G. (1984) *J. Bacteriol.* 157, 1–6.

Ralph, E.D., and Clarke, D.A. (1978) *Antimicrob. Agents Chemother.* 14, 377–383.

Rawlings, D.E., Jones, W.A., O'Neill, E.G., and Woods, D.R. (1987) *Gene* 53, 211–217.

Reitzer, L.J., and Magasanik, B. (1985) *Proc. Natl. Acad. Sci. USA* 82, 1979–1983.

Reitzer, L.J., and Magasanik, B. (1987) in Escherichia coli *and* Salmonella typhimurium; *Cellular and Molecular Biology* (Neidhardt, F.C., Ingraham, J.L., Low, K.B., et al., eds.), pp. 302–320, American Society for Microbiology, Washington, DC.

Rogers, P. (1986) *Adv. Appl. Microbiol.* 31, 1–60.

Rosenkranz, H.S., and Speck, W.T. (1975) *Biochem. Biophys. Res. Commun.* 66, 520–525.

Santangelo, J.D., Jones, D.T., and Woods, D.R. (1991) *J. Bacteriol.* 173, 1088–1095.

Schönheit, P., Brandis, A., and Thauer, R.K. (1979) *Arch. Microbiol.* 120, 73–76.

Schreier, H.J., Brown, S.W., Hirschi, K.D., Nomellini, J.F., and Sonenshein, A.L. (1989) *J. Mol. Biol.* 210, 51–63.

Schreier, H.J., Fisher, S.H., and Sonenshein, A.L. (1985) *Proc. Natl. Acad. Sci. USA* 82, 3375–3379.

Southern, J.A., Parker, J.R., and Woods, D.R. (1986) *J. Gen. Microbiol.* 132, 2827–2835.

Sonenshein, A.L. (1989) in *Regulation of Procaryotic Development: Structural and Functional Analysis of Bacterial Sporulation and Germination* (Smith, I., Slepecky, R.A., and Setlow, R., eds.), pp. 109–130, American Society for Microbiology, Washington, DC.

Stewart, V. (1988) *Microbiol. Rev.* 52, 190–232.

Strauch, M.A., Aronson, A.I., Brown, S.W., Schreier, H.J., and Sonenshein, A.L. (1988) *Gene* 71, 257–265.

Streicher, S.L.,and Tyler, B. (1980) *J. Bacteriol.* 142, 69–78.

Tally, F.P., Goldin, B.R., and Sullivan, N.E. (1981) *Scan. J. Infect. Dis.* (Suppl.) 26, 46–53.

Tally, F.P., Goldin, B.R., Sullivan, N.E., Johnston, J., and Gorbach, S.L. (1978) *Antimicrob. Agents Chemother.* 13, 460–465.

Tanaka, M., Haniu, M., Yasunobu, K.T., and Mayhew, S. (1974) *J. Biol. Chem.* 249, 4393–4396.

Tanaka, M., Haniu, M., Yasunobu, K.T., Mayhew, S., and Massey, V. (1973) *J. Biol. Chem.* 248, 4354–4366.

Thauer, R.K., Jungermann, K., and Decker, K. (1977) *Bacteriol. Rev.* 41, 100–180.

Tischer, E., DasSarma, S., and Goodman, H.M. (1986) *Mol. Gen. Genet.* 203, 221–229.

Tollin, G., and Edmondson, D.E. (1980) in *Methods in Enzymology*, vol. 69, (SanPietro, A., ed.), pp. 392–406, Academic Press, New York.

Uchino, F., Miura, K., and Doi, S. (1968) *J. Ferment. Technol.* 46, 188–195.

Ueno-Nishio, S., Backman, K.C., and Magasanik, B. (1983) *J. Bacteriol.* 153, 1247–1251.

Ueno-Nishio, S., Mango, S., Reitzer, L.J., and Magasanik, B. (1984) *J. Bacteriol.* 160, 379–384.

Usdin, K.P., Zappe, H., Jones, D.T., and Woods, D.R. (1986) *Appl. Environ. Microbiol.* 52, 413–419.

Wakabayashi, S., Kimura, T., Fukuyama, K., Matsubara, H., and Rogers, L.J. (1989) *Biochem. J.* 263, 981–984.

Wardman, P., and Clarke, E.D. (1976) *J. Chem. Soc. Faraday Trans. I* 72, 1377–1390.

Wong, H.C., and Chang, S. (1986) *Proc. Natl. Acad. Sci. USA* 83, 3233–3237.

Yerushalmi, L., Volesky, B., and Szczesny, T. (1985) *Appl. Microbiol. Biotechnol.* 22, 103–107.

Yeung, T.-C., Beaulieu, B.B., McLafferty, M.A., and Goldman, P. (1984) *Antimicrob. Agents Chemother.* 21, 131–134.

Young, M., and Mandelstam, J. (1979) *Adv. Microb. Physiol.* 20, 102–162.

Zhang, J., Strauch, M., and Aronson, A.I. (1989) *J. Bacteriol.* 171, 3572–3574.

CHAPTER 10

Transposons in Clostridia

Peter Dürre

The study of mutants represents the key to elucidation of metabolic pathways and gene regulation, as is evident from the impressive results obtained since the 1940s with *Escherichia coli*, the most prominent bacterial research object. Mutagenesis is also important for improving industrially useful organisms. Traditionally, chemical or physical treatments have been used to induce desired mutations. Applications of the respective methods to the genus *Clostridium* are detailed in Chapter 4, this volume. These techniques are limited by the possibility of inducing secondary mutations unrelated to the gene of interest that might obscure the expected phenotype and thus the selection method. A revolutionary development came with the use of so-called transposable elements. These mobile DNA sequences can integrate (mostly at random) into the chromosome of the target organism thereby creating an insertion that destroys the information conserved in the respective gene. Unlike chemical or physical treatments, transposon mutagenesis results in a single, unique physical alteration in the host DNA, thus a specific mutation. Genetic techniques can be used that test whether indeed only one copy of the transposon has inserted into the genome. Screening of the respective

Work reported from this laboratory has been supported by grants from the Deutsche Forschungsgemeinschaft and the Bundesminister für Forschung und Technologie.

mutants is greatly facilitated by the fact that transposons often carry genes coding for antibiotic resistance. Mutagenesis is, therefore, extremely efficient, most mutations lead to a total inactivation of the gene, and the level of secondary mutations is very low (Botstein and Shortle 1985).

Transposon-mutagenesis techniques have been developed mostly for use in the Gram-negative organisms *E. coli* and *Salmonella typhimurium*. The knowledge of respective elements in Gram-positive bacteria lagged behind. That is probably why the first reports on mobile genetic elements in clostridia and their use in mutagenesis appeared in the literature only a few years ago. The purpose of this chapter is to summarize our present knowledge on transposons in this group of anaerobic, spore-forming bacteria.

10.1 NONCONJUGATIVE TRANSPOSONS FROM GRAM-POSITIVE BACTERIA

During the last few years, many new transposable elements from Gram-positive bacteria have been discovered and described. Generally, they fall into two different groups. The first one contains relatively small transposons (up to 9-kilobase pair (kbp) length) that in many cases resemble the Tn*3* group of Gram-negative organisms. The second class is almost exclusively found in Gram-positive bacteria and consists of so-called conjugative transposons that are much longer (between 15- and 67-kbp length) and possess fertility potential. They will be dealt with in the following section.

Table 10–1 lists nonconjugative transposons from Gram-positive bacteria and some of their properties. It becomes evident that only three examples of naturally occurring transposons are known to date for the genus *Clostridium*. Tn*4451* and Tn*4452* were detected on two different plasmids from *C. perfringens* (Abraham and Rood 1987). Both elements encode a chloramphenicol resistance determinant that was still functional after cloning into *E. coli*. A high spontaneous excision rate could be observed in this organism. Transposition of Tn*4451* in the Gram-negative host was random although a preference for one chromosomal spot could be observed (three out of five clones investigated showed identical patterns in Southern hybridization). Restriction analyses indicated a precise excision of the element in *C. perfringens* and *E. coli* (Abraham and Rood 1987), which was later confirmed by sequence determinations (Abraham and Rood 1988). The termini of Tn*4451* shared some homology with the Tn*3* family of transposons. Both clostridial transposons showed an extensive homology with only three small regions of nonidentity that suggests a common origin (Abraham and Rood 1987). The nucleotide sequence of the chloramphenicol acetyltransferase-encoding gene has recently been determined (Steffen and Matzura 1989).

Another transferable element that carried resistance gene(s) to erythromycin and clindamycin has been found in *C. difficile* (Hächler et al. 1987a). It was detected on plasmid pBBB2 and also in the chromosome of various

TABLE 10–1 Nonconjugative Transposons from Gram-positive Bacteria

Transposon	*Natural Host*	*Length (kbp)*	*Resistance to*	*Comments*	*References*
Tn*551*	*Staphylococcus aureus*	5.2	Erythromycin	Similar to Tn*3*, termini sequenced	Novick et al. 1979; Khan and Novick 1980
Tn*552*	*Staphylococcus aureus*	6.5	Ampicillin	Similar to Tn*21* subgroup of Tn*3*, completely sequenced	Murphy and Novick 1979; Rowland and Dyke 1990
Tn*554*	*Staphylococcus aureus*	6.7	Erythromycin, Spectinomycin	Special transposition mechanism, completely sequenced	Phillips and Novick 1979; Murphy et al. 1985
Tn*917*	*Enterococcus faecalis*	5.3	Erythromycin	Similar to Tn*3* and Tn*551*, completely sequenced, transposition in *E. coli*	Tomich et al. 1980; Shaw and Clewell 1985; Kuramitsu and Casadaban 1986
Tn*3851*	*Staphylococcus aureus*	5.2	Gentamycin	Random transposition	Townsend et al. 1984
Tn*3852*	*Staphylococcus aureus*	7.3	Methicillin		Kigbo et al. 1985
Tn*3871*	*Enterococcus faecalis*	5.1	Erythromycin	Very similar to Tn*917*	Banai and LeBlanc 1984
Tn*4001*	*Staphylococcus aureus*	4.7	Gentamycin, tobramycin, kanamycin	Partially sequenced	Lyon et al. 1984; Byrne et al. 1989
Tn*4002*	*Staphylococcus aureus*	6.7	Penicillin	Very similar to Tn*552*, partially sequenced	Gillespie et al. 1988; Gillespie and Skurray 1989
Tn*4003*	*Staphylococcus aureus*	4.7	Trimethoprim	Completely sequenced	Rouch et al. 1989
Tn*4004*	*Staphylococcus aureus*	7.8	Mercury		Lyon and Skurray 1987
Tn*4031*	*Staphylococcus epidermidis*	5	Gentamycin	Similar to Tn*4001*	Thomas and Archer 1989
Tn*4201*	*Staphylococcus aureus*	6.7	Penicillin	Similar to Tn*552*	Weber and Goering 1988
Tn*4291*	*Staphylococcus aureus*	7.8	Methicillin		Trees and Iandolo 1988

(continued)

TABLE 10–1 (continued)

Transposon	*Natural Host*	*Length (kbp)*	*Resistance to*	*Comments*	*References*
Tn*4430*	*Bacillus thuringiensis*	4.2	—	Completely sequenced	Lereclus et al. 1986; Mahillon and Seurinck 1988
Tn*4451*	*Clostridium perfringens*	6.2	Chloramphenicol	Precise excision in *E. coli*, partially sequenced	Abraham and Rood 1988; Steffen and Matzura 1989
Tn*4452*	*Clostridium perfringens*	6.2	Chloramphenicol	Similar to Tn*4451*	Abraham and Rood 1987
Tn*4556*	*Streptomyces fradiae*	6.6	—	Similar to Tn*3*, completely sequenced	Chung 1987; Siemieniak et al. 1990
Tn*4560*	*Streptomyces lividans*	8.6	Viomycin	Constructed from Tn*4556*	Chung 1987
Tn*5096*	*Streptomyces griseofuscus*	3.0	Apramycin	Constructed from IS*493*	Solenberg and Baltz 1991
Tn*5097*	*Streptomyces griseofuscus*	3.0	Apramycin	As Tn*5096*, resistance gene in opposite orientation	Solenberg and Baltz 1991
Tn*5098*	*Streptomyces griseofuscus*	5.3	Apramycin	Derivative of Tn*5096*	Solenberg and Baltz 1991
No designation	*Clostridium difficile*	Unknown	Erythromycin, clindamycin	Antibiotic resistance gene similar to Tn*551*, possibly conjugative properties	Hächler et al. 1987a

C. difficile strains (by Southern hybridization) that resisted the respective drugs. Using the filter-mating technique, a transfer of the antibiotic resistance determinant from *C. difficile* to *Staphylococcus aureus* could be achieved at low frequencies ($\leq 10^{-8}$ per donor; hardly significantly above the rate of spontaneous resistant mutants that was 3×10^{-9} per cell). The recipients contained an additional 12.8-kbp *Hin*dIII DNA fragment. Homology studies indicated a close relationship of the clostridial erythromycin resistance gene (designated *erm*Z) with that from the staphylococcal transposon Tn*551* (*erm*B). No further data on this putative clostridial transposon are available. In the original description, the authors speculated on possible conjugative properties. If this assumption turns out to be correct, the new element would be one of the rare cases of conjugative transposons not carrying a tetracycline resistance determinant.

Tn*917* represents the only example of a nonconjugative transposon transferred to and expressed in clostridia. Yu and Pearce (1986) introduced this element into *C. acetobutylicum* by filter mating using the conjugative plasmid pVA797 as donor. Resistance to erythromycin was evidence of successful transfer of the transposon. This occurred with a frequency of about 10^{-4} per recipient and was the same as with the conjugative plasmid without the transposon (using, of course, a different drug selection). However, no data were reported on transposition. Thus, it is unclear whether the transposon still remained on the plasmid or whether it integrated into the chromosome of the new clostridial host. Bertram and Dürre (1989) also transferred Tn*917* to *C. acetobutylicum*. As a vector, the conjugative transposon Tn*925* (in the form of a tandem transposon) was used. Transfer frequencies as judged from acquisition of antibiotic resistance were in the range of 10^{-3} per recipient and donor. However, they dropped drastically when the transconjugants were selected simultaneously in the presence of tetracycline (resistance encoded on Tn*925*) and erythromycin (resistance encoded on Tn*917*). A possible reason for this phenomenon could be the inducible nature of the *erm* determinant of Tn*917* (Tomich et al. 1980). Thus, the transconjugants could have been unable to express the *erm* gene and the few surviving colonies might have acquired an additional resistance unrelated to Tn*917*, since the frequency of 10^{-8} is in the range of spontaneous mutations. The fact that erythromycin resistance of Tn*917* is not expressed in *E. coli* (Shaw and Clewell 1985) supports this assumption. *C. acetobutylicum* also shows a gradual response towards erythromycin. Whereas no growth of the wild type was found on solid media containing this drug, a clear adaptation occurred in exponentially growing liquid cultures (up to 1 mg/ml). It must be mentioned, however, that there seem to be enormous variations in this respect in the different strains of this organism.

Restriction analyses and Southern hybridizations of a transconjugant of *C. acetobutylicum* (selected in the presence of tetracycline) revealed that Tn*917* indeed was present on the chromosome (Bertram and Dürre 1989). It was still integrated into its vector (Tn*925*) as judged from identical bands after hybridization with the respective probes. No secondary transposition events of this element were observed. Further work is necessary to show whether this transposon will be useful for mutagenesis in clostridia.

10.2 CONJUGATIVE TRANSPOSONS FROM GRAM-POSITIVE BACTERIA

The second class of transposable elements in Gram-positive bacteria consists of the so-called conjugative transposons. Since a review on this topic was last published (Clewell and Gawron-Burke 1986) many additional elements have been described (see Table 10–2). Investigations with Tn*916*, the best studied transposon of this group, revealed that it had the capacity to promote its own transfer to a new bacterial host in the absence of any plasmid or phage. Prior to conjugation Tn*916* is excised from the donor chromosome (or plasmid) by a novel recombination event. A staggered cleavage yields molecules with single-stranded ends that form a covalently closed monomeric transposon circle with a heteroduplex at the joining site (Caparon and Scott 1989). The existence of such a supercoiled circular transposition intermediate has been shown (Scott et al. 1988). Integration into the new host chromosome probably is the reverse of excision. Since the circular intermediate contains a heteroduplex, no homologous target region is required (Caparon and Scott 1989). It has been shown that target regions are adenine-thymine (AT)-rich (Clewell et al. 1988) which makes this class of transposons attractive for use in the mostly AT-rich clostridia. Secondary transpositions seem to be rare and transfer of these elements is obviously not repressed by copies already present in the recipient (Norgren and Scott 1991). Tn*916* has also been used to introduce foreign genes into a new host (Norgren et al. 1989). The nucleotide sequences of the ends and *tetM* gene of Tn*916* have been determined (Clewell et al. 1988; Burdett 1990). The termini were not homologous to one another and contained imperfect inverted repeats. The *tetM* gene encoded a protein of 639 amino acid residues and showed a high degree of homology to previously sequenced tetracycline resistance determinants of class M, e.g., to that of Tn*1545* (Martin et al. 1986). This transposon has also been partially sequenced with respect to its termini and the erythromycin and tetracycline resistance genes (Martin et al. 1986; Caillaud and Courvalin 1987; Trieu-Cuot et al. 1990). The *erm* determinant proved to be almost identical to that from Tn*917*, although it was expressed constitutively and was not inducible. This was obviously due to several changes in the regulatory region (Trieu-Cuot et al. 1990). In agreement with the results obtained for Tn*916*, the ends of Tn*1545* were not homologous to one another and insertion occurred in AT-rich target DNA sequences (Caillaud and Courvalin 1987). The transposition mechanism, too, seems to be identical to that of Tn*916*. Excision and integration of Tn*1545* require two transposon-encoded proteins designated Xis-Tn and Int-Tn (Poyart-Salmeron et al. 1989, 1990). These proteins are able to *trans-*

TABLE 10–2 Conjugative Transposons from Gram-positive Bacteria[1]

Transposon	*Natural Host*	*Length (kbp)*	*Resistance to*	*Reference*
Tn*916*	*Enterococcus faecalis*	16.4	Tetracycline	Franke and Clewell 1981
Tn*918*	*Enterococcus faecalis*	15	Tetracycline	Clewell et al. 1985
Tn*919*	*Streptococcus sanguis*	15.8	Tetracycline	Fitzgerald and Clewell 1985
Tn*920*	*Enterococcus faecalis*	23	Tetracycline	Murray et al. 1988
Tn*925*	*Enterococcus faecalis*	14.5	Tetracycline	Christie et al. 1987
Tn*1545*	*Streptococcus pneumoniae*	25.3	Tetracycline, kanamycin, erythromycin	Carlier and Courvalin 1982
Tn*3701*	*Streptococcus pyogenes*	50	Contains Tn*3703*	Le Bouguénec et al. 1988
Tn*3702*	*Enterococcus faecalis*	18.5	Tetracycline, minocycline	Horaud et al. 1990
Tn*3703*	*Streptococcus pyogenes*	19.7	Tetracycline, erythromycin, minocycline	Le Bouguénec et al. 1988
Tn*3951*	*Streptococcus agalactiae*	67	Tetracycline, chloramphenicol, erythromycin	Inamine and Burdett 1985
Tn*5031*	*Enterococcus faecium*	Unknown	Tetracycline	Fletcher et al. 1989
Tn*5032*	*Enterococcus faecium*	Unknown	Tetracycline	Fletcher et al. 1989
Tn*5033*	*Enterococcus faecium*	Unknown	Tetracycline	Fletcher et al. 1989
Ω(cat-tet)	*Streptococcus pneumoniae*	65.5	Tetracycline, chloramphenicol	Vijayakumar et al. 1986
No designation	*Clostridium difficile*	Unknown	Tetracycline	Hächler et al. 1987b

[1] With respect to Gram-negative bacteria different types of conjugative transposons have been discovered in the genus *Bacteroides* (Shoemaker and Salyers 1988, 1990; Hecht et al. 1989; Halula and Macrina 1990).

complement in vivo a deletion derivative of Tn*1545* and show partial homology to the respective proteins from lambdoid phages (Poyart-Salmeron et al. 1989). Similar to Tn*916*, Tn*1545* forms a circular transposition intermediate with a heteroduplex between the joined termini (Poyart-Salmeron et al. 1990). The results obtained for both elements indicate that conjugative transposons from Gram-positive bacteria possess an integration/excision system similar to that of lambdoid phages of Gram-negative organisms (Poyart-Salmeron et al. 1989, 1990).

Only one of the transposons listed in Table 10–2 has been found to date to occur naturally in clostridia (Hächler et al. 1987b). This element showed a high degree of homology to Tn*916*. It could be transferred between different strains of *C. difficile* and *Bacillus subtilis* with low frequencies of 10^{-7} to 10^{-8} per donor (Mullany et al. 1990). This putative transposon is probably similar or identical to transferable-tetracycline resistance determinants described earlier in *C. difficile* by several groups (Ionesco 1980; Smith et al. 1981; Wüst and Hardegger 1983). Transfer occurred at low frequencies of 10^{-6} to 10^{-8} per donor and was not associated with plasmids present in donor strains. Insensitivity towards DNase treatment and the requirement of cell-to-cell contact indicated a conjugation-like mechanism (Smith et al. 1981). In one case, evidence for a different transferable element carrying clindamycin, erythromycin, and streptogramins resistance determinants was obtained (Wüst and Hardegger 1983). Such a resistance element was recently cloned from *C. perfringens* and the authors speculated on its location on a transposable element (Berryman and Rood 1989). *C. perfringens* also contained a resolvase probably necessary for plasmid stabilization. The resolvase structural gene most closely resembled that from Tn*917* (Garnier et al. 1987). A putative transposon-located erythromycin-clindamycin resistance determinant was found in *C. innocuum* that could be transferred to different strains of *C. perfringens* at rates of 10^{-6} to 10^{-8} per recipient (Magot 1983).

Whereas only few transposable elements of clostridial origin are known, several reports appeared recently that described efficient introduction of conjugative transposons into some clostridial species (Table 10–3). The transfer occurred via mating and will not be dealt with in this section since Chapter 5 of this volume covers conjugative gene transfer in clostridia. Frequencies of almost 10^{-2} per recipient were obtained when Tn*916* was located on a plasmid in the donor strain (Bertram and Dürre 1989). In general, higher transfer frequencies are observed when the donor contained the transposon on a plasmid rather than integrated in the chromosome (see Table 10–3). Tn*916* integrated randomly into the chromosome of its new host in most cases as shown by Southern hybridization (Volk et al. 1988; Bertram and Dürre 1989; Mattsson and Rogers 1989). Only one strain of *C. acetobutylicum* (NCIB 8052) possibly provided a hot spot for insertion. However, random transposition occurred when Tn*1545* was used (Woolley et al. 1989). In all cases where the transposon was located on a plasmid in the donor, no extrachromosomal elements could be detected in the recipients after mating. Although

TABLE 10–3 Frequencies of Transfer of Conjugative Transposons to Clostridia

				Transfer Frequencies		
Transposon	*Location*	*Donor*	*Recipient*	*Per Donor*	*Per Recipient*	*Reference*
Tn*916*	Plasmid	*Ec. faecalis*	*C. tetani*	1.7×10^{-5}	ND[1]	Volk et al. 1988
Tn*916*	Chromosome	*Ec. faecalis*	*C. tetani*	9.4×10^{-6}	ND	Volk et al. 1988
Tn*916*	Chromosome	*C. tetani*	*C. tetani*	3.9×10^{-4}	ND	Volk et al. 1988
Tn*916*	Plasmid	*Ec. faecalis*	*C. acetobutylicum*	7.3×10^{-4}	8.7×10^{-3}	Bertram and Dürre 1989
Tn*925*	Plasmid	*Ec. faecalis*	*C. acetobutylicum*	5.3×10^{-4}	1.5×10^{-2}	Bertram and Dürre 1989
Tn*925*::Tn*917*	Plasmid	*Ec. faecalis*	*C. acetobutylicum*	4.5×10^{-3}	1.3×10^{-3}	Bertram and Dürre 1989
Tn*916*	Plasmid	*Ec. faecalis*	*C. acetobutylicum*	ND	1.0×10^{-4}	Mattsson and Rogers 1989
Tn*916*	Chromosome	*Ec. faecalis*	*C. acetobutylicum*	ND	0.9×10^{-4}	Woolley et al. 1989
Tn*1545*	Chromosome	*Ec. faecalis*	*C. acetobutylicum*	ND	Up to 1.2×10^{-5}	Woolley et al. 1989
Tn*1545*	Chromosome	*B. subtilis*	*C. acetobutylicum*	ND	Up to 3.4×10^{-5}	Woolley et al. 1989
Tn*1545*	Chromosome	*C. acetobutylicum*	*C. acetobutylicum*	ND	2.0×10^{-6}	Woolley et al. 1989
Tn*916*	Plasmid	*E. coli*	*C. acetobutylicum*	1.4×10^{-5}	5.3×10^{-5}	Bertram et al. 1991
No designation	Chromosome	*C. innocuum*	*C. perfringens*	ND	Up to 7.1×10^{-6}	Magot 1983
No designation	Chromosome	*C. perfringens*	*C. perfringens*	ND	Up to 1.4×10^{-7}	Magot 1983
No designation	Chromosome	*C. difficile*	*C. difficile*	1.0×10^{-7}	ND	Mullany et al. 1990
No designation	Chromosome	*B. subtilis*	*C. difficile*	1.0×10^{-8}	ND	Mullany et al. 1990

[1] ND = not determined.

Tn*916* and Tn*925* are able to promote their own transfer, it cannot be excluded that plasmid transfer via conjugation also occurred that could have added to the observed transfer frequencies. Since no plasmids have been found after mating, this would mean that they were unstable in *C. tetani* and *C. acetobutylicum* and were lost after transposition of Tn*916* or Tn*925* to the chromosome. In most of the experiments reported, only one copy of the transposon used was found in the chromosome of the recipients (Volk et al. 1988; Bertram and Dürre 1989; Woolley et al. 1989). In some cases, however, two or more copies were present. This is in agreement with data of Tn*916* (Norgren and Scott 1991) mentioned before that the presence of one copy of the transposon in the recipient strain does not impede transfer of a second copy of the element. Recently, Tn*916* has also been introduced into *C. perfringens* (Allen and Blaschek, unpublished observations). An *E. coli* plasmid carrying this element (pAM120) was transferred to *C. perfringens* by electroporation. Transposition and subsequent integration into the clostridial genome occurred as a result of the inability of this plasmid to replicate in the Gram-positive host. The transposon inserted randomly and was present in multiple copies. A transformation frequency of about 10^2 transformants per μg of DNA was obtained. Transfer of Tn*916* from the same plasmid to *C. acetobutylicum* by electroporation has also been reported (Bertram 1989). Transconjugants contained the transposon on the chromosome as shown by Southern hybridization.

Usually, the Gram-positive bacterium *Enterococcus faecalis* has been used as a transposon donor in matings with Gram-positive clostridial species (see Table 10–3). This is in agreement with the old assumption that no natural DNA transfer can occur between Gram-positive and Gram-negative bacteria. Artificially constructed shuttle vectors were shown in the last few years to mediate conjugal transfer between *E. coli* and a variety of Gram-positive genera (Trieu-Cuot et al. 1987c, 1988; Mazodier et al. 1989; Schäfer et al. 1990). Recently, however, it was reported that conjugative transposons such as Tn*916* can *naturally* transfer between many Gram-positive (including *C. acetobutylicum*) and Gram-negative species, with subsequent expression in the new host (Bertram et al. 1991). Although the transfer frequencies were mostly lower compared to Gram-positive:Gram-positive exchange, the possibility of employing *E. coli* as a transposon donor opens up avenues for the use of these elements. Construction of special derivatives or insertion of foreign DNA is much easier in *E. coli* than in *Enterococcus* and the full repertoire of techniques and mutant strains for the former organism becomes available.

The ease with which this class of conjugative transposons can be introduced into clostridia from either the Gram-positive *Ec. faecalis* or even the Gram-negative *E. coli*, the fact that no sophisticated equipment is necessary for this process, and the mostly random integration mechanism makes them ideal tools for insertional mutagenesis. The efficiency of transfer might even be enhanced by using subinhibitory concentrations of tetracycline during the mating as has been shown recently for the transfer of Tn*1545* from *Ec. faecalis* to *Listeria monocytogenes* (Doucet-Populaire et al. 1991). Although until now only few clostridial mutants have been isolated and characterized this way, it is envisaged that this number will increase enormously in the future.

10.3 TRANSPOSON MUTAGENESIS IN CLOSTRIDIA

Many mutants of several clostridial species have been isolated and characterized. All of them either spontaneously arose or were obtained by chemical treatments. Since Chapter 4 of this volume deals with such strains and the strategies to isolate them details will not be given here. Only one report has been published to date on transposon-induced clostridial mutants (Bertram et al. 1990). It represents part of the efforts to eludicate the mechanism of regulation of solvent formation in *C. acetobutylicum*, an organism that was used for the industrial production of acetone and butanol until about the middle of this century when oil-based processes became more economical. *C. acetobutylicum* is a strictly anaerobic microorganism with the ability to switch its main fermentation products from acetic and butyric acids to acetone and butanol. Ethanol is also formed, but only in minor amounts. Onset of solvent formation requires a low pH, certain threshold concentrations of the aforementioned acids, and a suitable growth-limiting factor such as phosphate or sulfate (Bahl et al. 1982; Bahl and Gottschalk 1988). The regulatory mechanism of this shift to solventogenesis at the molecular level is still unknown.

Elucidation of this process requires the availability of well-characterized mutants with specific defects in either the solventogenic or acidogenic pathways. Several specific detection systems for the identification of such mutants have been described. Selection in the presence of allyl alcohol allows the isolation of strains that are defective in butanol synthesis (Dürre et al. 1986). In the wild type, allyl alcohol is oxidized to the toxic aldehyde acrolein that is highly reactive and leads to cell death by inactivation of proteins. It seems that the enzyme responsible for this oxidation in *C. acetobutylicum* is the butyraldehyde dehydrogenase. This is somewhat surprising since one would expect alcohol dehydrogenases to perform such a reaction. However, the data reported for *C. acetobutylicum* clearly point to the butyraldehyde dehydrogenase (Dürre et al. 1986). Similarly, acetone-negative mutants can be selected for in the presence of 2-bromobutyrate, another suicide substrate whose reaction mechanism in *C. acetobutylicum* is not quite clear yet (Janati-Idrissi et al. 1987; Junelles et al. 1987). Strains with lower acid production were isolated by a proton suicide method in the presence of bromide and bromate as selective agents (Cueto and Méndez 1990).

To obtain solvent-negative mutants *C. acetobutylicum* DSM 792 was mated with Tn*916*-containing *Ec. faecalis*, and transconjugants were isolated by tetracycline resistance (Bertram et al. 1990). These colonies were then picked and streaked on agar plates containing allyl alcohol or bromobutyrate. This was done to assure that transposition of Tn*916* was responsible for the mutation and not some other secondary effect. Any mutation in the solvent pathway not caused by insertion of the transposon would thus have shown up on the plates as one or a few single colonies growing on the streak. On the other hand, butanol- or acetone-negative mutants induced by Tn*916* transposition would grow on the whole streak. These colonies were picked, grown in minimal medium and characterized according to their fermentation products. Two main groups could be identified. One (designated I) contained strains that completely lost the ability to form acetone and butanol, but still produced ethanol. Members of the other group (designated II) were able to make all three solvents, but in clearly lower amounts compared to the parent. Phenotypically similar mutants have been obtained (also by transposition of Tn*916*) by D. Mattsson from the University of Minnesota, St. Paul, but have not yet been characterized further (P. Rogers, personal communication).

Representative mutants from the two groups mentioned above and the parent were analyzed for all solvent-forming enzymes (Table 10–4). It is obvious that members of group I do not show any activity of butyraldehyde dehydrogenase, NADH-dependent alcohol dehydrogenase, acetoacetyl-CoA:acetate/butyrate CoA transferase, and acetoacetate decarboxylase, concomitantly with the loss of butanol- and acetone-forming ability. However, they still possess a NADPH-dependent alcohol dehydrogenase and a specific acetaldehyde dehydrogenase activity that was formerly unknown. This represents the first genetic evidence for the in vivo function of the alcohol-forming enzymes. It had long been assumed that a NADPH-dependent alcohol dehydrogenase was responsible for butanol production (Jones and Woods 1986; Rogers 1986). Dürre et al. (1987) first showed the existence of two alcohol dehydrogenase activities with different coenzyme specificities in *C. acetobutylicum*. The NADPH-dependent enzyme showed a certain level of activity throughout the whole fermentation and was induced about threefold at the onset of solvent formation. The data obtained from group I mutants clearly indicate that the in vivo function of this NADPH-dependent alcohol dehydrogenase together with the specific acetaldehyde dehydrogenase is ethanol formation, since those are the only enzymes active in the mutants and ethanol is the only solvent still formed. The assumption that this alcohol dehydrogenase represents a constitutive enzyme is supported by the facts that: (1) enzymatic activity is found throughout the whole fermentation (Dürre et al. 1987), and (2) its structural gene is located next to the gene encoding β-hydroxybutyryl-CoA dehydrogenase (Youngleson et al. 1989), an enzyme involved in the synthesis of C_4-carbon intermediates in the pathway of butyrate *and* butanol formation and, thus, necessarily constitutive. However, the existence of an operon and a common regulatory system for these two genes has not been shown, and recent studies suggest that the two genes are independently regulated and transcribed (D. Woods, personal communication).

Enzymes responsible for butanol production according to the data obtained from the transposon-mutants are the butyraldehyde dehydrogenase

TABLE 10–4 Characterization of Tn*916*-Induced Mutants of *C. acetobutylicum*[1]

		Enzymatic Activity of						
Strain[2]	*Solvents Formed*[3]	*Acetaldehyde DH*[4] *(NADH)*	*Alcohol DH (NADPH)*	*Butyraldehyde DH (NADH)*	*Alcohol DH (NADH)*	*CoA Transferase*[5]	*Acetoacetate Decarboxylase*	*Copy No. of Tn*916
Parent (DSM792)	A,B,E	+	+	+	+	+	+	0
AA 2	e	+	+	–	–	–	–	1
AA 5	e	+	+	–	–	–	–	3
AA 1	a,b,e	0[6]	0	0	0	ND[7]	ND	1
AA 6	a,b,E	0	+	0	+	ND	ND	1
BB 3	a,b,E	0	+	0	+	ND	ND	1
BB 4	a,b,e	0	0	0	0	ND	ND	1

1 Adapted from Bertram et al. (1990).
2 AA strains were isolated from allyl alcohol plates, BB strains from bromobutyrate plates.
3 Acetone (A, a), butanol (B, b), ethanol (E, e); capitals indicate amounts similar to the parent (25% or more), lowercase letters stand for concentrations of 25% or less of those found in the parent.
4 Dehydrogenase, coenzyme specificity is indicated in brackets.
5 Acetoacetyl coenzyme A:acetate/butyrate:coenyme A transferase.
6 Less than 25% of the activity found in the parent.
7 ND = not determined.

and the NADH-dependent alcohol dehydrogenase. Both show some substrate nonspecificity and react with C_2- and C_4-carbon compounds, although the latter are clearly preferred (Palosaari and Rogers 1988; Welch et al. 1989). On the basis of kinetic studies performed with the purified enzyme it was already concluded that the NADH-dependent alcohol dehydrogenase was involved in butanol formation (Welch et al. 1989). The loss of the two enzymes mentioned and of CoA transferase and acetoacetate decarboxylase explains why no butanol and acetone can be produced by group I mutants. Strains of the other group (II) generally showed reduced levels of solvent-forming enzymes as was expected from their fermentation patterns (only reduced levels of all three solvents were formed, see Table 10–4). Mutants probably similar to group I have been obtained by nitrosoguanidine treatment and UV irradiation of different strains of *C. acetobutylicum* (Clark et al. 1989; Hayashida and Ahn 1990).

The Tn*916*-induced mutants AA 2 and AA 5 also provide evidence for a different regulation of acetone and butanol formation compared to that of ethanol. Production of ethanol seems to be an outlet for a surplus of NADPH that is needed for anabolic reactions. Regeneration of NADPH in clostridia is generally mediated by NADPH-ferredoxin oxidoreductase (Jungermann et al. 1973; Petitdemange et al. 1976). The existence of such an anabolic enzyme has been shown in *C. acetobutylicum* (Petitdemange et al. 1976, 1977). Thus, the physiological role of the NADPH-dependent alcohol dehydrogenase probably is to regulate the pool of reduced nicotinamide-adenine dinucleotide phosphate.

Production of acetone and butanol must be regulated by one genetic locus as indicated by mutant AA 2 that contains only one copy of Tn*916* in the chromosome and is unable to form both solvents. It seems to be a central regulatory gene or locus since the genes for the solvent-forming enzymes are arranged in different operons (see Chapter 8, this volume). The genes for the two enzymes of the acetone pathway, CoA transferase and acetoacetate decarboxylase, have been cloned and partially sequenced (Cary et al. 1990; Gerischer and Dürre 1990; Petersen and Bennett 1990). They are located in two different operons that share a common transcription termination structure, but are divergently arranged. Northern blots probed with oligonucleotides directed against the structural genes confirm these data (U. Gerischer and P. Dürre, unpublished observations). The metabolic pathway shown in Figure 10–1 summarizes the information obtained from analysis of the transposon-induced mutants of *C. acetobutylicum*.

The insertion of Tn*916* into a regulatory gene or locus of solventogenesis opens the possibility of its targeting and cloning. This transposon will be excised from cloned DNA in *E. coli* in the absence of tetracycline (Gawron-Burke and Clewell 1984). Thus, the fragment containing the regulatory region could be cloned in *E. coli* by selecting for the transposon-encoded resistance. After growth in the absence of the drug, excision of Tn*916* will result in exact restoration of the original sequence in some cases. This DNA can be analyzed

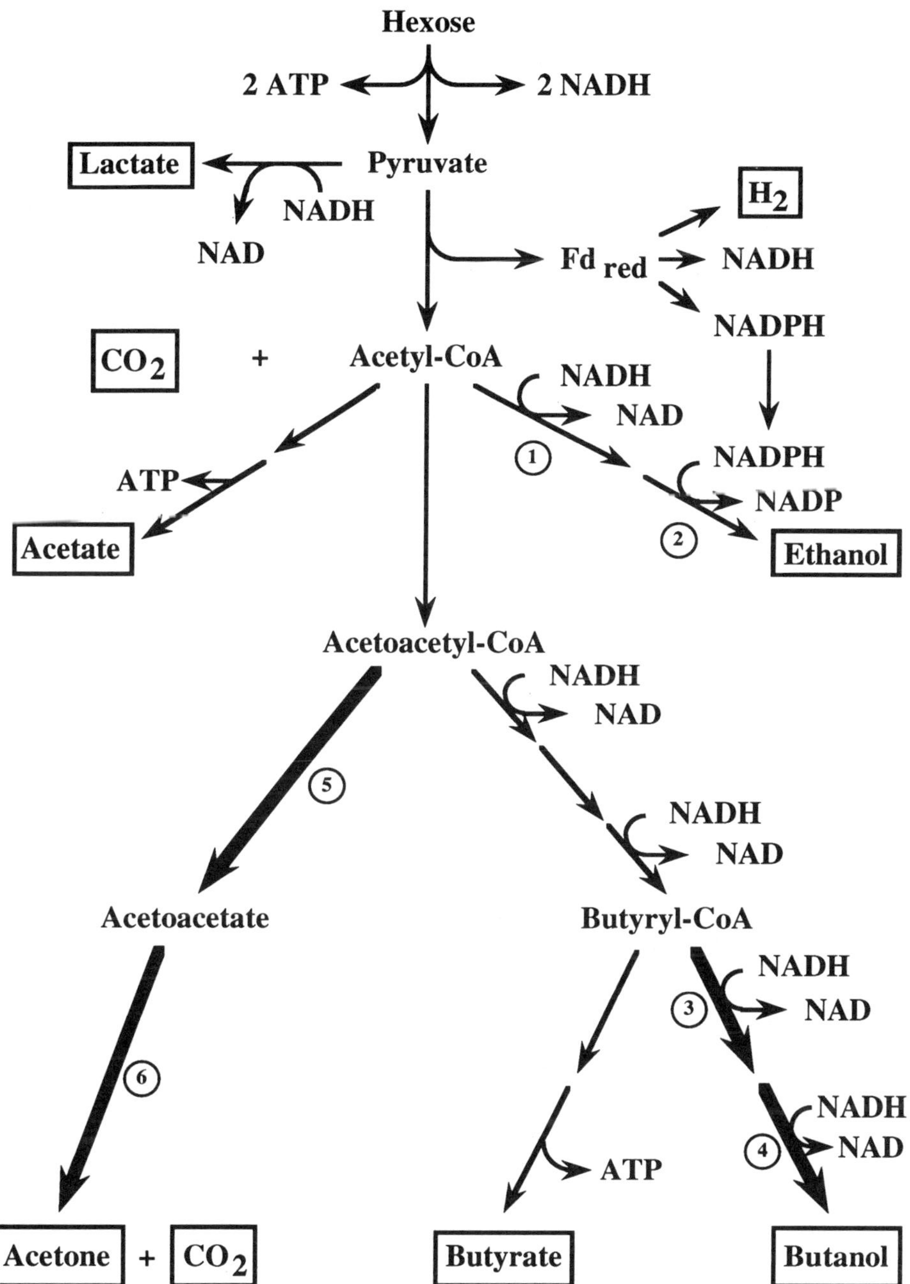

FIGURE 10–1 Metabolic pathway of *C. acetobutylicum*. Compounds shown in rectangles are fermentation end products. Thick arrows indicate reactions by enzymes induced at the onset of solventogenesis. Circled numbers refer to the solventogenic enzymes dealt with in the text: (1) acetaldehyde dehydrogenase; (2) NADPH-dependent alcohol dehydrogenase; (3) butyraldehyde dehydrogenase; (4) NADH-dependent alcohol dehydrogenase; (5) acetoacetyl coenzyme A:acetate/butyrate:coenzyme A transferase; (6) acetoacetate decarboxylase.

directly or used as a probe to isolate the respective fragments from the parent. Such a technique cannot be used with the conventionally obtained mutants, thus showing another advantage of transposon mutagenesis. A clone probably carrying the regulatory region has been obtained from *C. acetobutylicum* DSM 792 according to this method (U. Sauer and P. Dürre, unpublished observations) and is currently being analyzed.

The assumption that transposon mutagenesis in the genus *Clostridium* will become more important in the future is strengthened by a recent report that phospholipase C-minus mutants of *C. perfringens* have been obtained by transposition of Tn*916* (H.P. Blaschek, personal communication).

10.4 TRANSPOSONS AND EVOLUTION

Antibiotic resistance determinants from Gram-positive microorganisms are often expressed in Gram-negative bacteria, whereas the reverse is uncommon (Chang and Cohen 1974; Kreft et al. 1978; Trieu-Cuot et al. 1987a). Analysis of the nucleotide sequences and deduced amino acid sequences of macrolide and aminoglycoside resistance genes led to the conclusion that a horizontal transfer of genetic material from Gram-positive to Gram-negative bacteria must have occurred (Trieu-Cuot et al. 1985, 1987a, 1987b; Brisson-Noël et al. 1988). However, in most cases, plasmids from Gram-positive organisms cannot be stably maintained and replicated in Gram-negative bacteria, and vice versa, thus, some other transfer vehicle must exist. The conjugative transposons found in Gram-positive organisms fulfill the respective requirements. They can be transferred naturally between a variety of Gram-negative and Gram-positive genera, being more or less stably maintained in the new host after integration into its chromosome (Dybvig and Cassell 1987; Jones et al. 1987; Kathariou et al. 1987; Roberts and Kenny 1987; Ivins et al. 1988; Sen and Oriel 1990; Bertram et al. 1991). This class of elements shows a remarkable degree of homology (including the *tetM* resistance determinant that seems to be present in all of them), which indicates that these transposons originate from a common ancestor. Supporting this assumption is their broad host range in Gram-positive and even Gram-negative bacteria. Only a few organisms have been found to date that did not receive Tn*916* in mating experiments (Bertram et al. 1991). The fact that a similar element has been found in *C. difficile* suggests an early intergeneric transfer of genetic material (Hächler et al. 1987b). Recently, a Tn*916*-homologous *tetM* determinant has been identified in *Fusobacterium nucleatum*, a Gram-negative bacterium. It could be transferred to *Ec. faecalis* and some anaerobic species (Roberts and Lansciardi 1990). This represents the first report on natural occurrence of such a transposable element in Gram-negative bacteria.

It has been speculated that horizontal gene transfer plays a significant role in evolution (Syvanen 1986). Foreign genetic information, once introduced and stably maintained in the new bacterium, depends on the fate of its new host. If this material does not provide a selective advantage, chances for its getting lost again are relatively high. It has been shown, however, that the transposable element Tn*10* working as a mutator gene in evolution actually confers an advantage to the host bacterium (Chao et al. 1983). This is probably achieved by increasing the number of more fit mutants in the host population. Similarly, competition experiments in a chemostat with Tn*5*-containing and noncontaining *E. coli* strains showed more rapid growth rates and increases in frequency to 90% within the first 15 to 20 generations for the transposon-bearing strains (Biel and Hartl 1983). Thus, there is growing evidence for the beneficial effect of transposable elements not only as tools for molecular biologists and microbiologists, but also for the respective bacterial hosts in evolution.

REFERENCES

Abraham, L.J., and Rood, J.I. (1987) *J. Bacteriol.* 169, 1579–1584.

Abraham, L.J., and Rood, J.I. (1988) *Plasmid* 19, 164–168.

Bahl, H., Andersch, W., and Gottschalk, G. (1982) *Eur. J. Appl. Microbiol. Biotechnol.* 15, 201–205.

Bahl, H., and Gottschalk, G. (1988) in *Biotechnology* (Rehm, H.-J., Reed, G., ed.), vol. 6b, pp. 1–30, VCH Verlagsgesellschaft, Weinheim, Germany.

Banai, M., and LeBlanc, D.J. (1984) *J. Bacteriol.* 158, 1172–1174.

Berryman, D.I., and Rood, J.I. (1989) *Antimicrob. Agents Chemother.* 33, 1346–1353.

Bertram, J. (1989) *Entwicklung eines Systems zum DNA-Transfer und zur Transposon-Mutagenese für* Clostridium acetobutylicum, Reihe Biologie, vol. 2, pp. 72–74, Unitext Verlag, Göttingen, Germany.

Bertram, J., and Dürre, P. (1989) *Arch. Microbiol.* 151, 551–557.

Bertram, J., Kuhn, A., and Dürre, P. (1990) *Arch. Microbiol.* 153, 373–377.

Bertram, J., Strätz, M., and Dürre, P. (1991) *J. Bacteriol.* 173, 443–448.

Biel, S.W., and Hartl, D.L. (1983) *Genetics* 103, 581–592.

Botstein, D., and Shortle, D. (1985) *Science* 229, 1193–1201.

Brisson-Noël, A., Arthur, M., and Courvalin, P. (1988) *J. Bacteriol.* 170, 1739–1745.

Burdett, V. (1990) *Nucleic Acids Res.* 18, 6137.

Byrne, M.E., Rouch, D.A, and Skurray, R.A. (1989) *Gene* 81, 361–367.

Caillaud, F., and Courvalin, P. (1987) *Mol. Gen. Genet.* 209, 110–115.

Caparon, M.G., and Scott, J.R. (1989) *Cell* 59, 1027–1034.

Carlier, C., and Courvalin, P. (1982) in *Microbiology—1982* (Schlessinger, D., ed.), pp. 162–166, American Society for Microbiology, Washington, DC.

Cary, J.W, Petersen, D.J., Papoutsakis, E.T., and Bennett, G.N. (1990) *Appl. Environ. Microbiol.* 56, 1576–1583.

Chang, A.C.Y., and Cohen, S.N. (1974) *Proc. Natl. Acad. Sci. USA* 71, 1030–1034.

Chao, L., Vargas, C., Spear, B.B., and Cox, E.C. (1983) *Nature* (*London*) 303, 633–635.

Christie, P.J., Korman, R.Z., Zahler, S.A., Adsit, J.C., and Dunny, G.M. (1987) *J. Bacteriol.* 169, 2529–2536.

Chung, S.-T. (1987) *J. Bacteriol.* 169, 4436–4441.

Clark, S.W., Bennett, G.N., and Rudolph, F.B. (1989) *Appl. Environ. Microbiol.* 55, 970–976.

Clewell, D.B., An, F.Y., White, B.A., and Gawron-Burke, C. (1985) *J. Bacteriol.* 162, 1212–1220.

Clewell, D.B., Flannagan, S.E., Ike, Y., Jones, J.M., and Gawron-Burke, C. (1988) *J. Bacteriol.* 170, 3046–3052.

Clewell, D.B., and Gawron-Burke, C. (1986) *Annu. Rev. Microbiol.* 40, 635–659.

Cueto, P.H., and Méndez, B.S. (1990) *Appl. Environ. Microbiol.* 56, 578–580.

Doucet-Populaire, F., Trieu-Cuot, P., Dosbaa, I., Andremont, A., and Courvalin, P. (1991) *Antimicrob. Agents Chemother.* 35, 185–187.

Dürre, P., Kuhn, A., and Gottschalk, G. (1986) *FEMS Microbiol. Lett.* 36, 77–81.

Dürre, P., Kuhn, A., Gottwald, M., and Gottschalk, G. (1987) *Appl. Microbiol. Biotechnol.* 26, 268–272.

Dybvig, K., and Cassell, G.H. (1987) *Science* 235, 1392–1394.

Fitzgerald, G.F, and Clewell, D.B. (1985) *Infect. Immun.* 47, 415–420.

Fletcher, H.M., Marri, L., and Daneo-Moore, L. (1989) *J. Gen. Microbiol.* 135, 3067–3077.

Franke, A.E., and Clewell, D.B. (1981) *J. Bacteriol.* 145, 494–502.

Garnier, T., Saurin, W., and Cole, S.T. (1987) *Mol. Microbiol.* 1, 371–376.

Gawron-Burke, C., and Clewell, D.B. (1984) *J. Bacteriol.* 159, 214–221.

Gerischer, U., and Dürre, P. (1990) *J. Bacteriol.* 172, 6907–6918.

Gillespie, M.T., Lyon, B.R., and Skurray, R.A. (1988) *J. Gen. Microbiol.* 134, 2857–2866.

Gillespie, M.T., and Skurray, R.A. (1989) *Nucleic Acids Res.* 17, 8854.

Hächler, H., Berger-Bächi, B., and Kayser, F.H. (1987a) *Antimicrob. Agents Chemother.* 31, 1039–1045.

Hächler, H., Kayser, F.H., and Berger-Bächi, B. (1987b) *Antimicrob. Agents Chemother.* 31, 1033–1038.

Halula, M., and Macrina, F.L. (1990) *Rev. Infect. Dis.* 12 (suppl. 2), S 235–S 242.

Hayashida, S, and Ahn, B.K. (1990) *Agric. Biol. Chem.* 54, 343–351.

Hecht, D.W., Thompson, J.S., and Malamy, M.H. (1989) *Proc. Natl. Acad. Sci. USA* 86, 5340–5344.

Horaud, T., Delbos, F., and de Cespédès, G. (1990) *FEMS Microbiol. Lett.* 72, 189–194.

Inamine, J.M., and Burdett, V. (1985) *J. Bacteriol.* 161, 620–626.

Ionesco, H. (1980) *Ann. Microbiol. Inst. Pasteur* 131 A, 171–179.

Ivins, B.E., Welkos, S.L., Knudson, G.B., and LeBlanc, D.J. (1988) *Infect. Immun.* 56, 176–181.

Janati-Idrissi, R., Junelles, A.M., El Kanouni, A., Petitdemange, H., and Gay, R. (1987) *Ann. Inst. Pasteur/Microbiol.* 138, 313–323.

Jones, D.T., and Woods, D.R. (1986) *Microbiol. Rev.* 50, 484–524.

Jones, J.M., Yost, S.C., and Pattee, P.A. (1987) *J. Bacteriol.* 169, 2121–2131.

Junelles, A.-M., Janati-Idrissi, R., El Kanouni, A., Petitdemange, H., and Gay, R. (1987) *Biotechnol. Lett.* 9, 175–178.

Jungermann, K., Thauer, R.K., Leimenstoll, G., and Decker, K. (1973) *Biochim. Biophys. Acta* 305, 268–280.

Kathariou, S,. Metz, P., Hof, H., and Goebel, W. (1987) *J. Bacteriol.* 169, 1291–1297.

Khan, S.A., and Novick, R.P. (1980) *Plasmid* 4, 148–154.

Kigbo, E.P., Townsend, D.E., Ashdown, N., and Grubb, W.B. (1985) *FEMS Microbiol. Lett.* 28, 39–43.

Kreft, J., Bernhard, K., and Goebel, W. (1978) *Mol. Gen. Genet. 162*, 59–67.

Kuramitsu, H.K., and Casadaban, M.J. (1986) *J. Bacteriol.* 167, 711–712.

Le Bouguénec, C., de Cespédès, G., and Horaud, T. (1988) *J. Bacteriol.* 170, 3930–3936.

Lereclus, D., Mahillon, J., Menou, G., and Lecadet, M.-M. (1986) *Mol. Gen. Genet.* 204, 52–57.

Lyon, B.R., May, J.W., and Skurray, R.A. (1984) *Mol. Gen. Genet.* 193, 554–556.

Lyon, B.R., and Skurray, R. (1987) *Microbiol. Rev.* 51, 88–134.

Magot, M. (1983) *FEMS Microbiol. Lett.* 18, 149–151.

Mahillon, J., and Seurinck, J. (1988) *Nucleic Acids Res.* 16, 11827–11828.

Martin, P., Trieu-Cuot, P., and Courvalin, P. (1986) *Nucleic Acids Res.* 14, 7047–7058.

Mattsson, D.M., and Rogers, P. (1989) *Abstr. Annu. Meet. Am. Soc. Microbiol. 1989*, O-39, p. 311.

Mazodier, P., Petter, R., and Thompson, C. (1989) *J. Bacteriol.* 171, 3583–3585.

Mullany, P., Wilks, M., Lamb, I. et al. (1990) *J. Gen. Microbiol.* 136, 1343–1349.

Murphy, E., and Novick, R.P. (1979) *Mol. Gen. Genet.* 175, 19–30.

Murphy, E., Huwyler, L., and Bastos, M.C.F. (1985) *EMBO J.* 4, 3357–3365.

Murray, B.E, An, F.A., and Clewell, D.B. (1988) *Antimicrob. Agents Chemother.* 32, 547–551.

Norgren, M., Caparon, M.G., and Scott, J.R. (1989) *Infect. Immun.* 57, 3846–3850.

Norgren, M., and Scott, J.R. (1991) *J. Bacteriol.* 173, 319–324.

Novick, R.P., Edelman, I., Schwesinger, M.D., et al. (1979) *Proc. Natl. Acad. Sci. USA* 76, 400–404.

Palosaari, N.R., and Rogers, P. (1988) *J. Bacteriol.* 170, 2971–2976.

Petersen, D.J., and Bennett, G.N. (1990) *Appl. Environ. Microbiol.* 56, 3491–3498.

Petitdemange, H., Cherrier, C., Bengone, J.M., and Gay, R. (1977) *Can. J. Microbiol.* 23, 152–160.

Petitdemange, H. Cherrier, C., Raval, G., and Gay, R. (1976) *Biochim. Biophys. Acta* 421, 334–347.

Phillips, S., and Novick, R.P. (1979) *Nature (London)* 278, 476–478.

Poyart-Salmeron, C., Trieu-Cuot, P., Carlier, C., and Courvalin, P. (1989) *EMBO J.* 8, 2425–2433.

Poyart-Salmeron, C., Trieu-Cuot, P., Carlier, C., and Courvalin, P. (1990) *Mol. Microbiol.* 4, 1513–1521.

Roberts, M.C., and Kenny, G.E. (1987) *J. Bacteriol.* 169, 3836–3839.

Roberts, M.C., and Lansciardi, J. (1990) *Abstr. Annu. Meet. Am. Soc. Microbiol. 1990*, A-63, p. 11.

Rogers, P. (1986) *Adv. Appl. Microbiol.* 31, 1–60.

Rouch, D.A., Messerotti, L.J., Loo, L.S., Jackson, C.A., and Skurray, R.A. (1989) *Mol. Microbiol.* 3, 161–175.

Rowland, S.-J., and Dyke, K.G.H. (1990) *Mol. Microbiol.* 4, 961–975.

Schäfer, A., Kalinowski, J., Simon, R., Seep-Feldhaus, A.-H., and Pühler, A. (1990) *J. Bacteriol.* 172, 1663–1666.

Scott, J.R., Kirchman, P.A., and Caparon, M.G. (1988) *Proc. Natl. Acad. Sci. USA* 85, 4809–4813.

Sen, S., and Oriel, P. (1990) *FEMS Microbiol. Lett.* 67, 131–134.

Shaw, J.H., and Clewell, D.B. (1985) *J. Bacteriol.* 164, 782–796.

Shoemaker, N.B., and Salyers, A.A. (1988) *J. Bacteriol.* 170, 1651–1657.
Shoemaker, N.B., and Salyers, A.A. (1990) *J. Bacteriol.* 172, 1694–1702.
Siemieniak, D.R., Slightom, J.L., and Chung, S.-T. (1990) *Gene* 86, 1–9.
Smith, J.C., Markowitz, S.M., and Macrina, F.L. (1981) *Antimicrob. Agents Chemother.* 19, 997–1003.
Solenberg, P.J., and Baltz, R.H. (1991) *J. Bacteriol.* 173, 1096–1104.
Steffen, C., and Matzura, H. (1989) *Gene* 75, 349–354.
Syvanen, M. (1986) *Trends Genet.* 2, 63–66.
Thomas, W.D., and Archer, G.L. (1989) *Antimicrob. Agents Chemother.* 33, 1335–1341.
Tomich, P.K., An, F.Y., and Clewell, D.B. (1980) *J. Bacteriol.* 141, 1366–1374.
Townsend, D.E., Ashdown, N., Greed, L.C., and Grubb, W.B. (1984) *J. Antimicrob. Chemother.* 14, 115–124.
Trees, D.L., and Iandolo, J.J. (1988) *J. Bacteriol.* 170, 149–154.
Trieu-Cuot, P., Arthur, M., and Courvalin, P. (1987a) in *Streptocccal Genetics* (Ferretti, J.J., and Curtiss, R., III, ed.), pp. 65–68, American Society for Microbiology, Washington, D.C.
Trieu-Cuot, P., Arthur, M., and Courvalin, P. (1987b) *Microbiol. Sci.* 4, 263–266.
Trieu-Cuot, P., Carlier, C., and Courvalin, P. (1988) *J. Bacteriol.* 170, 4388–4391.
Trieu-Cuot, P., Carlier, C., Martin, P., and Courvalin, P. (1987c) *FEMS Microbiol. Lett.* 48, 289–294.
Trieu-Cuot, P., Gerbaud, G., Lambert, T., and Courvalin, P. (1985) *EMBO J.* 4, 3583–3587.
Trieu-Cuot, P., Poyart-Salmeron, C., Carlier, C., and Courvalin, P. (1990) *Nucleic Acids Res.* 18, 3660.
Vijayakumar, M.N., Priebe, S.D., and Guild, W.R. (1986) *J. Bacteriol.* 166, 978–984.
Volk, W.A. Bizzini, B., Jones, K.R., and Macrina, F.L. (1988) *Plasmid* 19, 255–259.
Weber, D.A., and Goering, R.V. (1988) *Antimicrob. Agents Chemother.* 32, 1164–1169.
Welch, R.W., Rudolph, F.B., and Papoutsakis, E.T. (1989) *Arch. Biochem. Biophys.* 273, 309–318.
Woolley, R.C., Pennock, A., Ashton, R.J., Davies, A., and Young, M. (1989) *Plasmid* 22, 169–174.
Wüst, J., and Hardegger, U. (1983) *Antimicrob. Agents Chemother.* 23, 784–786.
Youngleson, J.S., Jones, D.T., and Woods, D.R. (1989) *J. Bacteriol.* 171, 6800–6807.
Yu, P.-L., and Pearce, L.E. (1986) *Biotechnol. Lett.* 8, 469–474.

CHAPTER

11

Heat Shock Response and Onset of Solvent Formation in *Clostridium acetobutylicum*

Hubert Bahl

All organisms, from bacteria to plants and animals, respond to several environmental stresses by inducing the synthesis of a set of proteins. The heat shock proteins (hsp) are the best-known members of these so-called stress proteins. The heat shock response in bacteria (in *Escherichia coli*) was discovered long after the phenomenon had been studied in eucaryotic systems (Lemaux et al. 1978; Yamamori et al. 1978). The presence of the heat shock response in *E. coli* made it evident that this cellular response to a temperature upshift is universal in living organisms. It is now known that some of the components of the heat shock response belong to the most conserved genetic elements known (e.g., Bardwell and Craig 1984).

The function of the heat shock response or hsp is still a matter of debate. Besides the view that the heat shock response serves as a mechanism to protect

Work reported from this laboratory has been supported by the Deutsche Forschungsgemeinschaft. I express my appreciation to the members of my group, Andreas Pich, Franz Narberhaus, Katharina Giebeler, and Susanne Behrens, for their enthusiasm and valuable contributions.

the cells from damage at high lethal temperatures (acquired thermotolerance) (Yamamori and Yura 1982), there is growing evidence that hsp, which are present also in the unstressed cell under normal physiological conditions, play a central role in cell physiology. The involvement in assembly and disassembly of protein complexes and in protein-translocation processes are two recently discovered functions of hsp (Pelham 1986; Ellis 1987; Goloubinoff et al. 1989; Parsell and Sauer 1989). Neidhardt and VanBogelen (1987) have reviewed the heat shock response in *E. coli*. Lindquist and Craig (1988) and Watson (1990) have presented recent overviews on the heat shock response in different microorganisms.

Although *E. coli*, with its well-known genetic properties, will continue to be the experimental system of choice to study the heat shock response, investigations of other bacteria, especially those with unusual metabolic capacity, will make substantial contributions to our understanding of the stress response. The obviously central role of hsp in cell physiology makes the heat shock response an important research area also for clostridia with their high potential in biotechnology (see Chapter 1, this volume). When used in industrial processes, the cells are often grown under stressful conditions to improve the yield of the desired product. In this chapter, the presently limited knowledge on the heat shock response in a prominent member of the clostridia, *Clostridium acetobutylicum*, is summarized.

11.1 HEAT SHOCK PROTEINS IN *CLOSTRIDIUM ACETOBUTYLICUM*

The hsp can be easily identified by comparing autoradiograms of two-dimensional gels (O'Farrell 1975) of cell extracts that have been pulse-labeled (e.g., with [^{35}S]methionine) before and shortly after a shift to high temperature. By this method, at least 10 hsp were recognized in *C. acetobutylicum* ATCC 4259 (Terracciano et al. 1988), whereas 15 hsp were found in strain DSM 792 (Pich et al. 1990). In Figure 11–1, the effect of heat on protein synthesis in *C. acetobutylicum* is depicted. The temperature upshift from 30–42°C has obviously two consequences: (1) enhanced synthesis of *C. acetobutylicum* hsp and (2) decreased synthesis of several other cellular proteins. The molecular masses of the hsp range from 14–73 kDa (16–83 kDa in strain ATCC 4259). In accordance with convention, the heat shock-inducible proteins are referred to as hsp72, hsp67, hsp21, etc. The number after hsp corresponds to the molecular mass, which is usually estimated by sodium dodecyl sulfate-polyacrylamide gel electrophoresis (SDS-PAGE). Despite slightly different molecular masses, several hsp of the examined two *C. acetobutylicum* strains to date are considered as probably the equivalent proteins, as based on their behavior on two-dimensional gels (Pich et al. 1990; see also Table 11–1).

To date two hsp of *C. acetobutylicum* were identified by use of specific heterologous antibodies. The hsp67 of *C. acetobutylicum* is immunoreactive

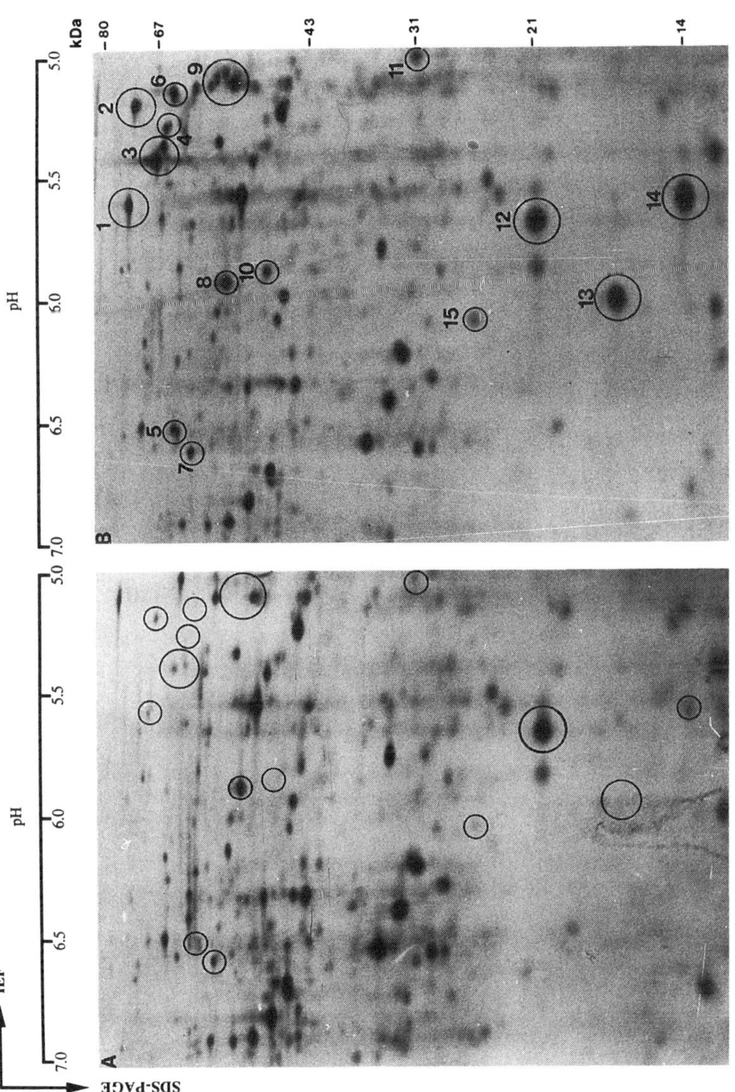

FIGURE 11-1 Heat shock proteins (hsp) in *C. acetobutylicum* DSM 792. Proteins of cells growing at 30°C (A) and heat shocked for 8 min at 42°C (B) were pulsed-labeled with [^{35}S]methionine, extracted, fractionated by sodium dodecyl sulfate-polyacrylamide gel electrophoresis (SDS-PAGE) in two-dimensional gels, and visualized by autoradiography. Proteins induced by the temperature shift are circled: 1, hsp73; 2, hsp72 (DnaK); 3, hsp67 (GroEL); 4, hsp66; 5, hsp65a; 6, hsp65b; 7, hsp61; 8, hsp53; 9, hsp50; 10, hsp45; 11, hsp31; 12, hsp21; 13, hsp17; 14, hsp14; 15, hsp25. IEF, isoelectric focusing. Reproduced with permission from Pich et al. (1990).

TABLE 11–1 Heat Shock Proteins (hsp) in *E. coli* and *C. acetobutylicum*

E. coli[1]		C. acetobutylicum[2]
Lon (94 kDa)	↔	hsp83*
hsp84.1 (ClpB)		
hsp71	↔	hsp73 [1]
sigma-70		
DnaK (69 kDa)	⇔	DnaK (72 kDa, 74 kDa*) [2]
GroEL (63 kDa)	⇔	GroEl (67 kDa, 68 kDa*) [3]
Lysyl-tRNA synthethase		hsp66 [4]
form II (60.5 kDa)		hsp65a [5]
		hsp65b/hsp62* [6]
		hsp61 [7]
hsp48.5	↔	hsp53 [8]
DnaJ (41 kDa)		hsp50/hsp49a,b* [9]
hsp33.4		hsp45 [10]
		hsp36*
GrpE (25 kDa)	↔	hsp31 [11]
hsp21.5		hsp25 [15]
hsp21		hsp21/hsp22* [12]
		hsp18b*
GroES (11 kDa)		hsp17/hsp18a* [13]
hsp14.7	↔	hsp14/hsp16* [14]
hsp13.5		
hsp10.1		

[1] Data taken from Neidhardt and VanBogelen (1987).
[2] Data taken from Pich et al. (1990) (*C. acetobutylicum* DSM 792) and from Terracciano et al. (1988) (*C. acetobutylicum* ATCC 4529, marked by asterisks). Numbers in brackets [ ] refer to the same numbers in Figure 11–1. ⇔, indicates homologous proteins, as based on their immunological relationship; ↔, indicates tentatively equivalent proteins, as based on their behavior after two-dimensional gel electrophoresis.

to antibodies raised against GroEL, a hsp of *E. coli* (Pich et al. 1990). Antibodies against the DnaK hsp of *E. coli* identified the hsp72 (hsp74 in strain ATCC 4259) as a similar protein (Terracciano et al. 1988; Pich et al. 1990). Although none of the hsp of *C. acetobutylicum* reacted to polyclonal antibodies raised against the hsp DnaJ, GrpE, and σ^{70} of *E. coli*, *dnaJ* and *grpE* homologous genes of *C. acetobutylicum* were cloned and sequenced as well as the *dnaK*, *groEL*, and *groES* homologous genes (Narberhaus and Bahl 1991; F. Narberhaus et al., unpublished observations). These results indicate that *C. acetobutylicum* induces several similar hsp after a temperature upshift as *E. coli*. The hsp of both microorganisms are listed in Table 11–1.

Since it is well established that hsp are also induced when the cells are exposed to other environmental stresses than heat, it was not surprising that several hsp of *C. acetobutylicum* were induced by solvents or air (Terracciano et al. 1988) and changes of the pH or the growth rate in continuous culture (Pich et al. 1990) (Table 11–2).

TABLE 11–2 Induction of Heat Shock Proteins (hsp) in *C. acetobutylicum* by Environmental Stresses Other than Heat

Environmental Stress	*Heat Shock Proteins*	*Reference*
Butanol[1]	hsp83, DnaK; GroEL hsp22, hsp18, hsp16	Terracciano et al. 1988
Air[1]	GroEL, hsp22	Terracciano et al. 1988
pH change[2]	hsp73, DnaK, GroEL	Pich et al. 1990
Growth rate change[2]	DnaK, GroEL; hsp21	Pich et al. 1990

[1] Strain ATCC 4259 was used.
[2] Strain DSM 1731 was used.

11.2 INDUCTION OF HEAT SHOCK PROTEINS AT THE ONSET OF SOLVENT FORMATION

C. acetobutylicum ferments sugars to acids such as acetate and butyrate in addition to CO_2 and H_2. At the end of the exponential growth phase, the culture undergoes the so-called shift and residual sugar and preformed acids are converted to butanol and acetone. For almost a century, microbial physiologists have been studying the mechanism underlying solvent formation (for recent reviews see Jones and Woods 1986; Bahl and Gottschalk 1988; and several chapters of this book). A tremendous amount of information has been obtained by several approaches. One approach studies the physiological conditions, which favor solvent production (Bahl and Gottschalk 1984), other approaches investigate the enzymes involved (Zerner et al. 1966; Palosaari and Rogers 1988; Wiesenborn et al. 1989; Welch et al. 1989) and clone and characterize the corresponding genes (Youngleson et al. 1989; Gerischer and Dürre 1990; Petersen and Bennett 1990; Petersen et al. 1991). Pich et al. (1990) have recently described a more global strategy. The idea is to identify sets of genes or proteins that respond to certain environmental conditions, e.g., conditions which initiate the shift from acid to solvent production in *C. acetobutylicum*. The final goal is to determine the function of the proteins affected and uncover the regulatory components that combine the different groups of proteins. Descriptions of such a global system approach for different multigene systems are given by Gottesman (1984), and VanBogelen and Neidhardt (1990).

Since the factors necessary for the initiation of solvent formation by *C. acetobutylicum* (e.g., low pH, low growth rate, and excess of carbohydrates; Bahl and Gottschalk 1984) might provide a stressful situation for this organism, it is likely that some kind of stress response is involved in this metabolic shift. Further evidence that several regulatory circuits are involved in the shift comes from the observation that sporulation of *C. acetobutylicum* and solvent formation are linked (Long et al. 1984). A schematic possible regulatory network in *C. acetobutylicum* is presented in Figure 11–2. The first observation

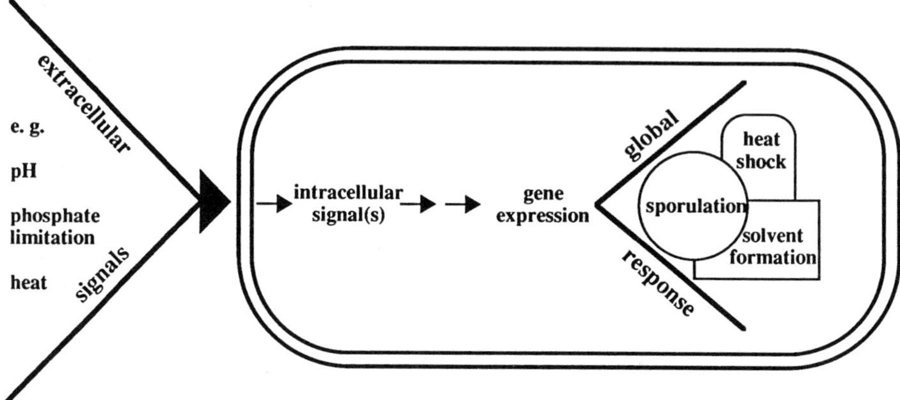

FIGURE 11–2 Schematic view of a regulatory network in *C. acetobutylicum*.

that hsp might be involved in the shift was made by Terracciano et al. (1988). In the absence of a heat shock, the synthesis rate of several hsp was found to vary with the *C. acetobutylicum* ATCC 4259 growth phase. The DnaK protein and the hsp22 were most prominent in the late acid-producing phase, and the GroEL protein was produced at maximum rates in the solvent-producing phase. However, the question of whether initiation of solvent formation and the heat shock response are, in fact, related could not be answered with these batch culture experiments. Several parameters change at the same time, which could be responsible for the induction of heat shock/stress proteins (pH, product concentration, growth rate, etc.). The increased synthesis of stress proteins during different growth phases has been shown in other bacteria (Goodman et al. 1985; Watt and North 1987; Nelson and Killen 1986). Therefore, a chemostat was used by Pich et al. (1990) to be independent from the growth phase and determine the effect of conditions needed for the initiation of solvent formation on the synthesis of hsp. The shift towards solvent formation by *C. acetobutylicum* can be reproducibly achieved by a pH change from 6.0 to 4.3 under phosphate limitation, if a threshold butyrate concentration is present (Bahl and Gottschalk 1984). Thus, if the chemostat is running at pH 6.0 (temperature, 37°C; dilution rate, 0.1 h^{-1}; phosphate concentration, 0.38 mM), acids are produced. Lowering the pH below 5.0 results in a shift to solvent formation (Figure 11–3A). Protein synthesis before, during, and after the shift can then be examined by pulse-labeling at different time points and two-dimensional gel electrophoresis (Figure 11–3D). Since it is known that enzyme induction occurs during the shift, a change in the protein pattern is expected. Interestingly, some of the proteins whose synthesis rate increases are hsp of *C. acetobutylicum*. Hsp14, hsp17, and GroEL show a striking increase, and hsp73 and DnaK are slightly induced during the shift (Figures 11–3B and 11–3D). In addition to hsp, other cellular proteins show altered synthesis rates, which could not be identified with the exception of one protein. One group of these proteins (P38, P34.5, and P28; designated according to their molecular masses) was synthesized at elevated levels during and after the shift to pH 5.0 (Figures 11–3C and 11–3D). These proteins are considered to be solvent phase-associated proteins. The P28 protein was identified as acetoacetate decarboxylase (cited as unpublished result in Pich et al. 1990), the final enzyme in the pathway of acetone formation. The synthesis rate of another group of proteins is reduced, and these proteins are obviously acid phase-associated (Figures 11–3C and 11–3D). Important is the time point when the synthesis rates of hsp increase. Inducting specific hsp immediately after the initiation of the shift, together with inducting solvent phase-associated proteins (e.g., acetoacetate decarboxylase) and several hours before solvents appear in the medium, indicates that this phenomenon is not a response to low amounts of solvents produced by the cells. In addition, pH changes under acid-producing conditions elicited the synthesis of only a few hsp in *C. acetobutylicum* (Table 11–2). These results are the first evidence that the conditions responsible for the metabolic shift in *C. acetobutylicum* are also able to trigger the heat shock/stress response.

The link between these two processes has still to be uncovered. For this purpose, the isolation and characterization of suitable mutants will be helpful, together with the analysis and comparison of the regulation of genes encoding hsp and enzymes involved in solvent formation. In the last part of this chapter, an overview on data and speculations about the regulation of the heat shock/stress response in *C. acetobutylicum* is given.

11.3 REGULATION OF THE HEAT SHOCK/STRESS RESPONSE IN *CLOSTRIDIUM ACETOBUTYLICUM*

In *E. coli*, the heat shock response has become one of the best-known model systems for studying global control of gene expression, and the mechanism of hsp induction has been elucidated. The coordinate expression of the heat shock genes is based on the fact that they all are transcribed by RNA polymerase holoenzyme that contains an alternate sigma-factor (σ^{32}). Following temperature upshift the concentration of σ^{32} transiently increases. As a result, the rate of transcription of heat shock genes transiently rises. This simple mechanism is complicated by the fact that the DnaK, DnaJ, and GrpE hsp negatively regulate the activity of σ^{32}. The amount of σ^{32} present in the cell is regulated at the transcriptional, posttranscriptional, and posttranslational levels (Straus et al. 1987; Bahl et al. 1987; Erickson et al. 1987).

In *C. acetobutylicum*, the heat shock response is also transient (Figure 11–4) (Pich et al. 1990). Whether this fact points to a similar regulatory mechanism can only be speculated. As outlined previously, heat shock response, solvent formation, and sporulation are somehow connected in *C. acetobutylicum*. Since sporulation in *Bacillus subtilis* and heat shock in *E. coli*

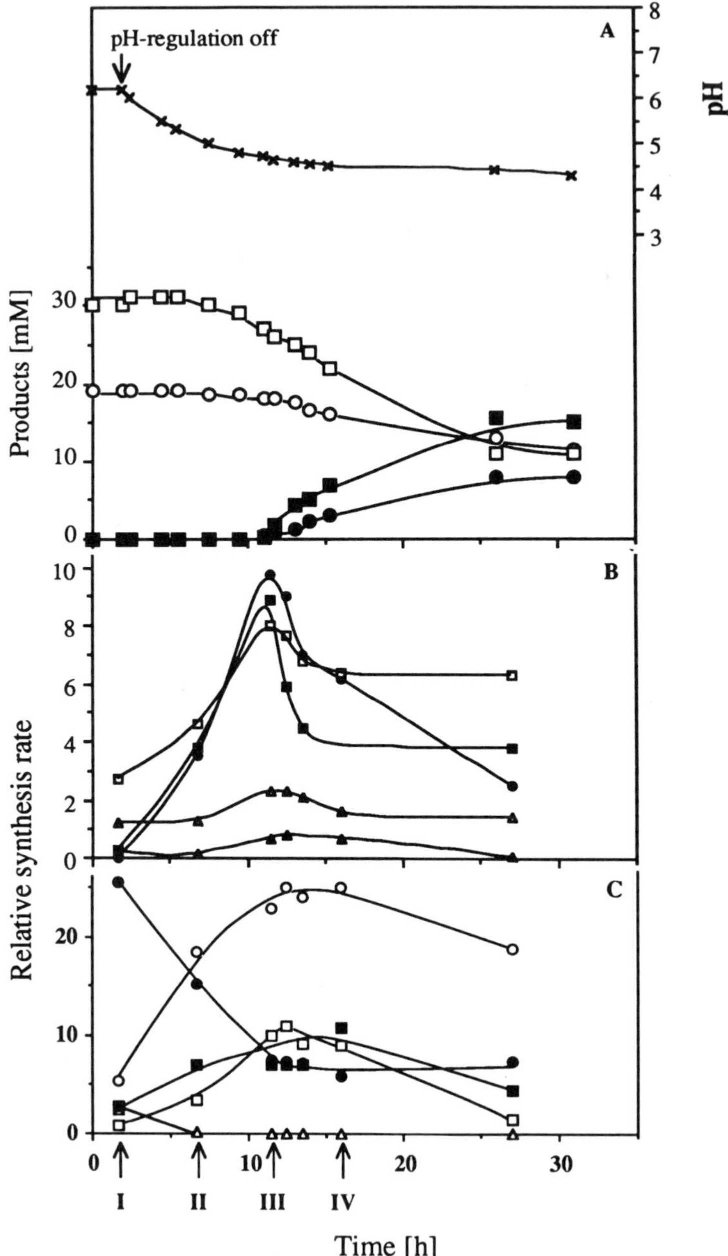

FIGURE 11–3 Synthesis of heat shock proteins (hsp) in *C. acetobutylicum* DSM 1731 during the onset of solvent formation. Cells were grown in a phosphate-limited chemostat (temperature 37°C, dilution rate 0.1 h^{-1}, 0.38 mM phosphate, pH 6.4). Solvent formation was induced by lowering the pH to 4.3 (A): x, pH; ●, acetone; ○, acetate; ■, butanol; □, butyrate. Samples were withdrawn at different time points, pulse-labeled with [^{35}S]methionine, and analyzed by two-dimensional gel electrophoresis.

(*continued on next page*)

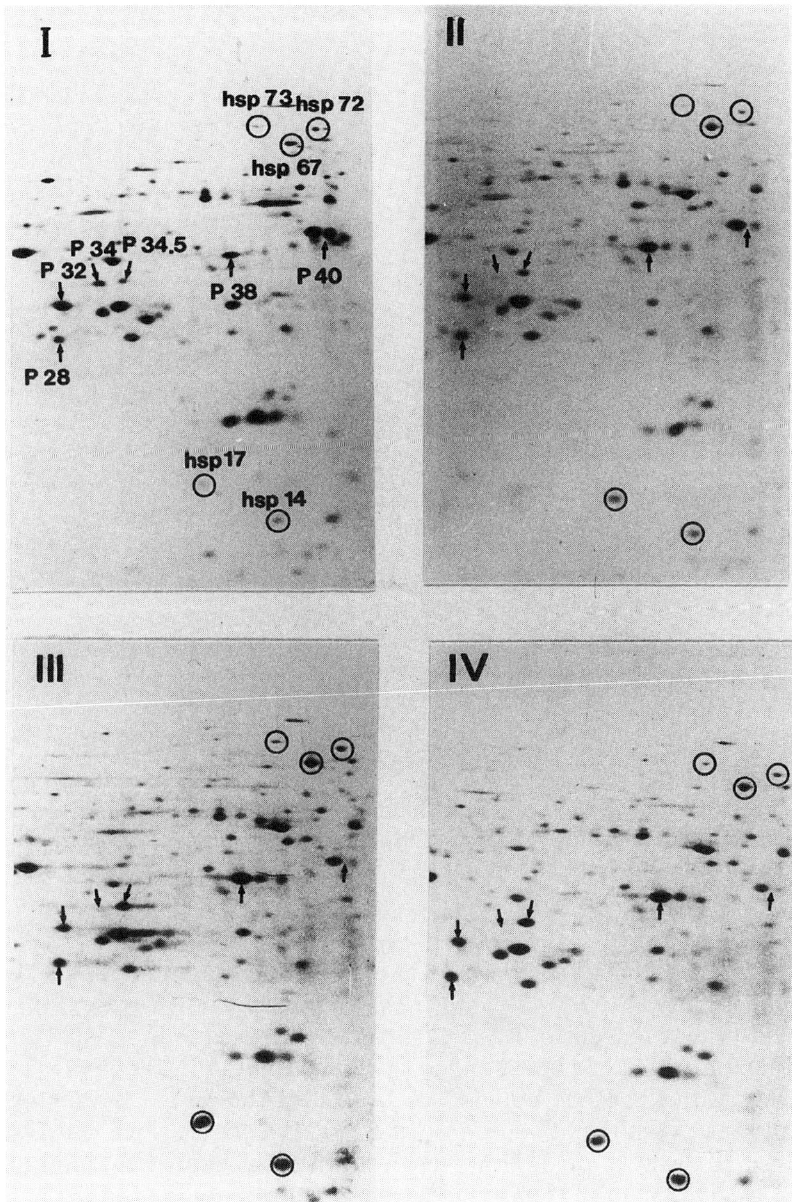

FIGURE 11–3 (continued) The synthesis rates of hsp and other cellular proteins have been determined from densitometer tracings of the two-dimensional gels. (B): ▲, hsp73; △, hsp72 (DnaK); □, hsp67 (GroEL); ●, hsp17; ■, hsp14. (C): ●, P40; ○, P38, □, P34.5; △, P34; ■, P28 (acetoacetate decarboxylase). Representative autoradiographs [samples taken at time points indicated by arrows I, II, III, and IV in (C)] are shown in (D). Figures 11–3A, 11–3B, and 11–3D reproduced with permission from Pich et al. (1990).

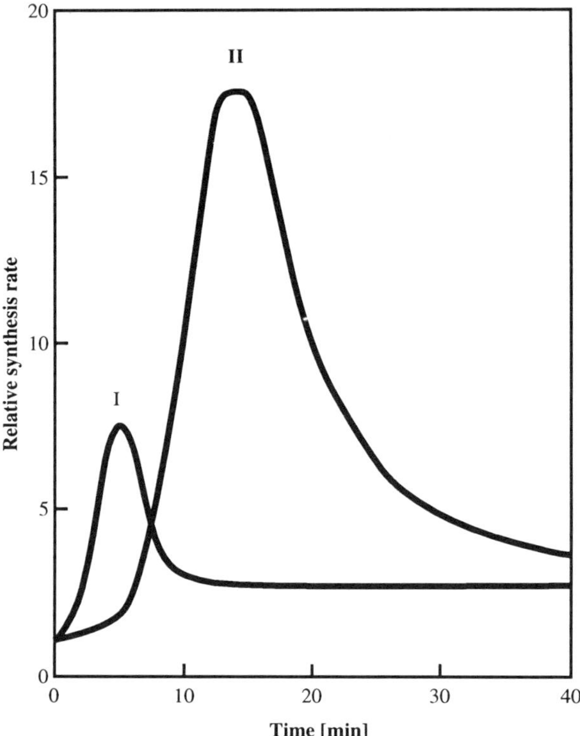

FIGURE 11–4 Time course (schematic) of heat shock protein (hsp) induction in *C. acetobutylicum*. Two temporal classes can be distinguished. (I): hsp61, hsp53, hsp21; (II): all other hsp including DnaK and GroEL. The synthesis rates of hsp in group (I) increase three- to sevenfold, in group (II) 4- to 18-fold.

involves alternate sigma-factors (Doi and Wang 1986), it is tempting to assume such a possibility also in *C. acetobutylicum*. The RNA polymerase from growing, acid-producing cells of *C. acetobutylicum* has been purified and a major sigma-factor (σ^{46}) could be identified (Pich and Bahl 1991). A 66-kDa protein copurifies with the RNA polymerase from heat-shocked *C. acetobutylicum* cells and reacts with antibodies against σ^{32} of *E. coli*. (A. Pich and H. Bahl, unpublished observations). However, it is not clear yet, whether this protein acts as a sigma-factor in *C. acetobutylicum*. Analysis of the expression of heat shock genes in *C. acetobutylicum* gave no evidence to date that an alternate sigma-factor is involved (F. Narberhaus and H. Bahl, unpublished observations). Furthermore, the RNA polymerase of solvent-producing cells is indistinguishable from the enzyme of acid-producing cells (A. Pich and H. Bahl, unpublished observations). Therefore, a common regulatory unit for the different stress responses (heat shock, solventogenesis, and sporulation) in *C. acetobutylicum* seems to be on another level.

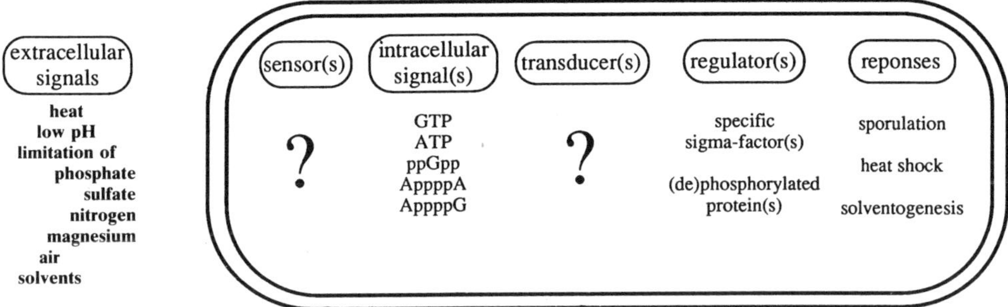

FIGURE 11–5 Speculative and simplified view of signalling in *C. acetobutylicum* during the stress response (sporulation, heat shock, and solventogenesis). It has been shown that a relationship between the three above-mentioned processes exists. However, it is not known at which point signalling is divided leading to the different responses.

In Figure 11–5 a simplified view of signalling during stress response in *C. acetobutylicum* is illustrated. The extracellular signals that lead to the different stress responses are mostly known. However, how *C. acetobutylicum* senses these signals and what are the intracellular signals is not known. Whereas no data are available on the sensor, several possible internal signals have been discussed (see also Chapter 2, this volume, with special emphasis on signals for the metabolic shift).

Two of the candidates examined are adenylated nucleotides, such as diadenosine 5′,5‴-P^1,P^4-tetraphosphate (Ap$_4$A) and adenosine 5′-P^1,P^4-tetraphospho-5′-guanosine (Ap$_4$G) (Balodimos et al. 1988). These two compounds accumulate in batch culture in the absence of imposed stresses, when *C. acetobutylicum* switches from the acidogenic to the solventogenic phase. However, they are extracellular and reach their maximum level in the late solvent phase after sporulation has initiated. Therefore, these nucleotides do not seem directly involved in stress response regulation in *C. acetobutylicum*. The role of guanosine triphosphate (GTP) and adenosine triphosphate (ATP) in the differentiation of *C. acetobutylicum* and during the onset of solvent formation has also been investigated (Santangelo et al. 1989; Meyer and Papoutsakis 1989; Grupe and Gottschalk 1990). Just before the initiation of solvent production and sporulation, the nucleoside triphosphate levels go down. Since a low GTP content seems to be one of the most crucial cellular signals controlling the adaptational network in *B. subtilis* (Hecker and Völker 1990), this nucleotide may be also important for signalling in *C. acetobutylicum*. Furthermore, Ahmed et al. (1988) have argued that the expression of the solventogenic pathway in a closely related bacterium, *C. beijerinckii*, is under the influence of guanosine tetraphosphate (ppGpp), a derivative of GTP and an indicator for certain starvation conditions in *E. coli* (Cashel and Rudd 1987) and *B. subtilis* (Hecker and Völker 1990).

After the signal has been created, whatsoever its nature may be like, the information has to be transferred to a regulator. The question arises: is there only one or are there different regulators for the different stress responses (heat shock, sporulation, and solventogenesis). The possible involvement of stress specific sigma-factor(s) has been discussed above. Balodimos et al. (1990) investigated whether a reversible protein phosphorylation mechanism participates in the regulation of the stress response in *C. acetobutylicum*. The extent of protein phosphorylation is increased by heat stress, and the phosphorylation of a 50-kDa protein is enhanced in extracts prepared from heat-stressed cells. Purified DnaK of *C. acetobutylicum* acts as a kinase catalyzing the phosphorylation of this protein. A more detailed characterization of the 50-kDa protein will be necessary, before a definite answer about its possible regulatory role can be given.

In conclusion, there are many problems that need further investigations for their solution. However, it should be emphasized that hsp in *C. acetobutylicum* are interesting proteins, because inducing these proteins may be an important adaptive strategy during the transition from acid to solvent formation and a growing to a nongrowing state, and during other stressful environmental conditions. Furthermore, the analysis of molecular details of the heat shock/stress response in *C. acetobutylicum* is helpful for the understanding of global gene expression in this organism.

REFERENCES

Ahmed, I., Ross, R.A., Mathur, V.K., and Chesbro, W.R. (1988) *Appl. Microbiol. Biotechnol.* 28, 182–187.

Bahl, H., Echols, H., Straus, D.B., et al. (1987) *Genes Dev.* 1, 57–64.

Bahl, H., and Gottschalk, G. (1984) *Biotechnol. Bioeng. Symp.* 11, 215–223.

Bahl, H., and Gottschalk, G. (1988) in *Biotechnology*, vol. 6b, (Rehm, H.J., and Reed, G., eds.), pp. 1–30, VCH Verlagsgesellschaft, Weinheim, Germany.

Balodimos, I.A., Kashket, E.R., and Rapaport, E. (1988) *J. Bacteriol.* 170, 2301–2305.

Balodimos, I.A., Rapaport, E., and Kashket, E.R. (1990) *Appl. Environ. Microbiol.* 56, 2170–2173.

Bardwell, J.C.A. and Craig, E.A. (1984) *Proc. Natl. Acad. Sci. USA* 81, 848–852.

Cashel, M., and Rudd, K.E. (1987) in Escherichia coli *and* Salmonella typhimurium: *Cellular and Molecular Biology, vol. 2*, (Neidhardt, F.C., Ingraham, J.L., Low, K.B., et al. eds.), pp. 1410–1438, American Society for Microbiology, Washington, DC.

Doi, R.H., and Wang, L.-F. (1986). *Microbiol. Rev.* 50, 227–243.

Ellis, R.J. (1987) *Nature* (*London*) 328, 378–379.

Erickson, J.W., Vaughn, V., Walter, W.A., Neidhardt, F.C., and Gross, C.A. (1987) *Genes Dev.* 1, 419–432.

Gerischer, U., and Dürre, P. (1990) *J. Bacteriol.* 172, 6907–6918.

Goloubinoff, P., Christeller, J.T., Gatenby, A.A., and Lorimer, G.H. (1989) *Nature* (*London*) 342, 884–889.

Goodman, H.J.K., Strydom, E., and Woods, D.R. (1985) *Arch. Microbiol.* 142, 362–364.
Gottesman, S. (1984) *Annu. Rev. Genet.* 18, 415–441.
Grupe, H., and Gottschalk, G. (1990) in *Sixth International Symposium on Genetics of Industrial Microorganisms, GIM 90, Proceedings, vol. 2*, (Heslot, H., Davies, J., Florent, J., et al. eds), Aug. 12–18, 1990, Strasbourg, France, pp. 715–729, Société Francaise de Microbiologie, Strasbourg, France.
Hecker, M., and Völker, U. (1990) *FEMS Microbiol. Ecol.* 74, 197–214.
Jones, D.T., and Woods, D.R. (1986) *Microbiol. Rev.* 50, 484–524.
Lemaux, P.G., Herendeen, S.L., Bloch, P.L, and Neidthardt, F.C. (1978) *Cell* 13, 427–434.
Lindquist, S., and Craig, E.A. (1988) *Annu. Rev. Genet.* 22, 631–677.
Long, S., Jones, D.T., and Woods, D.R. (1984) *Biotechnol. Lett.* 6, 529–534.
Meyer, C.L., and Papoutsakis, E.T. (1989) *Appl. Microbiol. Biotechnol.* 30, 450–459.
Narberhaus, F., and Bahl, H. (1991) *Bioforum* 14, 112.
Neidhardt, F.C., and VanBogelen, R.A. (1987) in Escherichia coli *and* Salmonella typhimurium: *Cellular and Molecular Biology, vol. 2*, (Neidhardt, F.C., Ingraham, J.L., Low, K.B., et al. eds.), pp. 1334–1345, American Society for Microbiology, Washington, DC.
Nelson, D.R., and Killeen, K.P. (1986) *J. Bacteriol.* 168, 1100–1106.
O'Farrell, P.H. (1975) *J. Biol. Chem.* 250, 4007-4021.
Palosaari, N.R., and Rogers, P. (1988) *J. Bacteriol.* 170, 2971–2976.
Parsell, D.A., and Sauer, R.T. (1989) *Genes Dev.* 3, 1226–1232.
Pelham, H.R.B. (1986) *Cell* 46, 959–961.
Petersen, D.J., and Bennett, G.N. (1990) *Appl. Environ. Microbiol.* 56, 3491–3498.
Petersen, D.J., Welch, R.W., Rudolph, F.B., and Bennett, G.N. (1991) *J. Bacteriol.* 173, 1831–1834.
Pich, A., and Bahl, H. (1991) *J. Bacteriol.* 173, 2120–2124.
Pich, A., Narberhaus, F., and Bahl, H. (1990) *Appl. Microbiol. Biotechnol.* 33, 697–704.
Santangelo, J.D., Jones, D.T., and Woods, D.R. (1989) *J. Gen. Microbiol.*, 135, 711–719.
Straus, D.B., Walter, W.A., and Gross, C.A. (1987) *Nature (London)* 329, 348–351.
Terracciano, J.S., Rapaport, E., and Kashket, E.R. (1988) *Appl. Environ. Microbiol.* 54, 1989–1995.
VanBogelen, R.A., and Neidhardt, F.C. (1990) *FEMS Microbiol. Ecol.* 74, 121–128.
Watson, K. (1990) *Adv. Microbial Physiol.* 31, 183–223.
Watt, P.W., and North, M.J. (1987) *FEBS Lett.* 215, 295–299.
Welch, R.W., Rudolph, F.B., and Papoutsakis, E.T. (1989) *Arch. Biochem. Biophys.* 273, 309–318.
Wiesenborn, D.P., Rudolph, F.B., and Papoutsakis, E.T. (1989) *Appl. Environ. Microbiol.* 55, 323–329.
Yamamori, T., Ito, K., Nakamura, Y., and Yura, T. (1978) *J. Bacteriol.* 134, 1133–1140.
Yamamori, T., and Yura, T. (1982) *Proc. Natl. Acad. Sci. USA* 79, 860–864.
Youngleson, J.S., Jones, W.A., Jones, D.T., and Woods, D.R. (1989) *Gene* 78, 355–364.
Zerner, B:, Coutts, S.M., Lederer, F., Waters, H.H., and Westheimer, F.H. (1966) *Biochemistry* 5, 813–816.

CHAPTER

12

Molecular Biology and Genetics of Substrate Utilization in Clostridia

Karin Bronnenmeier
Walter L. Staudenbauer

Solventogenic clostridia are the focus of increased interest because of their potential for converting carbohydrate substrates from surplus agricultural crops and industrial waste into solvents and fuels. Thus, extensive research is carried out on the acetone-butanol fermentation of *Clostridium acetobutylicum* (Jones and Woods 1986), while thermophilic clostridia are evaluated for the fermentative production of ethanol (Lynd 1989). Possible carbohydrate feed stocks for these fermentations include starch, cellulose, xylans, and lactose.

Research on clostridial fermentations has been greatly stimulated by the prospect of applying recombinant DNA technology to saccharolytic clostridia for elucidating the genetic structures and mechanisms involved in substrate utilization and product formation. Moreover, genetic engineering and related techniques offer great promise for the development of novel production strains with enhanced fermentative capabilities and extended substrate range.

It should be pointed out that the taxonomy of saccharolytic clostridia is still in a state of confusion. In particular, the ability to sporulate cannot be considered a primary taxonomic criterium. Thus, the nonsporulating rumen bacterium *Butyrivibrio fibrisolvens* was shown to be closely related to *C. sphenoides* and *C. aminovalericum* (Bryant 1986). Furthermore, comparative analysis of 16S ribosomal RNA sequences (Bateson et al. 1989) and DNA/DNA hybridization studies (Zeikus et al. 1990) have shown that various nonsporulating thermoanaerobic bacteria designated as either *Thermoanaerobium*, *Thermobacteroides* or *Thermoanaerobacter* are closely related to *C. thermohydrosulfuricum* and *C. thermosulfurogenes*, respectively. The thermoanaerobe *Caldocellum saccharolyticum* (Lüthi et al. 1990) also appears to be related to this group of organisms.

Carbohydrate utilization in the clostridia can be divided into three spatially separated stages: (1) polysaccharide degradation by either secreted or cell-associated extracellular enzymes, (2) uptake of mono- and oligosaccharides by PEP-dependent phosphotransferase systems (Mitchell et al. 1991) or by ion-linked transport systems, and (3) intracellular sugar catabolism. The principal route of hexose catabolism in the saccharolytic clostridia is the Emden-Meyerhof glycolytic pathway. Utilization of pentoses is accomplished by the pentose-phosphate pathway and subsequent metabolism via glycolysis.

The molecular mechanisms by which clostridial gene expression is controlled in response to substrate availability are still unknown. In general, bacterial genes concerned with substrate utilization are subject to induction and catabolite repression. The catabolite repression system of enterobacteria involves a positively acting cyclic AMP (cAMP) receptor protein (CRP), which activates transcription of a variety of catabolite-responsive operons. Clostridia and related Gram-positive bacteria such as bacilli apparently lack cAMP. It is, therefore, questionable whether a positive effector of gene expression analogous to the cAMP/CRP complex exists.

This chapter summarizes recent progress in characterizing the genes and proteins concerned with carbohydrate utilization from both sporulating and nonsporulating members of the *Clostridium* group. Possible modes of gene regulation are discussed with reference to the much more advanced genetics of *Bacillus*. Thereby we will employ the approach of molecular biology, that means attempting to understand biological function from molecular structure.

12.1 UTILIZATION OF STARCH

12.1.1 Substrate Structure and Catabolism

Starch granules can be separated into two different high molecular mass fractions, amylose and amylopectin. Amylose is a linear polysaccharide that is formed by α-1,4-linked D-glucose residues and has a tendency to adopt a coiled helical configuration. Amylopectin has a highly branched structure in which linear chains of α-1,4-linked glucose residues are interlinked by α-1,6-glucosidic bonds. The enzymatic hydrolysis of amylose and amylopectin is catalyzed by several enzymes, which have been classified according to the linkages hydrolyzed and their mode of action.

α-Amylases are endo-acting enzymes, which hydrolyze α-1,4-glucosidic linkages in a random fashion. A distinction is commonly made between liquefying and saccharifying enzymes. Hydrolysis of amylose by liquefying α-amylases produces a mixture of linear oligosaccharides. Saccharifying α-amylases further degrade the initially formed oligomers to maltotriose, maltose, and glucose. As neither type of enzyme is capable of cleaving α-1,6-glucosidic bonds, branched oligomers are released from amylopectin in addition to linear degradation products.

β-Amylase and glucoamylase are exo-acting enzymes attacking α-glucans from the nonreducing end of the chains. β-Amylase hydrolyzes alternate α-1,4-linkages releasing maltose units. It is incapable of hydrolyzing or bypassing α-1,6-linkages. Glucoamylase splits α-1,4-glucosidic bonds consecutively to yield glucose. α-1,6-Glucosidic linkages are also attacked, but with a rate of hydrolysis significantly lower than that of α-1,4-linkages. The activity of glucoamylase decreases with decreasing chain length of the substrate. It differs in this respect from α-glucosidase (maltase), which preferentially hydrolyzes maltose and short-chain oligosaccharides. Like glucoamylase, α-glucosidase shows reduced activity against α-1,6-glucosidic linkages. Thus, the α-1,6-linkages act as a barrier to exo-acting amylolytic enzymes.

Efficient degradation of amylopectin requires the action of "debranching enzymes." Pullulanases are endo-acting enzymes capable of hydrolyzing the α-1,6-linkages of pullulan, a fungal polysaccharide composed of maltotriosyl units linked by α-1,6-glucosidic bonds. Most pullulanases also cleave α-1,6-linkages in amylopectin thus exhibiting debranching activity. In general, pullulanases show no activity towards α-1,4-linkages in pullulan and amylose.

Most saccharolytic clostridia are able to use starch as carbon source due to the production of extracellular or cell-associated amylolytic enzymes. *C. acetobutylicum* produces both an α-amylase and a glucoamylase, originally referred to as maltase (Paquet et al. 1991). A β-amylase has been purified from *C. thermosulfurogenes* (Shen et al. 1988). A glucoamylase was recently isolated from *C. thermosaccharolyticum* (Specka et al. 1991). Production of a glucoamylase, which was later designated as α-glucosidase has also been reported for *C. thermohydrosulfuricum* (Hyun et al. 1985; Melasniemi 1987a; Saha and Zeikus 1991).

Several thermophilic anaerobes produce novel type of endo-acting amylolytic enzyme exhibiting both α-amylase and pullulanase activity (Coleman et al. 1987; Plant et al. 1987a; Saha et al. 1990). An amylolytic pullulanase has also been isolated from *C. thermohydrosulfuricum* (Melasniemi 1987b; Saha et al. 1988; Melasniemi 1988). Dual specificity towards α-1,6- and α-1,4-linkages was shown to reside at a single active site (Plant et al. 1987b; Mathupala et al. 1990). Therefore, the designation *amylopullulanase* appears more appropriate for this type of enzyme than the name *α-amylase-pullulan-*

ase, suggesting a bifunctional enzyme structure composed of two separate catalytic domains.

12.1.2 Protein Structure and Function

Cloned clostridial genes concerned with starch utilization are listed in Table 12–1. For most of these genes, the sequence of the encoded protein has been deduced from the nucleotide sequence. Sequence comparison reveals that the clostridial amylases can be grouped into two distinct protein families A and B (Figure 12–1).

Family A consists of several subfamilies and comprises all α-amylases and pullulanases, whose sequence has been determined. It can be seen that *C. thermosulfurogenes* α-amylase is highly homologous to *B. circulans* α-amylase (70% sequence identity). The same extent of similarity is observed between the amylopullulanases of *C. thermohydrosulfuricum* and *C. thermosulfurogenes*.

Family B encompasses a group of closely related bacterial β-amylases (sequence identity >50%) including *C. thermosulfurogenes* β-amylase. It should be noted that *B. polymyxa* α- and β-amylase are derived by proteolytic cleavage from a single precursor protein containing both amylolytic activities (Uozumi et al. 1989). In contrast, the α- and β-amylases of *B. circulans* and *C. thermosulfurogenes* are encoded by separate genes. The N-terminal amino acid (aa) sequence of mature *C. thermosulfurogenes* β-amylase deduced from the DNA sequence was found to be identical with that of the enzyme purified from *C. thermosulfurogenes* cultures (Kitamoto et al. 1988; Shen et al. 1988).

Although the various subfamilies of enzyme family A show little overall sequence homology, several highly conserved sequence elements, designated region I to IV, have been identified (Nakajima et al. 1986). As expected, these sequence elements are also present in the clostridial enzymes (Figure 12–2). Conservation of these regions indicates a common structural design and catalytic mechanism for all family A amylases. The three-dimensional structures of three family A enzymes, Taka-amylase A from *Aspergillus oryzae* (Matsuura et al. 1984), pig pancreas α-amylase (Buisson et al. 1987), and *B. circulans* cyclodextrin glycosyltransferase (Hofmann et al. 1989) have been determined. All three enzymes were found to contain a central catalytic domain with a parallel-stranded $(\alpha/\beta)_8$ barrel structure. X-ray structure analysis combined with in vitro mutagenesis (Vihinen et al. 1990; Holm et al. 1990) indicated a catalytic function of amino acid residues Arg-204, Asp-206, Glu-230, and Asp-297 (numbering of Taka-amylase). Asn-121 and His-210 are implicated in calcium (Ca^{2+}) ion binding. The presence of a single set of conserved regions in clostridial amylopullulanases is in line with a single active center catalyzing the hydrolysis of both α-1,4- and α-1,6-linkages. Furthermore, Podkovyrov et al. (1993) have shown that the cyclomaltodextrinase (Cdx) of *C. thermohydrosulfuricum* uses the same catalytic mechanism as the other amylolytic enzymes of family A. Thus, single Asp → Asn or Glu → Gln

TABLE 12–1 *Clostridium* Genes Concerned with Starch Degradation

Organism	*Gene*	*Molecular Mass of Protein (kDa)*	*Reference*
Bu. fibrisolvens	α-Amylase	106.9	Rumbak et al. 1991
C. acetobutylicum	α-Amylase	53.9	Verhasselt et al. 1989; Verhasselt 1992
Clostridium sp. G005	Glucoamylase (*cga*)	76.3	Ohnishi et al. 1992
C. thermosulfurogenes	β-Amylase	57.2	Kitamoto et al. 1988
	α-Amylase (*amyA*)	75.1	Bahl et al. 1991a
	Amylopullulanase (*amyB*)	130.5	Burchhardt et al. 1991
	Transport protein (*amyC*)	31.0	Bahl et al. 1991b
	Transport protein (*amyD*)	32.9	Bahl et al. 1991b
C. thermohydrosulfuricum	Amylopullulanase (*apu*)	162.2	Melasniemi et al. 1990
	Cyclomaltodextrinase (*cdx*)	68.0	Podkovyrov and Zeikus 1992
T. brockii	Amylopullulanase	105	Coleman et al. 1987

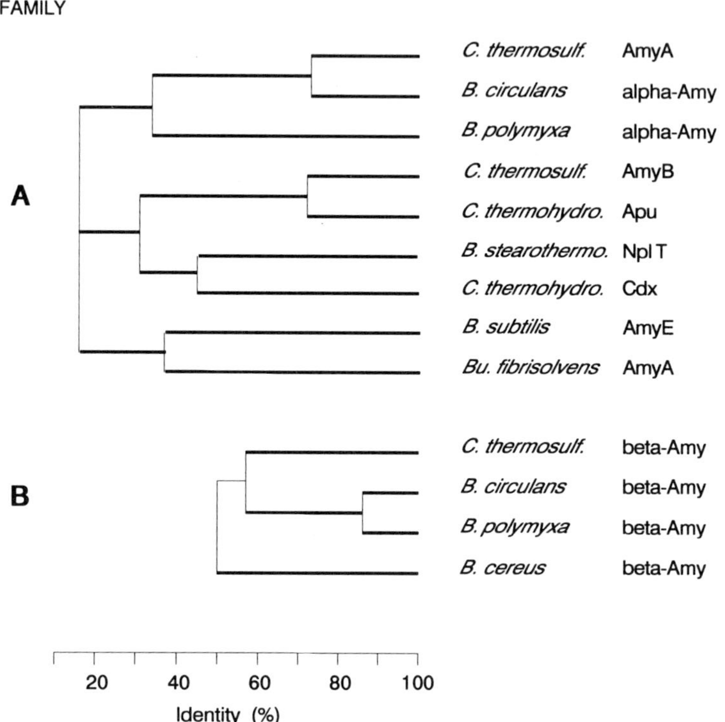

FIGURE 12–1 Amylase families. The dendrogram is built from pairwise identity scores of protein sequence alignments. References not included in Table 12–1: *B. cereus* β-amylase (Nanmori et al. 1993), *B. circulans* α-amylase (Nishizawa et al. 1985), *B. circulans* β-amylase (Siggens 1987), *B. polymyxa* α-amylase (Uozumi et al. 1989), *B. polymyxa* β-amylase (Kawazu et al. 1987), *B. stearothermophilus* neopullanase NplT (Kuriki and Imanaka 1989), *B. subtilis* AmyE (Yang et al. 1983).

replacements of Asp-325, Asp-421, and Glu-354 resulted in a complete loss of enzymatic activity.

Rumbak et al. (1991) noted the presence of additional conserved sequence motifs in the N-terminal half of type A amylases. One of these motifs, region A, shows characteristic sequence differences between the various subfamilies and might thus be employed as a signature sequence for the classification of type A amylases. The corresponding region of *C. acetobutylicum* α-amylase has been sequenced by Paquet et al. (1991) and was found to be closely related to that of *B. subtilis* α-amylase (see Figure 12–2). This finding suggests that both enzymes belong to the same subfamily of saccharifying α-amylases. It should be noted that the molecular mass (84 kilodalton, kDa) of the α-amylase purified from *C. acetobutylicum* ATCC 824 culture supernatants is significantly larger than that reported for the product of the amylase gene

```
                                A              I              II                 III            IV

A. oryzae       α-Amy   -----IYFLLTDRF--------LMVDVVANH--------IDGLRIDTVKHV-------IGEVLDG--------FVENHDN--
                             11               114               200                 228            292

C. thermosulf.  AmyA    -----IYQIVTDRF--------VIIDFAPNH--------IDGIRLDAVKHM-------FGEWFLG--------FIDNHDM--
                             17               131               222                 254            322
B. circulans    α-Amy   -----IYQIVTDRF--------VVIDFAPNH--------IDGIRVDAVKHM-------FGEWFLG--------FLDNHDM--
                             17               133               224                 256            324
B. polymyxa     α-Amy   -----IYFIMTDRF--------VMVDVVVNH--------IDGLRLDTVKHV-------MGEIFHG--------FIDNHDV--
                             714              817               900                 928            990

C. thermosulf.  AmyB    -----MYQIFPDRF--------IILDGVFNH--------ADGWRLDVENEV-------IAENWGD--------LLGSHDT--
                             369              496               600                 633            707
C. thermohydr.  Apu     -----MYQIFPDRF--------VIKDGVFNH--------ADGWRLDVANEV-------VAENWND--------LLGSHDT--
                             357              485               592                 625            699
B. stearother.  NplT    -----WYQIFPERF--------VMLDAVFNH--------IDGWRLDVANEI-------LGEIWHD--------LLGSHDT--
                             137              239               322                 355            419
C. thermohydr.  Cdx     -----VYQIFPERF--------VILDAVFNH--------IDGWRLDVANEI-------VGEVWHD--------LIGSHDT--
                             134              235               319                 352            416

B. subtilis     AmyE    -----ILHAWNWSF--------VIVDAVINH--------ADGFRFDAAKHI-------YGEILQD--------WVESHDT--
                             11               94                170                 206            264
B. fibrisclv.   AmyA    -----ILHAFCWSF--------VIVDILPNH--------ADGFRIDTAKHI-------YGEVLQG--------WVESHDN--
                             117              203               284                 340            409
C. acetobutyl.  α-Amy   -----MLHAFDWSF
                             11
```

FIGURE 12–2 Conserved regions of amylases. Regions are designated according to Nakajima et al. (1986) and Rumbak et al. (1991). Catalytic residues are denoted in enlarged case. Numbers refer to aa positions of the mature proteins. References not included in Table 12–1 and Figure 12–1: *A. oryzae* α-amylase (Matsuura et al. 1984), *C. acetobutylicum* α-amylase (Paquet et al. 1991).

cloned by Verhasselt et al. (1989). The relationship between these two enzymes remains, therefore, to be clarified. A truncated open reading frame (ORF) showing significant sequence similarity to the α-amylase of *B. subtilis* was detected upstream of the acetoacetate decarboxylase gene of *C. acetobutylicum* (Gerischer and Dürre 1990). It is, therefore, highly likely that this ORF encodes the α-amylase described by Paquet et al. (1991).

Sequence inspection to date has provided no clue why amylase secretion in clostridia is markedly affected by growth conditions and accompanied by the formation of membrane vesicles (Antranikian et al. 1987). The N termini of the deduced amino acid sequences of the amylolytic enzymes of thermophilic clostridia showed the typical features of secretory signal peptides. Moreover, membrane binding of clostridial amylases via hydrophobic anchor sequences seems unlikely since no hydrophobic regions of sufficient length could be identified in the deduced amino acid sequences of these enzymes.

Sequence analysis of clostridial amylases furthermore reveals a composite structure consisting of catalytic and noncatalytic domains (Figure 12–3). Thus, the otherwise unrelated α- and β-amylases of *C. thermosulfurogenes* contain a highly homologous C-terminal region of approximately 100 aa. A closely related sequence was originally identified as a starch-binding domain in *Aspergillus* glucoamylase and is also present in several bacterial amylases and cyclodextrin glycosyltransferases (Svensson et al. 1989). Similarly, the C-terminal half of the clostridial amylopullulanases contains a sequence that is present in three copies in a recently isolated alkalophilic amylase and has been implicated in starch binding (Candussio et al. 1991).

Recently a glucoamylase gene (*cga*) has been cloned and sequenced from the newly isolated thermophilic *Clostridium* sp. G0005 (Ohnishi et al. 1992). The deduced amino acid sequence indicates that the encoded enzyme has no

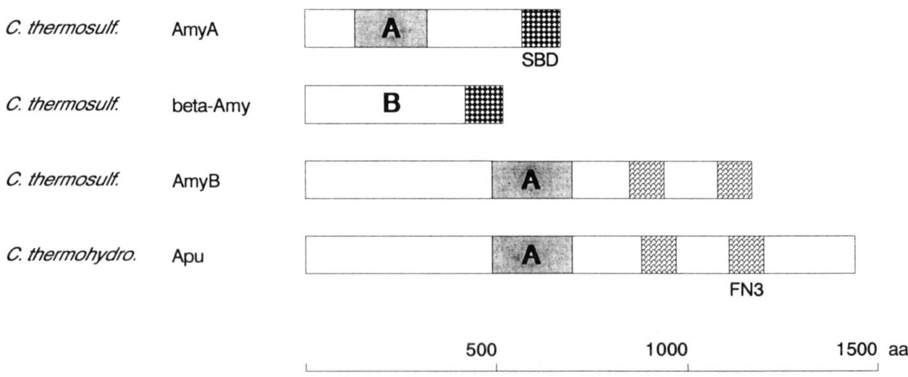

FIGURE 12–3 Domain structure of clostrial amylases. Homologous domains are indicated by the same patterns. Bold uppercase letters indicate membership of a α-glycanase catalytic domain family (see Figure 12–1). Abbreviation: SBD, starch-binding domain; FN3, fibronectin type III-like sequence (Hansen 1992).

similarity to other clostridial amylases, but is structurally related to eukaryotic glucoamylases. X-ray crystallography of the enzyme from *A. awamori* revealed an α/α barrel structure with a core of six, mutually parallel α-helices (Aleshin et al. 1992). These helices are connected to each other through a set of six peripheral α-helices. The active site of glucoamylase is comprised of loops connecting the N termini of the inner helices to the C termini of the peripheral helices. Within these loop regions, four sequence blocks containing the catalytically active amino acids are highly conserved among fungal enzymes and are also present in the thermophilic *Clostridium* glucoamylase.

12.1.3 Gene Structure and Regulation

Expression of the *C. thermohydrosulfuricum apu* gene is regulated by catabolite repression as well as induction (Hyun and Zeikus 1985a). Putative transcription control sites have been identified in the 5′-noncoding region of the *apu* gene by Melasniemi et al. (1990). As shown in Figure 12–4, this region contains canonical −35 and −10 promoter elements with an optimal spacing of 17 base pairs (bp). The putative *apu* promoter is flanked by palindromic sequences of 13 bp (site 1) and 7 bp (site 2). These palindromes could represent regulatory sites mediating either induction or catabolite repression of *apu* expression.

Interestingly, site 2 exhibits structural similarity to the operator regions of the *Escherichia coli lac* and *gal* operons and the operator *amyO* of the *B. subtilis amyE* gene (Figure 12–5). Putative operator sites similar to the *amyO* locus have been identified downstream from the promoters of several *B.*

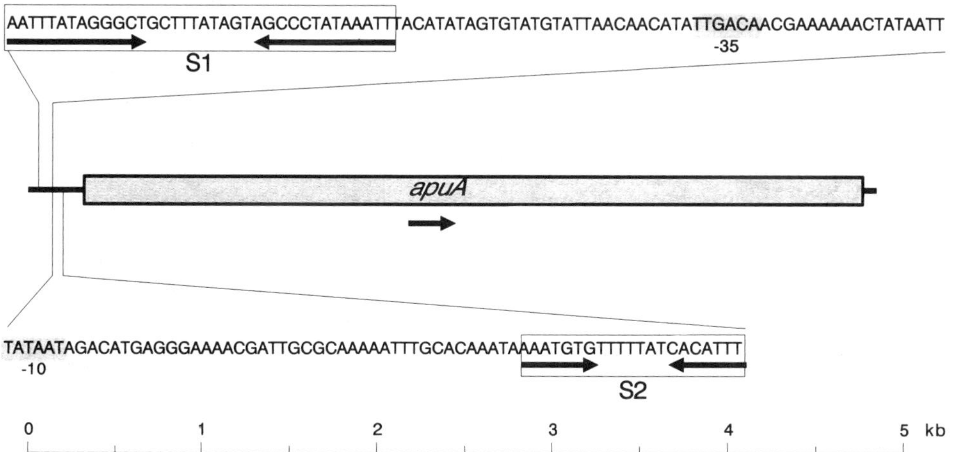

FIGURE 12–4 Structure of the *C. thermohydrosulfuricum apu* gene. Palindromic sequences are underlined by facing arrows. Putative −35 and −10 promoter elements are shaded. Sequence data from Melasniemi et al. (1990).

C. thermohydrosulfuricum	*apu* S2	AAAA	TGT	GTTTTTATC	ACA	TTT	- 112
E. coli	*lacO*	GAAT	TGT	GAGCGGATA	ACA	ATT	- 18
	galO$_E$	CTTG	TGT	AAAC GATT	CCA	CTA	- 87
	galO$_I$	CCAA	TGT	AACC GCTA	CCA	CCG	+ 17
B. subtilis 168	*amyO*	TAAA	TGT	AAGC GTTA	ACA	AAA	- 108
B. subtilis subsp. *natto*	*amyO*	AAAA	TGT	AAGC GTGA	ACA	AAA	- 108

FIGURE 12–5 Sequence comparison of amylase operators. Numbers indicate the position of the 3′-nucleotide relative to the start codon. Sequence data of *B. subtilis* operator sequences are from Weickert and Chambliss (1989).

subtilis genes subject to catabolite repression (Laoide et al. 1989). The *B. subtilis amyO* locus was shown to mediate glucose repression of *amyE* gene expression (Weickert and Chambliss 1989). This operator is recognized by a regulatory protein homologous to the *E. coli lacI* and *galR* repressors (Henkin et al. 1991). Thus, contrary to catabolite regulation in *E. coli*, glucose repression of *B. subtilis* α-amylase operates by a negative control mechanism. It is tempting to speculate, that a similar regulatory system is also operative in *C. thermohydrosulfuricum*.

The *apu* gene product purified from *C. thermohydrosulfuricum* is tightly associated with equimolar amounts of a 24-kDa protein of unknown function (Melasniemi 1988). The coding sequence of this "satellite protein" was found to be located about 500 bp downstream of the *apu* gene (Melasniemi et al. 1990). It is, therefore, likely that both genes are coordinately expressed.

Contrary to what one might expect, there is no evidence to date for clustering of the genes encoding the amylolytic enzymes of *C. thermosulfurogenes* (Burchhardt et al. 1991). However, the *amyA* gene is preceded at a short distance by two open reading frames (Figure 12–6) encoding highly hydrophobic proteins, which show homologies to the integral membrane components of binding protein-dependent transport systems of *E. coli*. These two proteins are presumably involved in the uptake of starch-degradation products and the corresponding genes have, therefore, been tentatively designated as *amyC* and *amyD* (Bahl et al. 1991b). Surprisingly, a truncated ORF nearly identical to *amyC* of *C. thermosulfurogenes* EM1 was located upstream of the β-amylase gene of strain ATCC 33743 sequence by Kitamoto et al. (1988). Thus, the two amylase genes apparently have been exchanged with each other at identical sites in the chromosomes of these two strains (Bahl et al. 1991a).

Due to their close vicinity, the genes *amyD*, *amyC*, and *amyA* could be expressed as a single transcription unit. A 15-bp palindrome located immediately upstream of *amyD* may represent an operator of the hypothetical *amyDCA* operon (see Figure 12–6). The 174-bp intergenic region between

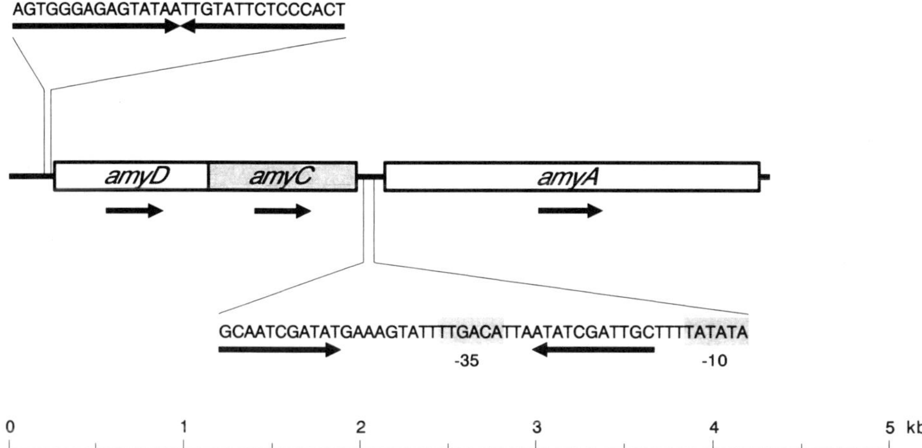

FIGURE 12–6 Structure of *C. thermosulfurogenes amyD-amyA* region. Sequence data are from Bahl et al. (1991a, 1991b).

amyC and *amyA* contains another dyad symmetry element, which overlaps the −35 region of a potential promoter sequence. It could form an RNA hairpin resembling a factor-independent transcription terminator of moderate strength ($\Delta G = -64.8$ kJ). Obviously, these regulatory DNA elements might play a role in uncoupling α-amylase synthesis from *amyDC* expression.

12.2 UTILIZATION OF CELLULOSE AND OTHER β-GLUCANS

12.2.1 Substrate Structure and Catabolism

Cellulose consists of linear chains of β-1,4-linked D-glucose residues, which adopt an extended, ribbon-like configuration stabilized by intrachain hydrogen bonds. Parallel chains are arranged in sheets due to the formation of interchain OH-6 to >OH-3 hydrogen bonds. Cellulose microfibrils consist of layers of hydrogen-bonded sheets, which are further stabilized by van der Waals interactions between successive sheets. This crystalline structure renders cellulose insoluble and highly resistant against enzymatic attack.

Cellulose-degrading enzymes are classified as endo-β-1,4-glucanases and exo-β-1,4-glucanases (cellobiohydrolases). Endoglucanases are characterized by their high activity towards substituted cellulose derivatives such as carboxymethyl cellulose (CMC). In contrast, exoglucanases have been defined by their ability to degrade unsubstituted, microcrystalline cellulose (Avicel), but not CMC.

The efficient hydrolysis of crystalline cellulose requires the synergistic interaction of at least two cellulases liberating cellobiose or short oligosaccharides in a processive fashion, i.e., by an exo-mode, from the nonreducing ends of the chains. It has been proposed that the initial hydrolysis of native cellulose occurs by the action of cellobiohydrolases and that the endoglucanases act in the later stages hydrolyzing the soluble cellodextrins (Enari and Niku-Paavola 1987).

It must be pointed out, however, that no clear distinction between endo- and exoglucanases can be made, since the mode of action may depend on the substrate. Thus, one and the same enzyme might degrade substituted cellulosics in a more or less random fashion, whereas crystalline cellulose is hydrolyzed in an exo-mode (Bronnenmeier et al. 1991). Moreover, the elucidation of the three-dimensional structure of cellobiohydrolase II of *Trichoderma reesei* has shown that no major structural differences exist between exo- and endoglucanases of the same cellulase family (Rouvinen et al. 1990).

Endo-β-1,4-glucanases are, in general, highly active towards mixed linkage β-glucans such as lichenan and barley β-glucan containing alternating β-1,3- and β-1,4 linkages. These substrates are also hydrolyzed by β-1,3-glucanases (laminarinases) and β-1,3-1,4-glucanases (lichenases), which specifically cleave β-1,4-linkages adjacent to β-1,3-glucosidic bonds. Neither laminarinases nor lichenases are capable of hydrolyzing the β-1,4 linkages of cellulose.

Several strains of *C. acetobutylicum* were found to produce an endoglucanase (CMCase), yet were unable to degrade crystalline cellulose presumably due to the lack of a cellobiohydrolase (Allcock and Woods 1981; Lee et al. 1985a). The CMCase activity of strain P270 was induced by growth on molasses, but apparently not repressed by glucose. Organisms that produce individual cellulolytic enzymes but do not possess a complete cellulase system have been termed *pseudo-cellulolytic* (Coughlan 1985).

C. stercorarium, a cellulolytic thermophile, produces a "low-complexity" cellulase system consisting only of an endoglucanase and exoglucanase (Creuzet et al. 1983; Bronnenmeier and Staudenbauer 1988a). Both enzymes are active towards Avicel and have, therefore, been designated avicelase I (Bronnenmeier and Staudenbauer 1990) and avicelase II (Bronnenmeier et al. 1991).

The endoglucanases of *C. thermocellum*, *C. cellulovorans*, and *C. cellulolyticum* are organized in multienzyme complexes termed *cellulosomes* containing a set of enzymes arranged around a cellulose-binding glycoprotein core (Lamed et al. 1987; Shoseyov and Doi 1990). The complex macromolecular organization is thought to be required for carrying out the coordinated hydrolysis of crystalline cellulose by the endoglucanase components of the cellulosome (Mayer et al. 1987). However, a cellulosome-associated cellobiohydrolase activity has recently been isolated that appears to be an essential component of the *C. thermocellum* cellulosomal complex (Morag et al. 1991).

The expression of *C. thermocellum* endoglucanases appears to be constitutive, since no striking differences were observed between the CMCase activities of cultures grown on either glucose, cellobiose, or cellulose. On the other hand, avicelase activity was fivefold enhanced upon growth on Avicel (Johnson et al. 1985). It will be interesting to see, whether this is due to an increased production of the cellobiohydrolase described by Morag et al. (1991).

Cellobiose and short-chain cellodextrins formed by cellulase action can be further degraded hydrolytically by β-glucosidases (cellobiases) or β-cellobiosidases (cellodextrinases) releasing glucose or cellobiose from the nonreducing ends of the oligosaccharides. β-Cellobiosidases differ from cellobiohydrolases by their lack of activity against crystalline cellulose. β-Glucosidases with cellobiase activity have been purified from *C. thermocellum* (Ait et al. 1982), *C. stercorarium* (Bronnenmeier and Staudenbauer 1988b), and *Ca. saccharolyticum* (Patchett et al. 1987).

Alternatively, both cellobiose and cellodextrins can be cleaved phosphorolytically by cellobiose phosphorylase and cellodextrin phosphorylase, respectively. Since this mode of cleavage conserves the energy of the β-glycosidic bond, it appears to be the preferential route of cellodextrin and cellobiose utilization by cellulolytic clostridia. Cellobiose and cellodextrin phosphorylases have been purified from both *C. thermocellum* (Alexander 1972) and *C. stercorarium* (K. Bronnenmeier, unpublished observations).

Substrate utilization by *C. thermocellum* is highly unusual in that cellobiose is preferred over glucose as carbon source (Shinmyo et al. 1979). When cellobiose-grown cells are transferred to a glucose medium, a lag period of several days is observed, which has been attributed to the need for glucokinase induction and induction of a glucose transport system (Hernandez 1982). Apparently, a genetic change is occurring during the lag period in response to the selection pressure exerted by the presence of glucose as the sole carbon source, since glucose-grown cells can be transferred to cellobiose and back to glucose without exhibiting a long lag period (Nochur et al. 1990). The target of this "adaptive mutation," which occurs at rather high frequency, remains to be identified.

12.2.2 Protein Structure

Clostridial cellulase, β-glucanase, and β-glucosidase genes, which have been sequenced or whose products have been biochemically characterized are compiled in Table 12–2. Additional endoglucanase genes have been cloned from *C. cellobioparum* (Shima et al. 1989), *C. cellulolyticum* (Faure et al. 1988), *C. cellulovorans* (Shoseyov et al. 1990), and *C. thermocellum* (Sakka et al. 1989; Bumazkin et al. 1990). A β-glucosidase gene has also been cloned from *C. acetobutylicum* (Zappe et al. 1988).

The large number of *C. thermocellum cel* genes reflects the complexity of the cellulolytic enzyme system. With the exception of the *celC* gene product, all *C. thermocellum* endo-β-1,4-glucanases are components of the cellulosome. A characteristic feature of cellulosomal enzymes is the presence of a conserved repetitive sequence of 25 aa (Béguin 1990). It is assumed that this

TABLE 12–2 ***Clostridium*** **and** ***Caldocellum*** **Genes Concerned with Cellulose Degradation**

Organism	*Gene Cloned*	*Molecular Mass of Protein (kDa)*	*Reference*
Bu. fibrisolvens	Endo-β-1,4-glucanase (*end1*)	61.0	Berger et al. 1989
	Endo-β-1,4-glucanase (*celA*)	48.9	Hazlewood et al. 1990
	β-Cellobiosidase (*ced1*)	61.0	Berger et al. 1990
	β-D-Glucosidase (*bglA*)	91.9	Lin et al. 1990
C. acetobutylicum	Endo-β-1,4-glucanase	49.4	Zappe et al. 1988
C. cellulovorans	Endo-β-1,4-glucanase (*engB*)	48.6	Foong et al. 1991
	Endo-β-1,4-glucanase (*engD*)	52.7	Hamamoto et al. 1990; Hamamoto et al. 1992
	Cellulose-binding protein A (*cbpA*)	189.0	Shoseyov et al. 1992
C. cellulolyticum	Endo-β-1,4-glucanase (*celCCA*)	50.7	Faure et al. 1989
	Endo-β-1,4-glucanase (*celCCC*)	47.2	Bagnara-Tardif et al. 1992
	Endo-β-1,4-glucanase (*celCCD*)	66.1	Shima et al. 1991
	Endo-β-1,4-glucanase (*celCCG*)	76.1	Bagnara-Tardif et al. 1992
C. josui	Endo-β-1,4-glucanase (*celA*)	38.4	Fujino and Ohmiya 1992
	Endo-β-1,4-glucanase 2 (*celB*)	42	Fujino and Ohmiya 1991
C. stercorarium	Avicelase I (*celZ*)	109.5	Jauris et al. 1990
	Avicelase II (*celY*)	103.0	W.H. Schwarz, unpublished observations
	β-D-Glucosidase (*bglZ*)	84.7	Schwarz et al. 1989

C. thermocellum	Endo-β-1,4-glucanase (*celA*)	52.6	Béguin et al. 1985
	Endo-β-1,4-glucanase (*celB*)	63.9	Grépinet and Béguin 1986
	Endo-β-1,4-glucanase (*celC*)	40.9	Schwarz et al. 1988a; Sakka et al. 1991
	Endo-β-1,4-glucanase (*celD*)	72.4	Joliff et al. 1986; Juy et al. 1992
	Endo-β-1,4-glucanase (*celE*)	90.2	Hall et al. 1988
	Endo-β-1,4-glucanase (*celF*)	82.1	Navarro et al. 1991
	Endo-β-1,4-glucanase (*celH*)	102.1	Yagüe et al. 1990
	β-Cellobiosidase (*cbh-4*)	83	Tuka et al. 1990
	β-Cellobiosidase (*cbh-9*)	69	Tuka et al. 1990
	β-1,3-Glucanase (*licA*)	148.2	Schwarz et al. 1988b; Schimming 1991
	β-1,3-1,4-Glucanase (*licB*)	37.9	Schimming et al. 1992
	β-D-Glucosidase A (*bglA*)	51.5	Gräbnitz et al. 1991
	β-D-Glucosidase B (*bglB*)	84.1	Gräbnitz et al. 1989
	Cellulose-binding protein S1	?	Romaniec et al. 1991; Fujino et al. 1992; Poole et al. 1992
Ca. saccharolyticum	Endo-β-1,4-glucanase (*celB*)	117.7	Saul et al. 1989; Saul et al. 1990
	β-D-Glucosidase	53.2	Love et al. 1988

highly conserved domain is involved in the attachment of the cellulosomal enzymes to the glycoprotein S1 core of the cellulosome. Thus, truncated forms of *C. thermocellum* CelD devoid of the 25-aa duplicated segment failed to bind to any cellulosomal protein (Tokatlidis et al. 1991). This reiterated domain is also present in the LicB lichenase and XynZ xylanase of *C. thermocellum*, in several endoglucanases (CelCCA, CelCCC, CelCCD, CelCCG) of *C. cellulolyticum*, and in the EngB endoglucanases of *C. cellulovorans*.

It is not known whether the β-cellobiosidases cloned by Tuka et al. (1990) are also components of the cellulosome. Their enzymatic properties differ from those of the cellobiohydrolase isolated by Morag et al. (1991) and resemble those of *Bu. fibrisolvens* cellodextrinase (Berger et al. 1990). Although the cloned *C. thermocellum* genes have been named cellobiohydrolase (*cbh*) genes, the encoded enzymes are more properly designated as β-cellobiosidases.

The confusing multiplicity of clostridial cellulases can be grouped into a limited number of cellulase families (Henrissat et al. 1989; Béguin 1990; Gilkes et al. 1991). As a rule, cellulase systems of cellulolytic microbes comprise at least two enzymes belonging to different cellulase families. On the other hand, a high degree of homology is frequently found between cellulases from distantly related organisms.

Most clostridial cellulases belong to family A, which constitutes the largest cellulase family and is further divided into several subfamilies (Figure 12–7). Endoglucanase A of *C. thermocellum* is a member of family D. It shares significant homology with an endoglucanase of *B. circulans* classified as β-1,3-1,4-glucanase (Bueno et al. 1990). The *C. thermocellum* enzymes CelD and CelF as well as the *C. stercorarium* celZ enzyme (avicelase I) are grouped into family E (see Figure 12–7). CelF and CelZ are closely related (>70% identity), whereas CelD belongs to a different subfamily. Avicelase II, an exoglucanase encoded by the *C. stercorarium celY* gene shows no obvious sequence similarity with other cellulolytic enzymes and might, therefore, represent the first member of an additional cellulase family.

The various β-glucosidases can also be grouped into two β-glycosidase families (Figure 12–8). Family A also includes *E. coli* phospho-β-glucosidase and the phospho-β-galactosidases of lactic acid bacteria and mammalian lactase-phlorizin hydrolase (Gräbnitz et al. 1991). In general, β-glucosidases of family A are rather nonspecific with respect to the C4 configuration of the hexose moiety and hydrolyze both β-gluosides and β-galactosides. β-Glucosidase A of *C. thermocellum* exhibits such a dual specificity and might also be considered a β-galactosidase (Gräbnitz and Saudenbauer 1988). Its properties closely resemble those of the *C. thermocellum* β-glucosidase purified by Ait et al. (1982).

β-Glucosidase family B comprises enzymes from thermophilic clostridia, rumen bacteria, yeasts, and filamentous fungi (Gräbnitz et al. 1989; Lin et al. 1990; Barnett et al. 1991). In case of the β-glucosidases from rumen anaerobes a region reshuffling has occurred moving the catalytic domain from

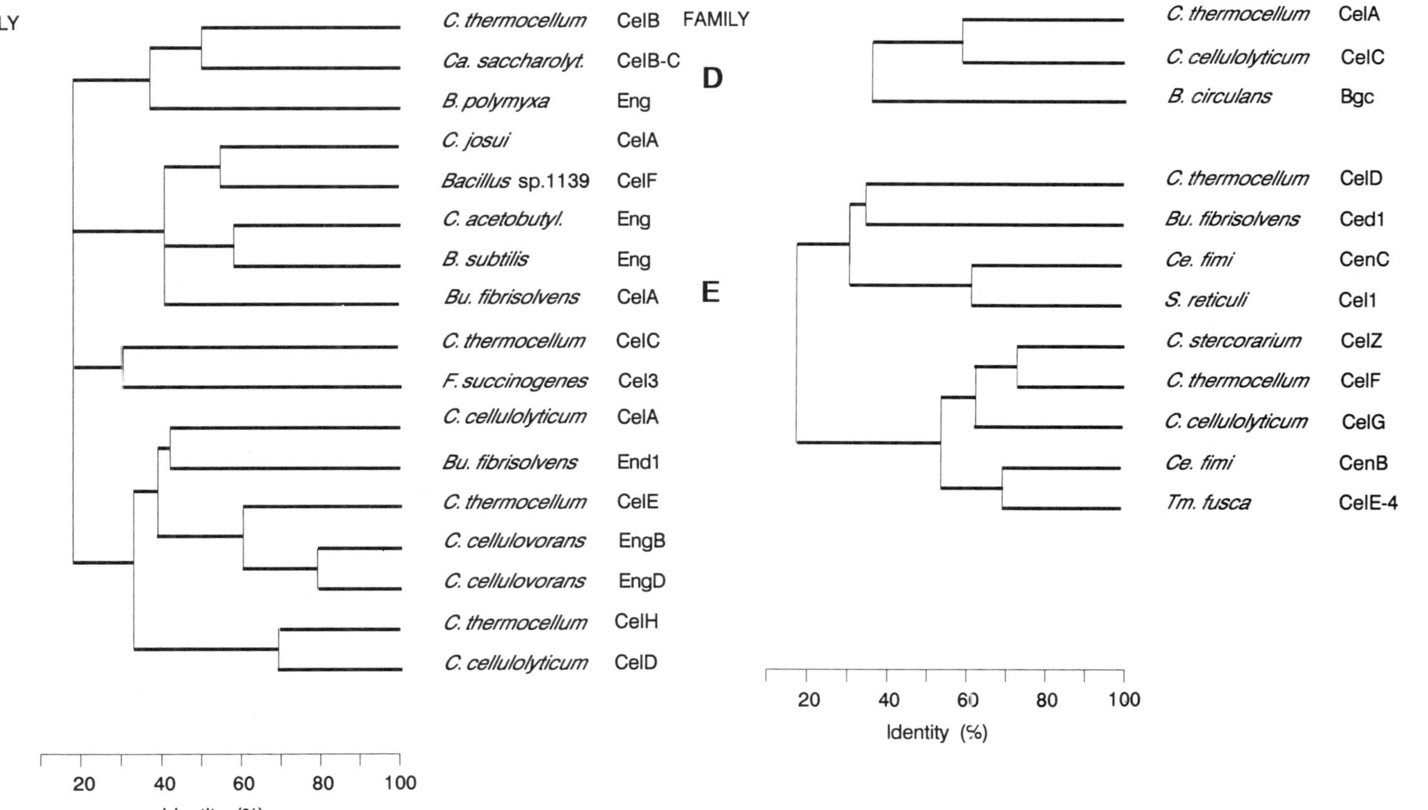

FIGURE 12–7 Cellulases families. CelB-C denotes the C-terminal catalytic domain C of *Ca. saccharolyticum* CelB (Saul et al. 1990). References not included in Table 12–2: *Bacillus* sp. 1139 (Fukumori et al. 1986), *B. circulans* Bgc (Bueno et al. 1990), *B. polymyxa* Eng (Baird et al. 1990b), *B. subtilis* Eng (Nakamura et al. 1987), *Ce. fimi* CenB (Meinke et al. 1991), *Ce. fimi* CenC (Coutinho et al. 1991), *F. succinogenes* Cel3 (McGavin et al. 1989), *S. reticuli* (Schlochtermeier et al. 1992), *Tm. fusca* (Lao et al. 1991).

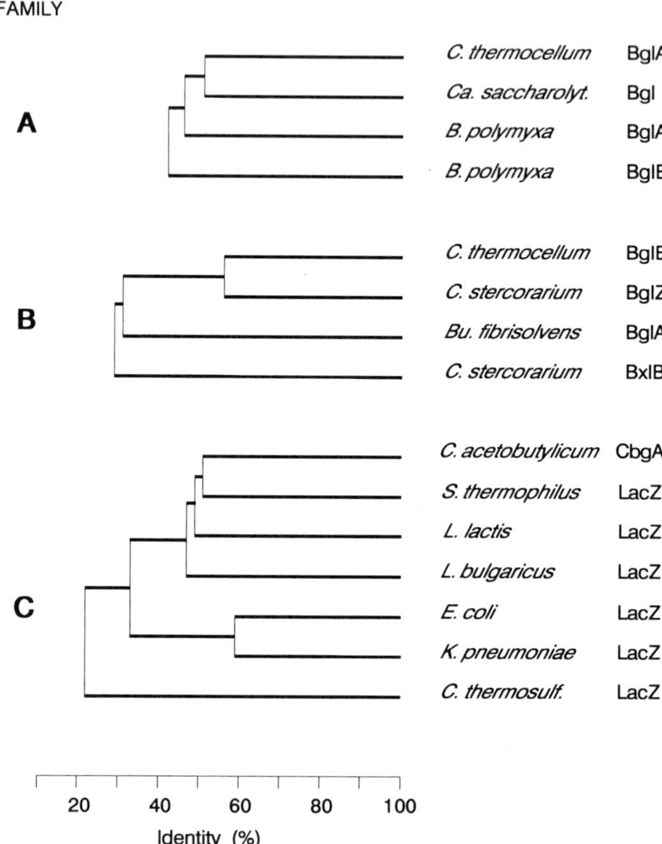

FIGURE 12–8 β-Glycosidase families. References not included in Table 12–2 and Table 12–4: *B. polymyxa* BglA and BglB (Gonzalez-Candelas et al. 1990), *E. coli* LacZ (Kalnins et al. 1983), *L. bulgaricus* (Schmidt et al. 1989), *L. lactis* (David et al. 1992), *K. pneumoniae* (Buvinger and Riley 1985), *S. thermophilus* (Schroeder et al. 1991).

the N-terminal to the C-terminal half of the enzymes. The *C. stercorarium bglZ* gene product corresponds to the *C. stercorarium* β-glucosidase that Bronnenmeier and Staudenbauer (1988b) described. It is closely related to BglB of *C. thermocellum*.

The catalytic domains of family A cellulases include four regions that are highly conserved among members of the different subfamilies (Figure 12–9). Region III contains a characteristic Asn-Glu-Pro motif, the Glu residue of which has been shown to be involved in catalysis (Baird et al. 1990a; Py et al. 1991). Similarly, the invariant His residue of region II was found to be essential for enzyme function (Py et al. 1991; Belaich et al. 1992). Furthermore, affinity labelling experiments indicate that the strictly conserved Glu

Organism	Enzyme	I	II	III	IV
C. thermocellum	CelB	-----GINVVRMPIAT----------VILDVHSP--------FDLKNEP--------LLGEWGG			
		93	151	199	360
Ca. saccharolyticum	CelB-C	-----GFNLLRVPISA----------IMLDIHSI--------FDLKNEP--------LIGEWGG			
		681	738	787	931
C. josui	CelA	-----GSNVIRLAMYV----------LFVDWHVL--------YELANEP--------FCSEWGT			
		100	140	179	284
C. acetobutylicum	Eng	-----GVNVIRAAMYT----------VIIDWHIL--------YEICNEP--------FVTEWGT			
		92	132	170	260
C. thermocellum	CelC	-----GFDHVRLPFDY----------LVLNTHHA--------FELLNEV--------YCGEFGV			
		41	85	135	277
C. cellulolyticum	CelA	-----GFNTVRIPVSW----------VILNTHHD--------FEGMNEP--------IIGECGA			
		99	142	190	329
C. thermocellum	CelE	-----GFNAVRVPVTW----------AIINLHHD--------FETMNEP--------IIGEFGT			
		99	143	188	313
C. cellulovorans	EngB	-----GFNTLRLPITW----------VIINLHHE--------FETMNEP--------VIGEMGT			
		85	129	184	302
C. thermocellum	CelH	-----GYKNVRIPVRW----------TIINSHHD--------FEIMNEP--------YFGEFAV			
		370	414	455	562
C. thermocellum	BglA	-----GIKSYRFSISW----------AITLYHWD--------WFTHNEP--------VISENGA			
		72	116	161	352
Ca. saccharolyticum	Bgl	-----GLKAYRFSIAW----------VVTLYHWD--------WITFNEP--------VITENGA			
		69	113	158	358
B. polymyxa	BglA	-----GIRTYRFSVSW----------FCTLYHWD--------WLTFNEP--------YITENGA			
		72	116	161	349
B. polymyxa	BglB	-----GFLHYRFSVAW----------MLTLYHWD--------WNTINEP--------LITENGA			
		74	117	162	354

FIGURE 12–9 Conserved regions of family A cellulases and β-glucosidases. Putative catalytic residues are denoted in enlarged case. Numbers refer to aa positions of the primary translation products. References: Baird et al. 1990a; Belaich et al. 1992; Macarron et al. 1993; Navas and Béguin 1992; Py et al. 1991; Trimbur et al. 1992; Tull et al. 1991.

residue within region IV constitutes the active site nucleophile (Macarron et al. 1993). Related sequence motifs have also been identified in type A β-glucosidases and members of celloxylanase family F (Tull et al. 1991). These three enzyme families might therefore constitute a widespread "superfamily" of β-glycosidases (Gräbnitz et al. 1991). All members of this proposed superfamily have been classified as "retaining" β-glycosidases, which hydrolyze the β-glycosidic bond with retention of the anomeric configuration (Gebler et al. 1992).

Cellulases of family E also contain three well-conserved sequence motifs, which presumably constitute the active center of this cellulase family (Figure 12–10). Invariant His and Glu residues of regions II and III, which are located close to the C terminus of the catalytic domain, have been implicated in catalytic activity of *C. thermocellum* endoglucanase D (CelD) by chemical modification and site-directed mutagenesis (Chauvaux et al. 1992; Tomme et al. 1991, 1992). The three-dimensional structure of CelD has recently been determined by X-ray diffraction techniques (Juy et al. 1992). The globular catalytic domain consists of 12 helices connected in an up-and-down pattern. This mode of folding results in a twisted α-barrel with six inner helices running in the same direction and six outer helices of opposite orientation (Figure 12–11). On one side of the barrel, long helix-connecting loops devoid of secondary structure form the active site of the enzyme. The highly conserved sequence motifs shown in Figure 12–10 are located within the loops preceding helices α2 and α12, respectively. Cleavage of glycosidic bonds catalyzed by CelD is thought to involve Glu-555 (region I) as a general acid catalyst and a nucleophilic water molecule whose ionization is promoted by Asp-201 (region III). His-516 and Arg-518 of region II are favorably oriented to contact polar groups of the substrate and are thus implicated in substrate binding.

On the basis of their sequence similarity, it can be assumed that all members of cellulase family E share the same catalytic domain structure. This protein fold is very similar to the α/α-barrel domain of glucoamylases (Aleshin et al. 1992), but clearly different from the $(\alpha\beta)_8$-like barrel topology determined for *Trichoderma reesei* cellobiohydrolase II (CBHII), a family B cellulase. The active site of CBHII is located at the C-terminal end of a parallel β-barrel (Rouvinen et al. 1990). Thus, the active sites of family B and family E cellulases lie at opposite ends of their respective barrels.

As shown in Figure 12–11, *C. thermocellum* CelD contains, in addition to the α/α-barrel domain, a smaller N-terminal immunoglobulin-like β-barrel. This domain consists of two Greek key motifs forming a three-stranded β-sheet packed against a four-stranded sheet. However, this antiparallel β-barrel is not directly involved in catalysis, since it is only found in cellulases of subfamily E1, but missing in enzymes of subfamily E2. Its function is unknown.

The presumptive cellulose-binding domain B (CBD-B) of the *C. stercorarium* cellulases, is also present in two copies in the C-terminal region of the *B. lautus* CelB endoglucanase (Jorgensen and Hansen 1990). Moreover, four

Organism	Enzyme	I	II	III
C. thermocellum	CelD	-----DSTKGWHDAG**D**YNKY---------- (191)	SYVTGLGINPPMNP**H**DR------ (502)	N**E**IAINWNAALIYALA-- (554)
B. fibrisolvens	Ced1	-----DVTGGWHDAG**D**YGRY---------- (138)	SYVTGNGEKAFKNP**H**LR------ (460)	N**E**ITIYWNSPLVFALS-- (528)
Ce. fimi	CenC	-----DVSGGWYDAG**D**HGKY---------- (496)	SYVTGWGEVASHQQ**H**SR------ (817)	N**E**LTVNWNSALSWVAS-- (890)
S. reticuli	Cel1	-----DVTGGWYDAG**D**HGKY---------- (332)	SYVTGYGEVASHNQ**H**SR------ (655)	N**E**TAINWNAALARMAS-- (725)
C. stercorarium	CelZ	-----DLTGGWYDAG**D**HVKF---------- (74)	SYVVGFGVNPPKRP**H**HR------ (386)	N**E**VACDYNAGFVGALA-- (446)
C. thermocellum	CelF	-----DLTGGWYDAG**D**HVKF---------- (74)	SFVVGFGKNPPRNP**H**HR------ (386)	N**E**VANDYNAGFVGALA-- (446)
C. cellulolyticum	CelG	-----DLTGGWYDAG**D**HVKF---------- (83)	SFVVGYGVNPPQHP**H**HR------ (394)	N**E**IACDYNAGFTGALA-- (454)
Ce. fimi	CenB	-----DLTGGWYDAG**D**HVKF---------- (81)	SYVVGFGANPPTAP**H**HR------ (396)	N**E**VATDYNAGFTSALA-- (457)
Tm. fusca	CelE4	-----DLTGGWYDAG**D**HVKF---------- (94)	SYVVGFGNNPPRNP**H**HR------ (406)	N**E**VATDYNAGFSSALA-- (467)

FIGURE 12–10 Conserved regions of family E cellulases. Putative catalytic residues are shown in enlarged case. Numbers refer to aa positions of the primary translation products. References: Chauvaux et al. 1992; Juy et al. 1992; Tomme et al. 1991; Tomme et al. 1992.

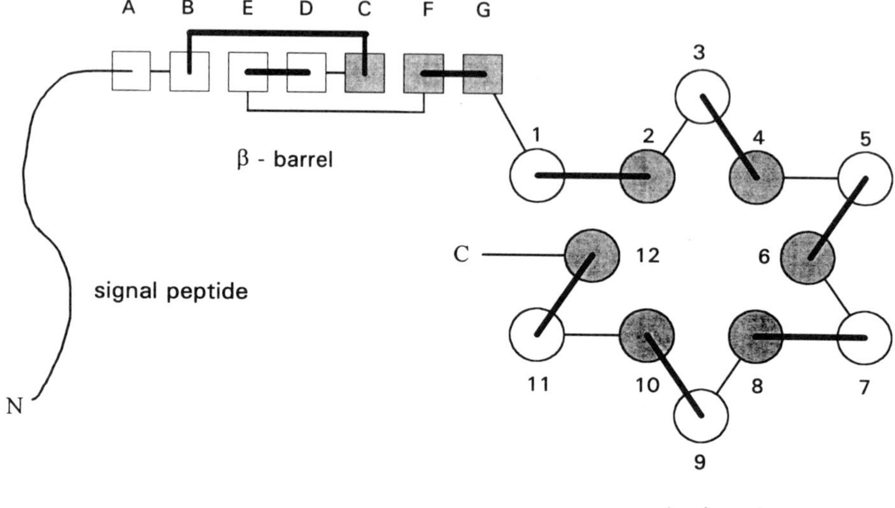

FIGURE 12–11 Topology diagram of *C. thermocellum* endoglucanase CelD. Helices are presented by circles, and β-strands by squares. The inner helices of the α/α-barrel and the strands forming the three-stranded β-sheet of the immunoglobulin-like domain are shaded. Labelling of helices (1–12 and β-sheets (A–G) is according to Juy et al. (1992).

copies of CBD-B and a single copy of CBD-C are found in the cellulose-binding protein CbpA of *C. cellulovorans* (Shoseyov et al. 1992). In addition to the cellulose-binding domains, CbpA contains nine copies of a 140-aa domain that most likely provides the anchoring sites for the various catalytic components of the cellulosome.

A reiterated domain homologous to the CbpA repeat has also been identified in truncated forms of the *C. thermocellum* S1 protein expressed in *E. coli* (Fujino et al. 1992; Poole et al. 1992). The fact, that the recombinant polypeptides retained the capacity to bind the CelD endoglucanase, suggests that the presence of the oligosaccharide moiety of S1 isolated from *C. thermocellum* (Gerwig et al. 1991) is not essential for this interaction. Comparison of the C-terminal regions of CbpA and S1 (Figure 12–12) reveals that in S1 the duplicated CBD-B has been replaced by a single copy of CBD-C. Furthermore, S1 carries an additional C-terminal domain with the duplicated 25-aa segment found in various cellulosomal enzymes. This finding intimates that S1 might be able to self-associate.

S1 can therefore be considered as a multifunctional protein composed of separate domains for cellulose-binding, anchoring of cellulolytic enzymes, and self-association (Salamitou et al. 1992). The molecular architecture of the CbpA and S1 proteins provides support for a model of the cellulosome according to which rows of catalytic components are positioned along the

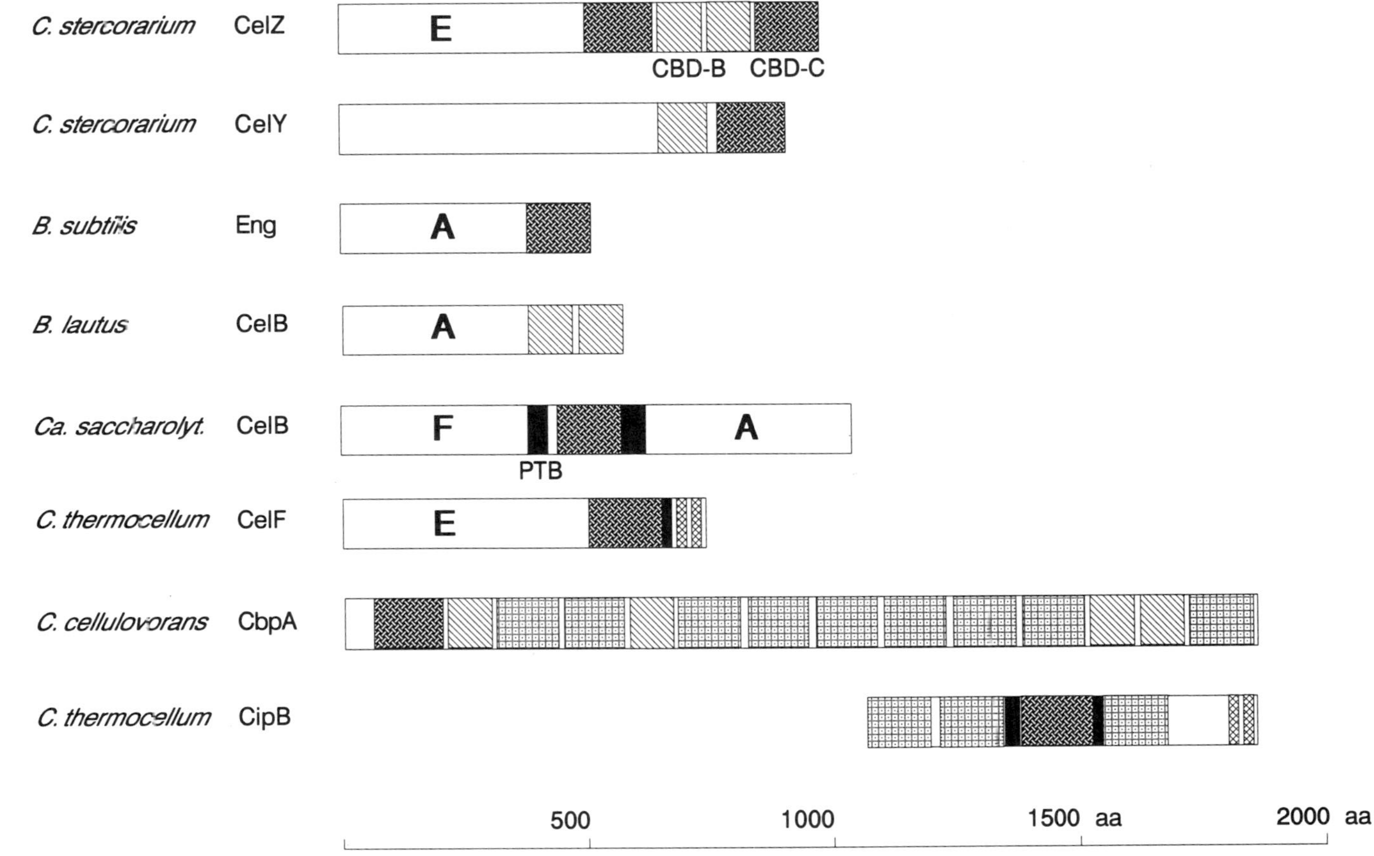

FIGURE 12–12 Domain structure of cellulases. Homologous domains are indicated by the same patterns. Bold uppercase letters indicate membership of a β-glycanase catalytic domain family (see Figure 12–7 and 12–18). Abbreviations: PTB, Pro-Thr box; CBD, cellulose-binding domain.

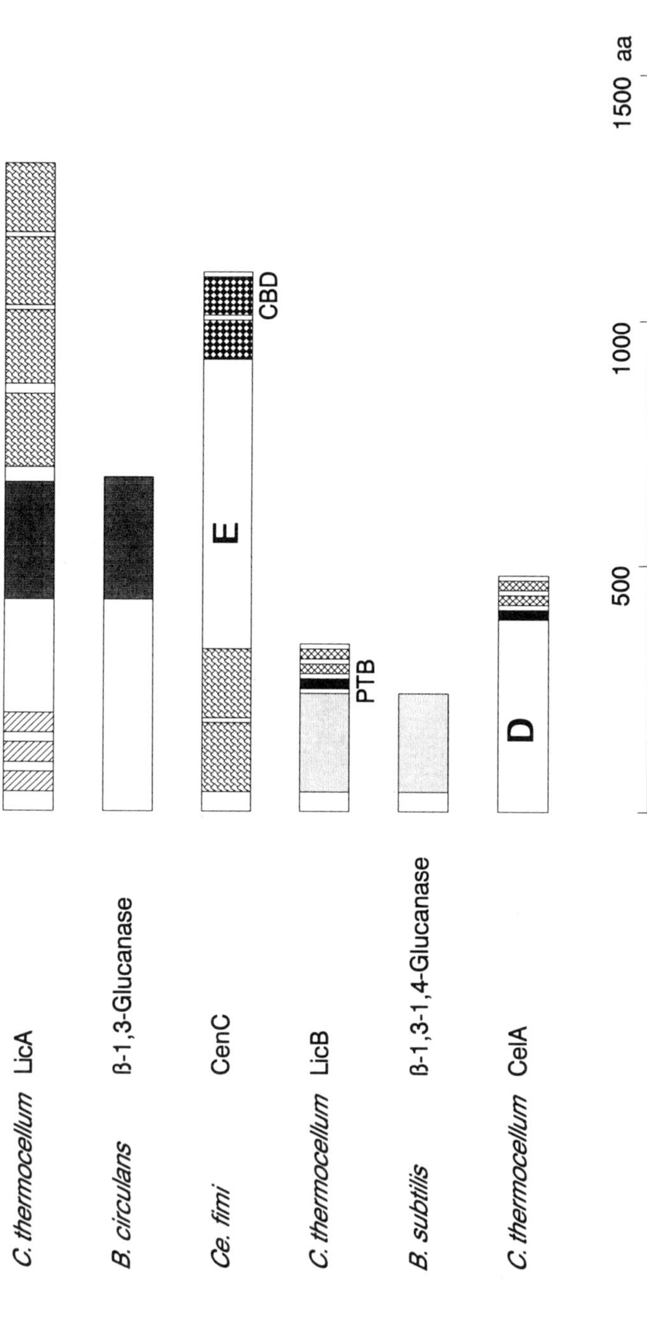

FIGURE 12–13 Domain structure of β-glucanases. Homologous domains are indicated by the same patterns. The catalytic domains of β-glucanases are shaded. Bold uppercase letters indicate membership of a β-glycanase catalytic domain family (see Figure 12–7). Abbreviations: PTB, Pro-Thr box; CBD, cellulose-binding domain.

cellulose chains by interaction with repetitive domains of a cellulose-bound scaffolding protein (Mayer et al. 1987; Béguin et al. 1992).

The avicelases of *C. stercorarium* contain, in addition to their unrelated catalytic domains, two distinct noncatalytic domains implicated in substrate binding (Jauris et al. 1990). These domains are present in single copies in CelY and duplicated in CelZ (Figure 12–12). The C-terminal cellulose-binding domain (CBD) of the *C. stercorarium* cellulases is also found in other cellulolytic enzymes including *B. subtilis* endoglucanase. This CBD is about 150 aa in length and includes three invariant Tyr residues that could be involved in substrate binding. In the bifunctional *Ca. saccharolyticum* enzyme CelB, the CBD forms the central domain B separating the N-terminal β-cellobiosidase/xylanase domain A from the C-terminal endoglucanase domain C (Saul et al. 1990). Multiple CBDs are present in the *Ce. fimi* CenB enzyme, which contains in addition to a central *Clostridium*-type CBD, a C-terminal *Cellulomonas*-type CBD (Meinke et al. 1991).

As shown in Figure 12–13, an unusual complexity of sequence organization is observed for the β-1,3-glucanase (laminarinase) encoded by the *C. thermocellum licA* gene (Schimming 1991). The C-terminal half of the enzyme consists of four copies of a noncatalytic domain of 150 aa, which is also iterated in the N-terminal portion of *Ce. fimi* endoglucanase C. The N-terminal half of LicA comprises three copies of a 35-aa sequence of unknown function and a central catalytic domain homologous to *B. circulans* β-1,3-glucanase (47% identity). Both enzymes are distantly related to the *Bacillus* β-1,3-1,4-glucanases and the lichenase encoded by the *C. thermocellum licB* gene (Schimming et al. 1992). Thus, this β-glucanase family consists of two subfamilies with different substrate specificities. The catalytic domain of LicB is joined by a Pro-Thr-rich linker sequence, to the reiterated domain of clostridial cellulases. This unexpected finding suggests a cellulosomal localization of the LicB enzyme.

12.2.3 Gene Structure and Regulation

The two avicelases of *C. stercorarium* are encoded by adjacent genes and presumably coordinately expressed (Figure 12–14). The transcription start point(s) of the putative *celYZ* operon has not yet been identified. A transcription terminator stem-loop ($\Delta G = -130$ kJ) is located immediately downstream of the *celZ* gene. The 138-bp nucleotide sequence that lies between the two stop codons of the *celY* and the initiation codon of *celZ* gene contains a 18-bp palindrome. Although this intercistronic sequence lacks the characteristic run of T residues found at the 3′ end of factor-independent transcription terminators, it could form a stable RNA hairpin ($\Delta G = -107$ kJ) and seems likely to play a role in regulating the rate of expression of *celZ* relative to the *celY* gene.

The *C. thermocellum cel* genes are apparently not clustered on the chromosome and are represented mostly as monocistronic transcription units. The

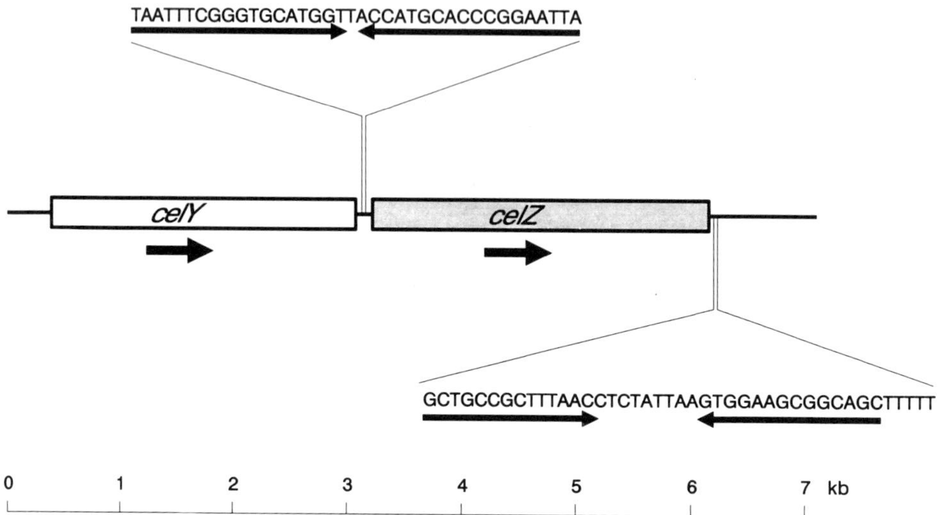

FIGURE 12–14 Structure of the *C. stercorarium celY-celZ* region. Sequence data are from Jauris et al. (1990) and W.H. Schwarz et al. (unpublished observations).

5′-termini of the *celA*, *celD*, and *celF* transcripts (Béguin et al. 1986; Mishra et al. 1991) have been mapped. Putative promoter elements located upstream of the transcription start points are shown in Figure 12–15 and compared with promoter sequences proposed on the basis of sequence similarity. The mapped transcription initiation sites of the *C. stercorarium xynA* gene (Adelsberger 1992) are also included in this comparison. It should be noted that promoter usage is host dependent. Thus, *celA* transcription in *C. thermocellum* starts predominantly at promoter P1, whereas transcription in *E. coli* initiates at promoter P2 (Béguin et al. 1986). The *celC307* promoter P2 was shown to be functional in *E. coli* (Sakka et al. 1991), but *celC* transcription start sites in *C. thermocellum* have not been determined.

Three distinct types of promoter structures can be distinguished (see Figure 12–15): The first type consists of the canonical promoter elements recognized by σ^A-associated form of *Bacillus* RNA polymerase. The presence of highly conserved TG motif at position -15 supports the notion of an extended -10 recognition sequence (Young et al. 1989). The second group of promoters contains a well-conserved -30 element with the consensus sequence TGGTAAA and somewhat variable -10 region resembling the -10 consensus of *B. subtilis* σ^D promoters (Béguin et al. 1986). The *celD* P1 promoter shows no obvious homologies to *Bacillus* promoter consensus sequences. The DNA sequence upstream of the transcription start site contains a nonanucleotide sequence, which is also found in the 5′-noncoding region of the *C. thermocellum celB* gene. It could conceivably represent the -10 recognition sequence of a third class of clostridial promoters. Clearly,

Organism	Gene		−35		−10		Position	
C. thermocellum	*celA* P2	AATTTGTA	TAAACA	TGACAAAATAAATATGA	TATAAT	GATTGTA	− 134	
	celC P2	ATAAAAAC	TGGACA	GAGAAGAAGAAAACGTGA	TATAAT	TAAATTA	− 153	???
	celD P2	ATTGAATA	TTGAAT	TATTTGCTATTTAGATGG	TATAAT	AAAATTA	− 294	
	celE	AATAAACA	GTAACA	TTACCGTTTAGTTTGT	TATAAT	GTTTTAT	− 118	???
C. stercorarium	*xynA* P1	TAATAAAT	TTAACA	AATAATAACACACTGC	TATCTT	CGACCGTAAAT	− 59	
C. thermocellum	*celA* P1	TGTTATTGGTT	TGGTAAA	TGTTTTTGGGTAA	CGATAT	TTATTTT	− 57	
	celC P1	GAATAACCCAG	TGTTAAA	TGGTTTCAGTTTA	CGATTT	CAATGTT	− 87	???
	celF	ATATTCAATTA	TGTTCAA	ATTTGATGTAAAT	CAATAG	AAATTAATC	− 218	
C. stercorarium	*xynA* P2	TTATGGGGTAC	TGGTAAA	GACGTGATAGTTAT	TAATAA	ATTTAACAA	− 91	
C. thermocellum	*celD* P1	GTTCAGTTCCAGCATACGTCTGTATTCAAAA	TGCCTGTAT			TTATAACTGC	− 124	
	celB	GACGGATTTAAAATTTTTAGAGAAAAGTATT	TGCCTGTAT			ATTTTAACCT	− 35	???

FIGURE 12–15 Sequence comparison of transcription start sites. Numbers indicate the position of the 3′-nucleotide relative to the start codon. References are given in the text. Sequences marked by question marks have not been mapped.

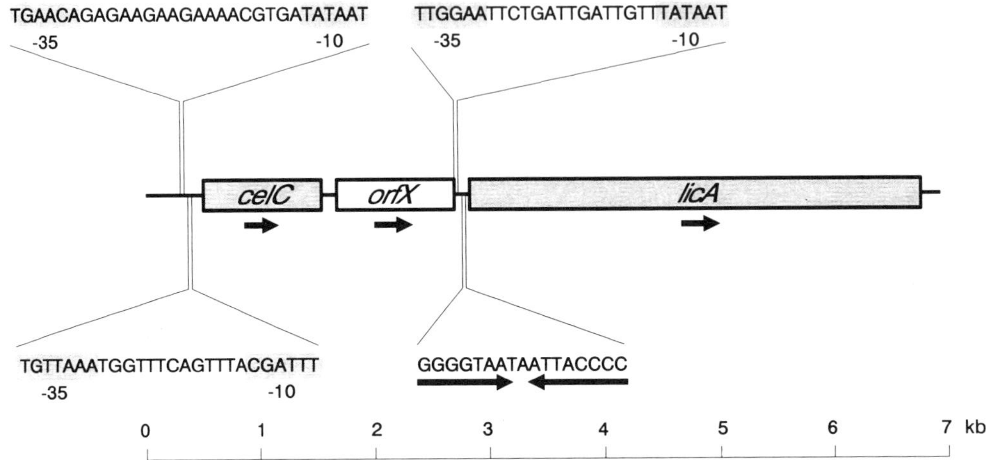

FIGURE 12–16 Structure of the *C. thermocellum celC-licA* region. Sequence data are from Schwarz et al. (1988a) and Schimming (1991).

additional mRNA starts have to be mapped before any solid conclusions can be made about the structure of clostridial promoters.

Expression of *C. thermocellum cel* genes appears to be controlled by the metabolic state of the cells (Mishra et al. 1991). Thus, during growth on cellobiose, transcripts of the *celA*, *celD*, and *celF* genes are formed predominantly in the late log phase. In contrast, *celC* expression is detectable only during early stationary phase. Expression of the *celD* gene is modulated by promoter switching. Transcripts are initiated predominantly at P1 during late log phase, but almost exclusively at the σ^A-type promoter P2 during early stationary phase. Promoter switching could result from alterations in the pattern of σ-factors at the onset of sporulation.

The *celC*-encoded endoglucanase differs with respect to domain structure, localization, and timing of expression from the other *cel* gene products. The finding that *celC* and *licA* are located in close proximity on the *C. thermocellum* chromosome furthermore suggested that these genes might constitute an operon concerned with β-glucan degradation (Young et al. 1989). The structure of *celC-licA* region (Figure 12–16) reveals that *celC* and *licA* are codirectionally expressed, but separated by an ORF (*orfX*) of unknown function (Schimming 1991). The absence of a discernible transcription terminator within the *celC-licB* region indicates that *licA* might, in part, be expressed by read-through of transcripts starting at the *celC* promoter(s).

Clustering was also observed for the *bglA* and *licB* genes of *C. thermocellum* (Figure 12–17). However, the DNA sequence of the *bglA-licB* region (Gräbnitz et al. 1991; Schimming et al. 1992) shows that these genes are convergently transcribed and flanked by symmetric sequences with a potential for RNA hairpin formation of $\Delta G = -162$ and -109 kJ, respectively. Since

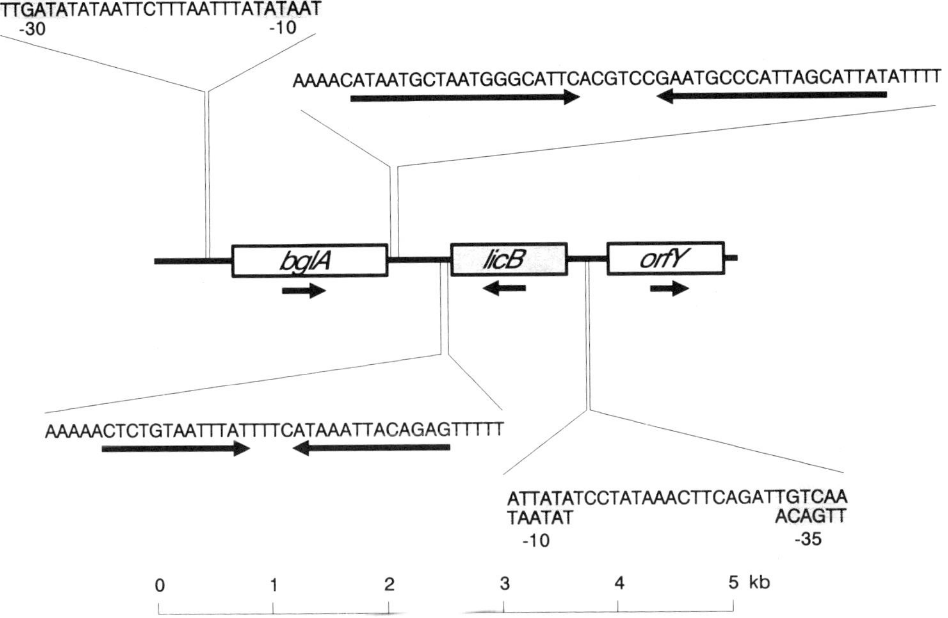

FIGURE 12–17 Structure of the *C. thermocellum bglA-licB* region. Sequence data are from Gräbnitz et al. (1989) and Schimming et al. (1992).

the oligo(T)-tracts of the stem-loop structures are part of the palindromic sequences, both could function as transcription terminator in either direction, thus preventing any read-through of convergently transcribed mRNA.

A cluster of cellulolytic genes has recently been identified in *C. cellulolyticum*. Bagnara-Tardif et al. (1992) have shown that the adjacent genes *celCCC* and *celCCG* are transcribed into a polycistronic mRNA. The size of the transcripts (5–6 kilobases, kb) indicates that additional genes might be present in the same transcription unit. Actually, the *celCCC-celCCG* region is flanked by two ORFs, which on the basis of sequence similarity could encode additional enzymes involved in cellulose degradation.

12.3 UTILIZATION OF HEMICELLULOSE

12.3.1 Substrate Structure and Catabolism

Hemicelluloses are a diverse group of complex polysaccharide components of the plant cell wall, in which they occur in close association with cellulose. Most hemicelluloses are branched heteropolysaccharides, containing different types of sugars in the backbone chain, and in the sidechains or branches. Major sugar components are D-xylose, D-mannose, D-glucose, D-galactose,

L-arabinose, and 4-O-methyl-D-glucuronic acid. Hemicelluloses are classified according to their principal monosaccharide constituents as xylans, mannans, etc.

Xylans, a major component of the hemicellulose fraction and, next to cellulose, the most abundant renewable resource, consist of a backbone of β-1,4-linked β-D-xylopyranosyl residues, which is more flexible than the cellulose chains, and adopts a twisted configuration. The frequency and composition of branches in xylans varies according to the source. Xylans from grasses are mainly arabinoxylans, in which the backbone is substituted by α-1,3-linked L-arabinofuranosyl residues. The xylan of hardwoods is 4-O-methylglucuronoxylan containing α-1,2-linked 4-O-methylglucuronic acid side groups. Hardwood xylans are highly acetylated mostly at the O-3 position of the xylose residues. Esterification by acetyl groups significantly increases the solubility of xylan in water. Softwood xylans differ from hardwood xylan in not being acetylated and in carrying both 4-O-methylglucuronic acid and L-arabinofuranose side groups.

Enzymatic hydrolysis of the xylan backbone involves endo-β-1,4-xylanases and β-D-xylosidases. Removal of side groups is catalyzed by α-L-arabinofuranosidases, α-D-glucuronidases and acetylxylan esterase. Utilization of xylose liberated from xylan involves xylose transport and intracellular conversion to D-xylulose catalyzed by xylose (glucose) isomerase. D-Xylulose is phosphorylated by xylulose kinase to D-xylulose-5-phosphate, which then enters the pentose phosphate pathway.

C. acetobutylicum ATCC 824 can grow on larch wood xylan as sole carbon source (Lee et al. 1985b), although the utilization of xylan is incomplete, possibly due to lack of an α-glucuronidase. Two distinct xylanases (Lee et al. 1987), a β-D-xylosidase (Lee and Forsberg 1987a), and an α-L-arabinofuranosidase (Lee and Forsberg 1987b) have been purified from culture supernatants.

Production of these enzymes is highest in cells grown on xylose. Glucose has a negative effect on the formation of xylanolytic enzyme presumably due to inducer exclusion, since xylose is not consumed in the presence of excess glucose (Lee et al. 1985b). Xylose utilization requires the induction of a xylose permease activity, which is repressed by glucose (Ounine et al. 1985).

C. stercorarium can use xylan or xylose as sole carbon source. Production of xylanases is induced by xylose, xylan, or cellulose and repressed by glucose and other hexoses (Berenger et al. 1985). Xylanolytic enzymes isolated from *C. stercorarium* culture supernatants include three xylanases, a β-xylosidase, and an α-L-arabinofuranosidase (Berenger et al. 1985; Bronnenmeier et al. 1990). Furthermore, two enzymes were purified that exhibited both β-cellobiosidase and xylanase activity. Because of its dual substrate specificity this novel type of enzyme has been termed *celloxylanase* (Bronnenmeier et al. 1990).

C. thermocellum is unable to use xylose and grows only poorly on hemicellulose (Wiegel et al. 1985). It produces several xylanolytic activities some of which are components of the cellulosome (Morag et al. 1990). Cellulosomal xylanases differ from their noncellulosomal counterparts in molecular mass and substrate specificity. Slow growth of *C. thermocellum* on xylans apparently results from utilization of xylooligomers formed by xylanase action (Wiegel et al. 1985).

Xylose isomerases have been purified from *C. thermosulfurogenes* and *Thermoanaerobacter saccharolyticus* (Lee and Zeikus 1991). Both enzymes were shown to catalyze also the isomerization of glucose to fructose. Because of their high thermostability, they seem superior to enzymes currently in use for industrial production of high-fructose corn syrup.

12.3.2 Protein Structure and Function

Table 12–3 lists hemicellulase genes that have been cloned from clostridia and thermoanaerobes. Nine distinct DNA fragments expressing genes related to xylan hydrolysis have also been isolated from *C. stercorarium* strain F-9 (Sakka et al. 1990). Comparison of restriction maps suggests that three of these clones might carry the genes *xynA*, *celW*, and *bxlA* listed in Table 12–3. Two additional *C. thermocellum* xylanase genes were cloned by MacKenzie et al. (1989) but not further characterized.

The xylanases encoded by the cloned genes can be assigned to two enzyme families (Figure 12–18) that have been designated β-glycanase family F and G, respectively (Gilkes et al. 1991). Family F includes several celloxylanases like *C. thermocellum* XynZ and *C. stercorarium* CelX that show both β-cellobiosidase (cellodextrinase) and xylanase activity. It is conceivable, that these enzymes participate both in xylan degradation and in the later stages of cellulose degradation. Family G is a group of closely related endo-β-xylanases from bacilli and clostridia, including XynA of *B. pumilus*, XynB of *C. acetobutylicum*, and XynA of *C. stercorarium*. The tertiary structure of *B. pumilus* XynA has recently been determined at 0.22-nm resolution (Arase et al. 1993). According to these data, the protein fold of family G xylanases is composed mostly of antiparallel β-sheets and is therefore entirely different from those of cellulase families B and E.

Sequence analysis reveals that the xylanases from thermophilic clostridia are composed of both catalytic and noncatalytic domains (Figure 12–19). Thus, the C-terminal two-thirds of *C. stercorarium* XynA consist of three copies of a CBD. This domain, which is unrelated to the CBD of the CelY and CelZ cellulases from the same organism, is also present in the otherwise unrelated XynZ enzyme of *C. thermocellum*. Consistent with its cellulosomal location, the central region of XynZ contains in addition to the CBD the reiterated 25-aa sequence found in most clostridial cellulases. The catalytic region of family F celloxylanases is located in different sections of the enzymes: C terminal in *C. thermocellum* XynZ, N terminal in *Ca. saccharolyticum* CelB, and central in *C. stercorarium* CelX. In case of XynZ and CelB, the various sections of the enzyme molecules are separated by Pro-Thr(PT)-

TABLE 12–3 ***Clostridium*** **and** ***Caldocellum*** **Genes Concerned with Hemicellulose Degradation**

Organism	*Gene Cloned*	*Molecular Mass of Protein (kDa)*	*Reference*
Bu. fibrisolvens	Xylanase (*xynA*)	46.7	Mannarelli et al. 1990
	Xylanase (*xynB*)	73.2	Lin and Thomson 1991
	α-L-Arabinofuranosidase (*xylB*)	62.0	Utt et al. 1991
C. acetobutylicum	Xylanase (*xynB*)	29.0	Zappe et al. 1990
C. stercorarium	Xylanase (*xynA*)	70.7	Adelsberger 1992
	Celloxylanase X (*celX*)	115.3	Jauris et al. 1990
	Celloxylanase W (*celW*)	42	Schwarz et al. 1990
	β-D-Xylosidase A (*bxlA*)	58.0	Adelsberger 1992
	β-D-Xylosidase B (*bxlB*)	79.3	Adelsberger 1992
	α-L-Arabinofuranosidase A (*arfA*)	54.9	Schwarz et al. 1990
	α-L-Arabinofuranosidase B (*arfB*)	52	Schwarz et al. 1990
C. thermocellum	Xylanase Z (*xynZ*)	92.3	Grépinet et al. 1988
	Xylanase	25	MacKenzie et al. 1989
C. thermohydrosulfuricum	Xylose isomerase	50.2	Dekker et al. 1991
C. thermosulfurogenes	Xylose isomerase (*xylA*)	50.2	Lee et al. 1990b
Ca. saccharolyticum	Celloxylanase (*celB*)	117.7	Saul et al. 1989; Saul et al. 1990
	Xylanase (*xynA*)	39.9	Lüthi et al. 1990
	β-D-Xylosidase (*xynB*)	55.6	Lüthi et al. 1990
	Acetylxylan esterase (*xynC*)	30.6	Lüthi et al. 1990
	β-Mannanase (*manA*)	38.9	Lüthi et al. 1991; Gibbs et al. 1992

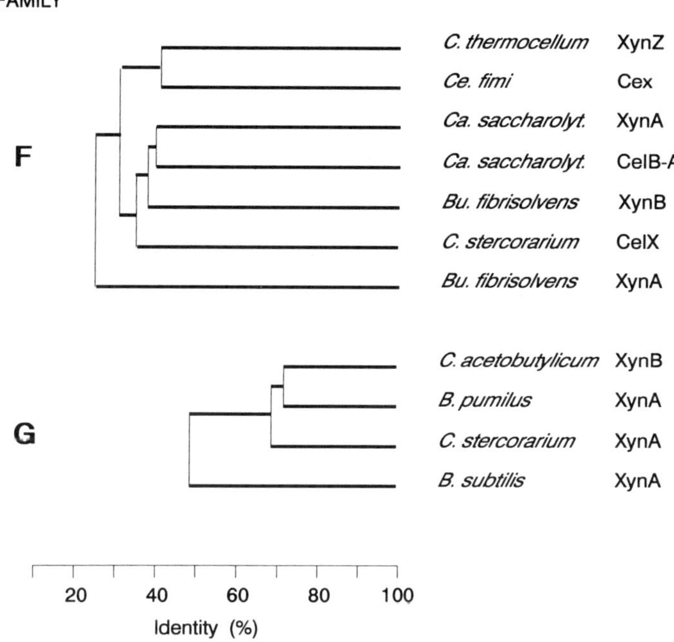

FIGURE 12–18 Xylanase families. CelB-A denotes the N-terminal catalytic domain of *Ca. saccharolyticum* CelB. References not included in Table 12–3: *B. pumilus* XynA (Fukusaki et al. 1984), *B. subtilis* XynA (Paice et al. 1986), *Ce. fimi* Cex (O'Neill et al. 1986).

rich linker sequences (PT boxes). No function has as yet been assigned to the N- and C-terminal regions of CelX or to the N-terminal section of XynZ.

Because of their composite domain structure, the products of the *C. stercorarium xynA* and *celX* genes were found to be susceptible to proteolytic processing in *E. coli*. The size of the enzymatically active cleavage products obtained from XynA is consistent with a stepwise removal of the iterated C-terminal CBDs. Thus, the immunologically related *C. stercorarium* xylanases A, B, and C (Berenger et al. 1985) are most likely derived from a single precursor enzyme. On the other hand, N-terminal sequence analysis indicates that the 80-kDa celloxylanase I isolated from *C. stercorarium* is not derived by proteolytic cleavage from the 115-kDa celloxylanase II encoded by the *celX* gene (F. Lottspeich, unpublished observations). It appears that these enzymes are protected from proteolytic degradation by glycosylation in their original host (K. Bronnenmeier, unpublished observations).

The β-xylosidase BxlA encoded by the *C. stercorarium bxlA* gene is homologous to the β-xylosidase produced by the *Ca. saccharolyticum* gene *xynB* (35% identity). These enzymes might be related to the β-glycosidases of family A. The β-xylosidase encoded by the *C. stercorarium* gene *bxlB*

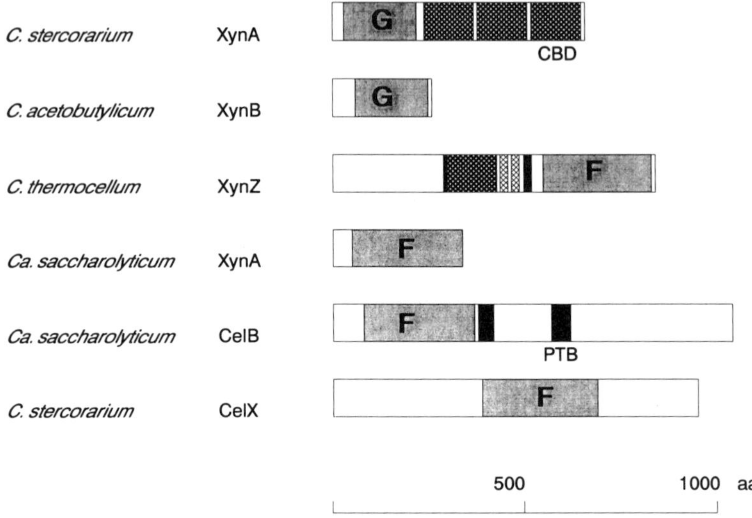

FIGURE 12–19 Domain structure of xylanases. Homologous domains are indicated by the same patterns. The catalytic domains are shaded. Bold uppercase letters indicate membership of a β-glycanase catalytic domain family (see Figure 12–18). Abbreviations: PTB, Pro-Thr box; CBD, cellulose-binding domain.

corresponds to the major β-xylosidase activity purified from *C. stercorarium* culture supernatants (Bronnenmeier et al. 1990). *C. stercorarium* BxlB is related to the family B β-glucosidases BglZ of *C. stercorarium* and BglB of *C. thermocellum* (see Figure 12–8). It should be noted that BglB exhibits both β-glucosidase and β-xylosidase activity (Gräbnitz and Staudenbauer 1988; Kateya et al. 1992), whereas the *C. stercorarium* enzymes BglZ and BxlX are rather specific for β-glucosides (BglZ) and β-xylosides (BxlB), respectively. Presumably, the corresponding genes arose by gene duplication and subsequent divergent evolution of a precursor gene encoding an enzyme with dual specificity like *C. thermocellum* BglB.

β-D-Xylosidases and α-L-arabinofuranosidases are, in general, not strictly specific for their respective substrates. Thus, β-xylosidase B of *C. stercorarium* shows significant α-L-arabinofuranosidase activity (Schwarz et al. 1990), whereas the α-L-arabinofuranosidases encoded by the *C. stercorarium arfA* and *Bu. fibrisolvens xylB* genes have similar activities towards aryl-α-L-arabinosides and aryl-β-D-xylosides (Schwarz et al. 1990; Utt et al. 1991). Deletion analysis has, to date, provided no evidence for the involvement of two distinct active sites. Therefore, these α-L-arabinosidases/β-D-xylosidases should not be considered as "bifunctional" proteins (Utt et al. 1991). Sequence comparison indicates that *C. stercorarium* ArfA, *Bu. fibrisolvens* XylB, and *B. pumilus* XynB (Xu et al. 1991) are structurally related and constitute a novel glycosidase family.

Contrary to an earlier report (Lüthi et al. 1991), the β-mannanase encoded by the *Ca. saccharolyticum manA* gene is part of a bifunctional multidomain enzyme (Gibbs et al. 1992). The *manA* gene product consists of an N-terminal β-mannanase domain, a central portion comprising two CBDs flanked by PT boxes, and a C-terminal endoglucanase/xylanase domain. The overall domain structure is therefore similar to that of the *Ca. saccharolyticum* CelB enzyme (see Figure 12–12). The β-mannanase domain contains the highly conserved sequence motifs of family A cellulases (Figure 12–9) and may therefore be considered as yet another member of this enzyme superfamily.

The sequences of the xylose isomerases of *C. thermohydrosulfuricum* and *C. thermosulfurogenes* are nearly identical. These enzymes are closely related to the glucose/xylose isomerases of *B. subtilis* (70% identity) and *E. coli* (50% identity). Glucose/xylose isomerases from diverse bacteria are homologous and the crystal structures of several enzymes have been determined. As in the case of α-amylases (see Section 12.1.2), the main domain of the enzyme is a parallel-stranded $(\alpha/\beta)_8$ barrel (Henrick et al. 1989). Lee et al. (1990a) have recently identified the invariant His-101 as the active-site residue of *C. thermosulfurogenes* xylose isomerase.

12.3.3 Gene Structure and Regulation

Clustering of *C. stercorarium* hemicellulase genes has not been detected to date. The size of the mapped *xynA* mRNA (see Figure 12–14) corresponds to a monocistronic transcript. In contrast, the genes encoding the xylan-degrading enzymes of *Ca. saccharolyticum* are clustered and presumably belong to a polycistronic transcription unit. The *xynACB* cluster is flanked by putative transcription control sequences (Figure 12–20). Two ORFs (ORF3 and ORF4) are located between the acetylxylan esterase gene *xynC* and the

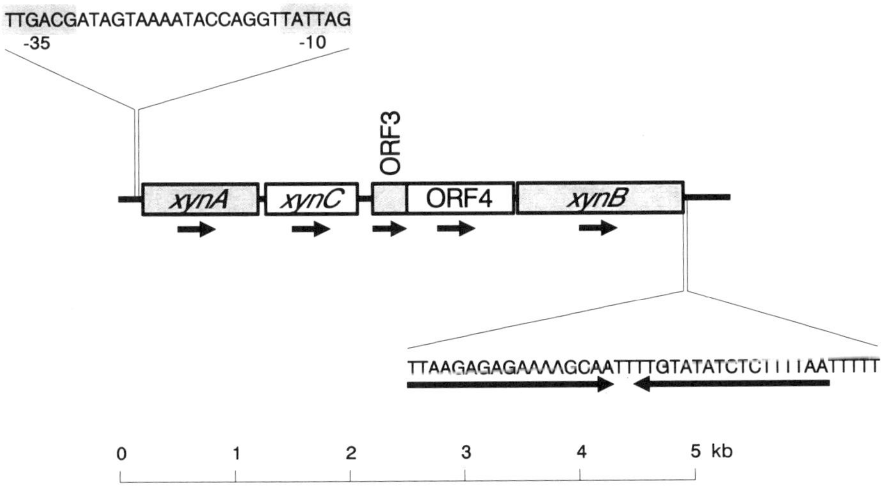

FIGURE 12–20 Structure of the *Ca. saccharolyticum xynA-xynB* region. Sequence data are from Lüthi et al. (1990).

C. thermosulfurogenes	xylA	AAT	GTTTGTT	GGACA	GACAAAC	GAA	- 18
B. subtilis 168	xyl	TTA	GTTTGTT	TAAAC	AACAAAC	TAA	- 74
B. subtilis 168	xyn	TTA	GTTTGTT	TGATC	AACAAAC	TAA	- 294
B. subtilis W23	xylA	TTA	GTTTGTT	TGGGC	AACAAAC	TAA	- 70
B. licheniformis	xylA	TTA	GTTTAAT	GGTTA	AACAAAC	ATT	- 43
B. megaterium	xylA	TTA	GTTTATT	GGATA	AACAAAC	TAA	- 60

FIGURE 12–21. Sequence comparison of *xyl* operators. Numbers indicate the position of the 3′-nucleotide relative to the start codon. References to *xyl* operator sequences: *B. subtilis* 168 (Hastrup 1988), *B. subtilis* W23 (Kreuzer et al. 1989), *B. licheniformis* (Scheler et al. 1991), *B. megaterium* (Rygus et al. 1991).

β-xylosidase gene *xynB*. The deduced amino acid sequences show significant homologies to the xylanase encoded by the *xynA* gene. Apparently, these pseudogenes originated from a genetic rearrangement introducing a frameshift into a former xylanase gene located between *xynC* and *xynB*. The putative transcription terminator is only of moderate strength ($\Delta G = -40.1$ kJ) and expected to allow read-through into a truncated ORF6 located 500 bp downstream of *xynB* (Lüthi et al. 1990).

In *B. subtilis*, the xylose regulon comprises six genes that are clustered within a 7.5-kb region of the chromosome (Klier and Rapoport 1988): *xynA* (xylanase), *xynB* (β-xylosidase), *xynC* (permease ?), *xylA* (xylose isomerase), *xylB* (xylulose kinase), and *xylR* (repressor). The operators for the *xynCB* and *xylAB* operons have been identified (Hastrup 1988). They correspond to a 10-bp palindromic sequence, which was shown to be the target for the *xylR* repressor (Kreuzer et al. 1989). It seems highly significant that the same dyad symmetry element is also present upstream of the *C. thermosulfurogenes xylA* coding region (Figure 12–21). This finding suggests that the organization and regulation of clostridial genes concerned with xylan utilization may be similar to the *Bacillus* xylose regulon.

12.4 UTILIZATION OF LACTOSE

12.4.1 Pathways of Lactose Catabolism

Whey, a waste product of the dairy industry, has attracted attention as an alternative substrate for solvent production because of its abundance and high-lactose content. Two major pathways exist for uptake and utilization of lactose in clostridia. One pathway involves transport of lactose by lactose permease and subsequent hydrolysis to glucose and galactose by β-galactosidase. An alternate pathway involves a PEP-dependent phosphotransferase system converting lactose to lactose-6-phosphate, which is then hydrolyzed by a phospho-β-galactosidase to yield glucose and galactose-6-phosphate.

Glucose formed by cleavage of lactose is phosphorylated to glucose-6-phosphate, which enters the glycolytic pathway. The further catabolism of galactose has not been rigorously elucidated in clostridia, but it appears likely that galactose is phosphorylated by galactokinase to galactose-1-phosphate, which is then converted via the Leloir pathway to glucose-6-phosphate. Galactose-6-phosphate formed by phospho-β-galactosidase action is presumably degraded to triose phosphates via the tagatose-6-phosphate pathway (Rosey et al. 1991).

It has been shown that several *C. acetobutylicum* strains can use whey as a substrate for solvent production (Ennis and Maddox 1985; Yu et al. 1987). Most strains produced both β-galactosidase and phospho-β-galactosidase. Gene expression was found to be developmentally regulated. Levels of phospho-β-galactosidase were highest during the acidogenic phase, whereas β-galactosidase induction was associated with solventogenesis and the onset of sporulation. Synthesis of both enzymes is subject to repression by glucose.

12.4.2 Protein Structure and Function

Two clostridial β-galactosidase genes have recently been cloned and sequenced (Table 12–4). The enzyme encoded by the *C. acetobutylicum cbgA* gene was found to be closely related to β-galactosidases of lactic acid bacteria (see Figure 12–8, family C). On the other hand, the *C. thermosulfurogenes* LacA β-galactosidase is only distantly related to this subfamily of β-galactosidases. The homology is limited to the N-terminal half of the LacA enzyme and apparently ends shortly after the putative active site region.

E. coli β-galactosidase is a tetrameric protein containing identical subunits of 1023 aa (Kalnins et al. 1983). Several regions of high-sequence similarity have been identified between the β-galactosidases of lactic acid bacteria and *E. coli* β-galactosidase (Schmidt et al. 1989; Schroeder et al. 1991). These regions include the active site residues Glu-461 and Tyr-503 (numbering of *E. coli* LacZ). The corresponding sequences are also highly conserved in the *C. acetobutylicum* β-galactosidase, except for the missing C-terminal segment (Figure 12–22). The sequence of the *C. acetobutylicum* CbgA ends 13 aa after homology region VII, corresponding to position 921 of *E. coli* LacZ. This discrepancy is puzzling, since it has been shown that the C-terminal region of LacZ is crucial for proper assembly of the enzyme. Thus, the mutant X90 protein lacking the C-terminal 10 aa is completely devoid of enzyme activity (Mandecki et al. 1981). However, X90 protein can be complemented to form active enzyme by addition of a C-terminal 32-aa peptide.

12.4.3 Gene Structure and Regulation

Expression of the *C. acetobutylicum* β-galactosidase gene in *E. coli* shows some perplexing features. Hancock et al. (1991) identified by transposon Tn5

TABLE 12–4 ***Clostridium*** **Genes Concerned with Lactose Utilization**

Organism	*Gene Cloned*	*Molecular Mass of Protein (kDa)*	*Reference*
C. acetobutylicum	β-Galactosidase (*cbgA*)	105.1	Hancock et al. 1991
C. pasteurianum	Galactokinase	40	Daldal and Applebaum 1985
C. thermosulfurogenes	β-Galactosidase (*lacZ*)	83.8	Burchhardt and Bahl 1991

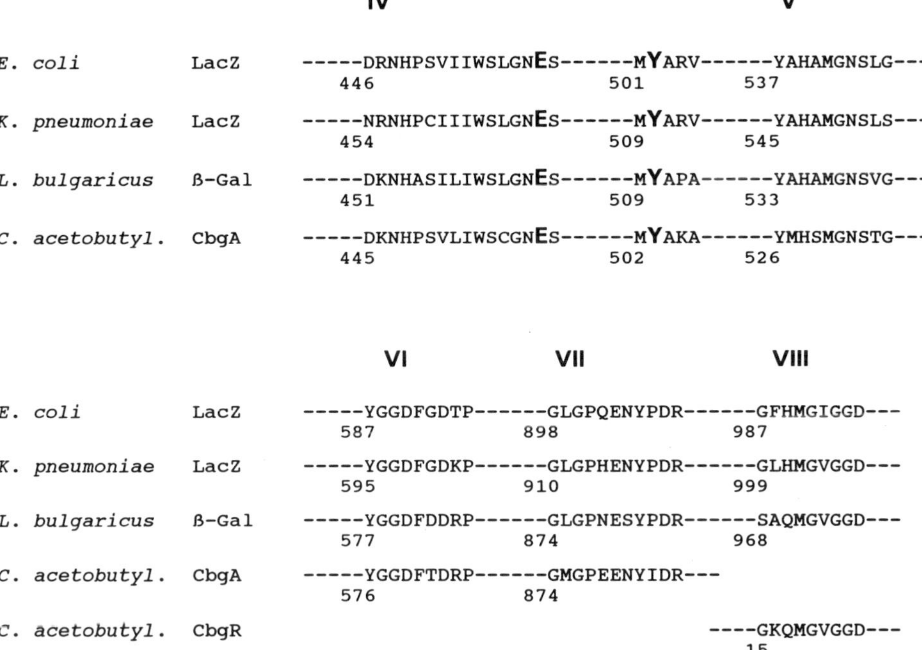

FIGURE 12–22 Conserved regions of β-galactosidases. Conserved regions are designated according to Schmidt et al. (1989). Catalytic residues are shown in enlarged case. For sequence references see Table 12–4 and Figure 12–8. CbgR denotes the deduced aa sequence of the fully translated ORF of the *cbgR* locus. Numbering of CbgR starts at the first codon of the ORF.

mutagenesis a second gene locus, designated *cbgR*, as necessary for the expression of the β-galactosidase encoded by the *cbgA* gene. The *cbgR* locus corresponds to an ORF of about 0.4 kb located downstream of *cbgA* (Figure 12–23). The fully translated ORF contains a sequence closely related to the C-terminal region VIII, which is highly conserved within β-glycosidase family C but missing in the *cbgA* gene product (see Figure 12–21). However, the interpretation that *cbgR* encodes a second subunit of the β-galactosidase holoenzyme (analogous to the C-terminal peptide complementing the LacZ X90 protein) seems unlikely since *cbgR* cannot complement *cbgA* in trans. Hancock et al. (1991), therefore, proposed that *cbgR* encodes a *cis*-acting positive regulatory protein necessary for *cbgA* expression. It is, however, unclear why such a protein should be required in a heterologous host like *E. coli*, where *cbgA* expression is neither subject to induction nor glucose repression. Assuming that the experimental data are correct, they seem to imply that β-galactosidase synthesis requires contiguous expression of the *cbgA*-*cbgR* loci possibly involving ribosomal frame-shifting. Such an unusual mode

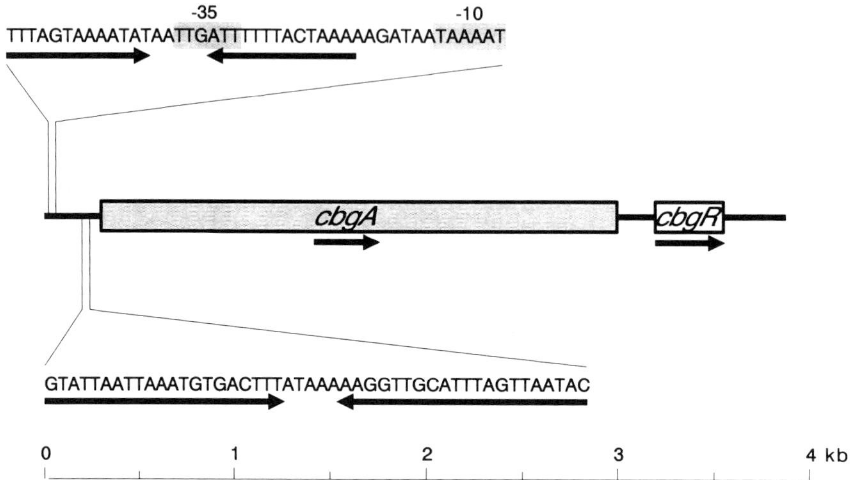

FIGURE 12–23 Structure of the *C. acetobutylicum cbgA* gene. Sequence data are from Hancock et al. (1991).

of translation has recently been proposed for the expression of the cellulase gene *endA* of *Fibrobacter succinogenes* (Cavicchioli et al. 1991).

This interpretation is supported by the recent finding that the *Leuconostoc lactis* β-galactosidase has a heterodimeric structure and is composed of 75- and 36-kDa subunits (David et al. 1992). These subunits are encoded by two partially overlapping genes, designated *lacL* and *lacM*. Expression of the *lacL* gene is essential for LacM production and both genes are required for β-galactosidase production in *E. coli*. Consistent with the absence of a ribosome-binding site for the *lacM* gene and its overlap with the *lacL* gene, it is assumed that the *lacM* gene is translationally coupled to the *lacL* gene. Obviously, this might also be the case for the *cbgA* and *cbgR* genes of *C. acetobutylicum*.

It should be noted that the 5′-noncoding region of *cbgA* contains two sites with extended dyad symmetry (see Figure 12–23). The first site is a 13-bp hyphenated palindrome that overlaps with the −35 region of a σ^A-type promoter structure. The second site is a 21-bp imperfect palindrome, which when transcribed into RNA could form a stable hairpin loop ($\Delta G = -74$ kJ) resembling those found in transcription termination signals. This raises the interesting possibility that control of *cbgA* expression might involve antitermination as has been observed for the phospho-β-glucosidase operon of *E. coli* (Houman et al. 1990) and the sucrose regulon of *B. subtilis* (Debarbouille et al. 1990). A possible role of CbgR in *cbgA* antitermination seems unlikely since its amino acid sequence shares no obvious homology with known antiterminator proteins. The apparent absence of putative transcription terminators downstream of the *cbgR* locus indicates that, like the *lacZ* gene of enteric bacteria, the *cbgA-cbgR* loci might be part of a larger transcription unit including additional genes concerned with lactose utilization.

12.5 GENETIC ENHANCEMENT OF SUBSTRATE UTILIZATION

12.5.1 Strain Improvement by Mutagenesis

The efficiency of substrate utilization by solventogenic clostridia may obviously be increased by classical genetic methods, i.e., mutagenesis and screening for hyperproductive mutants. Bowring and Morris (1985) have shown that *C. acetobutylicum* is susceptible to mutagenesis by the DNA-alkylating agents *N*-methyl-*N'*-nitro-*N*-nitrosoguanidine (MNNG) and ethyl methanesulfonate (EMS). On the other hand, indirect mutagens such as UV irradiation and mitomycin, which rely on error-prone DNA repair were found to be less effective.

Hyun and Zeikus (1985a, 1985b) employed MNNG mutagenesis to isolate catabolite-resistant mutants of *C. thermohydrosulfuricum* and *C. thermosulfurogenes*. Enrichment was achieved by incubating mutagenized cultures in starch medium in the presence of the nonmetabolizable catabolite repressor 2-deoxyglucose. Regulatory mutants were classified as either catabolite repression-resistant or constitutive and catabolite repression-resistant. Hyperproductive *C. thermohydrosulfuricum* mutants were obtained that produced about twofold more amylopullulanase and α-glucosidase than the wild type. A *C. thermosulfurogenes* mutant, which was both constitutive and catabolite repression-resistant showed an eightfold increase in β-amylase production. When grown on glucose, it produced nearly twice as much ethanol than the wild type as a result of switching from lactate to ethanol production.

Mutagenesis was also applied to increase cellulase production by *C. thermocellum*. Upon irradiation with UV light, Shinmyo et al. (1979) obtained a mutant that produced twice as much Avicelase and CMCase as the parent. Recently, Mori (1990a) reported the isolation of a naturally occurring hypercellulolytic *C. thermocellum* strain, which produced a 10-fold higher Avicelase activity than any other *C. thermocellum* strain. Cellulase production was further enhanced by two steps of mutagenesis employing UV treatment and gamma irradiation (Mori 1990b). A hyperproductive mutant was isolated that produced two to three times as much avicelase, xylanase, and cellobiase than the parent strain, whereas its CMCase activity was unchanged.

12.5.2 Applications of Recombinant DNA Technology

The advent of gene technology suggests new strategies for extending the substrate range of saccharolytic clostridia via introduction of heterologous, or suitably modified homologous, substrate utilization genes. Moreover, the properties of the encoded enzymes can be tailored for specific applications by site-directed mutagenesis and related techniques. The main obstacle for applying this promising new technology to clostridia was, until now, the lack of suitable host-vector systems. Only recently have techniques been worked out for efficient and reproducible gene transfer into *C. acetobutylicum* (Young et al. 1989). Genetic manipulation of thermophilic clostridia is still in an early stage.

In an attempt to increase the cellulolytic activity of *C. acetobutylicum* NCIB 8052, Minton et al. (1990) constructed an artificial operon consisting of the *celC* and *celA* genes of *C. thermocellum*. A recombinant plasmid carrying this *celCA* operon was introduced into *C. acetobutylicum* by electroporation and found to significantly enhance the growth of NCIB 8052 on soluble β-glucans. In future experiments, a complete cellulase system like the one encoded by the *celYZ* operon of *C. stercorarium* will be transferred to *C. acetobutylicum* to endow it with true cellulolytic properties.

An alternative strategy consists in supplementing fermentations with clostridial enzymes produced in a heterologous host. For this purpose, *Bacillus* strains employed in the large-scale production of technical enzymes appear as particularly suited. Clostridial enzymes successfully expressed in bacilli include: *C. thermosulfurogenes* β-amylase (Kitamoto et al. 1988), *T. brockii* amylopullulanase (Coleman et al. 1987), *C. thermocellum* cellulases CelA (Soutschek-Bauer and Staudenbauer 1987; Joliff et al. 1989), *C. stercorarium* cellulase CelZ (K. Bronnenmeier, unpublished observations), and *C. thermosulfurogenes* xylose isomerase (Lee et al. 1990b). Amylases and cellulases are efficiently secreted and apparently not subject to proteolytic degradation. The yields compare favorably with that obtained in the original host, but are still too low for industrial production. A further increase in stability and productivity can be achieved by insertion of the clostridial DNA into the *Bacillus* chromosome and subsequent gene amplification (Petit et al. 1990).

A glimpse of things to come was recently provided by Meng et al. (1991), who switched the substrate preference of *C. thermosulfurogenes* xylose isomerase from xylose to glucose by replacing Trp and Val residues that contact the C-6 OH group of the substrate with the smaller residues Phe and Val. Protein engineering combined with improved techniques for clostridial gene transfer offers nearly unlimited possibilities for redesigning clostridial enzymes and whole metabolic pathways.

REFERENCES

Adelsberger, H. (1992) Doctoral thesis, Technical University, Munich.

Ait, N., Creuzet, N., and Cattaneo, J. (1982) *J. Gen. Microbiol.* 128, 569–577.

Alexander, J.K. (1972) *Methods Enzymol.* 28, 944–948.

Aleshin, A., Golubev, A., Firsov, L.M., and Honzatko, R.B. (1992) *J. Biol. Chem.* 267, 19291–19298.

Allcock, E.R., and Woods, D.R. (1981) *Appl. Environ. Microbiol.* 41, 539–541.

Arase, A., Yomo, T., Urabe, I., Hata, Y., et al. (1993) *FEBS Lett.* 316, 123–127.

Antranikian, G., Zablowski, P., and Gottschalk, G. (1987) *Appl. Microbiol. Biotechnol.* 27, 75.

Bagnara-Tardif, C., Gaudin, C., Belaich, A., et al. (1992) *Gene* 119, 17–28.

Bahl, H. Burchhardt, G., Spreinat, A., et al. (1991a) *Appl. Environ. Microbiol.* 57, 1554–1559.

Bahl, H., Burchhardt, G., and Wienecke, A. (1991b) *FEMS Microbiol. Lett.* 81, 83–88.

Baird, S.D., Hefford, M.A., Johnson, D.A., et al. (1990a) *Biochem. Biophys. Res. Commun.* 169, 1035–1039.

Baird, S.D., Johnson, D.A., and Seligy, V.L. (1990b) *J. Bacteriol.* 172, 1576–1586.

Barnett, C.C., Berka, R.M., and Fowler, T. (1991) *Bio/Technology* 9, 562–567.

Bateson, M.M., Wiegel, J., and Ward, D.M. (1989) *Syst. Appl. Microbiol.* 12, 1–7.

Béguin, P. (1990) *Annu. Rev. Microbiol.* 44, 219–248.

Béguin, P., Cornet, P., and Aubert, J.P. (1985) *J. Bacteriol.* 162, 102–105.

Béguin, P., Racancourt, M., Chebrou, M.C., and Aubert, J.P. (1986) *Mol. Gen. Genet.* 202, 251–254.

Béguin, P., Millet, J., and Aubert, J.P. (1992) *FEMS Microbiol. Lett.* 100, 523–528.

Belaich, A., Fierobe, H.P., Baty, D., et al. (1992) *J. Bacteriol.* 174, 4677–4682.

Berenger, J.F., Frixon, C., Bigliardi, J., and Creuzet, N. (1985) *Can. J. Microbiol.* 31, 635.

Berger, E., Jones, W.A., Jones, D.T., and Woods, D.R. (1989) *Mol. Gen. Genet.* 219, 193–198.

Berger, E., Jones, W.A., Jones, D.T., and Woods, D.R. (1990) *Mol. Gen. Genet.* 223, 310–318.

Bowring, S.N., and Morris, J.G. (1985) *J. Appl. Bacteriol.* 58, 577–584.

Bronnenmeier, K., Ebenbichler, C., and Staudenbauer, W.L. (1990) *J. Chromatogr.* 521, 301–310.

Bronnenmeier, K., Rücknagel, K.P., and Staudenbauer, W.L. (1991) *Eur. J. Biochem.* 200, 379–385.

Bronnenmeier, K., and Staudenbauer, W.L. (1988a) *Appl. Microbiol. Biotechnol.* 27, 432–436.

Bronnenmeier, K., and Staudenbauer, W.L. (1988b) *Appl. Microbiol. Biotechnol.* 28, 380–386.

Bronnenmeier, K., and Staudenbauer, W.L. (1990) *Enzyme Microb. Technol.* 12, 431–436.

Bryant, M.P. (1986) in *Bergey's Manual of Systematic Bacteriology*, vol. 2 (Sneath, P.H.A., Mair, N.S., Sharpe, M.E., and Holt, J.G. eds.), pp. 1376–1379, The Williams & Wilkins Co., Baltimore.

Bueno, A., Vazquez de Aldana, C.R., Correa, J., and del Rey, F. (1990) *Nucleic Acids Res.* 18, 4248.

Buisson, G., Duee, E., Haser, R., and Payan, F. (1987) *EMBO J.* 6, 3909–3916.

Bumazkin, B.K., Velikodvorskaya, G.A., Tuka, K., Mogutov, M.A., and Strongin, A.Y. (1990) *Biochem. Biophys. Res. Commun.* 167, 1057–1064.

Burchhardt, G., and Bahl, H. (1991) *Gene*, 106, 13–19.

Burchhardt, G., Wienecke, A., and Bahl, H. (1991) *Curr. Microbiol.* 22, 91–95.

Buvinger, W.E., and Riley, M. (1985) *J. Bacteriol.* 163, 850–857.

Candussio, A., Schmid, G., and Böck, A. (1991) *Eur. J. Biochem.* 199, 637–641.

Cavicchioli, R., East, P.D., and Watson, K. (1991) *J. Bacteriol.* 173, 3265–3268.

Chauvaux, S., Béguin, P., and Auber, J.P. (1992) *J. Biol. Chem.* 267, 4472–4478.

Coleman, R.D., Yang, S.S., and McAlister, M.P. (1987) *J. Bacteriol.* 169, 4302–4307.

Coughlan, M.P. (1985) *Biotechnol. Genet. Eng. Rev.* 3, 39–109.

Coutinho, J.B., Moser, B., Kilburn, D.G., Warren, R.A.J., and Miller, R.C. (1991) *Mol. Microbiol.* 5, 1221–1233.
Creuzet, N., Berenger, J.F., and Frixon, C. (1983) *FEMS Microbiol. Lett.* 20, 347–350.
Daldal, F. and Applebaum, J. (1985) *J. Mol. Biol.* 186, 533–544.
David, S., Stevens, H., van Riel, M., Simons, G., and de Vos, W.M. (1992) *J. Bacteriol.* 174, 4475–4481.
Debarbouille, M., Arnaud, M., Fouet, A., Klier, A., and Rapoport, G. (1990) *J. Bacteriol.* 172, 3966–3973.
Dekker, K., Yamagata, H., Sakaguchi, K., and Udaka, S. (1991) *Agric. Biol. Chem.* 55, 221–227.
Enari, T.M., and Niku-Paavola, M.L. (1987) *Crit. Rev. Biotechnol.* 5, 67–87.
Ennis, B.M., and Maddox, I.S. (1985) *Biotechnol. Lett.* 7, 601–606.
Faure, E., Bagnara, C., Belaich, A., and Belaich, J.P. (1988) *Gene* 65, 51–58.
Faure, E., Belaich, A., Bagnara, C., Gaudin, C., and Belaich, J.P. (1989) *Gene* 84, 39–46.
Foong, F., Hamamoto, T., Shoseyov, O., and Doi, R.H. (1991) *J. Gen. Microbiol.* 137, 1729–1736.
Fujino, T., Béguin, P., and Aubert, J.P. (1992) *FEMS Microbiol. Lett.* 94, 165–170.
Fujino, T., Sasaki, T., Ohmiya, K., and Shimizu, S. (1990) *Appl. Environ. Microbiol.* 56, 1175–1178.
Fujino, T., and Ohmiya, K. (1991) *J. Ferment. Bioeng.* 72, 422–425.
Fujino, T., and Ohmiya, K. (1992) *J. Ferment. Bioeng.* 73, 308–313.
Fukumori, F., Kudo, T., Narahashi, Y., and Horikoshi, K. (1986) *J. Gen. Microbiol.* 132, 2329–2335.
Fukusaki, E., Panbangred, W., Shinmyo, A., and Okada, H. (1984) *FEBS Lett.* 171, 197.
Gebler, J, Gilkes, N., Claeyssens, M., et al. (1992) *J. Biol. Chem.* 267, 12559–12561.
Gerischer, U., and Dürre, P. (1990) *J. Bacteriol.* 172, 6907–6918.
Gerwig, G.J., Kamerling, J.R., Vliegenthart, J.F.G., et al. (1991) *Eur. J. Biochem.* 196, 115–122.
Gibbs, M.D., Saul, D.J., Lüthi, E., and Bergquist, P.L. (1992) *Appl. Environ. Microbiol.* 58, 3864–3867.
Gilkes, N.R., Henrissat, B., Kilburn, D.G., Miller, R.C., and Warren, R.A.J. (1991) *Microbiol. Rev.* 55, 303–315.
Gonzalez-Candelas, L., Ramon, D., and Polaina, J. (1990) *Gene* 95, 31–38.
Grépinet, O., and Béguin, P. (1986) *Nucleic Acids Res.* 14, 1791–1799.
Grépinet, O., Chebrou, M.C., and Béguin, P. (1988) *J. Bacteriol.* 170, 4582–4588.
Gräbnitz, F., Rücknagel, K.P., Seiß, M., and Staudenbauer, W.L. (1989) *Mol. Gen. Genet.* 217, 70–76.
Gräbnitz, F., Seiß, M., Rücknagel, K.P., and Staudenbauer, W.L. (1991) *Eur. J. Biochem.* 200, 301–309.
Gräbnitz, F, and Staudenbauer, W.L. (1988) *Biotechnol. Lett.* 10, 73–78.
Hall, J., Hazlewood, G.P., Barker, P.J., and Gilbert, H.J. (1988) *Gene* 69, 29–38.
Hamamoto, T., Shoseyov, O., Foong, F., and Doi, R.H. (1990) *FEMS Microbiol. Lett.* 72, 285–288.
Hamamoto, T., Foong, F., Shoseyov, O., and Doi, R.H. (1992) *Mol. Gen. Genet.* 231, 472–479.

Hancock, K.R., Rockman, E., Young, C.A., et al. (1991) *J. Bacteriol.* 173, 3084–3095.

Hansen, C.K. (1992) *FEBS Lett.* 305, 91–96.

Hastrup, S. (1988) in *Genetics and Biotechnology of Bacilli*, vol. 2 (Cranesan, A.T., and Hoch, J.A., eds.), pp. 79–84, Academic Press, New York.

Hazlewood, G.P., Davidson, K., Laurie, J.I., Romaniec, M.P.M., and Gilbert, H.J. (1990) *J. Gen. Microbiol.* 136, 2089–2097.

Henkin, T.M., Grundy, F.J., Nicholson, W.L., and Chambliss, G.H. (1991) *Mol. Microbiol.* 5, 575–584.

Henrick, K., Collyer, C.A., and Blow, D.M. (1989) *J. Mol. Biol.* 208, 129–157.

Henrissat, B., Claeyssens, M., Tomme, P., and Mornon, J.P. (1989) *Gene* 81, 83–95.

Hernandez, P.E. (1982) *J. Gen. Appl. Microbiol.* 28, 469–477.

Hofmann, B.E., Bender, H., and Schulz, G.E. (1989) *J. Mol. Biol.* 209, 793–800.

Holm, L., Koivula, A.K., Lehtovaara, P.M., Hemminiki, A., and Knowles, J.K.C. (1990) *Protein Eng.* 3, 181–191.

Houman, F., Diaz-Torres, M.R., and Wright, A. (1990) *Cell* 62, 1153–1163.

Hyun, H.H., Shen, G.J., and Zeikus, J.G. (1985) *J. Bacteriol.* 164, 1153–1161.

Hyun, H.H., and Zeikus, J.G. (1985a) *J. Bacteriol.* 164, 1146–1152.

Hyun, H.H., and Zeikus, J.G. (1985b) *J. Bacteriol.* 164, 1162–1170.

Jauris, S., Rücknagel, K.P., Schwarz, W.H., et al. (1990) *Mol. Gen. Genet.* 223, 258–267.

Johnson, E.A., Bouchot, F., and Demain, A.L. (1985) *J. Gen. Microbiol.* 131, 2303.

Joliff, G., Béguin, P., and Aubert, J.P. (1986) *Nucleic Acids Res.* 14, 8605–8613.

Joliff, G., Edelman, A., Klier, A., and Rapoport, G. (1989) *Appl. Environ. Microbiol.* 55, 2739–2744.

Jones, D.T., and Woods, D.R. (1986) *Microbiol. Rev.* 50, 484–524.

Jorgensen, P.L., and Hansen, C.K. (1990) *Gene* 93, 55–60.

Juy, M., Amit, A.G., Alzari, P.M., et al. (1992) *Nature* 357, 89–91.

Kalnins, A., Otto, K., Rüther, U., and Müller-Hill, B. (1983) *EMBO J.* 2, 593–597.

Kateya, I.A., Golovchenko, N.P., Chuvilskaya, N.A., and Akimenko, V.K. (1992) *Enzyme Microb. Technol.* 14, 407–412.

Kawazu, T., Nakanishi, Y., Uozumi, N., et al. (1987) *J. Bacteriol.* 169, 1564–1570.

Kitamoto, N., Yamagata, H., Kato, T., Tsukagoshi, N., and Udaka, S. (1988) *J. Bacteriol.* 170, 5848–5854.

Klier, A.F., and Rapoport, G. (1988) *Annu. Rev. Microbiol.* 42, 65–95.

Kreuzer, P., Gärtner, D., Allmansberger, R., and Hillen, W. (1989) *J. Bacteriol.* 171, 3840–3845.

Kuriki, T., and Imanaka, T. (1989) *J. Gen. Microbiol.* 135, 1521–1528.

Lamed, R., Naimark, J., Morgenstern, E., and Bayer, E.A. (1987) *J. Bacteriol.* 169, 3792–3800.

Lao, G., Ghangas, G.S., Jung, E.D., and Wilson, D.B. (1991) *J. Bacteriol.* 173, 3397–3407.

Laoide, B.M., Chambliss, G.H., and McConnell, D.J. (1989) *J. Bacteriol.* 171, 2435–2442.

Lee, C., Bagdasarian, M., Meng, M., and Zeikus, J.G. (1990a) *J. Biol. Chem.* 265, 19082–19090.

Lee, C., Bhatnagar, L., Saha, B.C., et al. (1990b) *Appl. Environ. Microbiol.* 56, 2638–2643.

Lee, C., and Zeikus, J.G. (1991) *Biochem. J.* 273, 565–571.

Lee, S.F., and Forsberg, C.W. (1987a) *Appl. Environ. Microbiol.* 53, 651–654.
Lee, S.F., and Forsberg, C.W. (1987b) *Can. J. Microbiol.* 33, 1011–1016.
Lee, S.F., Forsberg, C.W., and Gibbins, L.N. (1985a) *Appl. Environ. Microbiol.* 50, 220–228.
Lee, S.F., Forsberg, C.W., and Gibbins, L.N. (1985b) *Appl. Environ. Microbiol.* 50, 1068–1076.
Lee, S.F., Forsberg, C.W., and Rattray, J.B. (1987) *Appl. Environ. Microbiol.* 53, 644–650.
Lin, L.L., Rumbak, E., Zappe, H., Thomson, J.A., and Woods, D.R. (1990) *J. Gen. Microbiol.* 136, 1567–1576.
Lin, L.L., and Thomson, J.A. (1991) *Mol. Gen. Genet.* 228, 55–61.
Love, D.R., Fisher, R., and Bergquist, P.L. (1988) *Mol. Gen. Genet.* 213, 84–92.
Lüthi, E., Jasmat, N.B., Grayling, R.A., Love, D.R., and Bergquist, P.L. (1991) *Appl. Environ. Microbiol.* 57, 694–700.
Lüthi, E., Love, D.R., McAnulty, J., et al. (1990) *Appl. Environ. Microbiol.* 56, 1017–1024.
Lynd, L.R. (1989) *Adv. Biochem. Eng./Biotechnol.* 38, 1–52.
Macarron, R., van Beeumen, J., Henrissat, B., de la Mata, I., and Claeyssens, M. (1993) *FEBS Lett.* 316, 137–140.
MacKenzie, C.R., Yang, R.C.A., Patel, G.B., Bilous, D., and Narang, S.A. (1989) *Arch. Microbiol.* 152, 377–381.
Mandecki, W., Fowler, A.V., and Zabin, I. (1981) *J. Bacteriol.* 147, 694–697.
Mannarelli, B.M., Evans, S., and Lee, D. (1990) *J. Bacteriol.* 172, 4247–4254.
Mathupala, S., Saha, B.C., and Zeikus, J.G. (1990) *Biochem. Biophys. Res. Commun.* 166, 126–132.
Matsuura, Y., Kusunoki, M., Harada, W., and Kakudo, M. (1984) *J. Biochem.* 95, 697–702.
Mayer, F., Coughlan, M.P., Mori, Y., and Ljungdahl, L.G. (1987) *Appl. Environ. Microbiol.* 53, 2785–2792.
McGavin, M.J., Forsberg, C.W., Crosby, B., et al. (1989) *J. Bacteriol.* 171, 5587–5595.
Meinke, A., Braun, C., Gilkes, N.R., et al. (1991) *J. Bacteriol.* 173, 308–314.
Melasniemi, H. (1987a) *J. Gen. Microbiol.* 133, 883–890.
Melasniemi, H. (1987b) *Biochem. J.* 246, 193–197.
Melasniemi, H. (1988) *Biochem. J.* 250, 813–818.
Melasniemi, H., Paloheimo, M., and Hemiö, L. (1990) *J. Gen. Microbiol.* 136, 447–454.
Meng, M., Lee, C., Bagdasarian, M., and Zeikus, J.G. (1991) *Proc. Natl. Acad. Sci. USA* 88, 4015–4019.
Minton, N.P., Brehm, J.K., Oultram, J., et al. (1990) in *Proceedings Sixth International Symposium Genetics Industrial Microorganisms* (Heslot, H., Davies, J., Florent, J., et al., eds.), August 1990, Strassbourg, pp. 759–770, Société Francais de Microbiologie, Paris.
Mishra, S., Béguin, P., and Aubert, J.P. (1991) *J. Bacteriol.* 173, 80–85.
Mitchell, W.J., Shaw, J.E., and Andrews, L. (1991) *Appl. Environ. Microbiol.* 57, 2534–2539.
Morag, E., Bayer, E.A., and Lamed, R. (1990) *J. Bacteriol* 172, 6098–6105.
Morag, E., Halevy, I., Bayer, E.A., and Lamed, R. (1991) *J. Bacteriol.* 173, 4155–4162.

Mori, Y. (1990a) *Appl. Environ. Microbiol.* 56, 37–42.
Mori, Y. (1990b) *Agric. Biol. Chem.* 54, 825–826.
Nakajima, R., Imanaka, T., and Aiba, S. (1986) *Appl. Microbiol. Biotechnol.* 23, 355–360.
Nakamura, A., Uozumi, T., and Beppu, T. (1987) *Eur. J. Biochem.* 164, 317–320.
Nanmori, T., Nagai, M., Shimizu, Y., Shinke, R., and Mikami, B. (1993) *Appl. Environ. Microbiol.* 59, 623–627.
Navarro, A., Chebrou, M.C., Béguin, P., and Aubert, J.P. (1991) *Res. Microbiol.* 142, 927–936.
Navas, J., and Béguin, P. (1992) *Biochem. Biophys. Res. Commun.* 189, 807–812.
Nochur, S.V., Roberts, M.F., and Demain, A.L. (1990) *FEMS Microbiol. Lett.* 71, 199–204.
Ohnishi, H., Kitamura, H., Minowa, T., Sakai, H., and Ohta, T. (1992) *Eur. J. Biochem.* 207, 413–418.
O'Neill, G.P., Goh, S.H., Warren, R.A.J., Kilburn, D.G., and Miller, R.C. (1986) *Gene* 44, 325–330.
Ounine, K., Petitdemange, H., Raval, G., and Gay, R. (1985) *Appl. Environ. Microbiol.* 49, 874–878.
Paice, M.G., Bourbonnais, R., Desrochers, M., Jurasek, L., and Yaguchi, M. (1986) *Arch. Microbiol.* 144, 201–206.
Paquet, V., Croux, C., Goma, G., and Soucaille, P. (1991) *Appl. Environ. Microbiol.* 57, 212–218.
Patchett, M.L., Daniel, R.M., and Morgan, H.W. (1987) *Biochem. J.* 243, 779–787.
Petit, M.A., Joliff, G., Mesas, J.M., et al. (1990) *Bio/Technology* 10, 559–563.
Plant, A.R., Clemens, R.M., Daniel, R.M., and Morgan, H.W. (1987a) *Appl. Microbiol. Biotechnol.* 26, 427–433.
Plant, A.R., Clemens, R.M., Morgan, H.W., and Daniel, R.M. (1987b) *Biochem. J.* 246, 537–541.
Podkovyrov, S.M., and Zeikus, J.G. (1992) *J. Bacteriol.* 174, 5400–5405.
Podkovyrov, S.M., Burdette, D., and Zeikus, J.G. (1993) *FEBS Lett.* 317, 259–262.
Poole, D.M., Morag, E., Lamed, R., et al. (1992) *FEMS Microbiol. Lett.* 99, 181–186.
Py, B., Bortoli-German, I., Haiech, J., Chippaux, M., and Barras, F. (1991) *Protein Eng.* 4, 325–333.
Romaniec, M.P.M., Kobayashi, T., Fauth, U., Gerngross, U.T., and Demain, A.L. (1991) *Appl. Biochem. Biotechnology* 31, 119–134.
Rosey, E.L., Oskouian, B., and Stewart, G.C. (1991) *J. Bacteriol.* 173, 5992–5998.
Rouvinen, J., Bergfors, T., Teeri, T., Knowles, J.K.C., and Jones, T.A. (1990) *Science* 249, 380–385.
Rumbak, E., Rawlings, D.E., Lindsey, G.G., and Woods, D.R. (1991) *J. Bacteriol.* 173, 4203–4211.
Rygus, T., Scheler, A., Allmansberger, R., and Hillen, W. (1991) *Arch. Microbiol.* 155, 535–542.
Saha, B.C., Lamed, R., Lee, C.Y., Mathupala, S.P., and Zeikus, J.G. (1990) *Appl. Environ. Microbiol.* 56, 881–886.
Saha, B.C., Mathupala, S.P., and Zeikus, J.G. (1988) *Biochem. J.* 252, 343–348.
Saha, B.C., and Zeikus, J.G. (1991) *Appl. Microbiol. Biotechnol.* 35, 568–571.
Sakka, K., Furuse, S., and Shimada, K. (1989) *Agric. Biol. Chem.* 53, 905–910.

Sakka, K., Kojima, Y., Yoshikawa, K., and Shimada, K. (1990) *Agric. Biol. Chem.* 54, 337–342.

Sakka, K., Shimanuki, T., and Shimada, K. (1991) *Agric. Biol. Chem.* 55, 347–350.

Salamitou, S., Tokatlidis, K., Béguin, P., and Aubert, J.P. (1992) *FEBS Lett.* 304, 89–92.

Saul, D.J., Williams, L.C., Grayling, R.A., et al. (1990) *Appl. Environ. Microbiol.* 56, 3117–3124.

Saul, D.J., Williams, L.C., Love, D.R., Chamley, L.W., and Bergquist, P.L. (1989) *Nucleic Acids Res.* 17, 439.

Scheler, A., Rygus, T., Allmansberger, R., and Hillen, W. (1991) *Arch. Microbiol.* 155, 526–534.

Schimming, S. (1991) Doctoral thesis, Technical University, Munich.

Schimming, S., Schwarz, W.H., and Staudenbauer, W.L. (1991) *Biochem. Biophys. Res. Commun.* 177, 447–452.

Schimming, S., Schwarz, W.H., and Staudenbauer, W.L. (1992) *Eur. J. Biochem.* 204, 13–19.

Schlochtermeier, A., Walter, S., Schröder, J., Moorman, M., and Schrempf, H. (1992) *Mol. Microbiol.* 6, 3611–3621.

Schmidt, B.F., Adams, R.M., Requadt, C., Power, S., and Mainzer, S.E. (1989) *J. Bacteriol.* 171, 625–635.

Schroeder, C.J., Robert, C., Lenzen, G., McKay, L.L., and Mercenier, A. (1991) *J. Gen. Microbiol.* 137, 369–380.

Schwarz, W.H., Adelsberger, H., Jauris, S., et al. (1990) *Biochem. Biophys. Res. Commun.* 170, 368–374.

Schwarz, W.H., Jauris, S., Kouba, M., Bronnenmeier, K., and Staudenbauer, W.L. (1989) *Biotechnol. Lett.* 11, 461–466.

Schwarz, W.H., Schimming, S., Rücknagel, K.P., et al. (1988a) *Gene* 63, 23–30.

Schwarz, W.H., Schimming, S., and Staudenbauer, W.L. (1988b) *Biotechnol. Lett.* 10, 225–230.

Shen, G.J., Saha, B.C., Lee, Y.E., Bhatnagar, L., and Zeikus, J.G. (1988) *Biochem. J.* 254, 835–840.

Shima, S., Igarashi, Y., and Kodama, T. (1991) *Gene* 104, 33–38.

Shima, S., Kato, J., Igarashi, Y., and Kodama, T. (1989) *J. Ferment. Bioeng.* 68, 75–78.

Shinmyo, A., Garcia-Martinez, D.V., and Demain, A.L. (1979) *J. Appl. Biochem.* 1, 202–209.

Shoseyov, O., and Doi, R.H. (1990) *Proc. Natl. Acad. Sci. USA* 87, 2192–2195.

Shoseyov, O., Takagi, M., Goldstein, M.A., and Doi, R.H. (1992) *Proc. Natl. Acad. Sci. USA* 89, 3483–3487.

Siggens, K.W. (1987) *Mol. Microbiol.* 1, 86–91.

Soutschek-Bauer, E., and Staudenbauer, W.L. (1987) *Mol. Gen. Genet.* 208, 537–541.

Specka, U., Mayer, F., and Antranikian, G. (1991) *Appl. Environ. Microbiol.* 57, 2317–2323.

Svensson, B., Jespersen, H., Sierks, M.R., and MacGregor, E.A. (1989) *Biochem. J.* 264, 309–311.

Tokatlidis, K., Salamitou, S., Béguin, P., Dhurjati, P., and Aubert, J.P. (1991) *FEBS Lett.* 291, 185–188.

Tomme, P., Chauvaux, S., Béguin, P., et al. (1991) *J. Biol. Chem.* 266, 10313–10318.

Tomme, P., van Beeumen, J., and Claeyssens, M. (1992) *Biochem. J.* 285, 319–324.

Trimbur, D.E., Warren, R.A.J., and Withers, S.G. (1992) *J. Biol. Chem.* 267, 10248–10251.

Tuka, K., Zverlov, V.V., Bumazkin, B.K.,Velikodvorskaya, G.A., and Strongin, A.Y. (1990) *Biochem. Biophys. Res. Commun.* 169, 1055–1060.

Tull, D., Withers, S. G., Gilkes, N.R., et al. (1991) *J. Biol. Chem.* 266, 15621–15625.

Uozumi, N., Sakurai, K., Sasaki, T., et al. (1989) *J. Bacteriol.* 171, 375–382.

Utt, E.A., Eddy, C.K., Keshav, K.F., and Ingram, L.O. (1991) *Appl. Enivron. Microbiol.* 57, 1227–1234.

Verhasselt, P., Poncelet, F., Vits, K., Van Gool, A., and Vanderleyden, J. (1989) *FEMS Microbiol. Lett.* 59, 135–140.

Verhasselt, P. (1992) Doctoral thesis, Katholieke Universiteit te Leuven, Leuven, Belgium.

Vihinen, M., Ollikka, P., Niskanen, J., et al. (1990) *J. Biochem.* 107, 267–272.

Weickert, M.J., and Chambliss, G.H. (1989) *J. Bacteriol.* 171, 3656–3666.

Wiegel, J., Mothershed, C.P., and Puls, J. (1985) *Appl. Environ. Microbiol.* 49, 656–659.

Xu, W.Z., Shima, Y., and Urabe, I. (1991) *Eur. J. Biochem.* 202, 1197–1203.

Yagüe, E., Béguin, P., and Aubert, J.P. (1990) *Gene* 89, 61–67.

Yang, M., Galizzi, A., and Henner, D. (1983) *Nucleic Acids Res.* 11, 237–249.

Young, M., Minton, N.P., and Staudenbauer, W.L. (1989) *FEMS Microbiol. Rev.* 63, 301–326.

Yu, P.L., Smart, J.B., and Ennis, B.M. (1987) *Appl. Microbiol. Biotechnol.* 26, 254–257.

Zappe, H., Jones, W.A., Jones, D.T., and Woods, D.R. (1988) *Appl. Environ. Microbiol.* 54, 1289–1292.

Zappe, H., Jones, D.T., and Woods, D.R. (1986) *J. Gen. Microbiol.* 132, 1367–1372.

Zappe, H., Jones, W.A., and Woods, D.R. (1990) *Nucleic Acids Res.* 18, 2179.

Zeikus, J.G., Lee, C.Y., Lee, Y.E., Saha, B.C., and Bagdasarian, M. (1990) in *Proceedings Sixth International Symposium Genetics Industrial Microorganisms* (Heslot, H., Davies, J., Florent, J., et al., eds.), August 1990, Strassbourg, pp. 771–785, Société Francais de Microbiologie, Paris.

CHAPTER

13

Xylan and Cellulose Utilization by the Clostridia

G.P. Hazlewood
H.J. Gilbert

The structural polysaccharides cellulose and hemicellulose account for greater than 50% of plant biomass and are consequently the most abundant organic materials on earth. The value of plant biomass as a renewable and therefore potentially inexhaustible resource is immediately apparent if one considers that its estimated total energy content is equivalent to 640 billion tonnes of oil (Coughlan 1985). At present, the major proportion of plant biomass is recycled to carbon dioxide (CO_2) via natural decay or combustion, with lesser quantities being used as animal feed or in the production of paper and artificial building materials (Coombs 1987). Considerable potential exists for increasing the effective utilization of plant biomass in biological processes that involve the enzymatic hydrolysis of cellulose and hemicellulose, and the subsequent fermentation of liberated sugars to yield products ranging from fine chemicals to bulk organics and biofuels (Mandels 1985; Coombs 1987).

Known enzymes that hydrolyze cellulose and hemicellulose are mostly of microbial origin. Because these enzymes have a key role in the utilization of

We thank the Agricultural and Food Research Council (Grant no. LRG 138) for supporting part of this work.

TABLE 13–1 Clostridial Species Producing Cellulase or Xylanase Activity

Organism	*Cellulase*	*Xylanase*	*Reference*
Mesophilic spp.			
C. aerotolerans		+	van Gylswyk and van der Toorn 1987
C. acetobutylicum	+	+	Allcock and Woods 1981
C. cellobioparum	+	+	Hungate 1944
C. cellulovorans	+	+	Sleat et al. 1984
C. cellulolyticum	+		Petitdemange et al. 1984
C. josui	+		Sukhumavasi et al. 1988
C. lentocellum	+		Murray et al. 1986
C. longisporum	+	+	Hungate 1957
C. papyrosolvens	+		Madden et al. 1982
C. populeti	+		Sleat and Mah 1985
Thermophilic spp.			
C. stercorarium	+	+	Madden 1983
C. thermocellum	+	+	Viljoen et al. 1926

cellulosic biomass, cellulolytic microorganisms have been the subject of particularly intensive investigation since 1980.

The ability to partially or completely degrade cellulose or hemicellulose has been attributed to several clostridia (Table 13–1). In several instances, the identity and properties of the enzymes involved have not been revealed by detailed biochemical studies, and, in some cases, there is uncertainty regarding the phylogenetic affiliation of the strains concerned. Nevertheless, sufficient detailed studies have been published to confirm that the genus *Clostridium* includes several species that are active in degrading cellulose and xylan. The majority are mesophiles, but two of the most active cellulolytic species are the thermophiles *C. thermocellum* and *C. stercorarium*.

13.1 CELLULASES AND HEMICELLULASES

In order to use cellulose or xylan as sources of carbon and energy, microorganisms produce a repertoire of hydrolytic enzymes, with various specificities, which act cooperatively to convert these substrates to their constituent simple sugars (Coughlan and Ljungdahl 1988; Robson and Chambliss 1989). In fungi, cellulose hydrolysis is effected by the synergistic action of β-(1,4)-endoglucanases [EC 3.2.1.4] and β-(1,4)-exoglucanases (cellobiohydrolases) [EC 3.2.1.91] which act by randomly cleaving β-glycosidic bonds or releasing cellobiose from the nonreducing end of the cellulose molecule, respectively; a third enzyme, β-glucosidase [EC 3.1.1.21] hydrolyzes cellobiose and short-chain cellulooligosaccharides to glucose (Wood et al. 1988).

In bacterial cellulase systems, exoglucanase or cellobiohydrolase activity has been positively identified on only a few occasions (Creuzet et al. 1983; Gilkes et al. 1984; Gardner et al. 1987). Multiple isomeric forms of endoglucanase that differ subtly with respect to substrate specificity and individually have little effect on crystalline cellulose, appear to be the major components, although other activities such as β-glucosidase and cellodextrinase also form an integral part (Hazlewood et al. 1988; Robson and Chambliss 1989; Béguin 1990).

Xylan, the principal polymeric component of hemicellulose, is generally polydisperse, highly branched and requires a consortium of microbial enzymes of different specificities to effect its hydrolysis (Biely 1985; Wong et al. 1988). The linear backbone of the molecule comprising β-1,4-linked xylose residues is degraded by the combined actions of β-(1,4)-endoxylanase [EC 3.2.1.8] and xylosidase [EC 3.2.1.37], but other enzymes such as α-arabinofuranosidase [EC 3.2.1.55], α-glucuronidase [EC 3.2.1.4], and acetyl xylan esterase [EC 3.1.1.6] are necessary to remove side chain sugars and acetyl groups from highly branched and acetylated xylans, respectively, exposing the backbone to attack by xylanase. Because of the problems involved in dissociating multienzyme complexes while retaining the biological function of the individual components, purification of the single enzymes has proved to be difficult and biochemical analysis has not been wholly successful in revealing all of the complexities of microbial cellulases and hemicellulases. The cloning of genes using recombinant DNA methodology and their expression in *Escherichia coli* has in recent years provided an additional technique for dissecting cellulase and xylanase complexes and studying the function of the constituent proteins in a host that does not produce endogenous plant cell wall hydrolases. Such an approach has been applied successfully to several clostridia, most notably *C. thermocellum*, with the result that the cellulosome found in cultures of this organism has become the paradigm for cellulase in anaerobic bacteria of various genera.

13.2 *CLOSTRIDIUM THERMOCELLUM*

The ability to completely hydrolyze the most recalcitrant forms of cellulose at least as effectively as the commercially important fungus *Trichoderma reesei* (Ng and Zeikus 1981a; Johnson et al. 1982b), combined with the durability of thermostable enzymes and the possibility of continuous distillation of fermentation product (ethanol) during culture at elevated temperature has generated sustained commercial interest in this species. For the research scientist, the decomposition of a highly polymeric, substantially crystalline substrate by a group of extracellular functionally related enzymes that aggregate into an ordered complex has been of sufficient intrinsic interest to stimulate numerous studies dealing with topics ranging from the fermentation characteristics of the species (see for example: Freier et al. 1988; Lynd et al. 1989;

Tailliez et al. 1989) to the biochemistry and genetics of the enzymes involved in cellulolysis.

The structure and biochemical properties of the cellulase system of *C. thermocellum* have been described in detail recently (Lamed and Bayer 1988). For the purpose of this review, we intend to deal relatively briefly with these aspects in favor of a more comprehensive treatment of in vitro genetic studies.

13.2.1 Growth Physiology

C. thermocellum produces high levels of extracellular cellulase (measured as endoglucanase) when grown on either cellulose, cellobiose, or glucose (Ng et al. 1977; Ait et al. 1979b; Garcia-Martinez et al. 1980; Lamed and Zeikus 1980); it was subsequently shown that the more rapid growth that occurred with cellobiose, fructose, glucose, or sorbitol as carbon source was correlated with a lower level of Avicel-degrading activity (Johnson et al. 1985). In vitro, hydrolysis of Avicel by a crude extracellular preparation was inhibited by cellobiose and to a lesser extent glucose (Johnson et al. 1982a), and was dependent on the presence of calcium (Ca^{2+}) ions and a reducing agent such as dithiothreitol (Johnson et al. 1982b). The specific activities of endoglucanase and exoglucanase remained constant during batch culture (Ng et al. 1977) and relative to *T. reesei*, the ratio of endoglucanase/exoglucanase activity was higher for *C. thermocellum* (Ng and Zeikus 1981a). Xylan alone did not support growth but was rapidly degraded by extracellular enzymes produced during growth on cellobiose (Garcia-Martinez et al. 1980).

Most cellulolytic microorganisms convert the soluble cellulodextrins and cellobiose produced by endo- and exo-glucanase action to glucose, using β-glucosidase. *C. thermocellum* is apparently no exception since it produces a thermostable β-glucosidase with an M_r of 43 kilodalton (kDa) which is active against cellobiose, aryl-β-glucosides and cellulodextrins and is associated with the cell envelope of cellobiose-grown cells (Ait et al. 1979a, 1982). However, the presence of intracellular cellobiose phosphorylase and a cellulodextrin phosphorylase active against cellotriose, cellotetraose, and cellopentaose suggests that a further route is available for using the soluble products of cellulose hydrolysis (Sheth and Alexander 1967; Alexander 1968; Ng and Zeikus 1986).

Early biochemical studies established that the extracellular cellulase of *C. thermocellum* is composed of multiple proteins, some of which possess endoglucanase activity (Ait et al. 1979b; Béguin 1983). Two of these endoglucanases with molecular sizes of 56 kDa (Pétré et al. 1981) and approximately 94 kDa (Ng and Zeikus 1981b) were purified to homogeneity, but in general, attempts to resolve the individual cellulase components by conventional means were confounded by the difficulties involved in disaggregating the complex while retaining biological activity. In a recent solution to this problem, Wu et al. (1988) dissociated an extracellular Avicel-hydrolyzing aggregate composed of six major protein species, using sodium dodecyl sulfate, and purified two proteins with M_rs of 80 and 250 kDa that acted synergistically in degrading Avicel, albeit at a low rate compared with the crude aggregate. An alternative method employing lectin-affinity chromatography has since been used to purify the native aggregate, and based on its enzyme activities, it has been suggested that this complex of six main proteins may be the minimum subunit of the cellulase complex in *C. thermocellum* (Kobayashi et al. 1990).

13.2.2 The Cellulase Complex (The Cellulosome)

Our current perception of the topography of the *C. thermocellum* extracellular cellulase complex is the result of extensive biochemical and structural studies conducted since the early 1980s. It is not our intention to review each piece of evidence, but rather to summarize the consensus view that has emerged.

In batch cultures of *C. thermocellum* containing insoluble cellulose, around 80% of the total cellulase activity is located in a large cell-associated multiprotein aggregate (the cellulosome) that binds tightly to cellulose and, in the early stages of cellulolysis, mediates the attachment of cells to the substrate (Bayer et al. 1983; Lamed et al. 1983a, 1983b, 1985; Bayer et al. 1985). During the latter stages of culture, the cellulosome detaches from the cell but may remain attached to cellulose in its high molecular weight form (Bayer and Lamed 1986). Large cellulosome clusters ranging from 4.5–80 megadaltons (MDa) in size have been observed in certain strains (Coughlan et al. 1985; Mayer et al. 1987). The basic subunit in *C. thermocellum* strain YS appears to be a 2.1-MDa complex composed of a reproducible pattern of at least 14 different polypeptides ranging in size from 48–210 kDa (Lamed et al. 1983a, 1983b). Cellulosomes up to 3.5 MDa in size and composed of as many as 50 polypeptides, many of which are apparently glycosylated, have been described in other strains of *C. thermocellum* (Mayer et al. 1987; Kohring et al. 1990).

Multiple isomeric forms of endoglucanase (some with xylanase activity) and xylanases account for nearly all of the cellulosome components (Lamed et al. 1983a; Lamed and Bayer 1988; Kohring et al. 1990; Morag et al. 1990; Romaniec et al. 1992); several authors have presented evidence for exoglucanase-type activity, but an enzyme analogous to the cellobiohydrolases of aerobic fungi has only recently been definitively identified as a component of the cellulosome (Morag et al. 1991). Fungal cellobiohydrolase has been shown to interact synergistically with *C. thermocellum* cellulase in promoting the hydrolysis of cotton cellulose (Gow and Wood 1988).

Of the various polypeptides that make up the cellulosome, perhaps the greatest mystery has surrounded the largest. This subunit, designated S1, has an M_r of 210 kDa (in strain YS) and accounts for 25% of the total cellulosome protein; it is highly antigenic (Lamed et al. 1983a), contains 40% by weight of carbohydrate (Lamed and Bayer 1988), but has no demonstrable catalytic activity (Lamed et al. 1983a; Wu et al. 1988). Based on the original observation that the extracellular cellulase of an adherence-defective mutant of *C. ther-*

mocellum strain YS differed from the wild type in lacking the 210-kDa component (Bayer et al. 1983), a cellulose-binding role was proposed for S1 and was recently confirmed (Salamitou et al. 1992; Poole et al. 1992). It has also been shown that S1 serves as the template or framework upon which the other components of the cellulosome aggregate (Tokatlidis et al. 1991; Fujino et al. 1992), and it is abundantly clear that the S1 subunit is of pivotal importance in cellulosome structure and function. Partially denatured cellulosomes or preparations containing mixtures of cellulosome components only exhibited significant Avicel-hydrolyzing activity if the S1 subunit was retained (Lamed and Bayer 1988; Wu et al. 1988). Furthermore, there is little evidence that individual cellulosome components have significant activity against crystalline cellulose (Hazlewood et al. 1988; Lamed and Bayer 1988), or that dissociated and reaggregated cellulosomes regain the full catalytic activity of the starting material (Lamed and Bayer 1988). Together these facts suggest that the aggregation of polypeptide components to form an ordered complex, or the interaction of cellulosome endoglucanases with the S1 subunit is an essential prerequisite for the hydrolysis of crystalline cellulose. Indeed, the only model proposed to explain the action of *C. thermocellum* cellulase against crystalline substrates hinges on the retention of structural integrity by the cellulosome (Mayer et al. 1987).

13.2.3 Other Cellulase Components

Cellulase-related proteins that are not associated with the high molecular weight cellulosome complex of *C. thermocellum* may account for up to 20% of total activity (Lamed and Bayer 1988), but their origin and contribution to cellulolysis in vivo have not been determined. Included in this fraction are endoglucanases, xylanases, and a 250-kDa aggregate, composed of polypeptides 30–45 kDa in size, which combined synergistically with the cellulosome to hydrolyze crystalline cellulose (Shinmyo et al. 1979; Ng and Zeikus 1981a; Garcia-Martinez et al. 1980; Lamed and Bayer 1988; Morag et al. 1990). Also present is the yellow affinity substance that has been reported to enhance the binding of the cellulase complex to cellulose (Ljungdahl et al. 1983).

13.2.4 Molecular Studies

Investigation of the enzymes involved in cellulose and xylan hydrolysis by *C. thermocellum* using gene cloning was pioneered at the Pasteur Institute, Paris (Béguin et al. 1988; Béguin 1990) and has since been actively pursued by researchers in five other countries. As a result, in excess of 50 clones expressing cellulase and hemicellulase activities have now been described (Table 13–2). The precise number of unique genes that this represents has not been determined, but it is safe to assume that it exceeds the tally of 15 endoglucanase genes, two xylanase genes and two β-glucosidase genes revealed by the only comparative study carried out to date (Hazlewood et al. 1988). In

TABLE 13–2 ***Clostridium thermocellum*** **Cellulase and Hemicellulase Genes Cloned in *E. coli***

Source	*Main Substrates*[1] *Hydrolyzed (no. of clones)*	*Designation of Genes*	*Reference*
C. thermocellum NCIB 10682	CMC (2)	*celA*, *celB*	Cornet et al. 1983b
	CMC, MUC (5)	*celB*, *celC*, *celD*, *celG*, *celH*	Millet et al. 1985
	CMC (2)	*celA*, *celF*	
	MUC (3)	*bglB*, *xynZ*	
	CMC (2)	*celA*	Schwarz et al. 1985
	CMC, lich, lam, pNPC (1)	*celC*	
	lich, cellob, pNPG (2)	*bglA*	
	cellob, pNPG, pNPC (1)	*bglB*	
	CMC (9)	*celA*, *celF*, *celH*, *celI*	Romaniec et al. 1987a
	CMC, MUC (4)	*celB*, *celE*	
	pNPG (1)	*bglB*	Romaniec et al. 1987b
C. thermocellum ATCC 27405	MUG (1)	*bglB*	Kadam et al. 1988
C. thermocellum Strain F7	CMC (3)		Pirusyan et al. 1988
	CMC, lich (2)		Bumazkin et al. 1990
	CMC, lich, MUG (2)		
	CMC, lich, MUG, xylan (2)		
	CMC, lich, pNPC, xylan (2)		Tuka et al. 1990
Clostridium sp. F1	CMC (6)	*celF*	Sakka et al. 1989
	CMC, pNPC (2)	*celH*, *celC*	
	pNPC, pNPG (1)		
	CMC, xylan (2)		

[1] cellob = cellobiose; CMC = carboxymethylcellulose; lam = laminarin; lich = lichenan; MUC = 4-methylumbelliferyl-β-D-cellobioside; MUG = 4-methylumbelliferyl-β-D-glucoside; pNPC = *p*-nitrophenyl-β-D-cellobioside; pNPG = *p*-nitrophenyl-β-D-glucoside.

numerous instances, studies have progressed beyond the preliminary characterization of a gene and its encoded protein, and primary sequence data are available, revealing interesting details of molecular architecture and phylogeny.

13.2.4.1 Gene Sequences. Complete nucleotide sequences have been published for *celA* (Béguin et al. 1985), *celB* (Grépinet and Béguin 1986), *celC* (Schwarz et al. 1988a), *celD* (Joliff et al. 1986a), *celE* (Hall et al. 1988), *celF* (Navarro et al. 1991), *celH* (Yagüe et al. 1990), *bglB* (Grabnitz et al. 1989), and *xynZ* (Grépinet et al. 1988a); the sequence of *celI* (G.P. Hazlewood, K. Davidson and H.J. Gilbert, unpublished observations) has also been determined and awaits publication. The proteins encoded by these genes are endoglucanases A, B, C, D, E, F, H, β-glucosidase B, xylanase Z, and endoglucanase I, respectively.

The N-terminal regions of the translated sequence of each endoglucanase and xylanase Z conformed to the general pattern for the signal peptides of secreted prokaryotic proteins (a positively charged N terminus, followed by a central hydrophobic core and a flexible C-terminal region containing hydrophilic residues) and, in each case, potential cleavage sites were identified. All of these proteins are secreted by the normal host, but the efficacy of the signal peptide in directing translocation to the periplasmic space in *E. coli* was quite variable (Table 13–3).

Putative ribosome-binding sites (Shine-Dalgarno sequences) located between 5 and 12 base pairs (bp) upstream from the proposed translational start sites were typical of Gram-positive microorganisms (Table 13–4). The consensus sequence AGGAGG exhibited strong complementarity to the 3′-terminus of 16S rRNA from *Bacillus subtilis* (McLaughlin et al. 1981).

For each of the genes sequenced, the 5′ noncoding region contained a significantly higher percentage of A/T residues than the coding sequences, and within these regions putative promoters have been identified. Gene expression independent of orientation in plasmid vectors suggests that certain of these sequences do indeed function as promoters in *E. coli*, but with the exception of *celA*, *celD*, and *celF* their true status in *C. thermocellum* awaits confirmation. S1 nuclease mapping and primer extension have been used to determine the 5′ ends of *celA*, *celD*, and *celF* mRNA. Sequences reminiscent of *E. coli* and *B. subtilis* promoters were identified and their role in initiating transcription in *C. thermocellum* and *E. coli* was demonstrated (Béguin et al. 1986; Mishra et al. 1991).

The 3′ ends of *celA*, *celB*, *celD*, *celH*, and *xynZ* are flanked by palindromic sequences with the potential to form the mRNA hairpin loop structures which when followed by a succession of T residues, are characteristic of factor-independent termination. This observation suggests that in *C. thermocellum*, *cel* genes are transcribed singly and are not clustered or organized in operon structures. In general, this has been confirmed by hybridization experiments

TABLE 13–3 Comparison of the Signal Peptides of Endoglucanases and a Xylanase from *Clostridium thermocellum*

Enzyme[1]	*Signal Peptide Sequence*[2]	*Exported to Periplasm of* E. coli
EGA	M $\overset{+}{K}$ N V $\overset{+}{K}$ $\overset{+}{K}$ $\overset{+}{R}$ V G V V L L I L A V L G V Y M L A M P A N T V S A	+
EGB	M $\overset{+}{K}$ $\overset{+}{K}$ F L V L L I A L I M I A T L L V V P G V Q T S A	+
EGC	M V S F $\overset{+}{K}$ A G I N L G G W I S Q Y Q V F S	−
EGD	M S $\overset{+}{R}$ M T L $\overset{+}{K}$ S S M $\overset{+}{K}$ $\overset{+}{K}$ $\overset{+}{R}$ V L S L L I A V V F L S L T G V F P S G L I $\overset{-}{E}$ T $\overset{+}{K}$ V S A	−
EGE	M $\overset{+}{K}$ $\overset{+}{K}$ I V S L V C V L V M L V S I L G S F S V V A A S P V $\overset{+}{K}$ G F Q V	+
EGH	M $\overset{+}{K}$ $\overset{+}{K}$ $\overset{+}{R}$ L L V S F L V L S I I V G L L S F Q S L G N Y N S G L $\overset{+}{K}$ I G A W V G T Q P S $\overset{-}{D}$ S	ND[3]
EGI	M $\overset{+}{R}$ L V N S L G $\overset{+}{R}$ $\overset{+}{R}$ $\overset{+}{K}$ I L L I L A V I V A F S T V L L F A $\overset{+}{K}$ L W G $\overset{+}{R}$ $\overset{+}{K}$ T S S T L D E V G S $\overset{+}{K}$ T H G D L T A E	−
XYLZ	M S $\overset{+}{R}$ $\overset{+}{K}$ L F S V L L V G L M L M T S L L V T I S S T S A	ND

[1] EG = endoglucanase; XYL = xylanase.

[2] EGA (Béguin et al. 1985); EGB (Grépinet and Béguin 1986); EGC (Schwarz et al. 1988a); EGD (Joliff et al. 1986a); EGE (Hall et al. 1988); EGH (Yagüe et al. 1990); EGI (G.P. Hazlewood, K. Davidson, and H.J. Gilbert, unpublished observations); XYLZ (Gépinet et al. 1988a).

[3] ND = not determined.

TABLE 13–4 Comparison of Ribosome Binding Site Sequences of *C. thermocellum* *cel* and *xyn* Genes with the SD Sequences from *B. subtilis* and *E. coli*

Organism	*Sequence*[1]	*Position Relative to Translational Start*
E. coli	C T A G T G G A G G A A T	
B. subtilis	A G A A A G G A G G T G A	
C. thermocellum		
celA	A G G A G G	8 bp upstream
celB	A G G A G G	9 bp upstream
celC	A G G A G G	6 bp upstream
celD	A A G G G G G	12 bp upstream
celE	A G G A G A	7 bp upstream
celH	G G G A G G	6 bp upstream
celI	A G G A G G	5 bp upstream
xynZ	A G G A G G	6 bp upstream

[1] *celA* (Béguin et al. 1985); *celB* (Grépinet and Béguin 1986); *celC* (Schwarz et al. 1988a); *celD* (Joliff et al. 1986a); *celE* (Hall et al. 1988); *celH* (Yagüe et al. 1990); *celI* (G. P. Hazlewood, K. Davidson, and H. J. Gilbert, unpublished observations); *xynZ* (Grépinet et al. 1988a).

that showed that *cel* genes are scattered throughout the genome (Béguin et al. 1988). One exception is *celE*, which is located 300 bp downstream from the 3 ′ end of an open reading frame (ORF) coding for a protein that contains a sequence highly conserved in xylanase Z and all but two of the seven endoglucanases sequenced to date; the function of the second protein remains unclear (Hall et al. 1988).

Codon usage was similar for each of the sequenced genes; the marked bias towards adenine or thymine (A or T) nucleotides in the wobble position reflected the comparatively low guanine + cytosine (G + C) content of *C. thermocellum* DNA. Cloned *cel* and *xyn* genes from *C. thermocellum* failed to cross-hybridize, but comparison of their translated sequences revealed a highly conserved region containing duplicated segments of approximately 24 amino acids (aa) separated by 8 to 15 residues. This region occurred in xylanase Z (Grépinet et al. 1988a) and in all of the sequenced endoglucanases except for endoglucanases C and I, and was located at the C terminal of the proteins except in endoglucanase E and xylanase Z where it was positioned towards the center of the protein (Table 13–5). Gene-deletion studies have shown that the conserved domain is not required for the catalytic activity of endoglucanases E (Hall et al. 1988) and D (Chauvaux et al. 1990) or xylanase Z (Grépinet et al. 1988a). Neither is it part of the cellulose-binding domain (CBD) located in the C-terminal region of endoglucanase E (Durrant et al.

TABLE 13–5 Alignment of Conserved, Reiterated Regions in Cellulase-related Enzymes of *Clostridium thermocellum*[1]

Enzyme	Start	Sequence	End
Domain I			
Endoglucanase B	489	V T Y G D V N G D G R V N S S D V A L L K R Y L L G L V E N I N K E A A	533
Endoglucanase A	413	V V Y G D V N G D G N V N S T D L T M L K R Y L L K S V T N I N R E A A	448
Endoglucanase D	581	V L Y G D V N D D G K V N S T D L T L L K R Y V L K A V S T L P S S K A E K N A	620
Endoglucanase E	411	I L Y G D V N G D G K I N S T D C T M L K R Y I L R G I E E F P S	443
Endoglucanase X		D V N L D G Q V N S T D F S L L K R Y I L K V	
Xylanase Z	426	T G L G D L N G D G N I N S S D L Q A L K R H L L G I S P L T G E A L L R A	463
Endoglucanase H	828	I K H G D L N F D N A V N S T D L L M L K R Y I L K S L E L G T S E H E E K F K K A A	870
Domain II			
Endoglucanase B	534	D V N V S G T V N S T D L A I M K R Y V L R S I S E L P Y K — COOH	
Endoglucanase A	449	D V N R D G A I N S S D M T I L K R Y L I K S I P H L P Y — COOH	
Endoglucanase D	621	D V N R D G R V N S S D V T I L S R Y L I R V I E K L P I — COOH	
Endoglucanase E	451	D V N A D L K I N S T D L V L M K K Y L L R S I D K F P A E	480
Endoglucanase X		D M N N D G N I N S T D I S I L K R I L L R N — COOH	
Xylanase Z	464	D V N R S G K V D S T D Y S V L K R Y I L R I I T E F P G	492
Endoglucanase H	871	D L N R D N K V D S T D L T I L K R Y L L Y A I S E I P I — COOH	

[1] Source of sequences as listed in footnote to Table 13–3; endoglucanase X (Hall et al. 1988).

1991). Furthermore, despite the resemblance of the first 12 residues of the duplicated segments to the Ca^{2+}-binding sites of various Ca^{2+}-binding proteins, it is clear that, for endoglucanase D, Ca^{2+} binding and the effects of Ca^{2+} on substrate binding and thermostability are not mediated through the conserved regions (Chauvaux et al. 1990). Recent studies have shown conclusively that this domain interacts with a repeated region carried on the scaffolding protein S1 during the assembly of the cellulase complex (Tokatlidis et al. 1991; Fujino, et al., 1992).

13.2.4.2 Homology with Other Cellulases. Comparison of the deduced sequences of microbial cellulases encoded by cloned genes suggests that these enzymes are composed of domains that are more or less conserved, and whose order of assembly varies according to the source of the enzyme (Béguin 1990). A novel method for comparing protein sequences based on hydrophobic cluster analysis was proposed by Gaboriaud et al. (1987) and was subsequently used to assign the cellulases of prokaryotes and lower eukaryotes to six distinct families, some of which can be divided into subfamilies (Henrissat et al. 1989). Xylanase Z, and all of the endoglucanases encoded by genes cloned from *C. thermocellum* have been assigned to the established families (Béguin 1990; Gilkes et al. 1991). Endoglucanases B, C, E, and H are included in family A that contains endoglucanases from *C. cellulolyticum* and *T. reesei* and a large subfamily of endoglucanases which occur in *Bacillus* spp., *C. acetobutylicum*, *Schizophyllum commune*, and *Erwinia chrysanthemi*. With the exception of endoglucanase A (family D), the other endoglucanases (D, F, and I) occupy family E along with endoglucanases originating in *Pseudomonas fluorescens* subsp. *cellulosa* and in *Persea americana*. Xylanase Z, together with other true xylanases and the exoglucanase of *Cellulomonas fimi*, makes up family F. *C. thermocellum* β-glucosidase A (BGLA) is related to β-glucosidases from *Agrobacterium* sp. and *Caldocellum saccharolyticum*; BGLB is similar to a group of fungal β-glucosidases.

Among the *C. thermocellum* endoglucanases, the greatest degree of similarity was between endoglucanases E and H; the central 328 aa region of EGH was similar over 30.2% of its residues to the N-terminal catalytic region of endoglucanase E (Yagüe et al. 1990). One other noteworthy feature of the endoglucanases and xylanase of *C. thermocellum* that is common to other cellulase-related enzymes is the occurrence of sequences rich in hydroxy amino acids or proline. Similar sequences, often heavily glycosylated, are a recurring feature in cellulases of both bacterial and fungal origin (O'Neill et al. 1986; Penttilä et al. 1986; Wong et al. 1986; Knowles et al. 1987; Teeri et al. 1987; Hall and Gilbert 1988; Hall et al. 1989; Gilbert et al. 1990). These sequences often flank discrete functional domains and are assumed to function as hinges or linkers that maintain spatial separation between domains. Alternatively, they may have fulfilled a role analogous to that of the introns in eukaryotic genes during the evolution of cellulases and xylanases by the acquisition and shuffling of common ancestral sequences encoding functional domains (Ferreira et al. 1990). Pro-Thr-Ser-rich regions were especially evident in xylanase Z (Grépinet et al. 1988a) and endoglucanases A, B (Béguin et al. 1988), E (Hall et al., 1988), and H (Yagüe et al. 1990). In the latter case, the regions appeared to flank the centrally located catalytic domain.

13.2.4.3 Expression of Cloned *C. thermocellum* Genes. In each of the examples cited in Table 13–2, the cloned *C. thermocellum* sequences were isolated as a consequence of their ability to direct the synthesis of functional enzymes in *E. coli*. Subsequent studies have revealed that at least two of the *cel* genes may be efficiently expressed in a range of heterologous hosts including *Saccharomyces cerevisiae* (Sacco et al. 1984), *B. subtilis* (Soutschek-Bauer and Staudenbauer 1987; Mann 1988; Joliff et al. 1989), *B. stearorthermophilus* (Soutschek-Bauer and Staudenbauer 1987), *Streptomyces* spp. (Romaniec et al. 1987c), *Enterococcus faecalis* (Mann 1988) and *Lactobacillus plantarum* (Bates et al. 1989). Most interestingly, endoglucanase E was recently shown to be synthesized by stably transfected Chinese hamster ovary cells harboring *celE* from *C. thermocellum* (Hall et al. 1990). The endoglucanase, which was post-translationally modified (glycosylated), was secreted when fused to either prokaryotic or eukaryotic signal peptides and retained catalytic activity.

Immunological methods (immunoprecipitation or Western blotting) were used to show that endoglucanases B, C, D, E and I, and xylanase Z, encoded by cloned genes, are indeed produced and secreted by *C. thermocellum* (Table 13–6). Endoglucanase A synthesized by *E. coli* harboring *celA* had the same N-terminal amino acid sequence as the 56-kDa protein purified from *C. thermocellum* culture supernatant (Pétré et al. 1981). With the exception of endoglucanase C (Béguin 1990), each of the endoglucanases and xylanase Z (Grépinet et al. 1988b) are concluded to be part of the high molecular weight cellulase complex secreted by *C. thermocellum*.

C. thermocellum endoglucanases and xylanase Z synthesized by *E. coli* harboring cloned genes were, in general, smaller than the corresponding protein produced by the wild type (Table 13–6). This discrepancy in molecular size is attributable to post-translational proteolytic processing by *E. coli* proteinases and to the fact that *C. thermocellum* cellulases and hemicellulases are glycosylated in vivo (Gerwig et al. 1989). In other respects, it is likely that the products of *C. thermocellum* genes synthesized by *E. coli* are similar to the corresponding enzymes produced by the wild type; this fact has been confirmed experimentally for endoglucanase A (Schwarz et al. 1986; Béguin et al. 1988).

Hydrolysis of carboxymethylcellulose (CMC) by the products of the cloned *C. thermocellum cel* genes resulted in a rapid decrease in viscosity of the substrate solution indicating that the enzymes are principally endoglucanases. Individually the enzymes had little effect on crystalline cellulose (Hazlewood et al. 1988), but each exhibited a characteristically broad sub-

TABLE 13–6 Expression of *C. thermocellum cel* and *xyn* Genes

	Gene Product				
Gene	*Designation*[1]	*Size deduced from sequence*	*Size in* E. coli *kDa*	*Size in* C. thermocellum *kDa*	*Reference*
celA	EGA	52.5	49 52	56	Béguin et al. 1985 Pétré et al. 1981 Schwarz et al. 1986
celB	EGB	63.9	55 53	66	Cornet et al. 1983a Béguin et al. 1983 Grépinet and Béguin 1986
celC	EGC	40.4	38	40	Schwarz et al. 1988a Petre et al. 1986
celD	EGD	72.3	65	74	Joliff et al. 1986a Joliff et al. 1986b
celE	EGE	90.2	40 75	86	Hall et al. 1988 Hazlewood et al. 1990
celF	EGF	82.0	ND[2]	ND	Navarro et al. 1991
celH	EGH	102.3	ND	ND	Yagüe et al. 1990
celI	EGI	98.5	102 60	86	G.P. Hazlewood, K. Davidson, and H.J. Gilbert, unpublished observations
bglB	BGLB	84.1	86	ND	Grabnitz et al. 1989
xynZ	XYLZ	92.2	90	90	Grépinet et al. 1988a Grépinet et al. 1988b

[1] EG = endoglucanase; BGL = β-glucosidase; XYL = xylanase.
[2] ND = not determined.

strate specificity (Table 13–7). Endoglucanases C and I hydrolyzed a similar range of substrates displaying a marked preference for barley β-glucan. Endoglucanases E and H hydrolyzed both CMC and xylan. Xylanase Z was inactive against CMC, but hydrolyzed aryl glucoside and aryl cellobioside in the manner of some other prokaryotic xylanases. Of those investigated, endoglucanases A, B, and E were each shown to be secreted into the periplasm of *E. coli*; endoglucanases C and D were not (Table 13–3). Secretion of heterologous proteins that contain a signal peptide into the periplasm of *E. coli* is itself not unusual, but removal of the sequences coding for the signal peptide could be expected to prevent secretion. When *celE* was cloned with the signal sequence intact, 90% of endoglucanase E produced by *E. coli* was translocated to the periplasm. In the absence of a signal peptide, 50% of endoglucanase E was still secreted suggesting that an alternative region of the protein is able to direct secretion in *E. coli* (Hazlewood et al. 1990).

High-level expression of *C. thermocellum* endoglucanase genes *celA* and *celC* was obtained using an *E. coli* expression vector containing the phage lambda leftward promoter (Schwarz et al. 1987). Overproduction of endoglucanases E and D by *E. coli* was achieved by placing the respective genes under the transcriptional and translational control of *lacZ'* in pUC plasmids (Hazlewood et al. 1990; Joliff et al. 1986b). Endoglucanase D accumulated as intracellular cytoplasmic granules and was purified in sufficient quantity to allow structural studies by X-ray diffraction (Joliff et al. 1986c; Juy et al. 1992). Other studies with endoglucanase D, involving chemical modification and site-directed mutagenesis, have resulted in the identification of active site residues (Tomme et al. 1991; Chauvaux et al. 1992). High-level synthesis of xylanase Z occurred when the 3′ end of *xynZ*, coding for the catalytic domain, was fused in-frame with *lacZ'* in pUC19 (Grépinet et al. 1988a). These results suggest that the bias towards A or T nucleotides in the wobble position does not affect expression of the *C. thermocellum* genes in *E. coli*.

Severely truncated but catalytically active forms of endoglucanase E or xylanase Z were produced by *E. coli* harboring *celE* or *xynZ*, respectively. In the case of endoglucanase E, a truncated 40-kDa polypeptide comprising the catalytic domain was produced by proteolytic processing of the full-length protein (Hazlewood et al. 1990). This domain was encoded by the 5′ half of *celE* and a CBD located near the C terminus was encoded within the 3′ half of the gene (Hall et al. 1988; Durrant et al. 1991). Catalytically active xylanase Z around 40 kDa in size was encoded by the 3′ half of *xynZ*, and arose by internal initiation of translation from a point 470 bp downstream of the actual translational start (Grépinet et al. 1988a).

Identification of the C-terminal noncatalytic domain of endoglucanase E as a CBD raises the possibility that binding of the *C. thermocellum* cellulase complex to cellulose may not be the property of a single component such as the S1 subunit, but may result from the sum of the affinities for cellulose of individual enzymes like endoglucanase E. In this context, it is interesting that both xylanase Z (Grépinet et al. 1988a) and endoglucanase H (Yagüe et al.

TABLE 13–7 Substrate Specificity of *C. thermocellum* Enzymes Produced by *E. coli* from Cloned Genes

Enzyme		*Substrates Hydrolyzed*[1]	*Reference*
Endoglucanase	A	Acid-swollen cellulose, barley β-glucan, CMC, laminarin, lichenan	Schwarz et al. 1986
	B	CMC, lichenan, MUC	Hazlewood et al. 1988
	C	Barley β-glucan, cellulodextrins, CMC, lichenan, MUC, pNPC	Pétré et al. 1986 Hazlewood et al. 1988 Schwarz et al. 1988b
	D	CMC, lichenan, MUC, pNPC	Joliff et al. 1986b Hazlewood et al. 1988
	E	Barley β-glucan, CMC, lichenan, MUC, pNPC, xylan	Hazlewood et al. 1988
	F	CMC	Navarro et al. 1991
	H	Barley β-glucan, CMC, MUC, pNPC, xylan	Yagüe et al. 1990
	I	Barley β-glucan, CMC, lichenan, pNPC, pNPG	G.P. Hazlewood, K. Davidson, and H.J. Gilbert, unpublished observations
β-glucosidase	B	Cellobiose, gentiobiose, MUC, MUG, pNPC, pNPG, sophorose	Romaniec et al. 1987b Hazlewood et al. 1988
Xylanase	Z	MUC, pNPC, pNPG, xylan	Grépinet et al. 1988b

[1] CMC = carboxymethylcellulose; MUC = 4-methylumbelliferyl-β-D-cellobioside; MUG = 4-methylumbelliferyl-β-D-glucoside; pNPC = *p*-nitrophenyl-β-D-cellobioside; pNPG = *p*-nitrophenyl-β-D-glucoside.

1990) have extensive noncatalytic domains, whose function has yet to be ascertained. Removal of the CBD from full-length endoglucanase E resulted in a reduced affinity for cellulose, but produced a fourfold increase in specific activity against the soluble substrate barley β-glucan (Durrant et al. 1991). This modification effectively alters substrate specificity and may constitute an important mechanism for increasing the versatility of enzymes encoded by a limited number of genes, during the hydrolysis of insoluble substrates. Thus, early in cellulolysis when the concentration of soluble substrates is low, possession of a CBD increases the affinity of the enzyme for cellulose. As the concentration of soluble breakdown products increases, proteolytic removal of the CBD generates an enzyme with higher activity against these substrates.

13.3 *CLOSTRIDIUM STERCORARIUM*

The limited information available indicates that a second, more recently discovered, thermophilic species, *C. stercorarium*, differs markedly from *C. thermocellum* in its ability to use both xylan and cellulose as sources of fermentable sugars, and with respect to the biochemical properties of its cellulase/hemicellulase system.

13.3.1 Growth Physiology and Biochemistry

In batch cultures of *C. stercorarium* containing cellulose as substrate, cellulase and xylanase activities were secreted into the culture medium; the ratio of endoglucanase to exoglucanase was high relative to *C. thermocellum* (Creuzet and Frixon 1983). An endoglucanase with M_r of 91 kDa was purified to homogeneity; activity was maximal against CMC and strongly inhibited by cellobiose. The enzyme displayed significantly lower activity against ordered cellulose. A second extracellular enzyme that exhibited low level activity against crystalline cellulose, and released cellobiose from acid-swollen cellulose, was tentatively identified as a cellobiohydrolase (Creuzet et al. 1983). The enzyme combined synergistically with the endoglucanase in hydrolyzing filter paper; activity was further stimulated by the addition of β-glucosidase.

Production of endoglucanase and Avicel-hydrolyzing activities was not repressed during culture with cellobiose as carbon source (Bronnenmeier and Staudenbauer 1988). By using high-resolution anion exchange chromatography extracellular cellulase was resolved into five components; Avicelase I (M_r 100 kDa), Avicelase II (M_r 87 kDa), β-cellobiosidase I (M_r 82 kDa), β-cellobiosidase II (M_r 128 kDa), and β-glucosidase (M_r 85 kDa). Avicel-hydrolyzing activity of a reconstituted cellulase system composed of all five components was comparable to that of the crude extracellular cellulase, and was largely attributable to the combined activities of Avicelases I and II. Avicelase I hydrolyzed CMC and was tentatively judged to be the same endoglucanase as that described previously by Creuzet and Frixon (1983). Avicelase II was concluded to be an exoglucanase. Subsequent analysis of *C. stercorarium* extracellular cellulase confirmed the former point and revealed that endoglucanase (CMCase) and Avicel-hydrolyzing activities both reside in the 100-kDa monomeric protein previously designated Avicelase I (Bronnenmeier and Staudenbauer 1990). Activity against unsubstituted β-glucan was significantly higher than that against substituted cellulose. Hydrolysis of Avicel, the preferred insoluble substrate, was accompanied by the initial release of cellotetraose and cellotriose, with some subsequent hydrolysis to cellobiose; catalytic efficiency in the hydrolysis of cellulodextrins was 20 times higher than determined for endoglucanases I and II from *T. reesei*, and increased with the degree of polymerization up to cellohexaose. Specific activity towards Avicel was higher than reported for cellobiohydrolases I and II from *T. reesei*.

A study of the regulation of endoglucanase and xylanase production by *C. stercorarium* revealed that both activities were produced when culture medium contained either cellulose or xylan, but were highest when the organism was cultured on cellulose (Bérenger et al. 1985). Synthesis of both activities was partially repressed by xylose, cellobiose, lactose, and arabinose and was fully repressed by glucose, maltose, or galactose. During growth on cellulose, xylanase synthesis was strongly catabolite repressed by glucose, xylose, lactose, and cellobiose.

Three thermostable xylanases with M_rs of 44 kDa (A), 72 kDa (B), and 62 kDa (C) were purified from the supernatant of a culture containing cellulose as carbon source (Bérenger et al. 1985). Thermal stability of the three proteins differed, but in other respects, their properties were quite similar; all three contained covalently bound carbohydrate (from 3–19%), degraded xylan in a manner typical of β-(1,4)-endoxylanase and were apparently immunologically identical. The latter result is consistent with the three enzymes being derived from a single gene product by post-translational proteolysis.

Although published studies should not be regarded as exhaustive, it is evident that the cellulase/hemicellulase system of *C. stercorarium* is composed of relatively few components and lacks the structural complexity of the cellulosome of *C. thermocellum*. In general, bacterial cellulase systems tend to be composed of multiple isomeric forms of endoglucanase, which differ subtly with respect to substrate specificity, β-glucosidases and sometimes cellodextrinase; true exoglucanase (cellobiohydrolase), though an important component of fungal cellulase, has been described only occasionally in bacteria, and Avicelase I from *C. stercorarium* probably represents one of the best examples. Comparative studies indicate that cellulolytic bacteria can differ with respect to the nature and degree of organization of the enzymes that make up their cellulase systems. Most of the anaerobes appear to conform to the pattern established for the cellulosome of *C. thermocellum*, while others, such as *Pseudomonas fluorescens* subsp. *cellulosa*, *Cellulomonas fimi*, and *Thermomonospora fusca*, appear to mimic the aerobic fungus *T. reesei* by secreting fewer enzymes that act cooperatively but do not aggregate to form ordered complexes; *C. stercorarium* belongs to this second group.

13.3.2 Molecular Studies

Although in their infancy, molecular studies of the cellulases and hemicellulases of *C. stercorarium* support previous conclusions regarding the relative simplicity of these enzyme systems. Three genes (*celx*, *bglz*, and *celz*) encoding cellobiosidase, β-glucosidase, and endoglucanase activities, respectively, have been isolated from a *C. stercorarium* gene bank, and expressed in *E. coli* (Schwarz et al. 1989). Interestingly, clones harboring *celz* hydrolyzed both Avicel and CMC. In addition, activities encoded by *celz* and *celx* hydrolyzed cellopentaose with the production of cellobiose and cellotriose that were cleaved to glucose by the β-glucosidase encoded by *bglz*, suggesting that this minimum complement of three enzymes may account for the ability of *C. stercorarium* to hydrolyze cellulose to glucose. The combination of endoglucanase and Avicel-hydrolyzing activities elicited by the *celz* gene product led the authors to conclude that this enzyme corresponded to the protein previously purified from *C. stercorarium* culture supernatant and designated Avicelase I (Bronnenmeier and Staudenbauer 1988, 1990). In a subsequent study (Jauris et al. 1990), the complete nucleotide sequence of the *celz* gene was determined; codon usage was similar to that observed for *C. thermocellum cel* and *xyn* genes. The encoded polypeptide (M_r 109 kDa) was shown by N-terminal sequencing to be identical with Avicelase I purified from *C. stercorarium* culture supernatant (Bronnenmeier and Staudenbauer 1988, 1990). Catalytically active species varying in size from 100–48 kDa were produced by proteolytic processing in *E. coli* harboring *celZ*. The catalytic domain, which was contained in the N-terminal half of Avicelase I was homologous with endoglucanase D from *C. thermocellum*, endoglucanase A from *P. fluorescens* subsp. *cellulosa*, and an endoglucanase from *Persea americana* indicating that Avicelase I belongs in cellulase family E (Béguin 1990). The noncatalytic C-terminal half, retained the cellulose-binding properties of the full-length protein, and was composed of two duplicated domains, one of which was highly homologous with noncatalytic regions contained in endoglucanases from *Bacillus subtilis* and *Caldocellum saccharolyticum*. The reiterated sequences in the C-terminal half of Avicelase I were interpreted as multiple substrate-binding sites that could be expected to facilitate binding to and hydrolysis of cellulose and might explain the unusual ability of this enzyme to degrade Avicel (Jauris et al. 1990).

C. stercorarium enzymes involved in xylan utilization have also been investigated using a cloning approach and genes coding for one xylanase (*xynA*), two xylosidases (*bxlA* and *bxlB*), two arabinofuranosidases (*arfA* and *arfB*), and two so-called celloxylanases (*celw* and *celx*) have been isolated (Schwarz et al. 1990). Based on the size and properties of the recombinant form, xylanase encoded by *xynA* was suggested to be the precursor of a major xylanase previously purified from *C. stercorarium* culture supernatant (Bérenger et al. 1985). Similarly, the product of the *celx* gene (isolated previously and described as a cellobiosidase [Schwarz et al. 1989]) was concluded to be identical with a cellobiosidase isolated from a culture of the wild-type organism (Bronnenmeier and Staudenbauer 1988).

13.4 *CLOSTRIDIUM ACETOBUTYLICUM*

13.4.1 Growth Physiology and Biochemistry

Although apparently unable to grow in medium containing cellulose as sole source of fermentable sugars, *C. acetobutylicum* produces extracellular cellulase which, in vitro, hydrolyzes CMC, acid-swollen cellulose and, to a much lesser extent, Avicel (Lee et al. 1985a). In contrast, xylan is utilized as a source of fermentable sugars and will support the growth of the majority of *C. acetobutylicum* strains (Lee et al. 1985b; Lemmel et al. 1986), even though some of the enzymes necessary for the complete hydrolysis of this polysaccharide may be lacking (Lemmel et al. 1986).

Allcock and Woods (1981) first described the production by *C. acetobutylicum* of extracellular endoglucanase (CMCase) and cellobiase activities. Endoglucanase was induced by a dialyzable constituent of molasses and was not repressed or inhibited by glucose; cellobiase was produced constitutively. In a subsequent study involving six different strains of *C. acetobutylicum*, both extracellular and cell-bound endoglucanase and cellobiase were produced by only two strains, during culture with cellobiose (Lee et al. 1985a). Endoglucanase was mainly extracellular and cellobiase was mainly cell-associated; in continuous culture, production of both enzymes was severely repressed by glucose.

Both cellulolytic strains produced xylanase activity when cultured with cellobiose as fermentable substrate, but levels of xylanase, xylosidase, and arabinofuranosidase were maximal in medium containing xylose and significantly depressed in cultures containing glucose (Lee et al. 1985b). Extracellular xylanase hydrolyzed only 12% of larchwood xylan with the production of xylose, xylobiose, and xylotriose. In a separate study (Lemmel et al. 1986), up to 50% of the sugars contained in xylan were consumed during fermentation by *C. acetobutylicum* confirming that the organism probably does not produce the full complement of enzymes required to effect complete hydrolysis.

Two immunologically distinct extracellular xylanases (A and B) with M_rs of 65 and 29 kDa, respectively, were purified to homogeneity from *C. acetobutylicum* culture containing xylan as a carbon source (Lee et al. 1987). Both hydrolyzed xylan randomly in a manner typical of β-(1,4)-endoxylanase; xylanase A also hydrolyzed CMC and acid-swollen cellulose, while xylanase B, which was more efficient in hydrolyzing xylan than xylanase A, was equally active against lichenan. A xylosidase purified from culture supernatant of the same strain of *C. acetobutylicum* had an apparent M_r of 224 kDa and hydrolyzed *p*-nitrophenylxyloside and xylooligosaccharides from 2 to 6 units in length by removing xylose units; activity against CMC or xylan was negligible (Lee and Forsberg 1987).

13.4.2 Molecular Studies

Molecular cloning studies, showing that *C. acetobutylicum* has the genetic potential to produce endoglucanase and β-glucosidase activities, confirm the results of earlier biochemical analysis (Zappe et al. 1986). Both activities were encoded by genes carried on a 4.9 kilobase (kb) fragment of genomic DNA cloned in *E. coli*. Deletion of vector promoter sequences indicated that the cloned fragment contained a sequence with promoter activity in *E. coli*. Endoglucanase (CMCase) activity was largely secreted to the periplasm in *E. coli*. As with the native endoglucanase (Allcock and Woods 1981), there was little activity against acid-swollen cellulose or filter paper. The recombinant endoglucanase (calculated M_r 49.4 kDa) was encoded by an ORF of 1344 nucleotides (Zappe et al. 1988); the proposed translational start was preceded by a putative ribosome-binding site and sequences reminiscent of the promoters of Gram-positive bacteria. The first 350 residues of the *C. acetobutylicum* endoglucanase, containing the catalytic domain, exhibited substantial homology (up to 60%) with cellulase subfamily A2 endoglucanases from several *Bacillus* spp. and *Erwinia chrysanthemi* (Zappe et al. 1988; Béguin 1990). The nonhomologous C-terminal region contained approximately 40% of threonine and serine residues and was not required for catalytic activity.

A xylanase gene, carried on a 2-kb fragment of *C. acetobutylicum* DNA cloned in *E. coli*, directed the synthesis of a 28-kDa intracellular xylanase (Zappe et al. 1987). Antiserum prepared against the purified recombinant protein cross-reacted with an extracellular xylanase of similar size that had been purified from a *C. acetobutylicum* culture containing xylan and xylose as fermentable substrates. The deduced sequence of the encoded xylanase was homologous with other bacterial and fungal xylanases belonging to family G (Zappe et al. 1990; Gilkes et al. 1991).

13.5 *CLOSTRIDIUM CELLULOVORANS*

Although described only relatively recently (Sleat et al. 1984), *C. cellulovorans* has already generated much interest on account of the potency of its cellulase system and its apparent similarity to the well-characterized cellulosome of *C. thermocellum*.

Crystalline cellulose (Avicel) was completely degraded by *C. cellulovorans* cellulase and supported the growth of the organism when added to medium as the sole carbon source. Cellulase activity secreted into the culture supernatant and purified by ammonium sulfate precipitation followed by selective binding to Avicel, possessed endoglucanase (CMCase) and exoglucanase (4-methylumbelliferyl-β-D-cellobiosidase; MUCase) activities, and degraded filter paper cellulose in vitro (Shoseyov and Doi 1990). Xylanase activity was not reported, but a previous study (Sleat et al. 1984) had shown that xylan will support growth of *C. cellulovorans*. Chromatographic analysis of the extracellular cellulase revealed a multiprotein aggregate analogous to the cellulosome of *C. thermocellum*. The major subunits included an endoglucanase with M_r of 100 kDa and a 170-kDa noncatalytic protein with a high affinity for cellulose; at least four minor subunits also had endoglucanase activity. The large subunit was not required for binding of the catalytic subunits to cellulose, but was essential for the hydrolysis of crystalline cellulose in vitro, provoking the suggestion that it functions as the core protein to which the catalytic subunits bind, and that it somehow converts crystalline cellulose to a form which is amenable to attack by the endoglucanases (Shoseyov and Doi 1990).

Complementary genetic studies (Shoseyov et al. 1990; Hamamoto et al. 1990) have resulted in the isolation of genes coding for four apparently different endoglucanases (*engA*, *engB*, *engC* and *engD*), two of which have been sequenced (Foong et al. 1991; Hamamoto et al. 1992). Enzymes encoded by *engB* and *engC* hydrolyzed CMC only, but the *engA* product also displayed aryl-β-glucosidase activity. None of the recombinant proteins had activity against ordered cellulose and all appeared to be proteolytically processed by *E. coli* (Shoseyov et al. 1990). Enzyme activity expressed by *E. coli* containing *engD* had an apparent M_r of 50 kDa, hydrolyzed CMC, MUC and *p*-nitrophenyl-cellobioside and liberated some cellulodextrins from Avicel during prolonged incubation, presumably by attacking amorphous regions of the substrate (Hamamoto et al. 1990). The complete nucleotide sequence of *engD* comprises an ORF of 1545 bp, coding for a protein with M_r of 52.7 kDa, preceded by a putative ribosome-binding site and followed by sequences characteristic of factor-independent termination of transcription (Hamamoto et al. 1992). A noteworthy feature of the primary structure of EngD was a sequence rich in proline and threonine residues that separated the N-terminal portion of the protein from a 108 residue C-terminal domain that displayed significant homology with the CBD of *C. fimi* exoglucanase and endoglucanase (O'Neill et al. 1986; Wong et al. 1986). The sequence of a large cellulose-binding protein, analogous to the S1 subunit of the *C. thermocellum* cellulosome has also been reported (Shoseyov et al. 1992).

13.6 *CLOSTRIDIUM CELLULOLYTICUM*

The mesophilic anaerobe *C. cellulolyticum*, first isolated from decaying grass (Petitdemange et al. 1984), is yet another of the cellulolytic clostridia identified since 1980. Its cellulase system is relatively poorly characterized, despite the use of biochemical and molecular cloning techniques. A comprehensive study of growth physiology revealed that crystalline cellulose, added as sole source of fermentable sugars, was slowly hydrolyzed and supported growth of *C. cellulolyticum* albeit at a slower rate than when soluble sugars were present (Giallo et al. 1985). Glucose accounted for 30% of the sugars accumulating on cessation of growth suggesting that *C. cellulolyticum* has the full complement of enzymes necessary to completely hydrolyze cellulose. During growth on cellulose, endoglucanase (CMCase) activity was initially associated with cells and undegraded substrate but was subsequently released into the culture supernatant. Preliminary studies identified four endoglucanases (CMCases) with M_rs ranging from 38–71 kDa but in the absence of detailed biochemical analysis, the precise composition of the cellulase system of *C. cellulolyticum* remains unclear.

Cloning studies performed simultaneously by two independent groups working with the same strain have resulted in the isolation of genes coding for two or at most three different cellulase-related proteins. Two distinct genes (*celccA* and *celccB*) cloned and expressed in *E. coli* encoded endoglucanase activity which remained largely in the cytoplasm (Faure et al. 1988); enzyme encoded by *celccA* also hydrolyzed MUC. Both cloned fragments directed the synthesis of endoglucanase activity irrespective of their orientation in plasmid vectors suggesting that they each contain *E. coli*-like promoter sequences. Relatively strong hybridization was observed between the cloned fragment carrying *celccA* and the *celC* gene from *C. thermocellum*. In a separate report (Pérez-Martinez et al. 1988), another gene coding for an endoglucanase was isolated from a *C. cellulolyticum* gene bank. Although there were some differences in the respective physical maps of the cloned fragments, it seems probable that the gene coding for the latter endoglucanase is the same as *celccB*. Endoglucanase A, the product of *celccA*, consisted of 475 aa with an M_r of 53.6 kDa, and was encoded by an ORF of 1425 bp (Faure et al. 1989). The translational start site was preceded by a putative ribosome-binding site and sequences reminiscent of *B. subtilis* promoters; the 3′ untranslated region contained palindromic sequences that could function in transcription termination. The predicted N-terminus of endoglucanase A was typical for the signal sequence of extracellular proteins of Gram-positive bacteria, and codon usage was similar to that observed for *cel* genes from *C. acetobutylicum* (Zappe et al. 1988) and *C. thermocellum* (see Béguin 1990 for ref.). Interestingly, three distinct regions of endoglucanase A exhibited homology with sequences present in the cellulases of other organisms (Faure et al. 1989). An extended sequence located in the N-terminal region of endoglucanase A exhibited 40% sequence identity and 60% similarity with the N-terminal catalytic domain of endoglucanase E from *C. thermocellum* (Hall et al. 1988), suggesting that the catalytic core of the respective mesophilic and thermophilic proteins may have evolved from a common ancestral gene. A second region occurring at the C-terminus of endoglucanase A and comprising duplicated segments of 24 aa was highly homologous with a similar motif found in xylanase Z and in each of the endoglucanases from *C. ther-*

mocellum except endoglucanases C and I. Finally, the region between residues 127 and 147 of endoglucanase A was homologous with a short sequence conserved in several cellulase family A endoglucanases (Faure et al. 1989; Béguin 1990). Within the catalytic domain, the essential role of arginine and histidine residues was demonstrated by site-directed mutagenesis (Belaich et al. 1992).

13.7 OTHER CELLULOLYTIC CLOSTRIDIA

13.7.1 *Clostridium cellobioparum*

C. cellobioparum, isolated from rumen contents, was described by Hungate (1944) as producing extracellular cellulase activity that degraded cotton cellulose; hemicellulose was also fermented. In a more recent survey of the surface structures of cellulolytic bacteria, Lamed et al. (1987) noted a clear correlation between cellulolytic activity and the production of protuberant cell-surface structures. *C. cellobioparum* was one of the bacteria surveyed, and based on the observed surface structures, the ability of the cells to bind to labelled lectin and the cross-reaction that occurred between cell-free extract of *C. cellobioparum* and an antibody prepared against the cellulosome of *C. thermocellum*, it was suggested that *C. cellobioparum* extracellular cellulase might be analogous in structure to the *C. thermocellum* cellulosome.

13.7.2 *Clostridium longisporum*

C. longisporum, also isolated from rumen contents (Hungate 1957) completely hydrolyzed cellulose, and was recently shown to digest preparations of alfalfa cell walls to a greater extent than did strains of other anaerobic cellulolytic bacteria (Varel 1989). Although alfalfa (*Medicago sativa*) hemicellulose was degraded, xylan alone would not support growth of *C. longisporum* (Varel et al. 1989).

13.7.3 *Clostridium josui*

A moderately thermophilic anaerobe, isolated from compost and designated *C. josui* hydrolyzed crystalline cellulose in the form of Avicel and rice (*Oryza sativa*) straw (Sukhumavasi et al. 1988). Extracellular, thermostable endoglucanase (CMCase) with an M_r of 45 kDa was purified to homogeneity from the supernatant of a stationary phase culture containing ball-milled cellulose as source of fermentable sugars (Fujino et al. 1989). The enzyme was most active against CMC but released some reducing sugar from Avicel, and hydrolyzed cellulodextrins (from G4 to G6) with the production of cellobiose and cellotriose. A gene coding for endoglucanase activity was isolated and expressed in *E. coli* (Ohmiya et al. 1989). The recombinant protein, purified to homogeneity, had an M_r of 39 kDa, was thermostable, and based on its properties and N-terminal sequence differed from the endoglucanase previously purified from a culture of *C. josui* (Fujino et al. 1990).

Original descriptions include evidence that several other clostridia possess cellulase activity (see Table 13–1), but with the exception of the species already referred to, the authors are not aware of substantive work detailing the characteristics of the cellulase/hemicellulase systems of these other organisms.

13.8 CONCLUDING REMARKS

Within the context of biotechnology, the ability to use the plant structural polysaccharides cellulose and xylan as sources of fermentable carbon is an important feature of many clostridial species, since it provides the basis for using plant biomass as an industrial feedstock. Exploitation of this particular property of the clostridia becomes more likely as we extend our knowledge of the enzyme systems involved.

Among cellulolytic clostridia, the thermophile *C. thermocellum* has been examined most intensively but studies of other species have been sufficiently detailed to allow comparisons to be made. Based on biochemical and ultrastructural evidence it has been concluded that the multiprotein cellulase complex produced by *C. thermocellum* is a realistic pattern for the cellulases of a variety of anaerobic bacteria (Lamed et al. 1987; Lamed and Bayer 1988). Within the genus *Clostridium*, such an assertion may be true for *C. cellulovorans* (Shoseyov and Doi 1990) and possibly *C. cellobioparum* (Lamed et al. 1987), but it is almost certainly not the case for a second thermophile, *C. stercorarium*, or for other clostridia whose cellulase/hemicellulase systems appear to be composed of relatively fewer proteins.

Since 1980, the use of conventional biochemical techniques to study cellulases and hemicellulases has been augmented by rDNA methodologies, with the result that the primary structure of several clostridial cellulases and xylanases has been determined revealing interesting features of molecular architecture and contributing to our understanding of the catalytic process.

Molecular studies indicate that multiple cellulases and xylanases produced by clostridia are encoded by multiple genes. These genes exhibit a high degree of conformity with prokaryotic consensus sequences controlling initiation and termination of transcription and translation, and with consensus sequences regulating protein processing for membrane translocation. It could be argued that this is to be expected since the experimental approaches employed have selected only genes that are readily expressed in *E. coli*. There is of course no guarantee that regulatory sequences which function in *E. coli* will be active in the normal host. Given the large number of cellulase and xylanase genes that express in *E. coli* and other prokaryotic hosts, it is probably reasonable to conclude that in the clostridia there is no major deviation from consensus sequences with respect to cellulase and xylanase genes.

With regard to molecular architecture and the conservation of primary structure, clostridial cellulases and xylanases individually exhibit considerable similarity with enzymes of similar function from other clostridia and unrelated organisms. The homology that exists between these proteins suggests that they have evolved through the acquisition of several ancestral genes encoding functional domains that are conserved among a variety of cellulolytic species.

REFERENCES

Ait, N., Creuzet, N., and Cattaneo, J. (1979a) *Biochem. Biophys. Res. Commun.* 90, 537–546.

Ait, N., Creuzet, N., and Cattaneo, J. (1982) *J. Gen. Microbiol.* 128, 569–577.

Ait, N., Creuzet, N., and Forget, P. (1979b) *J. Gen. Microbiol.* 113, 399–402.

Alexander, J.K. (1968) *J. Biol. Chem.* 243, 2899–2904.

Allcock, E.R., and Woods, D.R. (1981) *Appl. Environ. Microbiol.* 41, 539–541.

Bates, E.E.M., Gilbert, H.J., Hazlewood, G.P., et al. (1989) *Appl. Environ. Microbiol.* 55, 2095–2097.

Bayer, E.A., Kenig, R., and Lamed, R. (1983) *J. Bacteriol.* 156, 818–827.

Bayer, E.A., and Lamed, R. (1986) *J. Bacteriol.* 167, 828–836.

Bayer, E.A., Setter, E., and Lamed, R. (1985) *J. Bacteriol.* 163, 552–559.

Béguin, P. (1983) *Anal. Biochem.* 131, 333–336.

Béguin, P. (1990) *Annu. Rev. Microbiol.* 44, 219–248.

Béguin, P., Cornet, P., and Aubert, J.-P. (1985) *J. Bacteriol.* 162, 102–105.

Béguin, P., Cornet, P., and Millet, J. (1983) *Biochimie* 65, 495–500.

Béguin, P., Millet, J., Grépinet, O., et al. (1988) in *Biochemistry and Genetics of Cellulose Degradation, FEMS Symposium No. 43*, 7–9 Sept. 1987, Paris, pp. 267–282, Academic Press, San Diego.

Béguin, P., Rocancourt, M., Chebrou, M.-C., and Aubert, J.-P. (1986) *Mol. Gen. Genet.* 202, 251–254.

Belaich, A., Fierobe, H.-P., Baty, D., et al. (1992) *J. Bacteriol.* 174, 4677–4682.

Bérenger, J.-F., Frixon, C., Bigliardi, J., and Creuzet, N. (1985) *Can. J. Microbiol.* 31, 635–643.

Biely, P. (1985) *Trends Biotechnol.* 3, 286–290.

Bronnenmeier, K., and Staudenbauer, W.L. (1988) *Appl. Microbiol. Biotechnol.* 27, 432–436.

Bronnenmeier, K., and Staudenbauer, W.L. (1990) *Enzyme Microb. Technol.* 12, 431–436.

Bumazkin, B.K., Velikodvorskaya, G.A., Tuka, K., Mogutov, M.A., and Strongin, A.Y. (1990) *Biochem. Biophys. Res. Commun.* 167, 1057–1064.

Chauvaux, S., Béguin, P., Aubert, J.-P., et al. (1990) *Biochem. J.* 265, 261–265.

Chauvaux, S., Béguin, P., and Aubert, J.-P. (1992) *J. Biol. Chem.* 267, 4472–4478.

Cornet, P., Millet, J., Béguin, P., and Aubert, J.-P. (1983a) *Bio/Technology* 1, 589–594.

Cornet, P., Tronik, D., Millet, J., and Aubert, J.-P. (1983b) *FEMS Microbiol. Lett.* 16, 137–141.

Coombs, J. (1987) *Phil. Trans. R. Soc. London* A321, 405–422.

Coughlan, M.P. (1985) *Biochem. Soc. Trans.* 13, 405–406.
Coughlan, M.P., Hon-Nami, K., Hon-Nami, H., et al. (1985) *Biochem. Biophys. Res. Commun.* 130, 904–909.
Coughlan, M.P., and Ljungdahl, L.G. (1988) in *Biochemistry and Genetics of Cellulose Degradation, FEMS Symposium No. 43*, 7–9 Sept. 1987, Paris, (Aubert, J.-P., Béguin, P., and Millet, J., eds.), pp. 11–30, Academic Press, San Diego.
Creuzet, N., Bérenger, J.-F., and Frixon, C. (1983) *FEMS Microbiol. Lett.* 20, 347–350.
Creuzet, N., and Frixon, C. (1983) *Biochimie* 65, 149–156.
Durrant, A.J., Hall, J., Hazlewood, G.P., and Gilbert, H.J. (1991) *Biochem. J.* 273, 289–293.
Faure, E., Bagnara, C., Belaich, A., and Belaich, J.-P. (1988) *Gene* 65, 51–58.
Faure, E., Belaich, A., Bagnara, C., Gaudin, C., and Belaich, J.-P. (1989) *Gene* 84, 39–46.
Ferreira, L.M.A., Durrant, A.J., Hall, J., Hazlewood, G.P., and Gilbert, H.J. (1990) *Biochem. J.* 269, 261–264.
Foong, F., Hamamoto, T., Shoseyov, O., and Doi, R.H. (1991) *J. Gen. Microbiol.* 137, 1729–1736.
Freier, D., Mothershed, C.P., and Wiegel, J. (1988) *Appl. Environ. Microbiol.* 54, 204–211.
Fujino, T., Béguin, P., and Aubert, J.-P. (1992) *FEMS Microbiol. Lett.* 94, 165–170.
Fujino, T., Sasaki, T., Ohmiya, K., and Shimizu, S. (1990) *Appl. Environ. Microbiol.* 56, 1175–1178.
Fujino, T., Sukhumavasi, T., Sasaki, T., Ohmiya, K., and Shimizu, S. (1989) *J. Bacteriol.* 171, 4076–4079.
Gaboriaud, C., Bissery, V., Benchetrit, T., and Mornon, J.-P. (1987) *FEBS Lett.* 224, 149–155.
Garcia-Martinez, D.V., Shinmyo, A., Madia, A., and Demain, A.L. (1980) *Eur. J. Appl. Microbiol. Biotechnol.* 19, 189–197.
Gardner, R.M., Doerner, K.C., and White, B.A. (1987) *J. Bacteriol.* 169, 4581–4588.
Gerwig, G.J., de Waard, P., Kamerling, J.P., et al. (1989) *J. Biol. Chem.* 264, 1027–1035.
Giallo, J., Gaudin, C., and Belaich, J.-P. (1985) *Appl. Environ. Microbiol.* 49, 1216–1221.
Gilbert, H.J., Hall, J., Hazlewood, G.P., and Ferreira, L.M.A. (1990) *Mol. Microbiol.* 4, 759–767.
Gilkes, N.R., Henrissat, B., Kilburn, D.G., Miller, R.C., and Warren, R.A.J. (1991) *Microbiol. Rev.* 55, 303–315.
Gilkes, N.R., Langsford, M.L., Kilburn, D.G., Miller, R.C., and Warren, R.A.J. (1984) *J. Biol. Chem.* 259, 10455–10459.
Gow, L.A., and Wood, T.M. (1988) *FEMS Microbiol. Lett.* 50, 247–252.
Grabnitz, F., Rücknagel, K.P., Seib, M., and Staudenbauer, W.L. (1989) *Mol. Gen. Genet.* 217, 70–76.
Grépinet, O., and Béguin, P. (1986) *Nucleic Acids Res.* 14, 1791–1799.
Grépinet, O., Chebrou, M.-C., and Béguin, P. (1988a) *J. Bacteriol.* 170, 4582–4588.
Grépinet, O., Chebrou, M.-C., and Béguin, P. (1988b) *J. Bacteriol.* 170, 4576–4581.
Hall, J., and Gilbert, H.J. (1988) *Mol. Gen. Genet.* 213, 112–117.
Hall, J., Hazlewood, G.P., Barker, P.J., and Gilbert, H.J. (1988) *Gene* 69, 29–38.

Hall, J., Hazlewood, G.P., Huskisson, N.S., Durrant, A.J., and Gilbert, H.J. (1989) *Mol. Microbiol.* 3, 1211–1219.

Hall, J., Hazlewood, G.P., Surani, M.A., Hirst, B.H., and Gilbert, H.J. (1990) *J. Biol. Chem.* 265, 19996–19999.

Hamamoto, T., Foong, F., Shoseyov, O., and Doi, R.H. (1992) *Mol. Gen. Genet.* 231, 472–479.

Hamamoto, T., Shoseyov, O., Foong, F., and Doi, R.H. (1990) *FEMS Microbiol. Lett.* 72, 285–288.

Hazlewood, G.P., Davidson, K., Clarke, J.H., et al. (1990) *Enzyme Microb. Technol.* 12, 656–662.

Hazlewood, G.P., Romaniec, M.P.M., Davidson, K., et al. (1988) *FEMS Microbiol. Lett.* 51, 231–236.

Henrissat, B., Claeyssens, M., Tomme, P., Lemesle, L., and Mornon, J.-P. (1989) *Gene* 81, 83–95.

Hungate, R.E. (1944) *J. Bacteriol.* 48, 499–513.

Hungate, R.E. (1957) *Can. J. Microbiol.* 3, 289–311.

Jauris, S., Rücknagel, K.P., Schwarz, W.H., Kratzsch, P., Bronnenmeier, K., et al. (1990) *Mol. Gen. Genet.* 223, 258–267.

Johnson, E.A., Bouchot, F., and Demain, A.L. (1985) *J. Gen. Microbiol.* 131, 2303–2308.

Johnson, E.A., Reese, E.T., and Demain, A.L. (1982a) *J. Appl. Biochem.* 4, 64–71.

Johnson, E.A., Sakajoh, M., Halliwell, G., Madia, A., and Demain, A.L. (1982b) *Appl. Environ. Microbiol.* 43, 1125–1132.

Joliff, G., Béguin, P., and Aubert, J.-P. (1986a) *Nucleic Acids Res.* 14, 8605–8613.

Joliff, G., Béguin, P., Juy, M., et al. (1986b) *Bio/Technology* 4, 896–900.

Joliff, G., Béguin, P., Millet, J., et al. (1986c) *J. Mol. Biol.* 189, 249–250.

Joliff, J., Edelman, A., Klier, A., and Rapoport, G. (1989) *Appl. Environ. Microbiol.* 55, 2739–2744.

Juy, M., Amit, A.G., Alzari, P.M., et al. (1992) *Nature (London)* 357, 89–91.

Kadam, S., Demain, A.L., Millet, J., Béguin, P., and Aubert, J.-P. (1988) *Enzyme Microb. Technol.* 10, 9–13.

Knowles, J., Lehtovaara, P., and Teeri, T. (1987) *Trends Biotechnol.* 5, 255–261.

Kobayashi, K., Romaniec, M.P.M., Fauth, U., and Demain, A.L. (1990) *Appl. Environ. Microbiol.* 56, 3040–3046.

Kohring, S., Wiegel, J., and Mayer, F. (1990) *Appl. Environ. Microbiol.* 56, 3798–3804.

Lamed, R., and Bayer, E.A. (1988) *Adv. Appl. Microbiol.* 33, 1–46.

Lamed, R., Kenig, R., Setter, E., and Bayer, E.A. (1985) *Enzyme Microb. Technol.* 7, 37–41.

Lamed, R., Naimark, J., Morgenstern, E., and Bayer, E.A. (1987) *J. Bacteriol.* 169, 3792–3800.

Lamed, R., Setter, E., and Bayer, E.A. (1983a) *J. Bacteriol.* 156, 828–836.

Lamed, R., Setter, E., Kenig, R., and Bayer, E.A. (1983b) *Biotechnol. Bioeng. Symp.* 13, 163–181.

Lamed, R., and Zeikus, J.G. (1980) *J. Bacteriol.* 144, 569–578.

Lee, S.F., and Forsberg, C.W. (1987) *Appl. Environ. Microbiol.* 53, 651–654.

Lee, S.F., Forsberg, C.W., and Gibbins, L.N. (1985a) *Appl. Environ. Microbiol.* 50, 220–228.

Lee, S.F., Forsberg, C.W., and Gibbins, L.N. (1985b) *Appl. Environ. Microbiol.* 50, 1068–1076.
Lee, S.F., Forsberg, C.W., and Rattray, J.B. (1987) *Appl. Environ. Microbiol.* 53, 644–650.
Lemmel, S.A., Datta, R., and Frankiewicz, J.R. (1986) *Enzyme Microb. Technol.* 8, 217–221.
Ljungdahl, L.G., Peterson, B., Eriksson, K.-E., and Wiegel, J. (1983) *Curr. Microbiol.* 9, 195–200.
Lynd, L.R., Grethlein, H.E., and Wolkin, R.H. (1989) *Appl. Environ. Microbiol.* 55, 3131–3139.
Madden, R.H. (1983) *Int. J. Syst. Bacteriol.* 33, 837–840.
Madden, R.H., Bryder, M.J., and Poole, N.J. (1982) *Int. J. Syst. Bacteriol.* 32, 87–91.
Mandels, M. (1985) *Biochem. Soc. Trans.* 13, 414–416.
Mann, S.P. (1988) *Lett. Appl Microbiol.* 7, 119–122.
Mayer, F., Coughlan, M.P., Mori, Y., and Ljungdahl, L.G. (1987) *Appl. Environ. Microbiol.* 53, 2785–2792.
McLaughlin, J.R., Murray, C.L., and Rabinowitz, J.C. (1981) *J. Biol. Chem.* 256, 11283–11291.
Millet, J., Pétré, D., Béguin, P., Raynaud, O., and Aubert, J.-P. (1985) *FEMS Microbiol. Lett.* 29, 145–149.
Mishra, S., Béguin, P., and Aubert, J.-P. (1991) *J. Bacteriol.* 173, 80–85.
Morag, E., Bayer, E.A., and Lamed, R. (1990) *J. Bacteriol.* 172, 6098–6105.
Morag, E., Halevy, I., Bayer, E.A., and Lamed, R. (1991) *J. Bacteriol.* 173, 4155–4162.
Murray, W.D., Hofmann, L., Campbell, N.L., and Madden, R.H. (1986) *System. Appl. Microbiol.* 8, 181–184.
Navarro, A., Chebrou, M.-C., Béguin, P., and Aubert, J.-P. (1991) *Res. Microbiol.* 142, 927–936.
Ng, T.K., Weimer, P.J., and Zeikus, J.G. (1977) *Arch. Microbiol.* 114, 1–7.
Ng, T.K., and Zeikus, J.G. (1981a) *Appl. Environ. Microbiol.* 42, 231–240.
Ng, T.K., and Zeikus, J.G. (1981b) *Biochem. J.* 199, 341–350.
Ng, T.K., and Zeikus, J.G. (1986) *Appl. Environ. Microbiol.* 52, 902–904.
Ohmiya, K., Fujino, T., Sukhumavasi, J., and Shimizu, S. (1989) *Appl. Environ. Microbiol.* 55, 2399–2402.
O'Neill, G., Goh, S.H., Warren, R.A.J., Kilburn, D.G., and Miller, R.C. (1986) *Gene* 44, 325–330.
Pérez-Martinez, G., González-Candelas, L., Polaina, J., and Flors, A. (1988) *J. Ind. Microbiol.* 3, 365–371.
Penttilä, M., Lehtovaara, P., Nevalainen, H., Bhikhabhai, R., and Knowles, J. (1986) *Gene* 45, 253–263.
Petitdemange, E., Caillet, F., Giallo, J., and Gaudin, C. (1984) *Int. J. Syst. Bacteriol.* 34, 155–159.
Pétré, D., Longin, R., and Millet, J. (1981) *Biochimie* 63, 629–639.
Pétré, D., Millet, J., Longin, R., et al. (1986) *Biochimie* 68, 687–695.
Pirusyan, E.S., Mogutov, M.A., Velikodvorskaya, G.A., and Pushkarskaya, T.A. (1988) *Genetics (USSR)* 24, 204–209.
Poole, D.M., Morag, E., Lamed, R., et al. (1992) *FEMS Microbiol. Lett.* 99, 181–186.

Robson, L.M., and Chambliss, G.H. (1989) *Enzyme Microb. Technol.* 11, 626–644.
Romaniec, M.P.M., Clarke, N.G., and Hazlewood, G.P. (1987a) *J. Gen. Microbiol.* 133, 1297–1307.
Romaniec, M.P.M., Davidson, K., and Hazlewood, G.P. (1987b) *Enzyme Microb. Technol.* 9, 474–478.
Romaniec, M.P.M., Fauth, U., Kobayashi, T., et al. (1992) *Biochem. J.* 283, 69–73.
Romaniec, M.P.M., Hazlewood, G.P., and Keiser, T. (1987c) *Abstracts of FEMS Symposium No. 43*, abstr. P3-14, p. 73.
Sacco, M., Millet, J., and Aubert, J.-P. (1984) *Ann. Inst. Pasteur Microbiol.* 135A, 485–488.
Sakka, K., Furuse, S., and Shimada, K. (1989) *Agric. Biol. Chem.* 54, 905–910.
Salamitou, S., Tokatlidis, K., Béguin, P., and Aubert, J.-P. (1992) *FEBS Lett.* 304, 89–92.
Schwarz, W., Bronnenmeier, K., and Staudenbauer, W.L. (1985) *Biotechnol. Lett.* 7, 859–864.
Schwarz, W.H., Gräbnitz, F., and Staudenbauer, W.L. (1986) *Appl. Environ. Microbiol.* 51, 1293–1299.
Schwarz, W.H., Schimming, S., and Staudenbauer, W.L. (1987) *Appl. Microbiol. Biotechnol.* 27, 50–56.
Schwarz, W.H., Schimming, S., Rücknagel, K.P., et al. (1988a) *Gene* 63, 23–30.
Schwarz, W.H., Schimming, S., and Staudenbauer, W.H. (1988b) *Appl. Microbiol. Technol.* 29, 25–31.
Schwarz, W.H., Jauris, S., Kouba, M., Bronnenmeier, K., and Staudenbauer, W.L. (1989) *Biotechnol. Lett.* 11, 461–466.
Schwarz, W.H., Adelsberger, H., Jauris, S., et al. (1990) *Biochem. Biophys. Res. Commun.* 170, 368–374.
Sheth, K., and Alexander, J.K. (1967) *Biochem. Biophys. Acta* 148, 808–810.
Shinmyo, A., Garcia-Martinez, D.V., and Demain, A.L. (1979) *J. Appl. Biochem.* 1, 202–209.
Shoseyov, O., and Doi, R.H. (1990) *Proc. Natl. Acad. Sci. USA* 87, 2192–2195.
Shoseyov, O., Hamamoto, T., Foong, F., and Doi, R.H. (1990) *Biochem. Biophys. Res. Commun.* 169, 667–672.
Shoseyov, O., Takagi, M., Goldstein, M.A., and Doi, R.H. (1992) *Proc. Natl. Acad. Sci. USA* 89, 3483–3487.
Sleat, R., and Mah, R.A. (1985) *Int. J. Syst. Bacteriol.* 35, 160–163.
Sleat, R., Mah, R.A., and Robinson, R. (1984) *Appl. Environ. Microbiol.* 48, 88–93.
Soutschek-Bauer, E., and Staudenbauer, W.L. (1987) *Mol. Gen. Genet.* 208, 537–541.
Sukhumavasi, J., Ohmiya, K., Shoichi, S., and Ueno, K. (1988) *Int. J. Syst. Bacteriol.* 38, 179–182.
Tailliez, P., Girard, H., Longin, R., Béguin, P., and Millet, J. (1989) *Appl. Environ. Microbiol.* 55, 203–206.
Teeri, T.T., Lehtovaara, P., Kauppinen, S., Salovuori, I., and Knowles, J. (1987) *Gene* 51, 43–52.
Tokatlidis, K., Salamitou, S., Béguin, P., Dhurjati, P., and Aubert, J.-P. (1991) *FEBS Lett.* 291, 185–188.
Tomme, P., Chauvaux, S., Béguin, P., et al. (1991) *J. Biol. Chem.* 266, 10313–10318.

Tuka, K., Zverlov, V.V., Bumazkin, B.K., Velikodvorskaya, G.A., and Strongin, A.Y. (1990) *Biochem. Biophys. Res. Commun.* 169, 1055–1060.
van Gylswyk, N.O., and van der Toorn, J.J.T.K. (1987) *Int. J. Syst. Bacteriol.* 37, 102–105.
Varel, V.H. (1989) *Arch. Microbiol.* 152, 209–214.
Varel, V.H., Richardson, A.J., and Stewart, C.S. (1989) *Appl. Environ. Microbiol.* 55, 3080–3084.
Viljoen, J.A., Fred, E.B., and Peterson, W.H. (1926) *J. Agric. Sci.* 16, 1–17.
Wong, W.K.R., Gerhard, B., Guro, Z.M., et al. (1986) *Gene* 44, 315–324.
Wong, K.K.Y., Tan, L.U.L., and Saddler, J.N. (1988) *Microbiol. Rev.* 52, 305–317.
Wood, T.M., McCrae, S.I., Wilson, C.A., Bhat, K.M., and Gow, L.A. (1988) in *Biochemistry and Genetics of Cellulose Degradation, FEMS Symposium No. 43*, 7–9 Sept. 1987, Paris, (Aubert, J.-P., Béguin, P., and Millet, J., eds.), pp. 31–52, Academic Press, San Diego.
Wu, J.H.D., Orme-Johnson, W.H., and Demain, A.L. (1988) *Biochemistry* 27, 1703–1709.
Yagüe, E., Béguin, P., and Aubert, J.-P. (1990) *Gene* 89, 61–67.
Zappe, H., Jones, D.T., and Woods, D.R. (1986) *J. Gen. Microbiol.* 132, 1367–1372.
Zappe, H., Jones, D.T., and Woods, D.R. (1987) *Appl. Microbiol. Biotechnol.* 27, 57–63.
Zappe, H., Jones, W.A., Jones, D.T., and Woods, D.R. (1988) *Appl. Environ. Microbiol.* 54, 1289–1292.
Zappe, H., Jones, W.A., and Woods, D.R. (1990) *Nucleic Acids Res.* 18, 2179.

CHAPTER 14

Utilization of Whey by Clostridia and Process Technology

I.S. Maddox
N. Qureshi
N.A. Gutierrez

Whey is a by-product of the dairy industry that is produced during the manufacture of cheese and casein. Each different process gives rise to a characteristic whey, and typical compositions are shown in Table 14–1. Sweet whey, derived from the manufacture of cheese or rennet casein, has a pH > 5.6, and is produced at a rate of 7.6 kg/kg cheese. Acid whey, derived from the manufacture of casein by either addition of sulfuric acid or production of lactic acid, has a pH < 5.1, and is produced at a rate of some 25 kg/kg casein (Zall et al. 1979). The exact composition of each type of whey varies with the location and season. Total world production approximates 85 million tonnes per annum (Zall 1984).

In the past, whey was considered to be a waste product. Now, however, it is treated as a by-product since a variety of applications are available for it including lactose manufacture, land irrigation, and protein recovery. The main means of protein recovery is by the application of ultrafiltration, and the liquid material remaining is known as whey permeate. A typical com-

TABLE 14–1 Composition of Whey[1]

	Cheese Whey *(g/kg)*	*Casein Wheys* *(g/kg)*	
		Lactic	*Sulfuric*
Total solids	67	64	59
Lactose	50	44	47
Protein	5.7	5.7	5.3
Nonprotein nitrogen	0.5	0.5	0.3
Fat	0.3	0.2	0.4
Ash	5.3	5.8	6.7

[1] Data from Short and Doughty (1977).

position of lactic whey permeate is shown in Table 14–2. Much of the emphasis in whey utilization is now on whey permeate rather than the original whey. For many years, a range of fermentation processes have been considered using whey as a substrate, either as a means of waste disposal or producing a higher-value product. Typical processes that have been investigated include

TABLE 14–2 Composition of Lactic Whey Permeate[1]

Component	*Concentration*
Total solids	56.8 g/kg
Lactose	44.8 g/kg
Total nitrogen	0.64 g/kg
Nonprotein nitrogen	0.46 g/kg
Ash	5.7 g/kg
Minerals	
Aluminium	5.1 mg/l
Calcium	835.1 mg/l
Copper	0.3 mg/l
Iron	1.2 mg/l
Potassium	881.6 mg/l
Magnesium	62.6 mg/l
Manganese	0.04 mg/l
Sodium	290.2 mg/l
Phosphorus	382.6 mg/l
Sulfur	43.1 mg/l
Zinc	43.5 mg/l

[1] Data from Kavanagh (1975) and Archer et al. (1982).

production of methane, ethanol, yeast, microbial fat, lactic acid, butanediol, citric acid, acetic acid, amino acids, microbial polysaccharides, lactase, cellulase, gibberellic acid, and a range of vitamins (Hobman 1984; Larsen and Maddox 1987). Unfortunately, as a fermentation substrate, whey suffers from two major disadvantages. First, the sugar concentration is relatively low, resulting in low product concentrations with subsequent high costs of product recovery. Prior concentration of the whey is technically feasible but the gains of reduced process volumes and increased product concentrations must be considered against the cost. Secondly, lactose is not a preferred carbon source for many microorganisms, resulting in relatively low reactor productivities when compared with raw materials containing glucose or sucrose. Despite these disadvantages, whey (permeate) is used in several countries as a raw material for the production of industrial and potable ethanol (Mawson 1987).

The purpose of this chapter is to consider the utilization of whey by clostridia, principally for the acetone-butanol-ethanol (ABE) fermentation process, and to investigate some new process technologies which are being applied to the ABE process using *Clostridium acetobutylicum* or *C. beijerinckii*.

14.1 THE ACETONE-BUTANOL-ETHANOL FERMENTATION

Although whey is a poor substrate for many fermentation processes, because of its low sugar content, this does not necessarily apply to those processes that suffer from product inhibition. The ABE fermentation is one such process, where product concentrations rarely exceed 20 g/l. This restricts the sugar utilization to the range 50–60 g/l, i.e., the approximate lactose content of whey. It is on these grounds that whey has proved to be a popular substrate for investigations into ABE production. Wix and Woodbine (1958), in a review concerning whey utilization, have described literature from the 1930s to 1940s regarding ABE production. Interestingly, ABE was often considered to be only a by-product of the fermentation, the vitamin riboflavin being the more valuable product (e.g., Meade et al. 1944).

14.1.1 Lactose as a Substrate

Various groups have used semi-defined media in attempts to evaluate lactose as a substrate for clostridia. The data are summarized in Table 14–3. Compere and Griffith (1979) used a meat-based medium containing various concentrations of lactose, but unfortunately their data do not allow calculation of reactor productivities or solvent (ABE) yields (total solvent produced per lactose used). Using a medium based on yeast extract, a reactor productivity of 0.10 g/(l · h) was recorded for *C. acetobutylicum* P262 at 30°C, with a yield of 0.38 (Ennis and Maddox 1985). In comparison, when glucose and galactose were used as substrates, productivity values were 0.22 and 0.24 g/(l · h),

TABLE 14–3 Lactose as a Substrate for Acetone-Butanol-Ethanol (ABE) Production

Strain	*Lactose* (*g*/*l*)	*Temperature* (°*C*)	*Total ABE* (*g*/*l*)	*Reference*
C. acetobutylicum ATCC 824	50	30	0.9	Ennis (1987)
C. acetobutylicum ATCC 824	50	37	8.9	Ennis (1987)
C. acetobutylicum P262	50	30	9.5	Ennis and Maddox 1985
C. acetobutylicum P262	50	37	1.5	Ennis and Maddox 1985
C. acetobutylicum NRRL B527	50	37	2.3	Compere and Griffith 1979
C. acetobutylicum NRRL B527	100	37	1.0	Compere and Griffith 1979
C. acetobutylicum NRRL B3179	50	37	10.1	Compere and Griffith 1979
C. acetobutylicum NRRL B3179	100	37	0.4	Compere and Griffith 1979
C. butylicum NRRL B592	50	37	5.8	Compere and Griffith 1979
C. butylicum NRRL B592	100	37	22.7	Compere and Griffith 1979
C. butylicum NRRL B593	50	37	1.3	Compere and Griffith 1979
C. butylicum NRRL B593	100	37	7.0	Compere and Griffith 1979
C. pasteurianum NRRL B598	50	37	1.8	Compere and Griffith 1979
C. pasteurianum NRRL B598	100	37	3.6	Compere and Griffith 1979
C. beijerinckii LMD 27.6	42	37	1.7	Schoutens et al. 1984

respectively, emphasizing the poorer nature of the disaccharide as a carbon source. One possible solution to this problem is to use a mixture of glucose and galactose as the substrate. In this situation, glucose is used preferentially to galactose, but overall results reveal that there is no real advantage in using the mixture of monosaccharides rather than the disaccharide (Schoutens et al. 1984; Ennis and Maddox 1985). An interesting feature of the results in Table 14–3 is the strong effect of temperature on the data for *C. acetobutylicum* P262 and ATCC 824. This will be discussed further in Section 14.5.5, but it suggests strongly that the screening experiments of the type described in Table 14–3 should be conducted at more than one temperature.

14.1.2 Whey as a Substrate

Different strains of clostridia have been screened for ABE production from whey, and some of the results are summarized in Table 14–4. All of these results were obtained from batch fermentation experiments without pH control or any other real attempt to optimize the fermentation conditions. Although the data demonstrate that several strains are capable of strong ABE production, the productivity values are all less than the 0.55 g/(l · h) observed using a substrate of molasses (Spivey 1978). From the data provided it appears that *C. acetobutylicum* P262 is the preferred strain for ABE production from whey.

Since lactose appears to be a difficult substrate compared to glucose or sucrose, Ennis and Maddox (1987b) performed experiments to compare the use of whey permeate with hydrolyzed whey permeate. The latter was prepared by treatment with a commercial β-galactosidase enzyme, and experiments were performed in batch fermentation using strain P262. The results for the nonhydrolyzed substrate were similar to those shown for this strain in Table 14–4. For the hydrolyzed substrate, ABE production commenced slightly earlier, but the final yield (0.31) and productivity [0.24 g/(l · h)] were not significantly different from the nonhydrolyzed control. Glucose and galactose were used simultaneously, but the former at a greater rate. Overall, there appears to be no advantage in using lactose-hydrolyzed whey as a substrate, since the slower galactose utilization negates the advantage conferred by faster utilization of glucose.

It is generally accepted that the ratio of products in the ABE fermentation approximates 1:2:0.3, acetone:butanol:ethanol (Prescott and Dunn 1959). However, Bahl et al. (1986) observed that when using a substrate of whey, the ratio of butanol to acetone is considerably higher. This could have profound economic implications as butanol is the preferred product if ABE is to be used as a fuel supplement. Bahl et al. (1986) demonstrated that the butanol/acetone ratio can be altered by varying the growth conditions, and they claimed that the conditions are optimal with whey. However, it is not clear as to which type of whey was used in their experiments, or why their results were so different from the others shown in Table 14–4. Nevertheless,

TABLE 14–4 Whey as a Substrate for Acetone-Butanol-Ethanol (ABE) Production

Strain	*Substrate*	*Temperature* (°C)	*Total ABE* (g/l)	*Productivity* [g/(l · h)]	*Ratio* (A:B:E)	*Reference*
C. acetobutylicum NCIB 2951	Sulfuric whey permeate + yeast extract	30	17.0	0.14	1:10:1	Maddox 1980
C. acetobutylicum ATCC 824	Acid whey	37	9.2	0.08	1:13:4	Welsh and Veliky 1984
C. acetobutylicum ATCC 824	Sulfuric whey permeate + yeast extract	30	3.9	0.04	1:4:0	Ennis 1987
C. acetobutylicum ATCC 824	Sulfuric whey permeate + yeast extract	37	4.4	0.08	1:4:0	Ennis 1987
C. acetobutylicum ATCC 4259	Cheese whey + yeast extract	37	9.4	0.10	1:4:0	Voget et al. 1985
C. acetobutylicum NRRL 596	Cheese whey + yeast extract	30	7.8	0.08	1:4:0	Voget et al. 1985
C. acetobutylicum DSM 792	Whey	37	10.8	—	1:128:24	Bahl et al. 1986
C. acetobutylicum P262	Sulfuric whey permeate + yeast extract	30	9.5	0.24	1:3:0	Ennis and Maddox 1985
C. beijerinckii LMD 27.7	Cheese whey permeate + yeast extract	30	5.0	0.05	1:4:0[1]	Schoutens et al. 1984
C. butylicum NRRL B592	Sulfuric whey permeate + yeast extract	30	11.2	0.09	1:4:0.2	Gapes et al. 1983

[1] Isopropanol is produced instead of acetone.

it is clear from all of the results that the butanol/acetone ratio is higher with whey than with other substrates. Bahl et al. (1986) have suggested that the effect is due to iron limitation or the presence of lactate in whey.

14.1.3 Enzyme Systems for Lactose Utilization

Microorganisms can use lactose by either or both of two metabolic pathways. The first pathway involves β-galactosidase and is common in Gram-negative bacteria. The second system is known as the phosphoenolpyruvate-dependent phosphotransferase (PEP:PTS) lactose system, and is more commonly found in Gram-positive bacteria (McKay et al. 1970; Saier 1985). The PEP:PTS system uses energy derived from the high-energy phosphate bond in phosphoenolpyruvate to translocate lactose through the cell membrane. The lactose phosphate is subsequently hydrolyzed by phospho-β-galactosidase to glucose and galactose-6 phosphate. Yu et al. (1987) investigated the activities of β-galactosidase and phospho-β-galactosidase in various strains of *C. acetobutylicum*, as shown in Table 14–5. All of the strains displayed both activities except strain ATCC 824 which had no detectable β-galactosidase activity. For the other strains, except strain P262, β-galactosidase activities were higher than those of phospho-β-galactosidase. When strains P262 and ATCC 824 were grown on glucose, neither of the enzyme activities were detected. Yu et al. (1987) went on to investigate the enzyme activities during the course of a typical ABE fermentation process on whey permeate. Using strain P262, there was early induction of phospho-β-galactosidase activity associated with the acidogenic phase of the fermentation, but the activity then decreased markedly. For β-galactosidase, induction occurred at a later stage, peaking during the solventogenic phase of the process. When using strain ATCC 824,

TABLE 14–5 Activities of β-Galactosidase and Phospho-β-Galactosidase in Different Strains of *C. acetobutylicum*[1]

Strain	*Substrate (10 g/l)*	*β-Galactosidase (U/mg protein)*	*Phospho-β-galactosidase (U/mg protein)*
NRRL 594	Lactose	0.19	0.015
NRRL 598	Lactose	0.23	0.066
NRRL 2490	Lactose	0.09	0.027
NCIB 2951	Lactose	0.39	0.205
P262	Lactose	0.16	0.40
	Glucose	0	0
	Galactose	0.015	0.32
ATCC 824	Lactose	0	0.15
	Glucose	0	0
	Galactose	0	0.25

[1] Data from Yu et al. (1987).

a similar early induction of phospho-β-galactosidase was observed, while β-galactosidase was not detected during the entire course of the fermentation. These data pose questions regarding the relative importance in vivo of these two enzyme systems for lactose utilization and solvent production.

Recently, a genomic library of *C. acetobutylicum* NCIB 2951 has been prepared in the broad host-range cosmid pLAFR1, and cosmids containing the β-galactosidase gene have been isolated following conjugal transfer of the library to a *lac* deletion derivative of *E. coli* (Hancock et al. 1991). Analysis of various subclones localized the β-galactosidase gene to a 5.1–kilobase (kb) *Eco*R1 fragment. Expression in *E. coli* was independent of the orientation of the fragment, suggesting expression from its own promoter, and was not subject to catabolite repression when the cells were grown in the presence of glucose. Six strains of *C. acetobutylicum* (ATCC 824, NRRL 594, NRRL 598, NRRL 2490, P262, and NCIB 2951; refer to Table 14–5) possessed a sequence highly homologous to the cloned β-galactosidase fragment (Hancock 1988). The result for strain ATCC 824 was rather surprising, given the data of Yu et al. (1987) shown in Table 14–5. However, repeated experiments have failed to demonstrate expression of this gene in strain ATCC 824, suggesting a mutation in either the structural or a regulatory gene (Hancock 1988). The β-galactosidase gene region of strain NCIB 2951 showed only low homology to the DNA from other Gram-positive bacteria, and no detectable homology to DNA from Gram-negative bacteria (Hancock 1988). To date, cosmids containing the phospho-β-galactosidase gene have not been isolated from the gene library.

14.2 BUTYRIC ACID PRODUCTION

Butyric acid has considerable use as an industrial chemical. In addition, it can be esterified with alcohols to produce carboxylic acid esters that have better properties as fuels than have alcohols alone. It is produced as a normal intermediate and by-product of the ABE fermentation, but very few studies have concentrated on its production in its own right, possibly because it is rarely produced as a pure product. Ennis and Maddox (1987a), conducted a study into the effect of pH on ABE production from sulfuric whey permeate using *C. acetobutylicum* P262. When the culture was controlled to be at not less than pH 5.6, butyric acid was produced at 13.3 g/l, accompanied by acetic acid (5.0 g/l) and butanol (2.4 g/l). Similar results have been described for *C. beijerinckii* growing on cheese whey (Alam et al. 1988). In this case, maximum butyric acid production (in excess of 12 g/l) was observed when the culture was controlled at pH 5. Acetic acid and butanol remained between 1–4 g/l. Stevens et al. (1988) have described the use of a mixed culture to increase the total product concentration without any sacrifice in the level of butyric acid formed. Thus, using cheese whey as substrate, a mixture of *C. beijerinckii* and *Bacillus cereus* fermented the lactose, at pH 5.5, to give a mixture of butanol and butyric acid at 7.0 and 11.5 g/l, respectively. Acetic acid remained at about 2 g/l. In contrast, Ennis and Maddox (1987a) used a pure culture of *C. acetobutylicum* P262, growing in sulfuric whey permeate at pH 5.2, to produce a mixture of butanol (6.5 g/l), butyric acid (9.9 g/l), and acetic acid (4.3 g/l).

14.3 ACETIC ACID PRODUCTION

Acetic acid is an important raw material in the chemical industry. It is normally produced by chemical synthesis since the traditional vinegar fermentation process suffers from low product yield and high energy cost (Tang et al. 1988). Recently, interest has revived in anaerobic fermentation processes since almost 100% of the substrate carbon can be recovered in the product.

As with butyric acid, acetic acid is a normal intermediate/by-product of the ABE process, but it is usually produced in lower concentrations. During their studies with *C. acetobutylicum* P262 into ABE production from whey permeate, Ennis and Maddox (1987a) reported a maximum acetic acid concentration of 5.9 g/l when the culture was maintained at pH 5.6. Butyric acid (11.8 g/l) was also present, but no butanol. Tang et al. (1988) have recently reported the use of a mixed culture of *Streptococcus lactis* and *C. formicoaceticum* for acetic acid production from sweet whey permeate. The former is a homolactic bacterium that can convert 90% of the lactose in whey to lactate. The latter cannot ferment lactose, but can produce acetate from lactate. When the mixed fermentation was controlled at pH 7.6, an acetic acid concentration of 20 g/l was produced within 20 h. Production then slowed due to product inhibition, but after 80 h acetate was present at 30 g/l, plus lactate at 20 g/l. Hence, this novel approach to acetic acid production from whey would seem to have promise.

14.4 ACRYLIC ACID PRODUCTION

Acrylic acid is an important chemical feedstock, whose production by fermentation from cheese whey has been demonstrated recently (O'Brien et al. 1990). The process involves two steps. In the first, lactose is converted into a stoichiometric mixture of acetic acid and propionic acid using a coculture of *Lactobacillus bulgaricus* and *Propionibacterium shermanii*. Then, in the second step, *C. propionicum*, which does not ferment sugars, oxidizes the propionate to acrylate in the presence of methylene blue as an electron acceptor. Although the reported concentration of acrylic acid was only 0.43 g/l, the concept of the two-stage fermentation warrants further investigation.

14.5 OPTIMIZATION OF BATCH ACETONE-BUTANOL-ETHANOL FERMENTATION

14.5.1 Strain Preservation and Inoculum Development

The most important aspect of any fermentation process is the organism that is used to perform the process. Hence, it is imperative that cultures be stored in such a way that they not only remain viable but also retain their desired physiological properties. Furthermore, owing to the effect of inoculum quality on the course of the subsequent fermentation, inocula must be developed from stock cultures in such a way that the main fermentation is vigorous, and that the maximum reactor productivities and yields are achieved. It is now well known that the manner of culture maintenance and inoculum development can affect the solventogenic properties of clostridia (Prescott and Dunn 1959). Gapes et al. (1983), studying ABE production from whey permeate using *C. butylicum* NRRL B592, demonstrated that heat shocking of cultures, and the number of subcultures performed, profoundly affected ABE production. Initially, progressive subcultures, accompanied by heat shocking at 75°C for 1.5 min, showed an improvement in solventogenic ability. After the third subculture, however, degeneration was rapid, and by the sixth no ABE was detected in the fermentation.

Gutierrez and Maddox (1987a) carried out a detailed investigation of some culture maintenance and inoculum development techniques for the production of ABE from whey permeate using *C. acetobutylicum* NCIB 2951. As expected, they observed that regular (monthly) transfer of cultures was detrimental to ABE production. The most effective method of culture maintenance proved to be storage as spores in nonnutrient media at −20°C. On revival of stock cultures, neither heat shocking (70°C for 1.5 min followed by immersion in ice-cold water) nor treatment with aqueous ethanol or butanol gave improvements in ABE production during the subsequent fermentation. This was rather surprising as a heat shock treatment has been variously described as stimulating spore germination (Spivey 1978) and destroying weaker spores that are poorly solventogenic (Prescott and Dunn 1959). The most-effective technique for ensuring a vigorous main fermentation, with maximum reactor productivity, was to ensure that all culture transfers during inoculum preparation were made at the time of maximum cell motility. The reason for this effect is not yet clear, although it is known that absence of motility during either inoculum development or the main fermentation process results in poor ABE production (Spivey 1978; Gutierrez 1985). In the present authors' laboratory, it is now standard procedure to heat shock cultures during the revival stage (to ensure absence of contamination) and to perform all transfers at the stage of maximum cell motility, as observed microscopically. However, it should be stressed that the optimum procedure may be strain specific, and vary with the substrate used.

14.5.2 Course of Fermentation

In general, the course of the fermentation in whey is similar to that in other substrates. Initially, the organism grows exponentially, at growth rates approaching 0.5 h^{-1}, and acetic acid and butyric acid are produced ("acidogenic" phase). Eventually, growth slows and the "solventogenic" phase commences.

Lactose is now converted to solvents, while there is also some uptake of the acids (Maddox 1980; Ennis and Maddox 1987a). Because of the importance of cell motility to ABE production, Gutierrez and Maddox (1987b) performed a detailed study of this feature during a typical batch fermentation process in whey permeate (Figure 14–1). Following inoculation, most cells quickly became nonmotile. Within 2 h, however, motility was restored, only to cease again by the fourth hour. By the sixth hour after inoculation, motility was

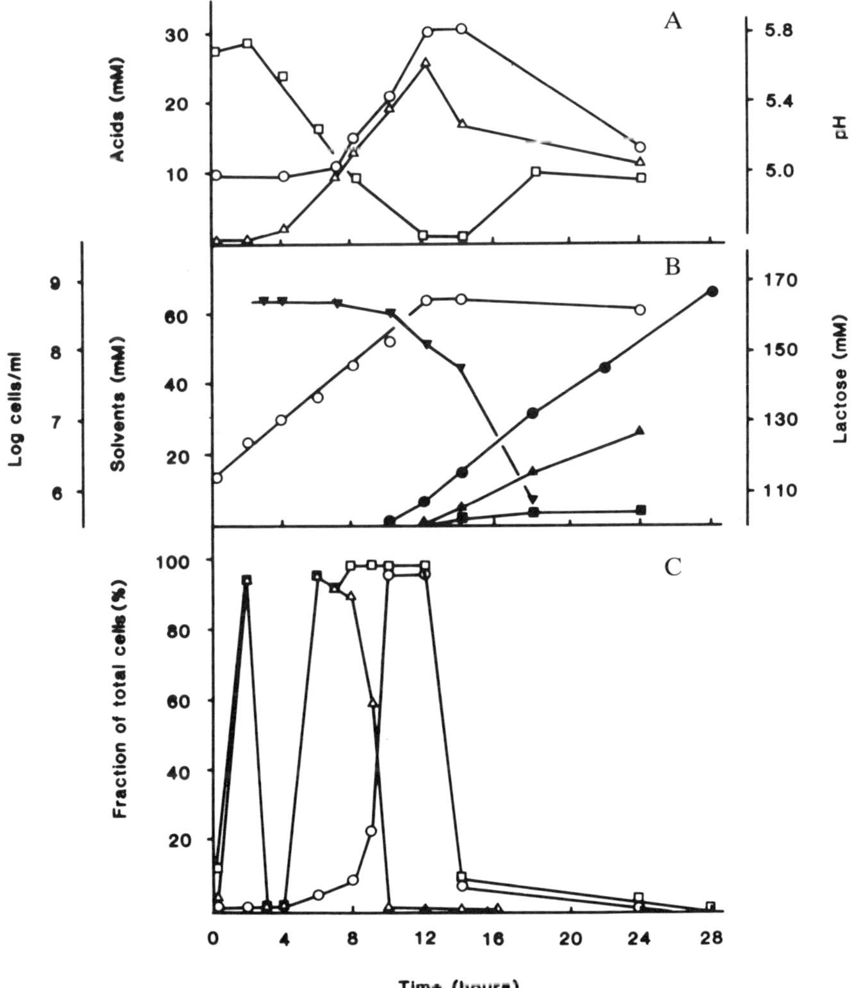

FIGURE 14–1 Time course of batch fermentation process. Symbols: (A) □, pH; ○, acetate; △, butyrate. (B) ○, log cells/ml; ▼, lactose; ●, butanol; ▲, acetone; ■, ethanol. (C) □, motile cells; △, running cells; ○, tumbling cells. Reproduced with permission from Gutierrez and Maddox (1987b).

fully restored, and the cells moved in long fast runs with little tumbling being observed. (The hiatus was probably due to a carbon source other than lactose being used initially.) The runs observed during this phase of the fermentation were associated with lactose utilization and acid production, and are characteristic of positive chemotaxis. The sugars and undissociated acetic and butyric acids are known to be attractants (Gutierrez and Maddox 1987b). As the fermentation proceeded, the motility became characterized by tumbles. This was associated with solventogenesis, and ABE are known to be repellents. The authors suggested that the motility of *C. acetobutylicum* during fermentation is a chemotactic response. This may help to explain the relationship between motility and solventogenesis. By 24 h after inoculation the majority of the cells were in the clostridial form.

The ABE fermentation process suffers from severe product inhibition, and this restricts the total ABE concentration to less than 20 g/l. This effect has been demonstrated in batch fermentation using whey permeate, and is similar to that in other substrates (Maddox 1982).

14.5.3 Effect of pH and Initial Sugar Concentration

The culture pH value is known to exert a profound effect on the ABE fermentation process, and a detailed study has been performed using *C. acetobutylicum* P262 growing in sulfuric whey permeate (Ennis and Maddox 1987a). Fermentations performed at relatively low pH values favored solventogenesis, but lactose utilization was relatively poor. At higher pH values growth and lactose utilization were much improved but acids rather than ABE were produced. To solve the problem, it was necessary to develop an optimum pH profile during the fermentation rather than simple pH control. The profile involved holding the pH near neutrality during the growth phase, and then allowing it to fall naturally to the optimum value for the ABE production rate (pH 5.1–5.5), at which point it was maintained by addition of ammonia. The authors stressed, however, that the optimum profile for a given substrate and bacterial strain is probably case specific, and should be evaluated for each situation.

The effect of the initial lactose concentration on ABE production appears slightly confusing. Compere and Griffith (1979), during their evaluation of different sugars as substrates for clostridia, showed that the concentration of lactose did have an effect on ABE concentration (Table 14–3). However, in some cases an increased lactose concentration was beneficial (e.g., *C. butylicum* NRRL B592) while in others it was detrimental (e.g., *C. acetobutylicum* NRRL B3179). Unfortunately, no data were provided regarding culture pH value so the results are difficult to interpret. Welsh and Veliky (1986), working with *C. acetobutylicum* in semisynthetic medium, reported that at lactose concentrations of 10 and 30 g/l, acid production predominated, while ABE production predominated at 50 g/l lactose concentration. Ennis and Maddox

(1987a) added extra lactose to whey permeate, and showed that when operating at high pH values, where acid production usually predominated, an increase in the initial lactose concentration resulted in lower rates of lactose utilization, and this was accompanied by increased ABE production and decreased acid production. Analysis of their data from a range of experiments revealed a strong inverse relationship between ABE yield and lactose utilization rate. It was suggested that conditions which minimize the lactose utilization rate, such as low culture pH values or high initial lactose concentrations, favor solventogenesis at the expense of acid production.

14.5.4 Other Nutrients

The compositions of whey and whey permeate are given in Tables 14–1 and 14–2. For a commercial operation, it would be economically advantageous if no additional nutrients needed to be added to ensure a successful fermentation process. Unfortunately, relatively few studies have been conducted on this point. When using whey permeate as substrate, most workers have routinely added yeast extract (Maddox 1980; Gapes et al. 1983; Schoutens et al. 1984; Ennis and Maddox 1985, 1987a), but it is not clear as to which particular nutrients require supplementation. For experiments conducted with whey, Welsh and Veliky (1984) provided no additional nutrients, but Voget et al. (1985) demonstrated that stronger solventogenesis occurred when yeast extract was present. Subsequently, Voget et al. (1987) demonstrated that corn steep or malt sprouts can replace yeast extract.

Kanchanatawee and Maddox (1990a) have described a study into the nutritional status of sulfuric whey permeate as a substrate for the ABE fermentation. They performed experiments to determine the effects of supplementary yeast extract (5 g/l) and a combination of vitamins (biotin 1 mg/l, thiamine hydrochloride 20 mg/l, and *p*-aminobenzoic acid 20 mg/l) on ABE production from whey permeate. In these experiments, the inocula were developed in such a way that there would be no ambiguity in the results due to inoculum carry-over. Various nutrients, including ammonium and phosphate ions, were analyzed and the results are shown in Table 14–6. Without supplementary yeast extract, lactose utilization and ABE production were poor. The combination of vitamins could not substitute for yeast extract, even though it is known to be appropriate in a fully defined medium (Kanchanatawee and Maddox 1990b). Since it is clear that whey permeate is not deficient in assimilable nitrogen or phosphate, these results indicated a deficiency of metal ion. Subsequent experiments provided evidence that there is iron deficiency (Kanchanatawee and Maddox 1990a), confirming the early claim of Meade et al. (1944), but it is possible that there are also some other unrecognized deficiencies. The possible role of iron in affecting the butanol/acetone ratio has been mentioned in Section 14.1.2 (Bahl et al. 1986).

TABLE 14–6 Effect of Yeast Extract and Vitamins on Acetone-Butanol-Ethanol (ABE) Production and Nutrient Utilization in Whey Permeate Using *C. acetobutylicum* P262[1]

	Supplement		
	None	*Yeast Extract*	*Vitamins*
NH_4^+ initial, g/l	0.26	0.39	0.23
NH_4^+ used, g/l	0.13	0.20	0.13
PO_4^{3-} initial, g/l[2]	1.96	1.71	1.71
Lactose used, g/l	14	35	6
ABE produced, g/l	0.2	6.9	0.1

[1] Based on Kanchanatawee and Maddox (1990a).

[2] Whey permeate, after autoclaving, contains a precipitate of mineral phosphate. There was a large excess of soluble phosphate at the completion of all fermentations.

14.5.5 Temperature

In Section 14.1.1, it was remarked that there was a strong effect of temperature on ABE production from lactose. Depending on the clostridial strain, similar effects have been observed using whey, but it is difficult to ascertain any general pattern. Schoutens et al. (1984), using *C. beijerinckii* LMD 27.7, observed far stronger solventogenesis at 30°C than at 37°C, and this has also been demonstrated for *C. butylicum* NRRL B592 by Voget et al. (1985). In contrast, *C. acetobutylicum* ATCC 4259 produced higher ABE yields at 37°C than at 30°C (Voget et al. 1985). For strains ATCC 824 and P262, little effect of temperature was observed when growing on whey permeate (Ennis 1987), in contrast to the results observed in a semisynthetic medium (Table 14–3). From these data, it appears that the effect of temperature is case-specific, and should be investigated for each clostridial strain and type of whey.

14.5.6 Agitation and Head-Space Pressure

The beneficial effect of maintaining a positive head-space pressure during the ABE fermentation process has been recognized for many years (Spivey 1978), and its importance has been demonstrated when fermenting whey permeate. Maddox et al. (1981) controlled the head-space pressure up to values of 105 kPa and observed a marked improvement in ABE productivity when compared with experiments performed at atmospheric pressure [0.2 g/(l · h) compared with 0.05 g/(l · h)]. It was also demonstrated that purging the culture with hydrogen gas at atmospheric pressure had no effect on solvent productivity. Closely linked to these observations, Welsh and Veliky (1984, 1986) have reported stronger ABE production when the fermentation is performed without agitation than with agitation. It is now believed that the beneficial effects of a positive head-space pressure, and non-agitation, are due to the maintenance in the culture of hydrogen supersaturation. This, in turn, favors the production of reduced products such as butanol rather than oxidized products such as butyrate.

14.6 CONTINUOUS ACETONE-BUTANOL-ETHANOL FERMENTATION

14.6.1 Freely Suspended Cells

Ennis et al. (1986a) and Maddox (1989) have reviewed the application of continuous culture to the ABE fermentation. However, little attention has been paid to the use of freely suspended cells in continuous culture using a substrate of whey. A major problem appears to be operational instability, probably reflecting the presence of two different physiological types of cells in the culture, i.e., acidogenic and solventogenic types (Clarke et al. 1988). Recently, a single-stage continuous reactor has been described for the utilization of whey permeate, but, as with other substrates, high-solvent productivities are attained only at low dilution rates (N. Qureshi and I.S. Maddox, unpublished observations). A maximum productivity of 0.16 g/(l · h) was recorded at a solvent concentration of about 8 g/l, but this fluctuated ±20%. It has been suggested that a successful continuous culture using freely suspended cells will be achieved only with a two- or multi-stage reactor (Maddox 1989).

14.6.2 Immobilized Cells

Cell immobilization is defined as a technique that confines cells within a reactor system, permitting their easy reuse. Compared to free-cell suspensions, greatly increased cell densities can be achieved, resulting in increased fermenter productivities. Further, in continuous culture systems, high dilution rates can be used without fear of washout, and it appears that these systems do not suffer from the same stability problems as do those employing free cells.

14.6.2.1 Alginate-Immobilized Cells. This is the most widely studied technique for application to the ABE fermentation, and Maddox (1989) has recently reviewed it.

Schoutens et al. (1985) used alginate-immobilized *C. beijerinckii* LMD 27.6 in a continuous stirred tank reactor (CSTR) using a substrate of whey permeate. They demonstrated the importance of using a low dilution rate during start-up, and that butanol can be produced continuously at reactor productivities 16 times those observed in batch culture. Ennis et al. (1986b) described similar work using *C. acetobutylicum* P262, but used a fluidized bed reactor in addition to a CSTR. The reactors could be routinely operated for up to 650 h at a productivity of 1.8 g/(l · h) and a yield of 0.3 g ABE per g lactose used. Unfortunately, lactose utilization remained incomplete, and this appears to be the sacrifice that is made for the achievement of high reactor productivities. In an attempt to solve this problem, Ennis et al. (1986b) applied pH regulation and a two-stage process, but both techniques proved detrimental to ABE production. The process was demonstrated to be controlled by the inhibitory product concentration, principally butanol. Ennis and Maddox (1987b) investigated the use of lactose-hydrolyzed whey permeate as a substrate for alginate-immobilized cells. Although a slightly improved reactor productivity was observed, the relatively poor utilization of galactose signified that there is no real operational advantage in lactose hydrolysis prior to the fermentation.

14.6.2.2 Immobilization by Adsorption. In contrast to alginate-immobilization, which is an entrapment technique, immobilization by adsorption of cells onto a solid surface should avoid any problems of nutrient/product diffusion in the bulk liquid to or from the cell. However, the accumulation of multicell layers may negate this advantage. Technically, immobilization of cells by adsorption is a cheap and mild procedure, and should be readily amenable to scale-up.

Qureshi and Maddox (1987) immobilized cells of *C. acetobutylicum* P262 on bonechar (as used in sugar refining), and used them in a packed bed reactor for continuous ABE production from whey permeate. A maximum productivity of 4.1 g/(l · h), representing a yield of 0.23 g ABE/g lactose used, was observed at a dilution rate of 1.0/h. The reactor was operated under stable conditions for 61 days, but then suffered from problems of blockage due to excess biomass production and gas hold-up. To solve these problems, various other reactor designs have been investigated including a two-stage packed bed reactor and a fluidized bed reactor (Qureshi and Maddox 1988). In the former, productivities approaching 6.5 g/(l · h) were achieved. Blockage was observed in the first reactor but not the second (probably due to solvent inhibition of growth or nutrient exhaustion), suggesting that occasional reversal of the order of reactors would be a solution to the blockage problem. The fluidized bed reactor provided an extremely stable system, with productivities up to 4.8 g/(l · h). Interestingly, based on the bonechar loading, the fluidized bed reactor was six times more productive than the packed bed, probably due to greater amounts of biomass being present.

The reactor productivities achieved using bonechar-immobilized cells in a packed bed reactor [approx. 6 g/(l · h)] are considerably higher than those using alginate-immobilized cells [1.8 g/(l · h)] or those observed in a traditional batch fermentation using freely suspended cells [0.25 g/(l · h)]. The reason for this is the large amount of biomass that is present in the system (in excess of 50 g/l). However, close examination of the data reveals that although volumetric productivities are high, the *specific* productivities of bonechar-immobilized cells are low [0.06–0.08 g/(g biomass · h) compared to 1.0 g/(g biomass · h) with freely suspended cells]. Qureshi et al. (1988) investigated the reasons for this since if the specific productivity can be improved, then even greater volumetric productivities should be achievable. A mathematical model was developed that postulated the presence of different morphological/physiological cell types within the reactor, similar to those described by Clarke et al. (1988) for a free-cell continuous fermentation. The majority of biomass in the reactor was demonstrated to be metabolically inert, while only small amounts actively grow, and even smaller amounts are solventogenic. This conclusion was supported by electron micrographs of the immobilized cells (Figure 14–2), showing many layers of cells adsorbed to the bonechar. Probably, of all the total biomass, only the outermost layers are involved in growth and solvent production. A similar conclusion has been reached following mathematical analysis of a fluidized bed reactor (Qureshi and Maddox 1990a).

14.6.3 Cell Recycle Techniques

Continuous culture (using freely suspended cells) with cell recycle is a technique by which the cells in the reactor effluent stream are separated and recycled to the reactor. In this way, the biomass concentration within the reactor is increased, with subsequent increases in productivity. Applications of this technique to the ABE fermentation have been summarized by Ennis et al. (1986a) and Maddox (1989). Originally, centrifugation was the technique used for cell separation but now the preferred technique is filtration, particularly using membranes in the "crossflow" mode (crossflow microfiltration, CFM). Ennis and Maddox (1989) have described the application of a cell recycle technique to continuous fermentation of whey permeate using *C. acetobutylicum* P262. Due to the presence of the precipitate that accompanies whey fermentations, regular backwashing of the CFM membrane was required, adding to the complexity of the system. Stable steady states were difficult to achieve, and the fermentations were characterized by cyclic solventogenic and acidogenic behavior, and ultimate degeneration to a completely acidogenic state. These poor results were in contrast to those described by other workers using synthetic media, and may reflect the nature of whey permeate as a substrate rather than the cell recycle technology (Maddox 1989). Ennis and Maddox (1989) explained their results by postulating the presence of several different morphological/physiological cell types within the reactor (Figure 14–3). The relative proportion of each cell type within the reactor varies with the environment (e.g., butanol concentration, butyric acid concentration), causing fluctuations in product concentrations which, in turn, feedback to affect the reaction rates, k, and thus the proportions of the cell types. Unfortunately, there is still insufficient knowledge concerning these reaction rates and their control mechanisms.

A

B

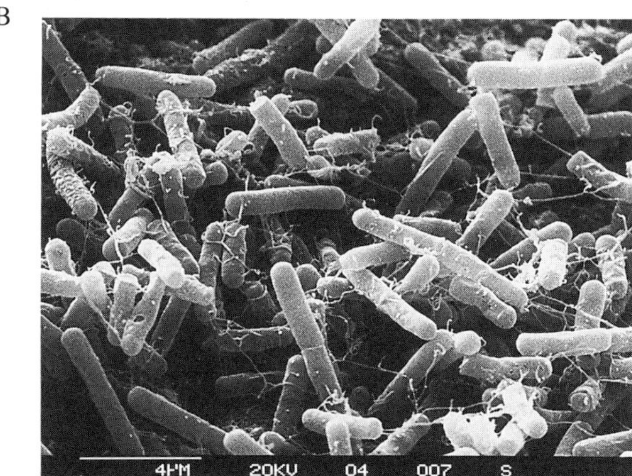

C

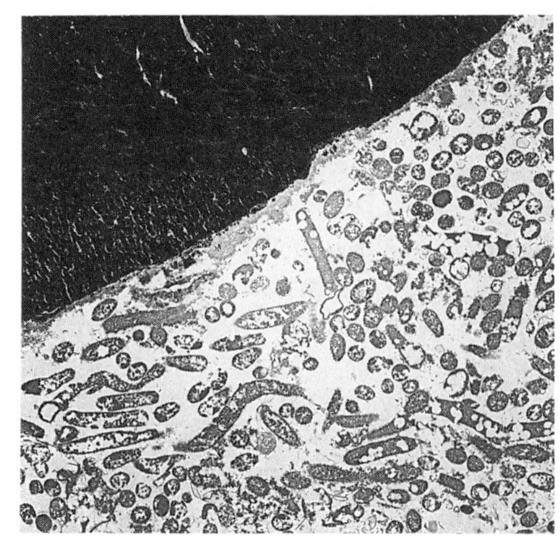

FIGURE 14–2 (A) Scanning electron micrograph of bonechar surface without immobilized cells (magnification ×3575). (B) Scanning electron micrograph of bonechar surface with immobilized cells (magnification ×3250). (C) Transmission electron micrograph of cells immobilized on bonechar (magnification ×2210). Reproduced with permission from Qureshi et al. (1988).

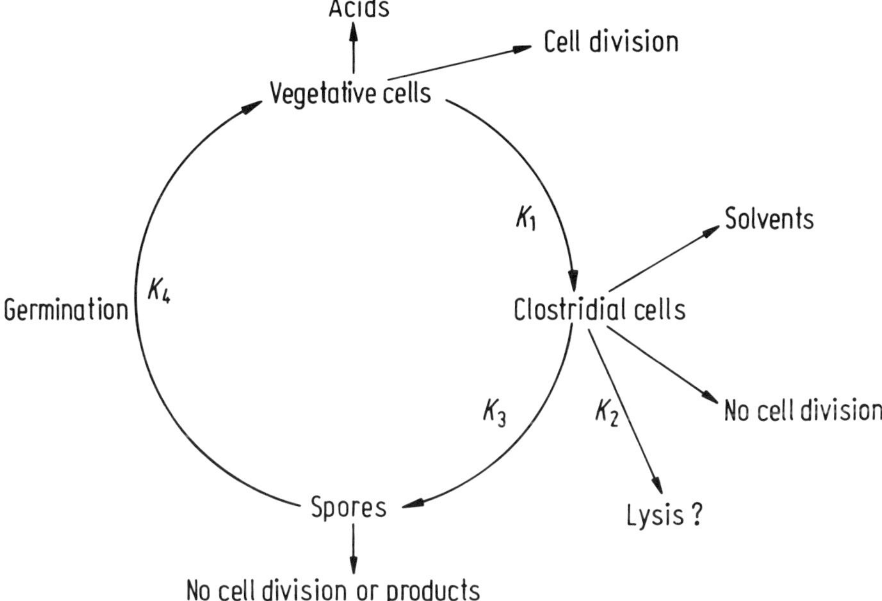

FIGURE 14–3 A schematic diagram of the morphological changes observed during continuous fermentation by *C. acetobutylicum* P262 fermenting whey permeate medium. K_1 designates the rate of conversion (or differentiation) of vegetative cells to clostridial forms. K_2 designates the rate of loss of clostridial forms due to lysis. K_3 designates the rate of sporulation of clostridial forms. K_4 designates the rate of spore germination to vegetative cells. Reproduced with permission from Ennis and Maddox (1989).

A summary of the various continuous fermentation techniques, as applied to the ABE fermentation using whey permeate as substrate, is shown in Table 14–7.

14.7 INTEGRATION OF ACETONE-BUTANOL-ETHANOL FERMENTATION WITH PRODUCT RECOVERY

The ABE fermentation process suffers from two major problems:

1. Low reactor productivity.
2. Product inhibition, resulting in low (<20 g/l) product concentrations in the fermentation broth.

As previously described, considerable progress has been made in the development of high-productivity continuous reactors, but these gains are often at the expense of reduced sugar utilization and lower-than-usual product

TABLE 14–7 Comparison of Continuous Fermentation Techniques for Acetone-Butanol-Ethanol (ABE) Production from Whey Permeate

Technique	*Reactor Productivity*[1] $[g(l \cdot h)]$	*Simplicity of Process*	*Stability of Operation*	*Reference*
Freely suspended cells	0.16	Good	Poor	N. Qureshi and I.S. Maddox, unpublished observations
Alginate-immobilized cells	1.8	Moderate	Good	Ennis et al. 1986b
Bonechar-immobilized cells	4.0	Good	Good	Qureshi and Maddox 1987
Cell recycle	2.9	Poor	Poor	Ennis and Maddox 1989

[1] All values were obtained at a dilution rate of $0.4\ h^{-1}$, except that for freely suspended cells which was at a dilution rate of $0.02\ h^{-1}$.

concentrations. One solution to this problem is to recycle the reactor effluent to allow further sugar utilization, but the effectiveness of this technique is limited by product inhibition. Thus, there is a requirement for a product removal/recovery technique that can be applied to the reactor effluent to remove inhibitory product, thus allowing subsequent recycle. Further, the traditional product recovery technique for the ABE fermentation process is distillation. This is an energy-intensive technique, hence considerable work has been conducted into the application of other technologies. These have been reviewed recently (Ennis et al. 1986a; Maddox 1989), so the following is restricted to reports where whey (permeate) is used as the substrate.

14.7.1 Adsorption

The desired characteristics of an adsorbent include a high adsorption capacity for ABE, but not for nutrients or reaction intermediates; favorable adsorption kinetics; and the availability of a simple adsorbent regeneration technique. Silicalite (a zeolite analogue) and XAD ion-exchange resins have been investigated for products removed from the effluent from a packed bed reactor operating with bonechar-immobilized *C. acetobutylicum* P262 (Ennis et al. 1987). The adsorption capacities were similar (0.08–0.10 g ABE/g adsorbent) but there was evidence in both cases that nutrients were also removed. These adsorption capacities are likely to be too low for any commercial application.

14.7.2 Liquid-Liquid Extraction

In this technique, a water-immiscible solvent is contacted with the fermentation broth; the inhibitory products dissolve in the solvent, and product inhibition is reduced. The products can subsequently be recovered from the solvent phase by distillation or back-extraction into another solvent. A wide range of solvents has been investigated (Maddox 1989) but there is little consensus on which is most appropriate. Those solvents with a suitably high partition coefficient are toxic to the bacterial cells. Conversely, those which are nontoxic have only low partition coefficients. In addition, most reports describe only removing butanol from the fermentation broth, with only scant attention to the other reaction products, ethanol and acetone, or to the reaction intermediates acetic acid and butyric acid. For successful commercial implementation, attention must be paid to these other metabolites.

N. Qureshi and I.S. Maddox (unpublished observations) have investigated the application of three solvents to ABE removal from continuous reactors operating with bonechar-immobilized *C. acetobutylicum* P262. Benzyl benzoate and dibutylphthalate were used to extract ABE from the effluent from a fluidized bed reactor. The effluent was then mixed with a concentrated whey-permeate feed medium and recycled to the reactor. The results suggested that the solvents exerted a toxic effect on the cells since reactor productivities decreased compared to control experiments. In addition, ABE yields were reduced, probably due to butyric acid removal by the solvents. The latter has been confirmed using model solutions. The use of oleyl alcohol has been studied in a similar manner, using the effluent from a packed bed reactor of bonechar-immobilized cells. In this case, the reactor productivity and the ABE yield were similarly reduced. Although oleyl alcohol appears to be the most promising of the three solvents tested, it is also the least effective for removing ethanol and acetone, as shown using model solutions (N. Qureshi and I.S. Maddox, unpublished observations).

Extraction using oleyl alcohol has also been applied to a traditional batch fermentation process, using freely suspended cells, where continuous feeding with a concentrated whey permeate medium was employed (Qureshi and Maddox 1990b). Continuous feeding is a widely used commercial fermentation technique. It differs from continuous fermentation in that there is no concomitant removal of culture. Its advantages over traditional batch fermentation include reductions in process volumes and reactor sizes, and increases in reactor productivity, sugar utilization, and product concentration. However, to achieve the latter in the ABE fermentation process, a product removal technique must be employed to alleviate product inhibition. Experiments were performed whereby the ABE concentration in the culture was allowed to reach 5 g/l. At this point, the culture was extracted with oleyl alcohol and the aqueous phase was returned to the reactor. After four such extractions, however, fermentation ceased, probably due to either the inhibitory effect of oleyl alcohol or salt accumulation in the broth. Nevertheless, compared with a standard batch fermentation process, the reactor productivity was doubled, and significant improvements were made in total ABE production and lactose utilization (Table 14–8).

14.7.3 Perstraction

As previously mentioned, direct contact of organic extractants with bacterial cultures can cause problems of cell inactivation. In addition, there can also be technical problems of sterilization of the extractant, and the formation of emulsions during separation of the aqueous and organic phases. The technique of perstraction has been developed in an attempt to solve these problems, in that a membrane is placed between the culture and the organic solvent. The membrane allows selective permeation of ABE followed by its instantaneous removal by the solvent.

Qureshi and Maddox (1990b) have performed experiments using perstraction with oleyl alcohol in a batch fermentation process using continuous feeding. The results (Table 14–8) show a productivity value comparable to that using liquid-liquid extraction, but the total lactose utilization and ABE production were markedly increased. Fermentation eventually ceased, however, either because of oleyl-alcohol diffusion into the culture or mineral salt accumulation. Nevertheless, the data demonstrate that perstraction may have

TABLE 14–8 Production of Acetone-Butanol-Ethanol (ABE) in a Batch Fermentation Operated with Continuous Feeding, and Coupled with a Product Removal Technique

Process	*Total ABE*[1] *(g/l Liquid in Reactor)*	*ABE Yield (g/g)*	*Lactose Utilized (g/l Liquid in Reactor)*	*ABE Productivity* $[g/(l \cdot h)]$
Control[2]	7.0	0.36	19.4	0.07
Continuous feeding				
+ Liquid-liquid extraction (oleyl alcohol)	21.0	0.31	68.6	0.13
+ Perstraction (oleyl alcohol)	38.0	0.24	157.5	0.13
+ Pervaporation	42.0	0.30	140.0	0.14
+ Gas stripping	45.2	0.25	182.5	0.17

[1] Calculated based on the ABE recovered.

[2] The control was a traditional batch fermentation process without continuous feeding.

potential as a product removal technique to allow the application of continuous feeding to the ABE process.

14.7.4 Pervaporation

Pervaporation is a membrane process whereby liquids selectively diffuse across a solid membrane, and are then evaporated and removed by a gas stream or by applying a vacuum (Ennis et al. 1986a). Important parameters to be considered for the technique include high-membrane selectivity for ABE, and high flux across the membrane. Ideally, there should be low selectivity for nutrients and acetic and butyric acids. Friedl et al. (1991) have applied the technique to a continuous packed bed reactor using bonechar-immobilized cells of *C. acetobutylicum* P262, in an attempt to improve the lactose utilization while retaining the high productivity of this fermentation system. ABE was removed from the reactor effluent that was then mixed with a concentrated whey permeate feed medium (lactose 130 g/l) and recycled to the reactor. A reactor productivity of 3.5 g/(l · h) was achieved, at a lactose utilization of 98%. Acid losses through the membrane were low, resulting in a high ABE yield of 0.39 g/g lactose used. The system was stable, and shows considerable promise for commercial application.

Pervaporation has also been investigated during a batch fermentation with continuous feeding of concentrated whey permeate, under conditions similar to those described above for liquid-liquid extraction and perstraction (Table 14–8). The reactor productivity was 0.14 g/(l · h), while the lactose used was 140 g/l, but fermentation ceased after 13 days of operation (N. Qureshi, A. Friedl, and I.S. Maddox, unpublished observations).

14.7.5 Gas Stripping

The use of gas stripping to remove inhibitory solvents during the ABE fermentation process, followed by condensation for product recovery, has been demonstrated by Ennis et al. (1986c). During a batch fermentation process, using nitrogen as the stripping gas, significant increases were achieved in sugar utilization rate and ABE productivity, presumably due to alleviation of product inhibition. Subsequently, Ennis et al. (1987) used gas stripping as a between-stages product removal technique in a two-stage continuous fermentation process employing packed bed reactors of bonechar-immobilized cells. The effluent from stage 1, which contained sugar but in which fermentation had ceased due to product inhibition, was gas stripped prior to being passed to stage 2, where further fermentation occurred. This product removal technique was suggested to be superior to others since only volatile products are removed, and not nutrients or reaction intermediates.

Recently, gas stripping has been integrated with a continuous fermentation process employing a fluidized bed reactor of bonechar-immobilized cells (Qureshi and Maddox 1991a). The reactor effluent was stripped of ABE using nitrogen gas, and was recycled to the reactor in a mixture with a concentrated whey permeate feed stream. At a reactor productivity of 5.1 g/(l · h), and a feed lactose concentration of 130 g/l, a lactose utilization of 75% was achieved. ABE in the stripping gas was condensed to give a concentration of 54 g/l, which could subsequently be recovered by distillation.

Gas stripping has also been applied to a batch fermentation process employing a continuous feeding technique, in a similar manner to that described above for liquid-liquid extraction, perstraction, and pervaporation (Qureshi and Maddox 1990b). The gas used in this case, however, was the mixture of hydrogen and carbon dioxide produced during the fermentation, rather than nitrogen gas. The process was commenced in the batch mode and was allowed to continue until ABE reached 4–5 g/l. At this stage, gas stripping was started and concentrated whey permeate was fed. The ABE concentration achieved was 45.2 g/l at a productivity of 0.17 g/(l · h) (Table 14–8). The fermentation ceased after 266 h, possibly due to accumulation of mineral salts in the culture.

Table 14–8 summarizes the experiments performed to develop a batch fermentation process coupled with continuous feeding. Of the four-product removal techniques investigated to alleviate product inhibition, gas stripping allowed the highest ABE production and productivity.

14.8 ECONOMICS

An economic study of ABE production from whey permeate has been described recently (Qureshi and Maddox 1991b). The conclusions drawn, based on a whey permeate price of $116 (U.S.) per tonne solids (New Zealand Dairy Board, personal communication), include:

1. A plant using traditional batch fermentation coupled with product recovery by distillation would not be economically viable.
2. A plant using a fluidized bed reactor of bonechar-immobilized cells, coupled with pervaporation to remove and concentrate ABE, would give a product price of $0.62 (U.S.) per liter for a plant capacity of 900 m^3 whey permeate per day. This price reduces to $0.20 (U.S.) per liter if whey permeate can be obtained at zero cost.

14.9 CONCLUSION

Whey is a relatively difficult substrate for industrial fermentation processes, both because of its low sugar concentration and because lactose is not a preferred carbon source for many clostridia. However, recent developments in the technology of continuous fermentation have allowed highly productive systems to be achieved for the ABE process. The integration of product removal techniques with these productive systems can overcome the problem of product inhibition while simultaneously reducing the costs of product recovery. The results achieved to date demonstrate that an economically viable process is technically possible.

REFERENCES

Alam, S., Stevens, D., and Bajpai, R. (1988) *J. Ind. Microbiol.* 2, 359–364.

Archer, R.H., Larsen, V.F., and McFarlane, P.N. (1982) *Comparison of Three High Rate Anaerobic Digester Designs for the Treatment of Whey, Annual Meeting*, Feb., Hamilton, New Zealand, Institute of Professional Engineers of New Zealand, Wellington, New Zealand.

Bahl, H., Gottwald, M., Kuhn, A., et al. (1986) *Appl. Environ. Microbiol.* 52, 169–172.

Clarke, K.G., Hansford, G.S., and Jones, D.T. (1988) *Biotechnol. Bioeng.* 32, 538–544.

Compere, A.L., and Griffith, W.L. (1979) *Dev. Ind. Microbiol.* 20, 509–517.

Ennis, B.M. (1987) Ph.D. thesis, Massey University, Palmerston North, New Zealand.

Ennis, B.M., and Maddox, I.S. (1985) *Biotechnol. Lett.* 7, 601–606.

Ennis, B.M., and Maddox, I.S. (1987a) *Biotechnol. Bioeng.* 29, 329–334.

Ennis, B.M., and Maddox, I.S. (1987b) *NZ J. Dairy Sci. Technol.* 22, 75–81.

Ennis, B.M., and Maddox, I.S. (1989) *Bioproc. Eng.* 4, 27–34.

Ennis, B.M., Gutierrez, N.A., and Maddox, I.S. (1986a) *Proc. Biochem.* 21, 131–147.

Ennis, B.M., Maddox, I.S., and Schoutens, G.H. (1986b) *NZ J. Dairy Sci. Technol.* 21, 99–109.

Ennis, B.M., Marshall, C.T., Maddox, I.S., and Paterson, A.H.J. (1986c) *Biotechnol. Lett.* 8, 725–730.

Ennis, B.M., Qureshi, N., and Maddox, I.S. (1987) *Enzyme Microbial Technol.* 9, 672–675.

Friedl, A., Qureshi, N., and Maddox, I.S. (1991) *Biotechnol. Bioeng.* 38, 518–527.

Gapes, J.R., Larsen, V.F., and Maddox, I.S. (1983) *J. Appl. Bacteriol.* 55, 363–365.

Gutierrez, N.A. (1985) M.Tech. thesis, Massey University, Palmerston North, New Zealand.

Gutierrez, N.A., and Maddox, I.S. (1987a) *Can. J. Microbiol.* 33, 82–84.

Gutierrez, N.A., and Maddox, I.S. (1987b) *Appl. Environ. Microbiol.* 53, 1924–1927.

Hancock, K.R. (1988) Ph.D. thesis, Massey University, Palmerston North, New Zealand.

Hancock, K.R., Rockman, E., Young, C.A., et al. (1991) *J. Bacteriol.* 173, 3084–3095.

Hobman, P.G. (1984) *J. Dairy Sci.* 67, 2630–2653.

Kanchanatawee, S., and Maddox, I.S. (1990a) in *Fermentation Technologies: Industrial Applications* (Yu, P.L., ed.), pp. 173–177, Elsevier Applied Science, London.

Kanchanatawee, S., and Maddox, I.S. (1990b) *J. Ind. Microbiol.* 5, 277–282.

Kavanagh, J.A. (1975) *NZ J. Dairy Sci. Technol.* 10, 132–136.

Larsen, V.F., and Maddox, I.S. (1987) in *Bioenvironmental Systems*, vol. IV, (Wise, D.L., ed.), pp. 37–74, CRC Press, Boca Raton, FL.

Maddox, I.S. (1980) *Biotechnol Lett.* 2, 493–498.

Maddox, I.S. (1982) *Biotechnol. Lett.* 4, 23–28.

Maddox, I.S. (1989) *Biotechnol. Genet. Eng. Rev.* 7, 189–220.

Maddox, I.S., Gapes, J.R., and Larsen, V.F. (1981) in *Chemeca 81: 9th Australasian Conference on Chemical Engineering*, 4 Sept., Christchurch, New Zealand, pp. 535–542, Institute of Professional Engineers of New Zealand, Wellington, New Zealand.

Mawson, A.J. (1987) *Australian J. Biotechnol.* 1 (3), 64–66.

McKay, L., Miller, A., Sandine, W.E., and Elliker, P.R. (1970) *J. Bacteriol.* 102, 804–809.

Meade, R.E., Rodgers, N.E., and Pollard, H.L. (1944) U.S. patent no. 2433232.

O'Brien, D.J., Panzer, C.C., and Eisele, W.P. (1990) *Biotechnol. Prog.* 6, 237–242.

Prescott, S.C., and Dunn, C.G. (1959) *Industrial Microbiology*, pp. 250–284, McGraw-Hill Book Co., New York.

Qureshi, N., and Maddox, I.S. (1987) *Enzyme Microb. Technol.* 9, 668–671.

Qureshi, N., and Maddox, I.S. (1988) *Bioproc. Eng.* 3, 69–72.

Qureshi, N., and Maddox, I.S. (1990a) *J. Chem. Technol. Biotechnol.* 48, 369–378.

Qureshi, N., and Maddox, I.S. (1990b) in *Proceedings of 18th Australasian Chemical Engineering Conference*, 27–30 Aug., pp. 659–665, Institute of Professional Engineers of New Zealand, Wellington, New Zealand.

Qureshi, N., and Maddox, I.S. (1991a) *Bioproc. Eng.* 6, 63–69.

Qureshi, N., and Maddox, I.S. (1991b) *Aust. J. Biotechnol.* 5, 56–59.

Qureshi, N., Paterson, A.H.J., and Maddox, I.S. (1988) *Appl. Microbiol. Biotechnol.* 29, 323–328.

Saier, Jr., M.H. (1985) *Mechanisms and Regulation of Carbohydrate Transport in Bacteria*, Academic Press, New York.

Schoutens, G.H., Nieuwenhuizen, M.C.H., and Kossen, N.W.F. (1984) *Appl. Microbiol. Biotechnol.* 19, 203–206.

Schoutens, G.H., Nieuwenhuizen, M.C.H., and Kossen, N.W.F. (1985) *Appl. Microbiol. Biotechnol.* 21, 282–286.

Short, J.L., and Doughty, R.K. (1977) *NZ J. Dairy Sci. Technol.* 12, 156–159.

Spivey, M.J. (1978) *Proc. Biochem.* 13, 2–5.

Stevens, D., Alam, S., and Bajpai, R. (1988) *J. Ind. Microbiol.* 3, 15–19.

Tang, I.C., Yang, S.T., and Okos, M.R. (1988) *Appl. Microbiol. Biotechnol.* 28, 138–143.

Voget, C.E., Mignone, C.F., and Ertola, R.J. (1985) *Biotechnol. Lett.* 7, 607–610.

Voget, C.E., Mignone, C.F., and Ertola, R.J. (1987) *Biotechnol. Lett.* 9, 421–424.

Welsh, F.W., and Veliky, I.A. (1984) *Biotechnol. Lett.* 6, 61–64.

Welsh, F.W., and Veliky, I.A. (1986) *Biotechnol. Lett.* 8, 43–46.

Wix, P., and Woodbine, M. (1958) *Dairy Sci. Abstr.* 20, 537–548.

Yu, P.L., Smart, J.B., and Ennis, B.M. (1987) *Appl. Microbiol. Biotechnol.* 26, 254–257.

Zall, R.R. (1984) *J. Dairy Sci.* 67, 2621–2629.

Zall, R.R., Kuipers, A., Muller, L.L., and Marshall, K.R. (1979) *NZ J. Dairy Sci. Technol.* 14, 79–85.

CHAPTER 15

Molecular Biology of Nitrogen Fixation in the Clostridia

Jiann-Shin Chen
John L. Johnson

Nitrogen fixation is the conversion of atmospheric nitrogen (N_2) to a form that plants can use, and nitrogen in forms useful to plants is a major limiting nutrient in crop production. In order to maintain the agricultural productivity to support the current world population, the application of industrial nitrogen fertilizer is essential. Nevertheless, over half of the input of fixed nitrogen into the world's soil and water still is supplied by nitrogen-fixing bacteria (Postgate 1987; Sprent and Sprent 1990). Biological nitrogen fixation is an ability restricted to some 50 genera of prokaryotes, which encompass both free-living organisms and symbioses or associations between nitrogen-fixing bacteria and higher plants (Sprent and Sprent 1990).

The symbiosis between *Rhizobium* and the legumes is the basis of some well-known farming practices that increase agricultural productivity. However, free-living nitrogen-fixing bacteria, which include clostridia, also make

This work was supported by competitive grants from the U.S. Department of Agriculture and by projects from the Commonwealth of Virginia.

an important contribution, especially in poor soils or where fertilizers are not used (Postgate 1987; Cannon et al. 1988).

Nitrogen-fixing bacteria belong to diverse taxonomic groups, ranging from obligate aerobes to obligate anaerobes. The anaerobic genus *Clostridium* contains several nitrogen-fixing species (Rosenblum and Wilson 1949; Jordan and McNicol 1979), which may be expected to play a role in the cycling of nitrogen in O_2-free environments (Rice and Paul 1972; Mishustin and Yemtsev 1973).

Clostridium pasteurianum, which was the first free-living nitrogen-fixing organism isolated in pure culture (Winogradsky 1895; McCoy et al. 1930; Witz et al. 1967) and with which the first demonstration of consistent cell-free nitrogen fixation was obtained (Carnahan et al. 1960; Burris 1988), has been extensively used in the study of the biochemistry and physiology of nitrogen fixation. The progress of modern studies on the biochemistry of nitrogen fixation parallels the work with this obligate anaerobe, and essentially all of the biochemical studies on *C. pasteurianum* have been done using strain W5 (ATCC 6013), which was originally obtained from Winogradsky (McCoy et al. 1930).

Although the nitrogenase of *C. pasteurianum* W5 has been extensively characterized (see references in Chen et al. 1986; Morgan et al. 1988; Bolin et al. 1990), genetic studies on the nitrogen-fixation system of this anaerobe were not actively pursued until the past several years after the nitrogen-fixation (*nif*) genes of *Klebsiella pneumoniae* had been cloned. Genetic studies with nitrogen-fixing clostridia are still hampered by a lack of effective gene transfer and gene replacement procedures. In this chapter, the structure and organization of the *nif* genes of *C. pasteurianum* W5, whose nitrogen-fixing system has several unique features, will be described.

15.1 NITROGENASE AND BIOLOGICAL NITROGEN FIXATION

Biological nitrogen fixation is catalyzed by the enzyme nitrogenase, which reduces N_2 to ammonia (NH_3) in the following reaction

$$N_2 + 8\,e^- + 8\,H^+ + n\,\text{MgATP} \longrightarrow 2\,NH_3 + H_2 + n\,\text{MgADP} + n\,\text{Pi}$$

Electrons for the reaction are transferred to nitrogenase from an electron donor with a low oxidation-reduction potential (e.g., ferredoxin or flavodoxin). Hydrolysis of MgATP to MgADP and Pi provides energy to drive this reaction. Hydrogen gas is an undesirable but apparently obligatory by-product of N_2 reduction by nitrogenase (Simpson and Burris 1984). H_2 production by nitrogenase and the inhibition of nitrogenase activity by H_2 affect the efficiency of biological nitrogen fixation and hence are potential targets for the improvement of this process. Nitrogenases from different organisms differ in their sensitivity toward H_2 inhibition, with the nitrogenase from *C.*

pasteurianum less sensitive to H_2 as an inhibitor than nitrogenases from other organisms (Guth and Burris 1983).

Three types of nitrogenase have been identified to date:

1. The molybdenum-containing nitrogenase, which is also called the conventional nitrogenase and the most extensively characterized among the three types. It will be referred to as nitrogenase in this chapter.
2. The vanadium-containing nitrogenase (Eady et al. 1990).
3. A third type that lacks significant amounts of molybdenum and vanadium (Bishop et al. 1990).

The presence of an alternative nitrogenase (molybdenum independent) in *C. pasteurianum* has been suggested (see Section 15.8).

15.2 NITROGEN-FIXATION (*nif*) GENES

The facultative anaerobe *K. pneumoniae* is a free-living nitrogen-fixing organism. Because *K. pneumoniae* can be manipulated genetically by many techniques developed with *Escherichia coli*, it was the first organism to be used in a thorough analysis of the genetics of nitrogen fixation (Streicher et al. 1971; Ausubel and Cannon 1980; Roberts and Brill 1981). Mutants that cannot grow with N_2 as a nitrogen source were isolated, and distinct genetic loci identified in these mutants were assigned as nitrogen-fixation (*nif*) genes. Later, additional *nif* genes of *K. pneumoniae* were established or inferred by reversion of *Mu* insertions, cloning, deletion, and DNA-sequence analysis (Roberts and Brill 1981; Brooks et al. 1985; Arnold et al. 1988).

nif genes of *K. pneumoniae* form a cluster spanning a region of about 24 kilobases (kb) near the operator end of the histidine (*his*) biosynthesis operon. Totally, 20 *nif* genes have been identified in *K. pneumoniae* (Figure 15–1),

K. pneumoniae

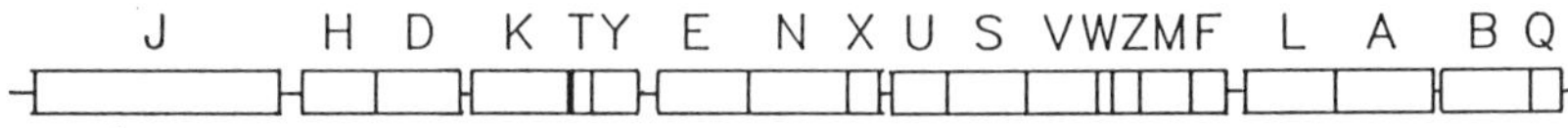

A. vinelandii

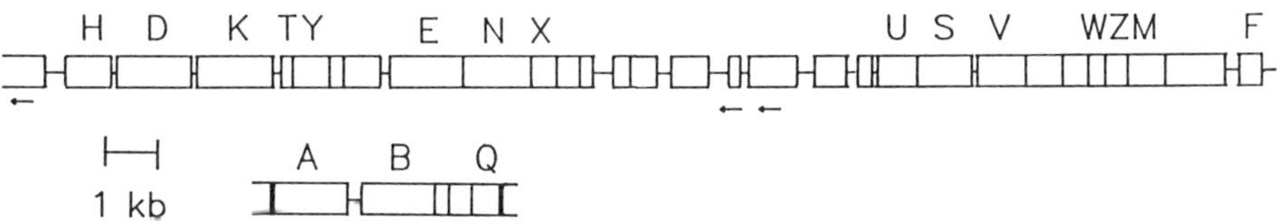

FIGURE 15–1 Organization of *nif* genes and flanking open reading frames (ORFs) (boxed) of *Klebsiella pneumoniae* and *Azotobacter vinelandii*. The direction of transcription is to the right except where indicated by an arrow.

and it is likely that all of the structural and regulatory genes that are exclusively required for nitrogen fixation are located in this cluster. Table 15–1 lists the proposed function of the *nif* genes, and a function is yet to be established for *nifT*, *nifY*, *nifU*, *nifS*, *nifW*, or *nifZ*. Some of the genes in the *nif* regulon seem to encode polypeptides that are not necessary to achieve significant levels of acetylene reduction or nitrogen-fixation activity, at least under standard conditions of growth in the laboratory (Roberts and Brill 1981). The function of these nonessential genes will be difficult to elucidate.

nif genes of other organisms are assigned on the basis of their sequence similarity to known *K. pneumoniae nif* genes. Thus, 18 *nif* genes have been assigned in the free-living, obligate aerobe *Azotobacter vinelandii* (see Figure 15–1). However, the organization of *nif* genes differ between *A. vinelandii* and *K. pneumoniae*, with the *nifABQ* genes occurring separately from the major *nif* cluster in *A. vinelandii* (Bennett et al. 1988; Joerger and Bishop 1988; Jacobson et al. 1989). Furthermore, additional open reading frames

TABLE 15–1 Proposed Function of Nitrogen-Fixation (*nif*) Genes Occurring in *Klebsiella pneumoniae*[1]

Gene	*Proposed Function*
J	Electron transport: pyruvate:flavodoxin oxidoreductase
H	Fe protein subunit of nitrogenase; synthesis of the iron-molybdenum cofactor
D	α-Subunit of the MoFe protein of nitrogenase
K	β-Subunit of the MoFe protein of nitrogenase
T	Unknown
Y	Unknown
E	Synthesis of the iron-molybdenum cofactor
N	Synthesis of the iron-molybdenum cofactor
X	Negative regulation
U	Unknown
S	Unknown
V	Synthesis of the iron-molybdenum cofactor: homocitrate synthase
W	Formation or accumulation of active MoFe protein
Z	Formation or accumulation of active MoFe protein
M	Maturation of the Fe protein of nitrogenase
F	Electron transport: flavodoxin
L	Transcriptional regulation: repression of *nif* genes other than *nifLA*
A	Transcriptional regulation: expression of *nif* genes other than *nifLA*
B	Synthesis of the iron-molybdenum cofactor
Q	Transport and processing of molybdenum

[1] General references: Arnold et al. (1988). Specific references: *nifX*, Gosink et al. (1990); *nifWZ*, Paul and Merrick (1989); *nifLA*, Buck (1990).

(ORFs) are present that separate *nif* genes in *A. vinelandii* (see Figure 15–1). Several of these ORFs have *nif*-related properties and have been numbered (Jacobson et al. 1989). It remains to be determined if the apparently *nif*-related ORFs found in *A. vinelandii* occur elsewhere on the *K. pneumoniae* chromosome.

15.3 CLONING OF NITROGENASE GENES FROM *CLOSTRIDIUM PASTEURIANUM*

Genetic studies with *C. pasteurianum* have been hampered by a lack of useable genetic markers and procedures for genetic manipulations. Since nitrogenase structural genes are conserved among nitrogen-fixing organisms (Mazur et al. 1980; Ruvkun and Ausubel 1980), the cloned *nifHDK* genes of *K. pneumoniae* became useful probes for the detection and isolation of *nif* genes from other organisms. The initial cloning of nitrogenase structural genes from *C. pasteurianum* used *K. pneumoniae nifHD* genes on fragment A3 as a probe (Chen et al. 1986). However, because of a significant difference in codon usage in nitrogenase genes between *C. pasteurianum* and *K. pneumoniae*, an exact match over 10 nucleotides in length is rare between *nifHDK* genes of these two organisms, that made the initial detection of *nif* genes in *C. pasteurianum* difficult (Chen et al. 1986).

Under more stringent hybridization conditions (50% formamide, 42°C), *Eco*RI-digested *C. pasteurianum* DNA showed only one very faint band around 3.8 kb when a DNA fragment containing *K. pneumoniae nifHDK* genes was used as the probe. Under less-stringent conditions, DNA fragments containing either *K. pneumoniae nifHDK* genes (10% formamide) or *nifHD* genes (25% formamide) detected six hybridizing bands (~10, ~7, 3.8, 2.6, 2.1, and 1.7 kb) in *Eco*RI-digested *C. pasteurianum* DNA (Chen et al. 1986). The less-stringent conditions also yielded a false positive band with *Hin*dIII-digested lambda phage DNA or *Eco*RI-digested DNA from the non-nitrogen-fixing *C. perfringens*.

Because the 3.8-kb fragment was the strongest among the weakly hybridizing *C. pasteurianum* bands, it was selected for cloning. DNA sequence analysis of the insert in recombinant plasmid pCP114 established the presence of the clostridial gene (*nifH1*) encoding the iron protein subunit and a portion of the gene (*nifD*) encoding the α subunit of molybdenum-iron (MoFe) protein. Surprisingly, another ORF (*nifH2*), which encodes a polypeptide similar to the *nifH1* gene product, was found upstream to the *nifH1* gene (Figure 15–2; see Section 15.8). Furthermore, another *nifH*-like sequence (*nifH3*) was identified on a 2.6-kb *Eco*RI fragment, which also hybridized to the *Klebsiella* nitrogenase genes (Chen et al. 1986). The presence of multiple *nifH*-like sequences in *C. pasteurianum* is an intriguing property of this organism (see Section 15.8).

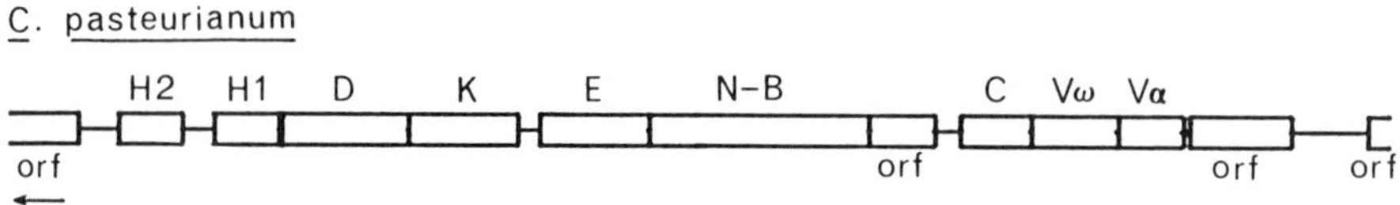

FIGURE 15–2 Organization of *nif* genes and flanking open reading frames (ORFs) (boxed) in a major *nif* cluster of *C. pasteurianum*. The direction of transcription is to the right except where indicated by an arrow.

Following the cloning of the *nifH1* gene and part of the *nifD* gene from *C. pasteurianum*, the remaining part of the *nifD* gene and additional *nif* genes have been cloned and sequenced by walking along the chromosome. The organization of the *Clostridium nif* genes is shown in Figure 15–2. The salient features of these *nif* genes are described below.

15.4 ORGANIZATION OF *nif* GENES IN *CLOSTRIDIUM PASTEURIANUM*

Identified *nif* genes and flanking ORFs that occur in a cluster in *C. pasteurianum* are shown in Figure 15–2. In a region spanning 11.6 kb of DNA, genes that correspond to the *nifH*, *nifD*, *nifK*, *nifE*, *nifN*, *nifB*, and *nifV* genes of *K. pneumoniae* and other organisms have been located. The nitrogenase structural genes (*nifH1*, *nifD*, and *nifK*) of *C. pasteurianum* were positively identified from a comparison of the deduced amino acid sequences with protein sequences (Chen et al. 1986; Wang et al. 1988b). Other *nif* genes (*nifE*, *nifN-B*, *nifVω*, and *nifVα*) of *C. pasteurianum* were identified through comparison of the deduced amino acid sequences with those of the *nif* gene products of *A. vinelandii*, *K. pneumoniae*, and other organisms. In addition, an ORF preceding the *nifVω* gene of *C. pasteurianum* was found to encode a sequence similar to the predicted *E. coli chlJ* gene product, possibly involved in molybdenum transport, and it has been designated the *nifC* gene (Wang et al. 1990).

Despite a conservation in the amino acid sequences encoded by the corresponding genes, the organization of *nif* genes in *C. pasteurianum* differs significantly from those found in two Gram-negative organisms (e.g., Arnold et al. 1988; Joerger and Bishop 1988; Jacobsen et al. 1989). The *nif* genes of *C. pasteurianum* seem to be organized according to their functions. Clustering of genes involved in the synthesis of the iron-molybdenum cofactor (FeMo-cofactor) is apparent. In *C. pasteurianum*, three consecutive groups (operons) of *nif* genes show the following arrangements: (1) the first group consists of structural genes (*nifH1DK*) for nitrogenase; (2) the second group contains genes (*nifEN-B*) required for FeMo-cofactor synthesis, where the *nifN* and the *nifB* genes are fused into one gene *nifN-B*; and (3) the third group contains a gene (*nifC*) possibly involved in molybdenum transport and additional genes (*nifVωVα*) for FeMo-cofactor synthesis, where the polypeptide encoded by

TABLE 15–2 Properties of *nif* Genes and Open Reading Frames (ORFs) in a Cluster in *C. pasteurianum* W5

		Deduced Polypeptide		
Gene	*Coding Region*	*No. of Amino Acid Residues*	M_r	*Estimated pI*
H2	1395–2213	272	29,580	4.93
H1	2623–3444	273	29,666	4.85
D	3486–5087	533	58,990	5.57
K	5086–6462	458	50,115	5.08
E	6747–8117	456	50,397	5.64
N–B	8140–10929	929	103,088	6.09
ORF	10892–11734	280	30,516	8.70
C	12084–12944	286	31,681	9.30
Vω	12962 14020	352	41,574	5.52
Vα	14030–14839	269	29,862	5.28
ORF	14953–16224	423	47,010	7.46

the *nifV* gene is encoded by two separate genes. ORFs with unknown functions are also present in groups 2 and 3. Properties of *nif* genes and ORFs in a contiguous region in *C. pasteurianum* (Figure 15–2) are listed in Table 15–2.

15.5 STRUCTURAL GENES FOR NITROGENASE

The three structural genes, *nifH1*, *nifD*, and *nifK*, constitute a transcription unit (Wang et al. 1988a). The *nifH1* gene encodes the iron protein, whereas the *nifD* and the *nifK* genes encode, respectively, the α and β subunits of the MoFe protein. The amino acid sequence deduced from the *nifH1* gene matches perfectly the protein sequence (Tanaka et al. 1977; Chen et al. 1986). Amino acid sequences deduced from the *nifD* and the *nifK* genes have several differences from the respective protein sequences (Hase et al. 1984; Wang et al. 1988b). *C. pasteurianum* also has five *nifH*-like genes (*nifH2* through *nifH6*) that have been cloned and sequenced (see Section 15.8).

15.5.1 The Iron Protein

The iron protein of *C. pasteurianum* has 273 amino acid (aa) residues, which is about 20 residues shorter than most *nifH*-encoded sequences studied to date (Normand and Bousquet 1989). The difference in the length of the polypeptides resides in their carboxyl termini (Chen et al. 1986). Recent results suggest that the inactivity of the heterologous complex between the *C. pasteurianum* iron protein and the *A. vinelandii* MoFe protein cannot be attributed solely to the difference in the carboxyl-terminal region (Jacobson et al. 1990).

The amino acid sequences encoded by *nifH* genes of diverse bacterial groups are highly related. It is thus interesting to consider the phylogeny of *nifH*—whether it reflects the ancestry of the bacteria that now carry the gene or it must involve lateral transfers of the gene between distant branches of the evolutionary tree. Phylogenetic trees based on the sequence of the 16S ribosomal RNA of major bacterial groups are presumed to represent the evolutionary relationships of the basic bacterial genomes (Fox et al. 1980; Woese et al. 1990). Analyses based on eight *nifH*-encoded sequences (Hennecke et al. 1985) or 28 *nifH*-encoded sequences (Young 1990) and their 16S rRNA suggest a concurrent evolution of *nifH* and 16S rRNA. It is possible to account for the present distribution of the *nifH* genes by postulating that they are all descended lineally from a gene in the common ancestor of all bacteria by means of several gene duplication events followed by divergence and loss in some lineages (Young 1990).

The *nifH* genes may be divided into four classes (Young 1990). Class A consists of *C. pasteurianum nifH3*, *A. vinelandii anfH*, and *Methanococcus thermolithotrophicus nifH1* (see Section 15.8). Class B encompasses most *nifH* genes, including the *vnfH* gene. Class C consists of the distinctive *nifH* genes (except *nifH3*) of *C. pasteurianum*. The first and second halves of the *nifH* of *Desulfovibrio gigas* seem to belong to, respectively, class C and class B. The *nifH* genes of methanogens form class D.

The *nifH*-encoded sequence of *D. gigas* (Kent et al. 1989a) is only one residue longer than that of *C. pasteurianum*. These two polypeptides have an association coefficient (S_{AB}) of 0.71, which is only slightly higher than the next highest value (0.69) found between the polypeptides encoded by *C. pasteurianum nifH1* and *A. vinelandii nifH* genes (with the extended carboxyl-terminal region of *A. vinelandii* NifH excluded from comparison). However, the *C. pasteurianum nifH1* and the *D. gigas nifH* genes encode long segments (up to 28 residues) of identical amino acid sequences, which include a region (residues 56 to 70) that contains a unique gap of two positions when the two sequences are compared with a *nifH* consensus sequence. There is an 80% identity in the first 160 residues between the two sequences. Besides their phylogenetic relationship, it should also be informative to compare properties of nitrogenases purified from the two organisms, as the *C. pasteurianum* nitrogenase has several unique properties (Chen et al. 1986; Wang et al. 1988b).

15.5.2 The Molybdenum-Iron Protein

The *nifD* gene is located 41 base pairs (bp) downstream from the *nifH1* gene. There are several remarkable features about the *nifD* gene and its product: (1) it uses the less common start codon GUG; (2) its stop codon (UAA) overlaps with the start codon (AUG) for the *nifK* gene by one base; and (3) there is an extra stretch (in comparison to the other *nifD*-encoded sequences) of about 50 aa residues in the 380 to 430 region and an apparent periodicity in amino acid sequence and hydrophobicity is observed in this stretch of residues (Wang et al. 1988b).

The *nifK* gene product of *C. pasteurianum* is about 60 aa residues shorter than the protein deduced from the *nifK* gene of other organisms, and the shortened region is mostly at the amino terminus (Wang et al. 1988b). The shortness of the *C. pasteurianum nifK* gene may be related to the proximity between the *nifD* and the *nifK* genes, which overlap by one base.

A comparison of the molecular weight of polypeptides encoded by the *nifD* and *nifK* genes of five organisms (Wang et al. 1988b) plus those of *K. pneumoniae* (Ioannidis and Buck 1987; Holland et al. 1987) shows that the *nifD* gene product (M_r, 58,900) of *C. pasteurianum* is the largest among the six, whereas the *nifK* gene product (M_r, 50,115) of *C. pasteurianum* is the smallest among the six. The molecular weight of the MoFe protein is between 220,000 to 230,000 for the six organisms. It thus appears that the overall size of the MoFe protein must be conserved. For the MoFe protein of *C. pasteurianum*, the extra stretch of about 50 aa residues in the *nifD* gene product and the shortened amino terminal region for about 50 aa residues in the *nifK* gene product may be related properties.

The *nifD* and *nifK* gene products share a significant degree of sequence similarity in the region surrounding three conserved cysteine residues (Ioannidis and Buck 1987; Wang et al. 1988b, and references cited therein), which are potential ligands for the iron-sulfur clusters (the P clusters) (Kent et al. 1989b; Dean et al. 1990; Kent et al. 1990). An analysis of the predicted secondary structure and hydrophobicity of the region in the two proteins suggests that (1) for the α subunit (the *nifD* gene product), the potential ligating thiols may be located near the surface of the subunit, but they will be within the subunit-subunit interface; and (2) for the β subunit (the *nifK* gene product), the potential ligating thiols may occur near the surface of the holo MoFe protein, and the iron protein interacting site is hence on the β subunit (Wang et al. 1988b). There are also unique structural features found in these regions (and elsewhere) in the *C. pasteurianum* polypeptides that may be related to the distinct properties of the *C. pasteurianum* nitrogenase.

The *nifK* gene is separated by 279 bp from the next ORF (*nifE*). There are thus no genes equivalent to the *nifTY* genes next to the *nifK* gene in *C. pasteurianum*. Between the *nifK* and the *nifE* genes, there are two regions of inverted repeats, involving 16 and 10 bp each, that may form stem-and-loop structures and serve as Rho-independent terminators of transcription.

15.6 GENES INVOLVED IN THE SYNTHESIS OF THE IRON-MOLYBDENUM-COFACTOR OR MOLYBDENUM TRANSPORT

In *K. pneumoniae* and *A. vinelandii*, at least six *nif* gene products are required for the synthesis of the FeMo-cofactor of nitrogenase (for reviews, see Shah et al. 1988; Hinton and Dean 1990). These genes include *nifH*, *nifE*, *nifN*, *nifV*, *nifQ*, and *nifB*. The *nifV* gene appears to code for homocitrate synthase, and homocitrate is a constituent of the FeMo-cofactor (Hoover et al. 1989).

The iron protein (the *nifH* gene product), but not its activity in nitrogen fixation, appears to be required for the synthesis of the FeMo-cofactor (Filler et al. 1986). The *nifE* and *nifN* gene products have amino acid sequences similar to, respectively, those of the *nifD* and *nifK* gene products (Brigle et al. 1987). They form a tetramer with a subunit structure similar to that of the MoFe protein (Paustian et al. 1989). It has been suggested that the NifEN complex serves as a scaffold for the synthesis of the FeMo-cofactor (Brigle et al. 1987) or the MoFe protein-like structure allows NifEN to have redox interactions with the iron protein during the biosynthetic cycle (Paustian et al. 1989). The *nifQ* gene product appears to function at an early step during the synthesis of the FeMo-cofactor, but it is not required for molybdenum uptake (Imperial et al. 1984). No clue is available as to the function of the *nifB* gene product in these two organisms. Thus, the specific functions of five of these gene products remain unknown.

From *C. pasteurianum*, genes corresponding to *nifH*, *nifE*, *nifN*, *nifB*, and *nifV* have been cloned and sequenced. Important differences in the spatial arrangement and in the structure of these genes in *C. pasteurianum* are discussed below.

15.6.1 The Group of *nifE-nifN-B-(ORF)*

The *C. pasteurianum nifE*-encoded polypeptide has 456 aa residues (Wang et al. 1989). It shows an S_{AB} of 0.33 with the *A. vinelandii nifE* gene product and an S_{AB} of 0.3 with the *C. pasteurianum nifD* gene product. Conserved residues are found over the entire length of the polypeptides. Cys-79, Cys-140, and Cys-255 of *C. pasteurianum* NifE correspond to Cys-79, Cys-144, and Cys-271 of *C. pasteurianum* NifD.

The *C. pasteurianum nifN-B* gene occurs 22 bp downstream of the *nifE* gene, with the sequence

(--- GTT TAG) AATATAAGTGGAGGTGAATATA (TTG AAT ---)
nifE *nifN-B*

occurring between the two genes (S.-Z. Wang, J.-S. Chen, and J.L. Johnson, unpublished observations). A potential ribosome-binding site (GGAG) occurs at −13 to −10 nucleotides from the predicted start codon UUG. The *nifN-B* gene has a coding capacity of 929 aa residues (M_r, 103,088); it is so named because the first 480 residues of the predicted polypeptide have sequence similarity to the predicted *nifN* gene product of other organisms, whereas the remaining 449 residues have sequence similarity to the predicted *nifB* gene product of other organisms. The NifN domain has an S_{AB} value of 0.29 with either *A. vinelandii* NifN or *C. pasteurianum* NifK. Cys-17, Cys-42, and Cys-100 of the NifN domain of NifN-B correspond to Cys-23, Cys-48, and Cys-106 of *C. pasteurianum* NifK. The NifB domain of *C. pasteurianum* NifN-B has an S_{AB} value of 0.385 with the predicted *A. vinelandii* NifB (Joerger and Bishop 1988). Twelve Cys residues are conserved between the *C. pasteurianum* NifB domain and *A. vinelandii* NifB.

Using antiserum raised against a 33,000-dalton (Da) polypeptide (corresponding to the carboxyl terminal 298 residues of *C. pasteurianum* NifN-B), western blot analysis detected a band with a M_r, 102,000 in nitrogen fixing but not in ammonia-grown *C. pasteurianum* cells, indicating the presence of an intact NifN-B product (S.-Z. Wang, J.-S. Chen, and J.L. Johnson, unpublished observations). The fusion of the *nifN* and the *nifB* genes into the *nifN-B* gene and the presence of the NifN-B product suggest that there is direct interaction between the *nifB* gene product and the NifEN complex in other organisms.

The carboxyl terminal 143 residues of the predicted NifN-B also has sequence similarity to the predicted *nifX* gene product of other organisms. The function of this *nifX*-like domain and the ORF downstream of *nifN-B* is not known.

15.6.2 The Group of *nifC-nifVω-nifVα-(ORF)*

The first ORF (*nifC*) of this cluster (see Figure 15–2) occurs 330 bp downstream of the ORF of the *nifE-nifN-B-(ORF)* cluster. Inverted repeats, which may form a stem-and-loop structure with a stem of 23 bp, are present 22–77 bp downstream of *nifEN-B-(ORF)*.

The predicted product (286 residues) of the *nifC* gene has sequence similarity to the predicted *E. coli chlJ* gene (Wang et al. 1990). Because the region of the *chlJ* gene cloned and sequenced covers only the carboxyl terminal 200 residues (Johann and Hinton 1987), the comparison is limited to this region, and it has an S_{AB} value of 0.325. The *chlJ* gene of *E. coli* is part of the *chlD* locus, and the *chlD* locus is involved in molybdenum transport. Molybdenum uptake by *C. pasteurianum* and *K. pneumoniae* is coregulated with nitrogen fixation (Elliott and Mortenson 1976; Pienkos and Brill 1981). Sequences similar to the proposed consensus *nif* promoter and upstream sequences of *C. pasteurianum* precede the *nifC* gene. We have thus proposed that the *nifC* gene is involved in molybdenum uptake and is coregulated with nitrogen fixation (Wang et al. 1990).

Although the *nif* cluster of *K. pneumoniae* does not contain a gene equivalent to the *nifC* gene, nitrogen fixation in *K. pneumoniae* requires the function of a *mol* locus, which is equivalent to the *chlD* locus of *E. coli* (Ugalde et al. 1985). Furthermore, Kennedy and Postgate (1977) showed that for the *nif* genes of *K. pneumoniae* to function in *E. coli* grown at low molybdenum concentrations, a functional *chlD* locus is required in *E. coli*. Thus, a *nifC*- or *chlJ*-like gene may be expected to exist in *K. pneumoniae*, although it is not present in the *nif* cluster and it may not be exclusively required for nitrogen fixation.

The two genes *nifVω* and *nifVα* are separated by 9 bp, with the *nifVω* gene occurring 17 bp downstream of the *nifC* gene. The *nifVω* gene has a coding capacity of 352 aa residues, whereas the *nifVα* gene has a coding capacity of 269 aa residues (Wang et al. 1991). These two genes are so named because (1) the carboxyl terminal 195 residues of NifVω are similar to the carboxyl-terminal region of the predicted NifV sequence of *A. vinelandii* and *K. pneumoniae*, and (2) the predicted sequence of NifVα is similar to the amino-terminal region of the predicted NifV of *A. vinelandii* and *K. pneumoniae*. In addition, a plasmid containing the *nifVω* and the *nifVα* genes, but not either alone, restored the NifV phenotype in an *nifV*-deleted strain of *A. vinelandii* (Wang et al. 1991). Studies of the enzymic properties of the products of the *nifVα* and the *nifVω* genes should further our understanding of the role of the *nifV* product in FeMo-cofactor synthesis. The function of the ORF following the *nifVα* gene is not known.

15.7 CODON USAGE IN *nif* GENES OF *CLOSTRIDIUM PASTEURIANUM*

Very biased codon usage in *C. pasteurianum* was first observed in the *nifH* genes (Chen et al. 1986), which is consistent with the fact that *C. pasteurianum* has a very low G + C content (26–28%; Cummins and Johnson 1971) and that nitrogenase is an abundant protein in the cell (Zumft and Mortenson 1973). Table 15–3 compares codon usage between nitrogenase genes (*nifH1DK*) and the other *nif* genes plus the two ORFs that belong to a major *nif* cluster in *C. pasteurianum* (see Figure 15–2).

In nitrogenase genes, codon usage is strongly biased toward codons with A or U at the third position, similar to what was found in the *nifH1* gene. The three codons AGA (arginine), UGU (cysteine), and GAA (glutamate) are each used over 90% of the time. Also, GCU/A (alanine), GGU/A (glycine), CCU/A (proline), ACU/A (threonine), and GUU/A (valine) are used 93.7–100% of the time. Nine codons (CGC, CGA, CGG, AGG, CUG, UCG, CCC, ACC, and ACG) are not used. The *nifH*-like genes (*nifH2* through *nifH6*; see next section) have a codon usage pattern similar to that of the *nifH1* gene (Chen et al. 1986; Wang et al. 1988a), and they are not included in Table 15–3.

The other five *nif* genes (*nifE*, *N-B*, *C*, *Vα*, *Vω*) and two ORFs in the major *nif* cluster are not expected to be as highly expressed as the nitrogenase genes. Nevertheless, codon usage in the other *nif* genes is also very biased. Codon usage for asparagine (AAU), aspartic acid (GAU), histidine (CAU), phenylalanine (UUU), and tyrosine (UAU) is more biased in these genes than in nitrogenase genes (Table 15–3).

The least frequently used codons in the *C. pasteurianum nif* genes covered in Table 15–3 (encompassing 4270 codons) are as follows (with the frequency of use among alternative codons and the number of times used indicated in parentheses): CGC (0.0%; 0), CGA (0.7%; 1), CGG (0.7%; 1), UCG (1.2%; 3), CUC (1.3%; 4), ACG (1.8%; 4), CUG (2.3%; 7), GCG (2.8%; 8), GUC (3.4%; 10), CGU (3.6%; 5), UCC (4.6%; 11), and ACC (5.0%; 11). These infrequently used codons in general agree with those found in *C. acetobu-*

TABLE 15–3 Codon Usage in *nif* Genes of *Clostridium pasteurianum* W5

Amino Acid	Codon	Usage (%)		Amino Acid	Codon	Usage (%)	
		(A)[1]	(B)			(A)	(B)
Arg	CGU	7.5	2.1	Val	GUU	58.2	37.5
	C	0.0	0.0		C	2.2	4.0
	A	0.0	1.0		A	37.4	49.0
	G	0.0	1.0		G	2.2	9.5
	AGA	92.5	83.5	Ile	AUU	26.5	36.3
	G	0.0	12.4		C	19.6	5.7
Leu	UUA	50.6	45.5		A	53.9	58.0
	G	4.7	15.2	Asn	AAU	56.3	89.3
	CUU	30.6	25.4		C	43.8	10.6
	C	1.2	1.3	Asp	GAU	75.0	89.5
	A	12.9	9.4		C	25.0	10.5
	G	0.0	3.1	Cys	UGU	95.5	86.5
Ser	UCU	21.9	27.7		C	4.5	13.5
	C	4.7	4.5	Glu	GAA	97.1	79.5
	A	35.9	26.0		G	2.9	20.5
	G	0.0	1.7	Gln	CAA	71.0	55.9
	AGU	18.8	28.2		G	29.0	44.1
	C	18.8	11.9				

(*continued*)

TABLE 15–3 (continued)

Amino Acid	*Codon*	*Usage (%)* (*A*)[1]	*Usage (%)* (*B*)	*Amino Acid*	*Codon*	*Usage (%)* (*A*)	*Usage (%)* (*B*)
				His	CAU	62.1	81.8
Ala	GCU	52.9	45.8		C	37.9	18.2
	C	1.1	10.8	Lys	AAA	76.6	71.6
	A	42.5	40.9		G	23.4	28.4
	G	3.4	2.5	Phe	UUU	42.9	87.5
Gly	GGU	50.8	34.0		C	57.1	12.5
	C	4.8	7.1	Tyr	UAU	78.8	92.4
	A	42.9	51.3		C	21.2	7.6
	G	1.6	7.6	Met	AUG	100	100
Pro	CCU	36.4	40.5	Trp	UGG	100	100
	C	0.0	10.7	Start	AUG	66.7	85.7
	A	61.8	39.3		GUG	33.3	0.0
	G	1.8	9.5		UUG	0.0	14.3
Thr	ACU	53.4	63.8	Stop	UAA	100	42.9
	C	0.0	7.4		UAG	0.0	42.9
	A	46.6	26.2		UGA	0.0	14.2
	G	0.0	2.7	No. of codons		1268	3002

[1] This compilation includes 4270 codons occurring in two groups of genes: column A (1268 codons), encompassing the three nitrogenase structural genes, *nifH1*, *nifD*, and *nifK*; and column B (3002 codons), encompassing the five genes, *nifE*, *nifN-B*, *nifC*, *nifV*ω, and *nifV*α, and two ORFs within the major *nif* cluster. An additional 1370 codons occurring in *nifH2* through *nifH6* are not included here, which have a codon usage pattern similar to that of *nifH1* (Chen et al. 1986).

tylicum, which has a similarly low G + C content (Youngleson et al. 1989). However, differences also exist. For example, the codon CAC is rarely used in the five *C. acetobutylicum* genes, but it is moderately used (26%) in the *C. pasteurianum* genes. A physiological explanation for such differences may be possible when the codon usage information is available for a wider range of genes covering different levels of expression in these organisms and when their tRNA population is examined. At present, the codon usage information from the two clostridial species is useful for the design of more effective oligonucleotide probes/primers based on amino acid sequence data of proteins isolated from clostridia with a low G + C content (Cary et al. 1990).

15.8 MULTIPLE *nifH*-LIKE GENES IN *CLOSTRIDIUM PASTEURIANUM*

In *C. pasteurianum*, there are five *nifH*-like genes (*nifH2* through *nifH6*) besides *nifH1*, which encodes the Fe protein (Wang et al. 1988a). *nifH2* is upstream of *nifH1* (see Figure 15–2), but the two genes belong to different transcriptional units. The genomic locations of *nifH3* through *nifH6* have not been determined. The length of polypeptides encoded by the *nifH*-like genes differs slightly: *nifH2* and *nifH6* each encodes 272 aa residues, *nifH3* encodes 275 residues, and *nifH1*, *nifH4*, and *nifH5* each encodes 273 residues.

The predicted amino acid sequences of *nifH1* through *nifH6* differ from one another to a different degree. Except for *nifH3* which differs from the others by 95–97 residues (S_{AB} = 0.65), the other five sequences have differences ranging from one (between *nifH2* and *nifH6*), two (between *nifH1* and *nifH5*) to 23 residues (S_{AB} values range from 0.99–0.92). When the nucleotide sequence of the coding region of the *nifH*-like genes is compared with that of *nifH1*, a sequence identity of 68% is found between *nifH3* and *nifH1*, whereas it ranges from 90–96% between the others and *nifH1*. High levels of sequence identity extend to the putative ribosome-binding sites in the sequences. Beyond that point little identity in the 5′ flanking regions is apparent, even between *nifH1* and *nifH5*, which have a very high sequence identity. However, *nifH2* and *nifH6* each occurs in a region of 2482 bp that differ only by two nucleotides. The mechanism(s) through which *C. pasteurianum* acquired the *nifH*-like genes is an intriguing question.

mRNAs for these *nifH*-like genes, with the exception of the *nifH3* gene, have been detected in nitrogen-fixing *C. pasteurianum* cells grown with normal amounts of molybdate (Wang et al. 1988a), but the function of these *nifH*-like genes remains to be determined. There is a striking identity (S_{AB} = 0.82; Figure 15–3) between the predicted amino acid sequence of the *nifH3* gene of *C. pasteurianum* and the sequence of the *anfH* gene of *A. vinelandii* (Joerger et al. 1989). The *anfH* gene encodes the iron protein-equivalent of the Mo- and V-independent nitrogenase of *A. vinelandii* (Bishop et al. 1990). Because *C. pasteurianum* can grow under nitrogen-fixing conditions with no

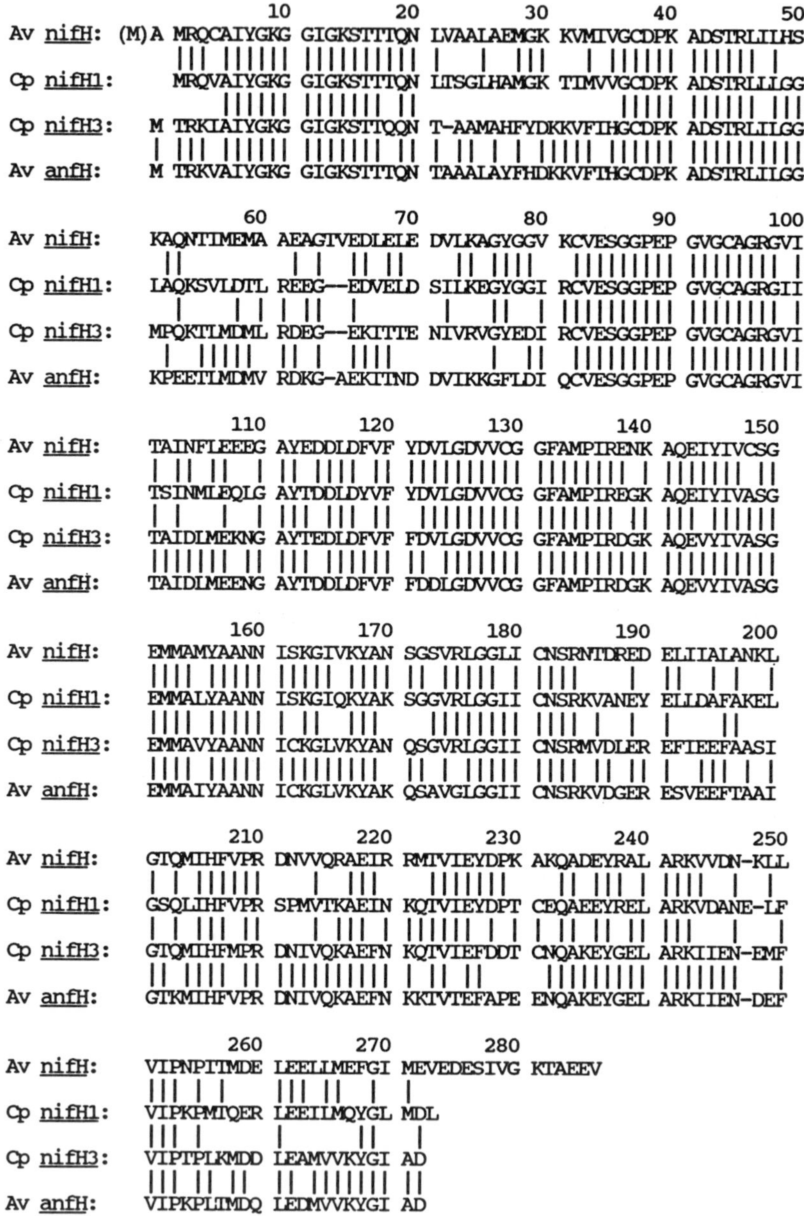

FIGURE 15–3 Comparison of amino acid sequences encoded by the *nifH1* gene (Tanaka et al. 1977; Chen et al. 1986) and the *nifH3* gene (Wang et al. 1988a) of *C. pasteurianum* (Cp) with those encoded by the *nifH* gene (Hausinger and Howard 1982; Brigle et al. 1985) and the *anfH* gene (Joerger et al. 1989) of *Azotobacter vinelandii* (Av). Amino acid residues are represented by the single letter code. Identical residues between neighboring sequences are connected by vertical bars.

or low molybdenum (Cardenas and Mortenson 1975; Dilworth et al. 1987; J.-S. Chen and J.L. Johnson, unpublished data), it is likely that the *C. pasteurianum nifH3* gene product is involved in a Mo-independent alternative nitrogenase system.

The deduced amino acid sequence of the *C. pasteurianum nifH3* gene has an S_{AB} value of 0.70 when compared with the corresponding region of the deduced sequence of the *nifH1* gene (encoding 284 aa residues) of *Methanococcus thermolithotrophicus* (Sibold and Souillard 1988). A separate class, which consists of the *anfH* gene of *A. vinelandii*, the *nifH3* gene of *C. pasteurianum*, and the *nifH1* gene of *M. thermolithotrophicus*, has been proposed when 28 *nifH* sequences are divided into four classes (Young 1990).

15.9 *nif* PROMOTER AND UPSTREAM SEQUENCES

The transcription start site for *nifH1*, *nifE*, and *nifH*-like genes (except *nifH3*) has been determined (Wang et al. 1988a). Apparently similar nucleotide sequences were observed in the −100 region relative to the transcription start. The proposed consensus sequence is ATCAatat-N_{6-10}-ATGGattc, which appears to be orientation-independent. The −35 and −10 regions are more heterogeneous, with sequences similar to TTG and TATAAT preceding some *nif* operons (Wang et al. 1988a). The promoter region and upstream sequence for the *nif* operons of *C. pasteurianum* are thus different from those found in *K. pneumoniae* and several other Gram-negative bacteria (Gussin et al. 1986). Further studies will be needed to define the structure of the control elements for the *nif* operons in *C. pasteurianum*.

15.10 GENETIC TECHNIQUES FOR THE STUDY OF NITROGEN FIXATION IN CLOSTRIDIA

Genetic techniques that are applicable to studies on clostridia are limited because few mutations that can serve as genetic markers are available for any species and effective methods for transferring DNA between *Clostridium* cells are still being developed. However, much progress has been made in recent years in the development of vectors and gene transfer techniques for clostridia (Young et al. 1989).

A mutant of *C. pasteurianum* defective in nitrogen fixation, apparently lacking an active MoFe protein, was obtained by mutagenesis with *N*-methyl-*N′*-nitro-*N*-nitrosoguanidine (Simon and Brill 1971). Mutants of *C. pasteurianum* altered in the biosynthesis of granulose (Robson et al. 1974), in sensitivity toward inhibitors of the membrane H^+-ATPase (Clarke et al. 1982), and in nutritional requirements, antibiotic resistance, sporulation, and resistance to UV light (Daldal 1985) have also been obtained by chemical and UV mutagenesis.

A procedure for the regeneration of *C. pasteurianum* protoplasts with frequencies of up to 10% reversion has been reported (Minton and Morris 1983). However, no transformation of *C. pasteurianum* has been reported, and the species appears recalcitrant in terms of gene transfers under laboratory conditions. Our recent results (unpublished observations) suggest that mobilization of DNA into *C. pasteurianum* cells is feasible, but a replication origin that is active in this species remains to be identified.

The genus *Clostridium*, consisting of more than 100 species, is one of the largest and more diverse genera of prokaryotes. Although several *Clostridium* spp. are capable of fixing nitrogen, only *C. pasteurianum* has been used extensively in the study of nitrogen fixation within this genus. Furthermore, nearly all of the studies on *C. pasteurianum* have been done using strain W5 (ATCC 6013). The use of a single strain in research tends to constrain the development of genetic techniques for the species. We have thus carried out a search for additional *C. pasteurianum* strains based on DNA homology measurements, and studies on their nitrogen-fixing properties and their amenability to genetic manipulations are in progress. The availability of additional strains and the identification of replication origins active in this species should accelerate the study of the genetics of nitrogen fixation in clostridia.

15.11 CONCLUDING REMARKS

The endospore-forming, obligately anaerobic, and usually Gram-positive clostridia represent a physiologically distinct group of nitrogen-fixing bacteria. The extensively characterized nitrogenase of *C. pasteurianum* has several unique properties, some of which may be useful for the design of a more-efficient nitrogenase. For example, the enzyme's relative insensitivity to inhibition by hydrogen gas is a useful feature that probably has evolved because this nitrogenase has to operate under a high concentration of H_2, which is a major metabolic product of this anaerobe. Thus, the nitrogen-fixation system of clostridia could represent a unique source of useful features for the understanding and improvement of biological nitrogen fixation.

Genetic studies of nitrogen fixation in clostridia have only been performed with the species *C. pasteurianum*. Because of a current lack of suitable genetic tools for this anaerobe, *nif* genes of *C. pasteurianum* have been identified on the basis of their sequence similarity to known *nif* genes of *K. pneumoniae* and *A. vinelandii*, with the exception of the *nifV* genes of *C. pasteurianum* that were also shown to function in *A. vinelandii*. The high degree of conservation in the amino acid sequence of *nif* gene products has allowed the identification of a major *nif* cluster in *C. pasteurianum*.

The organization of *nif* genes in *C. pasteurianum* is distinct. Unique features in the organization of *nif* genes in *C. pasteurianum* include:

1. An overlap between the *nifD* and the *nifK* genes.
2. The fusion of the *nifN* and the *nifB* genes into the *nifN-B* gene.

3. The division of the *nifV* gene into the *nifV*α and the *nifV*ω genes.
4. The presence of the *nifC* gene upstream of the *nifV*ω gene.
5. The presence of multiple copies of *nifH*-like genes.

The fused and split *nif* genes offer a unique opportunity for studies of the structure-function relationship in these gene products. Also, there appears to be a clustering of genes involved in the synthesis of the FeMo-cofactor and possibly in molybdenum uptake in *C. pasteurianum*, which, in conjunction with a proximity of these genes to the *nifH1DK* genes, suggests that in *C. pasteurianum* the *nif* genes are more tightly arranged than in some other diazotrophs. However, it remains to be determined how expression of *nif* genes is regulated in this anaerobe and whether the regulatory features are more primitive or more efficient than those operating in organisms that seem to contain a larger assembly of *nif* genes.

It is unknown how many genes are exclusively required for nitrogen fixation in *C. pasteurianum*, how different growth conditions might affect the expression of any specific set of *nif* genes, and whether or not all *nif* genes form a cluster or occur in nearby regions on the chromosome. To answer these questions, mutants of *C. pasteurianum* defective in nitrogen fixation will need to be isolated and their lesions mapped with respect to known genetic loci. Efficient methods for mobilizing DNA into *C. pasteurianum* cells and replicons active in this species will be needed for this pursuit, especially for a functional analysis of the *nif* genes and their control elements. The results should shed new light on the evolution and distribution of *nif* genes and will contribute to attempts to improve the efficiency of biological nitrogen fixation.

REFERENCES

Arnold, W., Rump, A., Klipp, W., Priefer, U.B., and Puhler, A. (1988) *J. Mol. Biol.* 203, 715–738.

Ausubel, F.M., and Cannon, F.C. (1980) *Cold Spring Harbor Symp. Quant. Biol.* 45, 487–499.

Bennett, L.T., Cannon, F., and Dean, D.R. (1988) *Mol. Microbiol.* 2, 315–321.

Bishop, P.E., MacDougal, S.I., Wolfinger, E.D., and Shermer, C.L. (1990) in *Nitrogen Fixation: Achievements and Objectives* (Gresshoff, P.M., Roth, L.E., Stacey, G., and Newton, W.E., eds.), pp. 789–795, Chapman & Hall, Ltd., New York.

Bolin, J.T., Ronco, A.E., Mortenson, L.E., et al. (1990) in *Nitrogen Fixation: Achievements and Objectives* (Gresshoff, P.M., Roth, L.E., Stacey, G., and Newton, W.E., eds.), pp. 117–124, Chapman & Hall, Ltd., New York.

Brigle, K.E., Newton, W.E., and Dean, D.R. (1985) *Gene* 37, 37–44.

Brigle, K.E., Weiss, M.C., Newton, W.E., and Dean, D.R. (1987) *J. Bacteriol.* 169, 1547–1553.

Brooks, S.J., Imperial, J., and Brill, W.J. (1985) in *Nitrogen Fixation and CO_2 Metabolism* (Ludden, P.W., and Burris, J.E., eds.), pp. 65–74, Elsevier, New York.

Buck, M. (1990) in *Nitrogen Fixation: Achievements and Objectives* (Gresshoff, P.M., Roth, L.E., Stacey, G., and Newton, W.E., eds.), pp. 451–457, Chapman & Hall, Ltd., New York.
Burris, R.H. (1988) in *Nitrogen Fixation: Hundred Years After* (Bothe, H., de Bruijn, F.J., and Newton, W.E., eds.), pp. 21–30, Fisher, Stuttgart, Germany.
Cannon, F.C., Beynon, J., Hankinson, T., et al. (1988) in *Nitrogen Fixation: Hundred Years After* (Bothe, H., de Bruijn, F.J., and Newton, W.E., eds.), pp. 735–740, Fisher, Stuttgart, Germany.
Cardenas, J., and Mortenson, L.E. (1975) *J. Bacteriol.* 123, 978–984.
Carnahan, J.E., Mortenson, L.E., Mower, H.F., and Castle, J.E. (1960) *Biochim. Biophys. Acta* 44, 520–535.
Cary, J.W., Petersen, D.J., Bennett, G.N., and Papoutsakis, E.T. (1990) *Ann. N.Y. Acad. Sci.* 589, 67–81.
Chen, K.C.-K., Chen, J.-S., and Johnson, J.L. (1986) *J. Bacteriol.* 166, 162–167.
Clarke, D.J., Kell, D.B., Morely, C.D., and Morris, J.G. (1982) *Arch. Microbiol.* 131, 81–86.
Cummins, C.S., and Johnson, J.L. (1971) *J. Gen. Microbiol.* 67, 33–46.
Daldal, F. (1985) *Arch. Microbiol.* 142, 93–96.
Dean, D.R., Setterquist, R.A., Brigle, K.E., et al. (1990) *Mol. Microbiol.* 4, 1505–1512.
Dilworth, M.J., Eady, R.R., Robson, R.L., and Miller, R.W. (1987) *Nature (London)* 327, 167–168.
Eady, R.R., Pau, R., Lowe, D.J., and Luque, F.J. (1990) in *Nitrogen Fixation: Achievements and Objectives* (Gresshoff, P.M., Roth, L.E., Stacey, G., and Newton, W.E., eds.), pp. 125–133, Chapman & Hall, Ltd., New York.
Elliott, B.B., and Mortenson, L.E. (1976) *J. Bacteriol.* 127, 770–779.
Filler, W.A., Kemp, R.M., Ng, J.C., et al. (1986) *Eur. J. Biochem.* 160, 371–377.
Fox, G.E., Stackebrandt, E., Hespell, R.B., et al. (1980) *Science* 209, 457–463.
Gosink, M.M., Franklin, M.M., and Roberts, G.P. (1990) *J. Bacteriol.* 172, 1441–1447.
Gussin, G.N., Ronson, C.W., and Ausubel, F.M. (1986) *Annu. Rev. Genet.* 20, 567–591.
Guth, J.H., and Burris, R.H. (1983) *Biochemistry* 22, 5111–5122.
Hase, T., Wakabayashi, S., Nakano, T., Zumft, W.G., and Matsubara, H. (1984) *FEBS Lett.* 166, 39–43.
Hausinger, R.P., and Howard, J.B. (1982) *J. Biol. Chem.* 257, 2483–2490.
Hennecke, H., Kaluza, K., Thoeny, B., et al. (1985) *Arch. Microbiol.* 142, 342–348.
Hinton, S.M., and Dean, D.R. (1990) *Crit. Rev. Microbiol.* 17, 169–188.
Holland, D., Zilberstein, A., Zamir, A., and Sussman, J.L. (1987) *Biochem. J.* 247, 277–285.
Hoover, T.R., Imperial, J., Ludden, P.W., and Shah, V.K. (1989) *Biochemistry* 28, 2768–2771.
Imperial, J., Ugalde, R.A., Shah, V.K., and Brill, W.J. (1984) *J. Bacteriol.* 158, 187–194.
Ioannidis, I., and Buck, M. (1987) *Biochem. J.* 247, 287–291.
Jacobson, M.R., Brigle, K.E., Bennett, L.T., et al. (1989) *J. Bacteriol.* 171, 1017–1027.
Jacobson, M.R., Cantwell, J.S., and Dean, D.R. (1990) *J. Biol. Chem.* 265, 19429–19433.

Joerger, R.D., and Bishop, P.E. (1988) *J. Bacteriol.* 170, 1475–1487.
Joerger, R.D., Jacobson, M.R., Premakumar, R., Wolfinger, E.D., and Bishop, P.E. (1989) *J. Bacteriol.* 171, 1075–1086.
Johann, S., and Hinton, S. (1987) *J. Bacteriol.* 169, 1911–1916.
Jordan, D.C., and McNicol, P.J. (1979) *Can. J. Microbiol.* 25, 947–948.
Kennedy, C., and Postgate, J.R. (1977) *J. Gen. Microbiol.* 98, 551–557.
Kent, H.M., Buck, M., and Evans, D.J. (1989a) *FEMS Microbiol. Lett.* 61, 73–78.
Kent, H.M., Ioannidis, I., Gormal, C., Smith, B.E., and Buck, M. (1989b) *Biochem. J.* 264, 257–264.
Kent, H.M., Baines, M., Gormal, C., Smith, B.E., and Buck, M. (1990) *Mol. Microbiol.* 4, 1497–1504.
Mazur, B.J., Rice, D., and Haselkorn, R. (1980) *Proc. Natl. Acad. Sci. USA* 77, 186–190.
McCoy, E., Fred, E.B., Peterson, W.H., and Hastings, E.G. (1930) *J. Inf. Dis.* 46, 118–137.
Minton, N.P., and Morris, J.G. (1983) *J. Bacteriol.* 155, 432–434.
Mishustin, E.N., and Yemtsev, V.T. (1973) *Soil Biol. Biochem.* 5, 97–107.
Morgan, T.V., Mortenson, L.E., McDonald, J.W., and Watt, G.D. (1988) *J. Inorg. Biochem.* 33, 111–120.
Normand, P., and Bousquet, J. (1989) *J. Mol. Evol.* 29, 436–447.
Paul, W., and Merrick, M. (1989) *Eur. J. Biochem.* 178, 675–682.
Paustian, T.D., Shah, V.K., and Roberts, G.P. (1989) *Proc. Natl. Acad. Sci. USA* 86, 6082–6086.
Pienkos, P.T., and Brill, W.J. (1981) *J. Bacteriol.* 145, 743–751.
Postgate, J. (1987) *Nitrogen Fixation*, 2nd ed., Edward Arnold, London.
Rice, W.A., and Paul, E.A. (1972) *Can. J. Microbiol.* 18, 715–723.
Roberts, G.P., and Brill, W.J. (1981) *Annu. Rev. Microbiol.* 35, 207–235.
Robson, R.L., Robson, R.M., and Morris, J.G. (1974) *Biochem. J.* 144, 503–511.
Rosenblum, E.D., and Wilson, P.W. (1949) *J. Bacteriol.* 57, 413–414.
Ruvkun, G.B., and Ausubel, F.M. (1980) *Proc. Natl. Acad. Sci. USA* 77, 191–195.
Shah, V.K., Hoover, T.R., Imperial, J., et al. (1988) in *Nitrogen Fixation: Hundred Years After* (Bothe, H., de Bruijn, F.J., and Newton, W.E., eds.), pp. 115–120, Fisher, Stuttgart, Germany.
Sibold, L., and Souillard, N. (1988) in *Nitrogen Fixation: Hundred Years After* (Bothe, H., de Bruijn, F.J., and Newton, W.E., eds.), pp. 705–710, Fisher, Stuttgart, Germany.
Simon, M.A., and Brill, W.J. (1971) *J. Bacteriol.* 105, 65–69.
Simpson, F.B., and Burris, R.H. (1984) *Science* 224, 1095–1097.
Sprent, J.I., and Sprent, P. (1990) *Nitrogen Fixing Organisms: Pure and Applied Aspects*, 2nd ed., Chapman & Hall, Ltd., London.
Streicher, S., Gurney, E., and Valentine, R.C. (1971) *Proc. Natl. Acad. Sci. USA* 68, 1174–1177.
Tanaka, M., Haniu, M., Yasunobu, K.T., and Mortenson, L.E. (1977) *J. Biol. Chem.* 252, 7093–7100.
Ugalde, R.A., Imperial, J., Shah, V.K., and Brill, W.J. (1985) *J. Bacteriol.* 164, 1081–1087.
Wang, S.-Z., Chen, J.-S., and Johnson, J.L. (1988a) *Nucleic Acids Res.* 16, 439–454.
Wang, S.-Z., Chen, J.-S., and Johnson, J.L. (1988b) *Biochemistry* 27, 2800–2810.
Wang, S.-Z., Chen, J.-S., and Johnson, J.L. (1989) *Nucleic Acids Res.* 17, 3299.

Wang, S.-Z., Chen, J.-S., and Johnson, J.L. (1990) *Biochem. Biophys. Res. Commun.* 169, 1122–1128.

Wang, S.-Z., Dean, D.R., Chen, J.-S., and Johnson, J.L. (1991) *J. Bacteriol.* 173, 3041–3046.

Winogradsky, M.S. (1985) *Arch. Sci. Biol. (St. Petersburg)* 3, 297–352.

Witz, D.F., Detroy, R.W., and Wilson, P.W. (1967) *Arch. Microbiol.* 55, 369–381.

Woese, C.R., Kandler, O., and Wheelis, M.L. (1990) *Proc. Natl. Acad. Sci. USA* 87, 4576–4579.

Young, J.P.W. (1990) in *Nitrogen Fixation: Achievements and Objectives* (Gresshoff, P.M., Roth, L.E., Stacey, G., and Newton, W.E., eds.), p. 840, Chapman & Hall, Ltd., New York.

Young, M., Minton, N.P., and Staudenbauer, W.L. (1989) *FEMS Microbiol. Rev.* 63, 301–326.

Youngleson, J.S., Jones, D.T., and Woods, D.R. (1989) *J. Bacteriol.* 171, 6800–6807.

Zumft, W.G., and Mortenson, L.E. (1973) *Eur. J. Biochem.* 35, 401–409.

CHAPTER
16

The Potential of Thermophilic Clostridia in Biotechnology

Francesco Canganella
Juergen Wiegel

Among thermophilic anaerobes, *Clostridium thermocellum* is the oldest validly published organism. It is also the one that has in the past and present drawn much attention for its potential in industrial applications. The interest in these thermophiles first arose in the late 1920s and surfaced again in the late 1970s due to the oil crisis. Since 1970, the number of thermophiles has increased tremendously. For instance, in 1979, there were five clostridial thermophiles known: *C. thermocellum* (since 1950 as pure culture), *C. thermosaccharolyticum*, *C. tartarivorum* (now regarded as a strain of *C. thermosaccharolyticum*) and *C. thermoaceticum*, all with a T_{max} around 65–69°C (moderate thermophiles), and *C. thermohydrosulfuricum* (with a T_{max} 75–78°C). About 12 years later, there were 16 thermophilic purified clostridial thermophiles (Table 16–1) [for a more detailed overview see Wiegel 1991].

The authors gratefully acknowledge the support of their work on thermophilic clostridia from the Dep. of Energy (DE-FG09-86ER13614 and DE-FG09-89ER-14059). F.C. is a recipient of a Doctorate of Research Fellowship on "Microbial Ecology" from the Italian Ministry of University and Scientific Research.

TABLE 16–1 Thermophilic Clostridia (1991)

Bacteria	*Temperature Optimum (°C)*	*pH Optimum*	*Isolated from*
C. thermohydrosulfuricum	69	6.9–7.5	Ubiquitous (thermobiotic + mesobiotic)
C. thermosaccharolyticum	55	7.0–8.5	Ubiquitous (soil, mud, canned food)
C. stercorarium	65	7.3	Compost (Ireland)
C. thermocopriae	60	6.5–7.3	Camel feces, compost, hot springs (Japan)
C. thermoaceticum	58	6.6–6.8	Horse manure (USA)
C. thermoautotrophicum	58	5.8–6.5	Ubiquitous
C. thermocellum	60	6.8–7.2	Ubiquitous (soil, mud, sewage)
C. thermolacticum	60–65	6.4–7.8	Cattle manure
C. thermobutyricum	57	6.8–7.1	Horse manure (USA)
C. fervidus	68	7.5–7.7	Geothermal springs (New Zealand)
C. josui	45	7.0	Compost (Thailand)
C. thermosulfurogenes	60	5.5–6.5	Algal mats (Yellowstone N.P., USA)
C. 'thermoamylolacticum'	NR[1]	NR	NR
C. thermosuccinogenes	58–72	7.6	Beet pulp, soil, cow manure
C. thermopalmarium	55	6.6	Palm wine (Senegal)
C. thermopapyrolyticum	59	NR	Riverside mud

[1] NR = Not reported.

Several of these species have been isolated only from hot springs, some only from mesobiotic environments or man-made environments and others appear to be truly ubiquitous (e.g., *C. thermohydrosulfuricum*). Interestingly, 8 out of the 16 species have been isolated at least once from manure or feces (Table 16–1, including those listed as ubiquitous; J. Wiegel, unpublished observations). However, besides the biotechnological application in sewage treatment plants, only a very few organisms have been promising enough to be specifically included in patents (Table 16–2). As evident from Table 16–2, most applications are either for alcohol production or as a source for thermostable hydrolytic enzymes. It should be stressed, despite all of the advantages these organisms offer, that none of them is used in an established large industrial production process. The reasons for this are manyfold; one of the reasons is that the renewed interest in their application is relatively recent, in some other cases the reason is that the concentration of products is too low and the process of concentration and purification (e.g., distillation) of the final product is often too energy demanding. One more reason is the lack of substantial molecular and genetic work and experiences with these organisms. This becomes more evident when the reader is comparing the various chapters of this book with the paragraph "Genetic studies" included in this chapter on thermophilic clostridia. This lack will surely disappear in the future. The authors are convinced that several thermophilic clostridial enzymes will be found increasingly important in industrial applications in the near future.

The genus *Clostridium* contains species representing a wide range of DNA base pair compositions, from a little above 20 up to 58 mol% guanidine and cytosine. Thermophilic clostridia range from about 32 mol% (using melting kinetics, *C. thermosulfurogenes*) to 58 mol% G + C (chemical analysis, *C. thermoautotrophicum*) (Wiegel 1991). Thus, it is no surprise that research is currently underway to reclassify again (more than 30 years ago the attempt by Prévot failed) these divergent microorganisms in more phylogenetic coherent groups. However, the taxonomic reclassification needs to include the nonspore-forming thermophiles. As discussed in more detail elsewhere (Wiegel 1991), several thermophiles, described as nonspore-forming and placed in new genera, according to the present definition of the genus *Clostridium* [spore-forming, nonsulfate-reducing (dissimilatory), strictly anaerobic, Gram-type positive organisms], might be just sporulation-"repressed" spore formers, and they should have been included in this genus. In fact, Cook et al. (1991) recently demonstrated that *Thermoanaerobium brockii* is a spore former. Another example is that, for instance, *C. thermohydrosulfuricum* strain 39E (isolated from Yellowstone National Park hot springs) is similar to *Thermoanaerobacter ethanolicus* (isolated from similar samples); presently, Zeikus et al. (personal communication) are proposing to transfer this spore-forming strain 39E into the so far nonspore-forming species *Thermoanaerobacter ethanolicus*. The genus *Thermoanaerobacter* already contains a spore-forming species, *T. finii*. However, the classification into the genus was based only on physiological and morphological properties; no DNA:DNA homology tests have been carried out between these two species.

TABLE 16–2 Some Patents Related to the Use of Thermophilic Clostridia in Processes of Biotechnological Interest (1977–1991)

Bacteria	*Patents*	*Applications*
C. thermocellum	Belg. 882,162	Ethanol production
	US 4,395,543	Ethanol, organic acids production
	US 4,400,470	Ethanol, enzymes production
	JP 60,224,495	Orange peels fermentation
	JP 61,128,898	Cellulose saccharification
	JP 61,224,983	Ethanol production
	JP 62,11,097	Alcohols, organic acids production
	JP 01,67,180	Ethanol production
C. thermoaceticum	EP 43,071	Acetate production
	EP 120,369	Acetate production
	EP 120,370	Acetate production
	DE 3,705,272	Bioconversions
	US 4,814,273	Acetate production
	JP 02,261,389	5-Aminolevulinic acid production
	JP 02,261,391	Corrinoids production
C. thermohydrosulfuricum	EP 258,050	α-Amylase, pullulanase production
	DE 3,603,348	α-Amylase production
	DE 3,727,987	Protein carriers production
	Fr. 2,549,854[1]	Ethanol production
	US 4,352,885	Oxidoreductase production
	US 4,400,470	Ethanol, enzymes production
	JP 61,224,983	Ethanol production
	WO 86 01,831	Pullulanase, glucoamylase production
	WO 86 01,833	Ethanol, enzymes production
C. thermosaccharolyticum	EP 276,806	Amylases, pullulanase production
	US 4,568,644	Ethanol production
	JP 61,209,584	Ethanol production
	JP 61,212,282	Butyric acid production
	JP 61,289,891	Alcohols production
	JP 62,00,267	Butyric acid production
	JP 62,11,097	Alcohols, organic acids production
	JP 62,166,882	Butanol production
	JP 62,289,189	Butanol production
	JP 63,03,794	Bioconversions
	JP 63,102,687	Butanol production
	JP 63,254,986	Alcohols production
	JP 01,23,890	Bioconversions
C. thermosulfurogenes	US 4,572,898	Pectinases production
	US 4,737,459	Amylolytic enzyme production
	US 4,814,267	Enzymes production
	WO 86 01,832	β-Amylase production
	WO 86 01,833	Ethanol, enzyme production
C. 'thermoamylolyticum'	US 4,536,477	Glucoamylase production
Clostridium sp.	EP 268,193	Amylolytic enzyme

[1] *C. thermohydrosulfuricum* 39E (soon to be renamed *Thermoanaerobacter ethanolicus*, see text).

In short, it is difficult at this time to distinguish between those thermophiles described as nonclostridial organisms and those described as clostridial species. Due to the aim of this book, related thermophiles were not included in this chapter. Thus, the reader interested in physiologically close-related thermophiles (which may or may not be a *Clostridium* sp.) is referred to a recent review on (eu)bacterial thermophiles (Wiegel 1991) and original literature cited therein.

16.1 LIQUID FUEL PRODUCTION

During the energy crisis of the early 1970s, the shortage of oil and increase in the price of its feedstocks led to intensified research of new energy sources and valuable substitutes for the more traditional liquid fuels and chemicals.

Short-chain alcohols produced via fermentation from cheap and renewable sources have been considered the most appropriate choices as alternatives to oil. Although the energy problem does not seem as urgent as it did 20 years ago, the concern is great, throughout the world, about the diminishing oil supply and need for new energy sources. For this reason, basic and applied research have been focusing on the potential use of microorganisms to produce liquid fuels, and in this respect, thermophilic microorganisms are considered useful and valuable candidates.

Ethanol can well be considered the most common fermentation product along with CO_2, H_2, acetate, and lactate in numerous species of thermophilic saccharolytic clostridia (Holdeman et al. 1977; Wiegel 1980; Gottshalk et al. 1981). The ability to produce alcohols (primarily ethanol) is quite wide spread among thermophilic clostridia. A possible reason for this ability is that alcohols are pH-neutral compounds and that at elevated temperatures their evaporation minimizes the accumulation of these products (boiling point of the azeotrope is 78.2°C).

All saccharolytic fuel-producing clostridia investigated to date use the fructose-biophosphate pathway (Embden-Meyerhof pathway) during glycolysis in the metabolism of hexose sugars (Reaction 16.1).

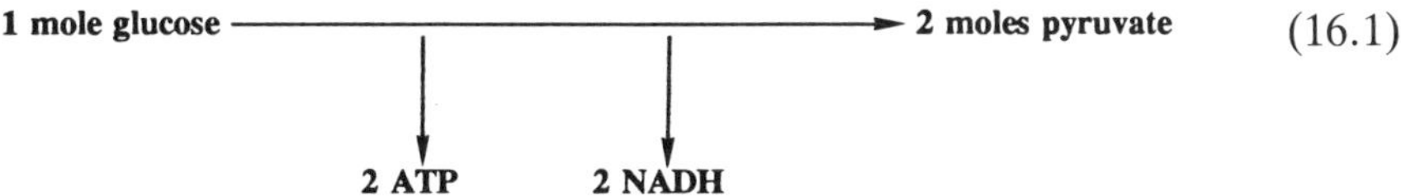

(16.1)

Some species (e.g., *C. thermohydrosulfuricum* and *C. thermobutyricum*) are also able to use pentose sugars via the hexose monophosphate pathway (Warburg-Dickens pathway) using the transaldolase and transketolase reactions (Reaction 16.2)

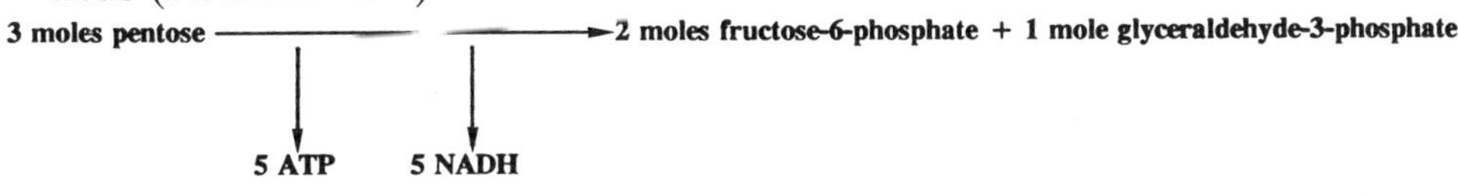

(16.2)

In all solvent-producing clostridia, the pyruvate generated by glycolysis can be cleaved into CO_2 and acetyl-coenzyme A (CoA) by means of a pyruvate ferredoxin (Fd) oxidoreductase with the production of a reduced Fd (Reaction 16.3). The oxidized Fd can then be regenerated by the action of a hydrogenase, with the resulting production of H_2. Most saccharolytic clostridia are able to use, at least to a small extent, the conversion of pyruvate to lactate involving a lactic dehydrogenase. Some mesophilic clostridia are also able to produce formate; in this case, pyruvate is cleaved to acetyl-CoA and formate by the pyruvate-formate lyase.

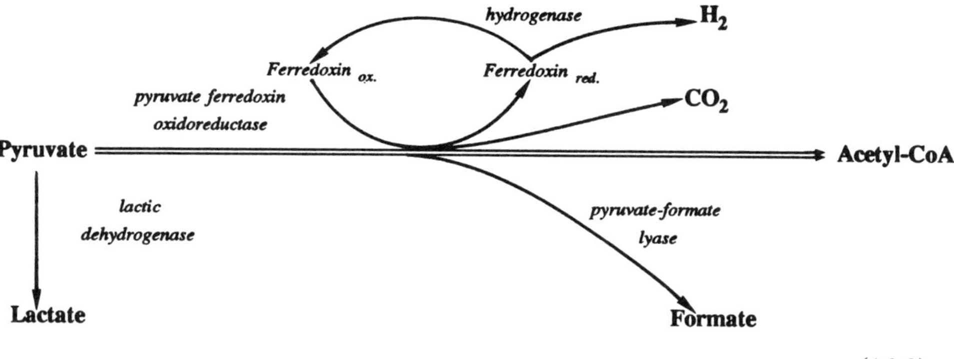

(16.3)

Among thermophilic clostridia, only *C. thermosuccinoges* (Drent et al. 1991) has been reported to produce formate as a fermentation end product. The pattern of product formation has not been described yet, but it is assumed to be similar to that of the mesophilic clostridia.

In general, fermentation strategies and type of end products obtained can be considered adaptations of the cell to sustain an effective growth under particular and selective environmental conditions. A direct relationship exists between the H_2 produced and the amount of ATP generated. Acetyl-CoA only can be used for energy formation if the reducing equivalents formed during the Embden-Meyerhof (EM) pathway are oxidized in some other way, e.g., by ethanol formation or after condensation to acetoacetyl-CoA forming butyrate or butanol. An elegant path for the oxidation of reducing equivalents is the formation of H_2 (2 H_2 per acetate and ATP formed) using an Fd-dependent hydrogenase reaction. In other words, the more H_2 released, the less-reduced final products such as ethanol and lactate are formed. If H_2 production is suppressed for any reason, a metabolic shift to the formation of reduced compounds such as ethanol or lactate may be observed.

16.1.1 Ethanol

16.1.1.1 Monocultures. In ethanol production, three microorganisms have been extensively characterized among thermophilic clostridia: *C. thermocel-*

lum, *C. thermohydrosulfuricum*, and *C. thermosaccharolyticum*. *C. thermocellum* has also been studied at the genetic level in relation to ethanol production. Here we just report some of its features in regard to ethanol formation, and the interested reader will find more details on the cellulase in Chapter 13 on cellulolytic microorganisms.

C. thermocellum has been isolated from both mesophilic and thermophilic environments (Wiegel et al. 1979). It is present on decaying materials and often associated with other anaerobes (Wiegel 1980). *C. thermocellum* is able to grow on cellulose as the only carbon source, and a mineral medium has been described (Johnson et al. 1981). The ability to grow and use other sugars has been also well established (Patni and Alexander 1971; Ng et al. 1977; Ng and Zeikus 1982; Brener and Johnson 1984). Final metabolic products are similar to *C. thermohydrosulfuricum* (acetate and ethanol) but different ratios and lower yields of ethanol are usually obtained (Ng and Zeikus 1982). *C. thermocellum* contains a complex of cellulose- and hemicellulose-degrading enzymes that have been well-characterized (Freier 1981; Su et al. 1981; Zertuche and Zall 1982; Bayer et al. 1983; Freier et al. 1988). The degradation of crystalline cellulose is carried out by extracellular cellulases (Lamed et al. 1983; Bayer et al. 1985; Hon-nami et al. 1986). This microorganism is considered a potential candidate for the direct conversion of cellulose to ethanol but the slow growth rate (using cellulose as substrate, the t_d is around 3–6 h), a limited ethanol tolerance, and final ethanol concentrations of less than 3% make its industrial application not attractive.

C. thermosaccharolyticum and *C. thermohydrosulfuricum* are two non-cellulolytic thermophilic clostridia that can produce above 1 mol ethanol/glucose used. Both organisms are versatile in respect to substrate utilization. *C. thermosaccharolyticum* can be forced to produce increased amounts of ethanol above 1 mol ethanol/1 mol glucose. The ratio of final products (ethanol, acetate, butyrate, lactate, CO_2, and H_2) are related to growth conditions and characteristics of the strain. During the vegetative phase of growth, acetate, butyrate, and lactate represent the main fermentation products, but a metabolic shift to ethanol can be observed if cell sporulation is induced (Hsu and Ordal 1970; Wang et al. 1979; Landuyt et al. 1983). Lee and Ordal (1967) showed that high yields of ethanol are produced during spore formation. Hsu (1976), using synchronized cells in the sporulation phase, was able to obtain yields of ethanol up to 2 mol ethanol per mol glucose used. New insights into the regulatory mechanisms of ethanol production have also been provided by biochemical studies. The fructose-1,6 diphosphate dependence of lactate dehydrogenase was recently demonstrated in *C. thermosaccharolyticum* (Vancanneyt et al., unpublished observations; Vancanneyt et al. 1990) and a similar effect has previously been reported in *C. thermohydrosulfuricum* (Germain et al. 1986; Turunen et al. 1987).

C. thermohydrosulfuricum has been known since 1965 (Klaushofer and Parkkinen 1965) but its detailed characterization began only a few years ago (Wiegel et al. 1979). Numerous data are now available on the isolation of

C. thermohydrosulfuricum from different environments (Wiegel et al. 1979; Zeikus 1979; Zeikus et al. 1980), on the range of substrates and fermentation products the optimal temperature and many other parameters such as ethanol production, pH, temperature and the effect of substrate concentration (Hyun and Zeikus 1985b; Parkkinen 1986; Wiegel et al. 1979; Zeikus et al. 1980). Besides the type strain, other isolates have been studied, e.g., *C. thermohydrosulfuricum* strain JW 102 (DSM 567) and *C. thermohydrosulfuricum* 39E (will be renamed as *Thermoanaerobacter ethanolicus* 39E; Zeikus et al., personal communication) strain 39E produces up to 1.9 mol ethanol per mol glucose used (Zeikus et al. 1981, 1983). For the type strain and strain JW 102 it was shown that a specific pH shift from above pH 7.2 to below pH 6.8 was required to obtain ethanol yields above 1 mol ethanol per mol glucose used (Wiegel and Ljungdahl 1979; Ljungdahl et al. 1981; Carreira and Ljungdahl 1983; Wiegel et al. 1983, 1984).

C. thermohydrosulfuricum is, like *C. thermosaccharolyticum*, a very versatile microorganism. It is able to metabolize hexoses, di- and polysaccharides, plus various pentoses such as ribose and xylose. The more the substrates are reduced compared to glucose, the more efficiently they are converted to ethanol, and vice versa (Harden 1901). All substrates metabolized to pyruvate, acetyl-CoA or acetyl phosphate and yielding high amounts of reducing equivalents can theoretically be converted to ethanol. If the H_2 pressure is increased, *C. thermohydrosulfuricum* does not show any metabolic shifts toward ethanol as the higher hydrogen concentration is inhibitory for the cell metabolism (Wiegel et al. 1979). In contrast, the addition of H_2 to a culture of *C. thermocellum* JW20, grown under shaking, increases the formation of ethanol, whereas the shaking of the culture under N_2 with a large head gas space increases the formation of acetate. This effect is due to shaking under an N_2-atmosphere that prevents the H_2-buildup in the culture and thus eases the release of H_2 and thus favors the formation of acetate plus ATP (Freier et al. 1988). A higher ethanol formation in nonshaking cultures degrading cellulose is similarly due to the H_2 produced that builds up in the settled cellulose containing the attached and entrapped cells (Wiegel and Dykstra 1984). Since *C. thermocellum* contains at least one bidirectional H_2-ase (Przybyla, personal communication) elevated H_2 concentrations cause a buildup of reducing equivalents in the cells, leading to the increased formation of the more reduced fermentation product ethanol (Freier et al. 1988).

Two different Fds {one highly thermostable with one [4Fe-4 (or 3)S]-cluster and one less thermostable with two [4Fe-4S]-clusters} were identified and two novel NADP-linked secondary alcohol dehydrogenases (alcohol-aldehyde/ketone oxidoreductase) were demonstrated (J. Wiegel, unpublished observations; Lamed and Zeikus 1981) in the type strain and strain 39E (*T. ethanolicus* 39E). These alcohol dehydrogenases appear to be similar to those characterized in *T. ethanolicus* JW 200 (Bryant et al. 1988) and in *Thermoanaerobacter brockii*. More details about these and other enzymes, suitable for industrial applications, are described below (see under Section 16.3).

16.1.1.2 Cocultures. Most of the research related to the production of liquid fuels by thermophilic microorganisms involves the fermentation of cellulolytic material by mixed cellulolytic cultures.

Among thermophilic clostridia, *C. thermocellum* represents the elective microorganism for ethanol production from cellulose. Numerous studies were conducted to test fermentation yields under different culture conditions (Zertuche and Zall 1982; Duong et al. 1983; Kundu et al. 1983) and to understand the regulatory mechanism of ethanol inhibition (Herrero and Gomez 1981; Herrero 1983). Novel cellulose degraders were isolated in the past (Wiegel et al. 1979; Wiegel 1980) and new species such as *C. stercorarium* (Madden 1983), *C. fervidus* (Patel et al. 1987), *C. josui* (Sukhumavasi et al. 1988), *C. thermolacticum* (Le Ruyet et al. 1985), *C. thermocopriae* (Jin et al. 1988), and *C. thermopapyrolyticum* (Mendez et al. 1991) have been described. *C. stercorarium* is able to grow at pH 5.5, but the doubling time of 16 h represents quite a disadvantage for industrial applications. *C. thermocopriae* is an acetic, butyric, and lactic acid producer but detailed fermentation balances have not been reported. *C. thermopapyrolyticum* is a new species of the thermophilic clostridia isolated from a river bank in Buenos Aires, Argentina, and produces several organic acids such as acetate, butyrate, and lactate in nearly equal amounts. The characterized strain was proposed as a new species due to some structural and metabolic features that were different from the ones previously observed in other cellulolytic and thermophilic clostridia.

In the effort to overcome some of the major problems associated with monocultures such as ethanol inhibition and slow growth rate, experimental cocultures were introduced. Various mixed cultures of thermophilic microorganisms have been studied to test ethanol production under several different conditions (Freier 1981). The general system involves a polymer degrader (or "primary producer") and one or more intermetabolic ("auxiliary") microorganisms, i.e., glucose/cellobiose fermentors (Wiegel and Ljungdahl 1979; Peck and Odom 1981); usually, the distribution of the final products resembles the one of the glucose/cellobiose fermentor. The cellobiose degradation increases the cellulose availability and makes its degradation, by the primary microorganism, more effective (Freier 1981; Wiegel et al. 1986).

Studies reporting the use of thermophilic clostridia in mixed cultures include: *C. thermocellum* LQR1 and *C. thermohydrosulfuricum* 39E (*T. ethanolicus* 39E) (Ng et al. 1981; Zeikus et al. 1981), *C. thermocellum* JW 20 and *T. ethanolicus* JW 200 (Wiegel and Ljungdahl 1979; Ljungdahl and Wiegel 1981), *C. thermohydrosulfuricum* JW 102, *C. thermosaccharolyticum* DMS 571, and *T. brockii* HTD4 (Freier 1981); *C. thermosaccharolyticum* HG-8 and *C. thermocellum* S7-19 (Avgerinos et al. 1981; Duong et al. 1983); and *C. thermocellum* and *Methanobacterium thermoautotrophicum* (Weimar and Zeikus 1977; Zeikus et al. 1983). Zeikus et al. (1981) showed that ethanol yields in different cocultures were usually higher compared to single monocultures, but a study comparing different cocultures of *C. thermocellum* JW 20 with *C. thermohydrosulfuricum*, *T. ethanolicus*, or *T. brockii* showed that the ethanol concentrations and yields can vary significantly (Freier 1981). Experimental cocultures can well represent a model for future fermentations at an industrial scale. Presently, cocultures seem to be the most effective systems for the microbial degradation of cellulose to ethanol at elevated temperatures.

16.1.2 Methanol

Methanol is usually not produced by microorganisms via fermentation. Among thermophilic clostridia, *C. thermosulfurogenes* is the only one to date reported to produce methanol in significant amounts (Schink and Zeikus 1983a, 1983b). In this case, methanol formation was observed only on pectin and not with any of the other carbohydrates tested as substrates. At present, methanol is industrially synthesized quite economically by chemical processes, thus there is no need or interest to produce methanol via fermentation. A potential exploitation for *C. thermosulfurogenes* is in the treatment of agricultural wastes where it might be used in mixed cultures with *C. thermoautotrophicum* and *Methanosarcina*.

16.1.3 Butanol and Higher Alcohols

Propanol, isobutanol, *n*- and iso-pentanol and hexanol are often found as final metabolic products of thermophilic clostridia. These alcohols are usually present in traces and usually formed from amino acids and peptides of the yeast extract or peptones added to the medium.

Isopropanol (1.2 mM) was produced by *C. thermosulfurogenes* after growth on pectin, in the presence of 0.3% yeast extract (Schink and Zeikus 1980, 1983a, 1983b). Among the thermophiles, butanol above 5 mM is produced only by *C. thermosaccharolyticum*; under selective conditions, yields as high as 45 mM butanol were obtained when starch was the carbon source (Freier et al. 1989). Butanol would be a valuable fermentation product since its energy content is higher than ethanol, and, as a liquid fuel, it is much easier to handle. However, at present the concentrations obtained are far too low to be economical. Specialized cell walls and membranes are apparently required in thermophiles to tolerate increased butanol concentrations due to the enhancing effect of elevated temperatures on the action of butanol on the lipids of the cell wall.

16.1.4 Future Research for Alcohol Production

Research on liquid fuels production by thermophilic microorganisms is currently focused on metabolic processes and the development of new fermentation systems. New insights will hopefully arise from the isolation of mutants with appropriate characteristics. Using gamma-ray mutagenesis, Duong et al. (1983) isolated a low-acid producing strain of *C. thermocellum* that formed ethanol with an ethanol to acetate ratio of 8:1 during the fermentation of cellobiose; moreover, spontaneous ethanol-tolerant mutants of *C. thermocellum* were isolated by Herrero and Gomez (1980). A mutant of *C. thermosaccharolyticum*, defective in acetate production and producing higher yields of ethanol than the parental strain, was isolated by Rothstein (1986). The ethanol tolerance is an important feature and further improvements are expected soon regarding this physiological trait. Some authors, in disagreement with the "phosphorylated sugars" hypothesis (Herrero and Gomez 1980; Herrero et al. 1982, 1985), explain the ethanol toxicity as a damage of the cell structure resulting in an inhibition of cellular processes (Ingram and Buttke 1984; Ingram 1986).

More effective fermentation systems are also expected from the improvement of extraction methods (Schlote and Gottschalk 1986; Ennis et al. 1987; Larrayoz and Puigjaner 1987; Jones and Woods 1989). The standard method presently used for the extraction of alcohols produced via fermentation is distillation which, despite recent developments in technical procedures, is still a high energy consuming process. Better extraction technologies will hopefully allow fermentation systems to be more economical in the future and more biotechnologically competitive.

16.2 ORGANIC ACID PRODUCTION

Thermophilic clostridia degrade different organic compounds to produce mixtures of acids, alcohols, CO_2, and H_2. Common organic acids are acetate, butyrate, and lactate, but formate (*C. thermosuccinogenes*), propionate (*C. josui*), and succinate (*C. thermosuccinogenes*) have also been recovered in some cases as final metabolic products. Moreover, valerate, isovalerate, isobutyrate, and caproate are sometimes produced, mainly from the metabolism of the amino acids but usually only in traces. At present, only a few organic acids are generally produced via fermentation in significant amounts (Wise 1983a, 1983b). In the food industry, acetate, citrate, butyrate, and lactate are commonly used but none of these are presently produced by microbial fermentations at elevated temperatures. Acetate, butyrate, and lactate are potentially valuable products of industrial fermentations. Some thermophilic clostridia such as *C. thermosaccharolyticum*, *C. thermoaceticum*, *C. thermoautotrophicum*, and *C. thermobutyricum* are suitable candidates for such applications. As mentioned before, organic acids are presently produced from petrochemicals, gas, or methanol. Since the oil crisis of 1972–1973, unstable prices and a considerable increase in the energy demand led to the reconsideration of using thermophilic microorganisms, especially clostridia, to produce organic acids from renewable sources (Zeikus 1980; Wise 1983a, 1983b; Busche 1985a; Rogers 1986; Wiegel and Ljungdahl 1986). A large reservoir of raw materials is potentially available worldwide from herbaceous crops, and near-term opportunities exist in the bioconversion of cellulosic material contained in urban, industrial, and agricultural wastes. Considering that cellulose and hemicellulose are the two most abundant renewable sources of carbon for industrial fermentations, an efficient utilization of both these lignocellulosic biomass constituents can be retained fundamentally for developing valuable fermentation processes. Thus, present research is trying to obtain more efficient organisms by isolation techniques as well as by genetic approaches.

16.2.1 Acetic Acid

Regarding the production of organic acids via microbial fermentations, more emphasis has been given to acetic acid, due to economical reasons and also to a better knowledge of acetate-producing microorganisms. Acetate is an important feedstock chemical and is used in different industrial processes, e.g., as a fungicide in grain storage, as a food additive, or as a road de-icer. Presently, it is produced from petrochemicals (in the USA, 1.3×10^9 kg per year, U.S. Int. Trade Comm. 1986) or by microbial fermentation from ethanol in a two-step process involving a mesophilic *Acetobacter* spp. or *Gluconobacter oxidans* (*Acetobacter suboxidans*) (Prescott and Dunn 1959; DeLey and Swings 1984).

The U.S. Federal Highway Administration found calcium magnesium acetate (CMA) to be a good, feasible and environmentally safe road de-icer to substitute for rock salt in environmentally sensitive areas (Chollar 1984). If this application is widely accepted, the acetic acid consumption will be at least doubled. Marynowsky et al. (1983), considering that such an increase of production is not possible via chemical routes, suggested the development of processes involving thermophilic clostridia for the anaerobic conversion of biomass to acetate.

Among thermophilic clostridia, homoacetogens are the catalysts of choice for the complete conversion of glucose and xylose to acetate as the only product (Ljungdahl 1983). This microbial group is important in the environment as it represents, in many cases, a link between glycolytic fermenters and methanogens during the mineralization of the biomass (Braun et al. 1979; Wiegel et al. 1981). These bacteria have been extensively studied and much data are available; furthermore, excellent reviews were recently published on the biochemistry of homoacetogens, especially about *C. thermoaceticum* (Ljungdahl 1986; Wood and Ljungdahl 1991). This bacterium and *C. thermoautotrophicum* are the only thermophilic clostridia producing acetic acid as the only fermentation product. The quantitative conversion of sugars to acetic acid is theoretically possible:

$$1 \text{ mol glucose} \longrightarrow 3 \text{ mol acetic acid or}$$

$$1 \text{ mol xylose} \longrightarrow 2.5 \text{ mol acetic acid}$$

This conversion is the result of the ability to fix CO_2; these microorganisms use CO_2 as an electron acceptor and thus are able to form nearly 3 mol of acetate per mol of hexose used with no production of other organic acids other than in trace amounts (Wood and Ljungdahl 1991). The hexose fermentation occurs via the Embden-Meyerhof pathway and leads to two acetate, two C_1 moieties, and four reducing equivalents. The formation of the third acetate from the C_1 units and the reducing equivalents is catalyzed by the tetrahydrofolate enzymes of the recently proposed Wood-Ljungdahl pathway (Wood and Ljungdahl 1991); this pathway is also responsible for the acetate formation from CO, methanol + CO_2, and H_2 + CO_2.

The ability of thermophilic homoacetogenic clostridia to grow on methanol has been described more recently (Wiegel et al. 1981; Wiegel and Garrison 1985); this substrate is a direct precursor of the methyl group of acetate and its incorporation may proceed either via the methyl-H_2 folate or by direct methylation of the corrinoid enzyme.

Different enzymes, associated with energy generation, have been isolated in thermophilic clostridia: H^+-ATPase (Ivey and Ljungdahl 1986; Mayer et al. 1986), hydrogenases (Hugenholtz and Ljungdahl 1988), cytochromes, and flavoproteins (Ivey 1987). The reader may refer to more specific reviews (Ljungdahl 1986; Wood and Ljungdahl 1991) for further details.

C. thermoaceticum is the homoacetogen most studied over the last 25 years (Wood 1982; Kerby and Zeikus 1983; Martin et al. 1983; Wiegel and Garrison 1985). The original strain was able to grow only heterotrophically, but Lundie and Drake (1984) developed a mineral medium with nicotinic acid as the only growth supplement. The optimum pH on glucose is 6.8 while on methanol it is 5.8; Wiegel et al. (1991) reported the successful adaptation of this microorganism to produce acetate from glucose at pH 4.5 when grown on a supporting material. *C. thermoaceticum* is capable of converting glucose and xylose to acetate, as the only product, in nearly stoichiometric yields (Ljungdahl 1983; Ljungdahl et al. 1986), but it does not possess the hydrolytic enzymes to digest lignocellulosic polysaccharides. This microorganism can, therefore, use these substrates only after their hydrolysis to component sugars (Ljungdahl 1983).

The other thermophilic homoacetogen, *C. thermoautotrophicum*, is a very ubiquitous microorganism. In this case, also, the Wood-Ljungdahl pathway has been demonstrated (Clarck et al. 1982) and a mineral medium developed (Savage and Drake 1986). The growth on H_2 + CO_2, CO or methanol + CO_2 is much faster compared to *C. thermoaceticum* (Wiegel et al. 1981; Wiegel 1991; Savage et al. 1987) and methanol can be tolerated up to a concentration of 6.5% (Wiegel 1991). The optimum pH differs according to the substrate used, e.g., 6.6 on sugars, 5.8 on glycerate or methanol (Wiegel et al. 1991).

Since 1980, both *C. thermoaceticum* and *C. thermoautotrophicum* have been used in laboratory fermentation systems to produce acetic acid. Reed and Bogdan (1985), using the cell-recycling system with the bacteria adsorbed on corn cob granules and activated carbon, obtained 14.3 g acetic acid l^{-1} h^{-1} and a concentration of 7.1 g/l. Wang and Wang (1984) used a batch system and incremental additions of glucose to overcome substrate inhibition;

by this method they were able to obtain a production rate of 0.8 g l^{-1} h^{-1} and a concentration of 45 g acetic acid/l. Moreover, Freier (1981) and later Wiegel et al. (1984), using cocultures of *C. thermocellum* plus *C. thermoautotrophicum* or *C. thermoaceticum* plus *C. thermoautotrophicum* and *Clostridium* sp. RB1, respectively, showed conversions above 2 mol of acetate per mol of glucose or xylose equivalent used, using as a substrate cellulosic and hemicellulosic material obtained by a steam-explosion process. In this respect, acetate production from hemicellulose seems more favorable than ethanol because some acetate is naturally produced from the acetylated sugars of hemicellulose during chemical or enzymatic deacylation.

In a recent work, Brownell and Nakas (1991) used *C. thermoaceticum* to ferment the carbohydrates released from pretreated oat-spelt xylan and hemicellulose isolated from hybrid poplar. Substrate hydrolysis was achieved with a selected concentration of sulfuric acid, and results showed that such a procedure, allowing a better utilization of xylan and hemicellulose by the microorganism, resulted in higher acetate production.

Wiegel et al. (1991) recently developed a new type of fermentor, the rotary fermentor, based on an idea from Clyde (1983) for ethanol production by yeast. The new system consists of a vertical tube with inserted disks of a white felt material (35% wool and 65% rayon) intermingled with disks of a stiffer Dupont Reemay 2033 material (Dupont, Wilmington, DE). Disks are fastened to a central rod and rotated during the fermentation; bacterial cells adhere to the disks and a large area is thus available for cell attachment and cell-to-substrate interaction due to the specific design of the system. CO_2 is bubbled through the fermentor from the bottom. The gassing fulfills several purposes: it pushes the outflow out of the vessel while keeping the fermentor anaerobic. With *C. thermoautotrophicum* and *C. thermoaceticum* (Wiegel et al. 1991), this system achieved acetate production rates of about 5 g l^{-1} h^{-1} at pH 5, a retention time of 1.3 h, and an acetate concentration of 100 mM. With a retention time of less than 30 min, a remarkable production rate of more than 11 g l^{-1} h^{-1} was obtained; however the acetate concentration was only about 50 mM and the pH 6.6. Higher acetate concentrations were obtained when two similar fermentors were run in a tail-head tandem mode. Generally, a conversion of 1 mol glucose into 2.7–2.9 mol acetate was achieved. It became clear that this fermentation process has good potential; however, a scaled-up version needs to be developed and run for a longer time to select better strains. It seems now that the success of future projects for the acetic acid production via microbial fermentations will depend on a further increase of petrochemical raw material prices and on the eventual overcoming of the existing problems for an economical fermentation technology. One of the limiting factors is that the increase of acetate in the medium also leads to a higher concentration of undissociated acetic acid and, as a consequence, cellular stress. The undissociated acetic acid diffuses into the cell where, due to the higher internal pH, it dissociates and changes the internal H^+ concentration. The attempt of the cell to maintain the ΔpH across the membrane resulted in a metabolic inhibition due to uncoupling (Baronowsky et al. 1984). Mutants of *C. thermoaceticum* able to grow, although slowly (doubling time above 35 h), at pH 4.5 were selected (Schwartz and Keller 1982a, 1982b; Keller et al. 1985), and a concentration of 75 mM acetic acid was obtained. The use of the previously mentioned rotary fermentor, using adapted cells adsorbed at different parts of the rotating fiber stack, is thus much more promising.

In addition, the use of novel substrates has been proposed. Since both *C. thermoaceticum* and *C. thermoautotrophicum* are able to synthesize acetate from CO (Wiegel 1982; Kerby and Zeikus 1983), Ljungdahl suggested that synthetic gas (syngas), produced from coal and water, might be converted to acetate using these microorganisms as biocatalysts. Further alternatives of cheap substrates for the production of CMA might be hydrolyzed corn syrup or soluble corn starch, not exposed to prior hydrolysis (Marynowsky et al. 1983; Wiegel et al. 1991).

Product recovery, product separation, and concentrating the product are the major cost in fermentation processes aside from the cost of the substrate. The most common method used to recover acetic acid is distillation after acidification of the broth, but such a procedure requires high energy (Busche 1985a, 1985b). More economical methods are the broth acidification via CO_2 followed by organic solvent extraction and final separation of the broth (Schimshick 1981; Yates 1981) or the use of membrane electrodialysis. For *C. thermoaceticum*, a possible recovery of the diluted acetic acid from the broth by a solvent membrane technique (Kuo and Gregor 1983) or an electrodialysis process (Busche 1983) has been suggested but further improvements are required.

16.2.2 Butyric Acid

Butyric acid is another organic acid that gives some expectations for a potential industrial application of thermophilic clostridia. Currently, it is used mainly as a food additive. In general, the most studied butyrate-producing microorganisms belong to the genus *Clostridium*. Fermentation processes for butyrate production have not yet been applied because of the low production yields. Currently, butyric acid is entirely produced by petrochemical means. However, the need for more biological processes and an increasing concern about chemical additives, should hopefully stimulate research to obtain increased yields of butyrate by fermentation systems.

Among thermophilic clostridia, butyric acid is produced in significant amounts by growing cultures (see under Section 16.1.1) of *C. thermosaccharolyticum* (Hollaus and Sleytr 1972) and by several recently isolated organisms (e.g., *C. thermopalmarium* and *C. thermopapyrolyticum*), but the elective microorganism for eventual industrial applications seems to be *C. thermobutyricum* (Wiegel et al. 1989). This microorganism ferments hexoses to 0.85 butyrate + 1.8 CO_2 + 1.9 H_2 + 0.2 lactate + 0.1 acetate, and after adaptation can grow in the presence of 15% glucose and 250 mM butyric acid (J. Wiegel et al., unpublished observations), although, with a significant increase of the doubling time. F. Canganella and J. Wiegel (unpublished observations) used the rotary fermentor system to cultivate *C. thermobutyricum* in a larger scale (500 ml). In this study, the highest butyrate concentration obtained in the medium was 210 mM and the best production rate was 1.1 g l^{-1} h^{-1}. Unfortunately, the amount of butyrate produced seemed strictly related to the percentage of yeast extract added, and, moreover, a metabolic shift to acetate production was observed at low volumetric retention times. Additional tests, carried out in liquid cultures and using ^{14}C-acetate, demonstrated that *C. thermobutyricum* is able to take up acetate and convert it to butyrate.

Recently, a new thermophilic species named *C. thermopalmarium* has been described (Soh et al. 1991) and reported to produce mainly butyrate from the fermentation of sugars. The representative strain, BVP, was isolated from palm wine in Senegal, and it closely resembles *C. thermobutyricum*, except for the ability to use sucrose and produce some ethanol. *C. thermobutyricum* did not produce any alcohol, even under an atmosphere of H_2.

16.2.3 Lactic Acid

Most of the lactic acid presently produced is used as Ca^{2+} vehicles in food preservatives. In Europe, about 30–40% of lactic acid production is carried out via fermentation whereas in the USA and Japan the synthesis is via chemicals. Considering the worldwide annual production, it would be really worthwhile to find or adapt a thermophilic microorganism for lactic acid production. Among thermophilic clostridia, *C. thermolacticum* (Le Ruyet et al. 1985) is the only species to produce mainly lactic acid plus some ethanol and CO_2. In addition, the strain H101-69 of *C. thermohydrosulfuricum* (Hollaus and Sleytr 1972) has been described to produce up to 1.2 mol lactate/mol glucose used (Wiegel et al. 1979). The potential of lactic acid for commercial application would significantly increase if new polymers, which are based on lactic acid, were to be marketed (Lipinsky 1981).

16.2.4 Other Organic Acids

As mentioned before, other organic acids are sometimes produced by different species of thermophilic clostridia, but, to date, they do not seem to have any biotechnological potential. Formate and succinate have been found to be produced by a new species, *C. thermosuccinogenes* (Drent et al. 1991). It produces mainly acetate and succinate when grown on inulin. Inulin has been previously suggested as a feasible cheap substrate in biotechnology for the microbial production of high-fructose syrups (Vandamme and Derycke 1983; Fuchs et al. 1985). Little information is available about the enzymes involved in the thermophilic breakdown of inulin. Indeed, it would be interesting to have more data on the metabolic pathway involved in such a degradation. Furthermore, it is remarkable that, although succinate is a common fermentation product among mesophilic anaerobes, it has been rarely found, to date, as a fermentation product of thermophilic anaerobes.

16.3 INDUSTRIAL ENZYMES

Since 1980, several enzymes have been considered for a wider application in the biotechnology industry. Some of these enzymes are used in the food industry and detergents. Others are also applied in different fields such as in the synthesis of pharmaceuticals, therapeutics, and clinical and chemical analysis. The most common enzymes in the market are hydrolases such as amylase, cellulase, pectinase, protease, lipase, and collagenase; two other important enzymes are glucose isomerase and glucose oxidase.

Thermophilic enzymes have, in general, some advantages for potential industrial exploitation because of their thermostability, i.e., they can operate at elevated temperatures at a reduced risk of contamination and increased reaction rates. In general, a thermostable enzyme should have the following properties to be considered as a potential candidate for industrial applications: (1) reasonable stability; (2) easy manipulation, storage, and handling (e.g., not O_2 sensitive); (3) high turnover number and stability at the temperature of application; (4) specific (or nonspecific) mode of action; (5) cheap and easy production from readily available sources (Wiegel and Ljungdahl 1986; Saha et al. 1989).

Thermophilic clostridia are a good potential source for thermostable enzymes of industrial interest since the various species are physiologically diverse and thrive in many different environments. Moreover, thermophilic clostridia probably evolved under energy-limited conditions, and this may have induced a selection of highly efficient catabolic enzymes. In Table 16–3, we report a list of thermophilic clostridia producing suitable enzymes for potential industrial applications. The technology of thermostable enzymes is not yet well exploited because thermophilic clostridia produce, in general, extremely low levels of extracellular enzymes, with the exception of *C. thermosulfurogenes* 4B (DSM 2229) (Hyun and Zeikus 1985d) and *C. thermohydrosulfuricum* (Klingerberg et al. 1990). The bulk of the enzyme activity is, in most cases, associated with the cell surface and not easily released in the culture medium. For instance, in *C. thermohydrosulfuricum* 39E (*T. ethanolicus* 39E), the amylolytic enzymes are completely cell bound (Hyun and Zeikus 1985d). A complete review on all the enzymes isolated from thermophilic clostridia is, of course, not possible within the page limitation. We will discuss here some general and novel catabolic features of some common commercialized thermostable enzymes in relation to the species/strains from which they have been isolated.

TABLE 16–3 Enzymes of Biotechnological Interest Present in Thermophilic Clostridia

Class-Enzyme	*Uses*	*Bacteria*	*References*
α-Amylase (EC 3.2.1.1)	Starch liquefaction	*C. thermohydrosulfuricum*	Hyun and Zeikus, 1985c, 1985d; Plant et al. 1987; Melasniemi et al. 1990
		C. thermosaccharolyticum	Klingerberg et al. 1990; Koch and Antranikian 1990; Spreinat et al. 1990
		C. thermosulfurogenes	Antranikian et al. 1987a, 1987b; Madi et al. 1987
		C. 'thermoamylolyticum'	Katkocin et al. 1985
β-Amylase (EC 3.2.1.2)	Maltose production	*C. thermosulfurogenes*	Hyun and Zeikus 1985a, 1985b
Glucoamylase (EC 3.2.1.2)	Glucose production	*C. thermoaceticum*	Ensley et al. 1975; Chojecki and Blashek 1986
		C. thermohydrosulfuricum	Hyun et al. 1985
		C. 'thermoamylolyticum'	Katkocin et al. 1985
		C. thermosaccharolyticum	Specka et al. 1991
Pullulanase (EC 3.2.1.41)	Debranching of starch in glucose and maltose syrups production	*C. thermosulfurogenes*	Hyun and Zeikus, 1985d; Madi et al. 1987; Antranikian 1989
		C. thermohydrosulfuricum	Melasniemi 1987b; Saha et al. 1988
		C. thermosaccharolyticum	Koch et al. 1987
Amylopullulanase	Starch liquefaction	*C. thermohydrosulfuricum*	Melasniemi 1988; Saha et al. 1988
		C. thermosulfurogenes	Madi et al. 1987; Antranikian 1989

Xylose/glucose isomerase (EC 5.3.1.5)	High-fructose syrups	*C. thermosulfurogenes*	Lee and Zeikus 1991
α-Glucosidase (EC 3.2.1.20)	Starch liquefaction	*C. thermohydrosulfuricum*	Melasniemi 1987a; Saha and Zeikus 1991
		C. thermosaccharolyticum	Koch and Antranikian 1990
		C. thermosulfurogenes	Antranikian et al. 1987a; Madi et al. 1987
Cellulases	Fruit and vegetable processing, ethanol production	*C. thermocellum*	Ljungdahl et al. 1983; Lamed and Bayer 1988; Saha et al. 1989
		C. stercorarium	Madden 1983
Xylanases (EC 3.2.1.8)	Fruit juice clarification, paper manufacture	*C. fervidus*	Patel et al. 1987
		C. thermocopriae	Jin et al. 1988
		C. thermolacticum	Debeire et al. 1990
		C. thermocellum	Ng et al. 1977; Wiegel et al. 1985
		C. stercorarium	Berenger et al. 1985
		C. thermohydrosulfuricum	Wiegel et al. 1985
		C. thermosaccharolyticum	Sjolander 1937; Wiegel et al. 1985
Pectinases	Fruit juice clarification	*C. thermosulfurogenes*	Schink and Zeikus 1983b
Oxidoreductases	Electron transfer reactions in stereoselective catalysis	*C. thermohydrosulfuricum*	Lamed and Zeikus 1981; Lamed et al. 1981
		C. thermosaccharolyticum	Vancanneyt et al. 1990

16.3.1 Amylolytic Enzymes

The majority of potentially industrial enzymes isolated from thermophilic clostridia are starch-hydrolyzing enzymes. They represent an enzymatic system regulated by substrate induction and catabolite repression. Enzyme expression is activated by maltose or maltose-containing carbohydrates where repression is generally due to the presence of glucose. Most of the starch-hydrolyzing enzymes in thermophilic clostridia show similar features such as insensitivity to oxygen and a nearly complete independence from metal ions; they also are extremely thermostable. When analyzed by gel electrophoresis (Spreinat and Antranikian 1990; Koch and Antranikian 1990; Klingerberg et al. 1990), these enzymes usually show several protein bands with enzymatic activity. Similar results have been generally obtained with other anaerobic and polysaccharide-degrading bacteria. Currently, starch-hydrolyzing thermostable enzymes are the subject of an intensive and specific research. Every year 18 billion sweeteners are produced in the USA, and the potential expansion of such a big market well justifies the increasing interest in these enzymes.

Several enzymes from mesophilic organisms are used in the manufacture of sugars from starch: α-amylase, glucoamylase, pullulanase, and β-amylase. The process is quite expensive and time consuming due to multi-step reactions and necessary removal of a large amount of salt. The isolation of more efficient and suitable thermostable enzymes should lead to better conversion processes, resulting in a lower cost of sugar syrup production.

16.3.1.1 α-Amylase. α-Amylase (α-1,4-D-glucan glucanohydrolase, EC 3.2.1.1, endoamylase) hydrolyzes internal α-1,4 starch linkages at random in an endo fashion producing oligosaccharides of various lengths. Generally, α-amylases cannot act on α-1,6 linkages of starch. This enzyme is used for the liquefaction of starch; its function is to lower the viscosity of the gelatinized starch and promote its hydrolysis into maltodextrins.

α-Amylase activity has been found in a variety of microorganisms (Fogarty 1983). This enzyme has been isolated from several species and strains of thermophilic clostridia: *C. thermohydrosulfuricum* (Hyun and Zeikus 1985c, 1985d; Melasniemi 1987a, 1987b; Melasniemi and Paloheimo 1989; Melasniemi et al. 1990; Plant et al. 1987), *C. 'thermoamylolyticum'* (Katkocin et al. 1985), *C. thermosaccharolyticum* (Spreinat and Antranikian 1990; Koch and Antranikian 1990; Klingerberg et al. 1990), and *C. thermosulfurogenes* EM1 (DSM 3896) (Antranikian et al. 1987a, 1987b; Madi et al. 1987). In general, maximal enzymatic activity has been detected at 70°C. The biosynthesis of α-amylase is regulated by substrate induction (e.g., maltose) and catabolite repression (e.g., glucose) except in *C. thermosulfurogenes* where it is constitutive (Antranikian et al. 1987a, 1987b). In this microorganism (strain EM1), Antranikian et al. (1987b) showed that the excretion of α-amylase in the medium is mediated by cytoplasmic membrane vesicles that are externally released during the renewal of the cell wall. To enhance the production of cell-free α-amylase, *C. thermosulfurogenes* EM1, *C. thermosulfurogenes* 4B, *C. thermosaccharolyticum*, and *C. thermohydrosulfuricum* have been cultivated in continuous cultures and varying parameters such as pH, dilution rate, and substrate (Antranikian et al. 1987a, 1987c; Koch et al. 1987; Madi et al. 1987; Antranikian 1989). An up to 80-fold increased concentration of enzyme was obtained when bacteria were grown under a limited starch supply, at optimum pH and dilution rate. Under these conditions, about 70% of the enzyme was released in the culture fluid. Similar observations were made with a mixed culture of *C. thermosulfurogenes* and *C. thermohydrosulfuricum* (Hyun and Zeikus 1985d); under starch limitation and optimum pH and dilution rates, an increase of the level of starch-hydrolyzing enzymes could be measured. It is likely that these positive results, obtained by continuous culture techniques, will lead to further potential applications with different thermophilic organisms.

16.3.1.2 β-Amylase. β-Amylase (α-1,4-D-glucan maltohydrolase, EC 3.2.1.2, saccharogenic amylase) hydrolyzes alternate α-1,4-glucosidic linkages of starch in an exo fashion from the nonreducing end, producing β-maltose. It cannot act on and bypasses α-1,6 linkages of starch.

Only one thermophilic clostridium, *C. thermosulfurogenes*, has been reported to produce an extracellular β-amylase (Hyun and Zeikus 1985a, 1985b). This enzyme was produced in large amounts at 65°C; it is inducible and regulated by catabolite repression. A hyperproductive mutant has been isolated that is able to produce eightfold more β-amylase than the wild type, as the synthesis of the enzyme is constitutive and not sensitive to catabolite repression (Hyun and Zeikus 1985b). The purified enzyme is a tetrameric glycoprotein with a molecular weight around 210,000; it is a sulphydryl enzyme with an optimum pH and temperature at pH 6.0 and 75°C, respectively (Shen et al. 1988).

Potential biotechnological applications of β-amylases are primarily in the production of high-maltose syrups at elevated temperatures from raw or soluble starch. Various maltose-containing syrups are presently used in brewing, baking, canning, and confectionery industries. In this regard, the β-amylase produced by *C. thermosulfurogenes* shows unique features and compatibilities with the current starch-processing conditions (elevated temperature and low pH). If produced economically, this enzyme might then have future exploitations in the manufacture of some maltose-containing syrups (Zeikus and Saha 1989).

16.3.1.3 Glucoamylase. Glucoamylase (1,4-α-D-glucan glucohydrolase, EC 3.2.1.3) is an exo-acting carbohydrase that consecutively cleaves glucose units from the nonreducing end of starch molecules.

Hyun and Zeikus (1985c) reported high levels of thermostable glucoamylase activity in *C. thermohydrosulfuricum*, but this enzyme was later described as a maltase or α-glucosidase activity (Hyun et al. 1985). A thermostable extracellular glucoamylase activity has been shown in *C. thermoamylolyticum* (Katkocin et al. 1985) and the enzyme was purified. It mainly hydrolyzes maltodextrin and soluble starch but also maltose; the molecular weight is 75,000; the maximal initial activity is at pH 5.0 and at a temperature of about 70–75°C.

Recently, a glucoamylase was isolated from *C. thermosaccharolyticum* (Specka et al. 1991). The enzyme has a "temperature optimum" of 70°C and a pH optimum of pH 5.0; about 50% of the glucoamylase activity is extracellular and no difference was observed when bacteria were cultivated in continuous culture. At present, glucoamylases used in industrial processes are obtained from fungal sources, and no thermostable enzymes of this kind have been commercially applied yet.

16.3.1.4 Pullulanase. Pullulanase (pullulan-6-glycanohydrolase, EC 3.2.1.41) is a debranching enzyme that cleaves specifically α-1,6 linkages in starch, amylopectin, pullulan, and other related oligosaccharides (Abdullah et al. 1966).

The biosynthesis of pullulanases follows the same pattern as that of amylases: maltose and carbohydrates containing maltose units induce the expression of the extracellular enzyme whereas glucose acts as the repressor (Hyun and Zeikus 1985d). Only in *C. thermosulfurogenes* has the synthesis been found to be constitutive (Antranikian et al. 1987a, 1987b; Madi et al. 1987). Higher levels of pullulanase were produced by the same mutant that Hyun and Zeikus (1985b) selected for the production of β-amylase. Moreover, the fermentation system developed to obtain larger amounts of cell-free amylases was similarly successful for the production of this enzyme (Antranikian et al. 1987a, 1987c; Koch et al. 1987; Madi et al. 1987; Antranikian 1989). The pullulanase of *C. thermohydrosulfuricum* 39E (*T. ethanolicus* 39E) has been purified to homogeneity (Saha et al. 1988). The enzyme is a glycoprotein with an apparent molecular weight of about 136,500; its maximal activity is at pH 5.5 and 90°C (Saha et al. 1988). Similar characteristics were also described for a pullulanase isolated from another strain of *C. thermohydrosulfuricum* (Melasniemi 1987b). Koch et al. (1987) reported the purification of a pullulanase in *C. thermosaccharolyticum*; this enzyme is maximally active at 75°C and a pH of 5.0–6.0. It is different than the one isolated from *C. thermohydrosulfuricum* because it shows a weak inhibition by β-cyclodextrin and also a complete insensitivity to α-cyclodextrin. Pullulanase is an important enzyme in industry, often used in combination with saccharifying amylases such as glucoamylases, fungal α-amylase or fungal β-amylase for the production of various sugar syrups. It has been shown to improve saccharification and product yields (Norman 1982). The main advantages of using pullulanases in starch-saccharification processes are: an increase in glucose yields (about 2% with the addition of glucoamylase) and maltose yields (about 20–25% with the addition of β-amylase), a reduction of saccharification time, an increase of substrate concentration, and a reduction in the addition of glucoamylase (up to 50%). Large-scale applications of thermostable pullulanases from thermophilic clostridia have not yet been published, but it is reasonable to expect that the newly studied enzymes will replace some of the current pullulanases in saccharification and starch-liquefaction processes.

16.3.1.5 Amylopullulanase. Amylopullulanase is a new class of enzyme that, like pullulanase, hydrolyzes α-1,6 linkages of pullulan but which is also able to cleave α-1,4 linkages of starch (Saha and Zeikus 1989).

A cell-bound amylopullulanase activity has been found in *C. thermohydrosulfuricum* 39E (*T. ethanolicus* 39E) as reported by Hyun and Zeikus (1985d). In this organism, synthesis of the enzyme was inducible and subject to catabolic repression. Purification of the cell-bound amylopullulanase was achieved by Saha et al. (1988); the reported molecular weight of the enzyme was 136,500 and the optimal temperature and pH were 90°C and pH 5.5, respectively. An amylopullulanase activity has been also demonstrated in *C. thermohydrosulfuricum* E101-69 (Melasniemi 1987b, 1988) and *C. thermosulfurogenes* (Madi et al. 1987; Madi and Antranikian 1989).

Potential applications of thermostable amylopullulanases in biotechnology might be in starch liquefaction (with or without α-amylase) or the single-step conversion of starch into various maltodextrins, with the advantage of a higher thermostability and larger glucose yields.

16.3.1.6 Xylose/Glucose Isomerase. Fructose, an isomer of D-glucose, is the sweetest natural sugar and its use has been previously limited by its availability; in this regard, one of the most significant achievements in the industrial application of enzymes can be considered the production of high-fructose corn syrups.

A specific glucose isomerase has not yet been found, and the chemical isomerization is an uneconomical process; however, the xylose isomerase was found to catalyze the glucose isomerization reaction and this fact led to an extensive use of this enzyme for the industrial production of high-fructose corn syrups, generally in fixed-bed reactors.

Most described xylose isomerases are from mesophilic microorganisms, but, recently, a xylose isomerase was purified to homogeneity from *C. thermosulfurogenes* (Lee and Zeikus 1991); this enzyme has a molecular weight of 200,000 and a maximal activity at pH 7.0 and 80°C. The interest lies in finding an enzyme that does not require Co^{2+} ions for activation and which is active and stable at a slightly acidic pH.

16.3.1.7 α-Glucosidase. α-Glucosidase (EC 3.2.1.20, α-D-glucoside glucohydrolase) hydrolyzes the terminal nonreducing α-1,4 linked glucose residues of various substrates, releasing α-D-glucose.

The ability to produce α-glucosidase is widely distributed among anaerobic bacteria. As is shown in Table 16–3, some thermophilic clostridia form α-glucosidase beside other amylolytic enzymes. An α-glucosidase activity was recently detected in *C. thermosaccharolyticum* (Koch and Antranikian 1990) and, previously, also in other species of thermophilic bacteria such as *C. thermohydrosulfuricum* (Hyun et al. 1985; Melasniemi 1987a) and *C. thermosulfurogenes* EM1 (Antranikian et al. 1987a; Madi et al. 1987). Purification of an α-glucosidase has not yet been achieved from a thermophilic clostridium. Little data is available about the optimal conditions for the production of the enzyme, and on whether it is primarily cell bound or extracellular. In *C. thermosulfurogenes* EM1 (Antranikian et al. 1987a; Madi et al. 1987) optimal conditions for α-glucosidase production were reported to be pH 5.9 and 60°C; in this case, about 60% of the enzymatic activity was released into the culture broth. A partial purification of a thermostable α-glucosidase has been recently accomplished by Saha and Zeikus (1991) from *C. thermohydrosulfuricum* 39E (*T. ethanolicus* 39E). The enzyme was reported to be cell bound; maximal activity was detected at pH 5.5 and a temperature of 75°C with a half-life of 35 min.

16.3.2 Xylanases

Xylans are branched and substituted polysaccharides of xylose comprising up to 10–40% of plant cell walls. They also represent a waste product of wood pulp manufacture and make up a significant component of agricultural lignocellulosic wastes.

Among thermophilic clostridia, a xylanolytic activity has been reported in different species: *C. fervidus*, *C. thermocopriae*, *C. thermolacticum*, *C. thermocellum*, *C. thermohydrosulfuricum*, *C. thermosaccharolyticum*, and *C. stercorarium*. Recently, xylanases were isolated from a new *C. stercorarium* isolate (F-9), and various xylanases-related genes, showing no activity against para-nitrophenylxyloside and some β-xylosidases, have been cloned in *Escherichia coli* (Sakka et al. 1990). Furthermore, an endoxylanase has been isolated and characterized in *C. thermolacticum* by Debeire et al. (1990) and, also, in taxonominally non-identified clostridial strains (De Blois and Wiegel 1991). Regulation of xylanases is generally considered to be constitutive and subject to catabolite repression (Brodel et al. 1990).

A potential exploitation of thermostable xylanases is in food and feed manufacture, for the clarification of fruit juices and wine, or to improve nutritional and physical properties of cereal fibers. Furthermore, an upcoming application is in pulp, paper, and fiber manufacture. One route is to convert xylan wastes to fermentable sugars as substrates for the production of valuable compounds (e.g., ethanol). To separate xylans from the lignocellulosic matrix of wood pulp and convert them into fermentable sugars, xylanases should be free of cellulase activity so as to minimize any possible damage to cellulose fibers. Thermostability would be an important feature for the industrial application. The other route is to use the xylanase treatment to reduce the use of chlorine. Such experiments are in progress (K.E. Ericksson et al., personal communication).

16.3.3 Pectinases

Pectin is a complex carbohydrate containing pectic acid esterified with methyl alcohol.

Pectin-degrading enzymes can be classified in two groups: esterases and depolymerases; such catalytic activities have been described in a variety of microorganisms, including clostridia (Fogarty and Kelly 1983). A polygalacturonate hydrolase and a pectin methylesterase have been isolated from *C. thermosulfurogenes* (Schink and Zeikus 1983b). The enzymes have maximal initial activity at pH 5.5 and 75°C (polygalacturonate hydrolase) and at pH 6.5 and 70°C (pectin methylesterase). Enzymatic activities were, in both cases, equally distributed between the cell and the culture fluid. Possible applications for thermostable pectinases are presently in the food industry (e.g., fruit juice clarification) or in the processing of agricultural and forest products where pectin gums often present a problem.

16.3.4 Oxidoreductases

This group includes many enzymes involved in electron transfer reactions. Most of these enzymes often require electron transfer mediators that are usually quite expensive and unstable.

C. thermohydrosulfuricum strain 39E (*T. ethanolicus* 39E) and *T. ethanolicus* type strain JW200 were found to contain a novel NADP-linked secondary alcohol dehydrogenase (alcohol-aldehyde/ketone oxidoreductase) (Lamed and Zeikus 1981; Lamed et al. 1981). The type strain produces also two different lactate dehydrogenases; they differ in molecular weight, isoelectric point, and requirement of 1,6-biphosphate activation; one enzyme is formed when cells are grown on starch, the other when cells are grown on sucrose (Turunen et al. 1987). For both enzymes, the highest initial activity is at 70°C. A different NADP-linked secondary alcohol dehydrogenase has been described in several strains of *C. thermosaccharolyticum* (R. Lamed, unpublished observations). From a different strain of *C. thermosaccharolyticum* (LMG 6564), cultivated in continuous culture, Vancanneyt et al. (1990) obtained a lactate and an ethanol dehydrogenase; both enzymes were insensitive to O_2 and showed maximal initial activity at pH 6.0–6.5 and at a temperature of 55–60°C.

Many chiral compounds are currently synthesized by microbial hydrogenation using hydrogen (or formate) and hydrogenase-containing microorganisms. Furthermore, the NADP-linked alcohol-aldehyde/ketone oxidoreductase from *C. thermohydrosulfuricum* 39E (*T. ethanolicus* 39E) has been patented (Zeikus and Lamed 1982).

Oxidoreductive enzymes are of significant interest for industrial-applied processes because the direct stereoselective reduction of acyclic ketones, previously carried out by horse liver or yeast alcohol dehydrogenases, is not suitable for linear secondary alcohols. Besides this consideration, the thermal stability and the solvent tolerance, shown by these enzymes, are valuable features making these oxidoreductases primary candidates for biotechnological applications.

16.3.5 Immobilized Enzymes

The isolation and purification of thermostable enzymes, suitable for industrial applications, is a main aspect of basic research, but further insights are also expected from the improvement of fermentation systems, cultivation techniques, and application techniques. In this regard, the immobilization of enzymes or whole cells has recently become the focus of intensive-applied research (Poulsen 1981; Bucke and Wiseman 1981; Hamilton 1984). Frequently, enzyme immobilization leads to a more stable enzyme activity, which, if the temperature can be elevated, allows an increase in product yields from the raw substrate and a reduction in the risk of microbial contamination. A restriction of a higher cell density to the beads makes the separation of the metabolites from the cells much more effective.

Generally, thermophilic microorganisms or their enzymes do not require any special technique for immobilization and different procedures can be used: covalent binding, entrapment, cross-linking, and adsorption. Immobilization techniques are expected to become even more competitive when applied to thermophilic enzymes. But, to date, only a few reports are available on immobilized enzymes or entrapped whole cells from thermophilic microorganisms. Regarding thermophilic clostridia, the calcium-alginate bead immobilization has been used to entrap cells of *C. thermosulfurogenes* EM1 and *C. thermosaccharolyticum*, strain Sol1 and Uel1 (Vorlop and Steinert 1987). In this case, full or hollow sphere beads were used, but most of the enzymes were released in the medium.

Further potential applications will probably arise from developing fermentation systems in which the metabolic state of the immobilized cells can be monitored and determined. To date, the only positive reports are the application of a fluorescence probe to monitor the intracellular NAD(P)H pool (Müller et al. 1988) and the use of antibodies to detect the pullulanase produced by *C. thermosulfurogenes* EM1 (Spreinat and Antranikian 1990; Freitag et al. 1991).

16.4 BIOCONVERSIONS AND BIOREDUCTIONS

Only a few studies have been reported, to date, regarding the use of thermophilic microorganisms to modify biological compounds like vitamins and steroids; the reason for this is probably a lack of research (Wiegel and Ljungdahl 1986). Also, little data are available in this respect on thermophilic clostridia, although many species of this genus are well known as useful agents of anaerobic bioconversions. Single-step bioconversions may be achieved, in vitro, using pure enzymes or whole cells, either in suspension or after immobilization on a support material. The required low Eh is usually maintained by the cells themselves, however, the production of CO_2 or H_2 can lead to the splitting of the polymer bead and thus present a problem. Although most of the thermophilic clostridia are presently reported to possess unique enzymes and carry out remarkable bioreactions, a large amount of data comes from basic research and no industrial applications have been reported to date.

In some bioconversions, fermentation pathways are "distorted" to produce a particular end product. One example is that *C. thermosaccharolyticum* (forming usually acetate, butyrate, and lactate from hexoses) produces, on sugars and under certain culture conditions (60°C, pH 6.0, H_2 atmosphere), significant amounts of R(−)-1,2-propanediol and acetol. Presumably this is due to routing of intermediates into the methylglyoxal bypass (Cameron and Cooney 1986; Sanchez-Riéra et al. 1987). The interest in R(−)-1,2-propanediol is due to its potential application in organic synthesis including the preparation of R-propylene oxide, optically active polymers, and chiral crown ethers. Another example is, that when viologen dyes are added to the growth medium, *C. thermoaceticum* produces methanol from CO (White et al. 1987). In the same organism, a potentially useful dehydrogenase has also been isolated and characterized; it is a formate-NADP oxidoreductase containing selenium and tungsten that is active at 45°C and capable of reacting with methylviologen and benzylviologen instead of the usual pyridine nucleotide (Ljungdahl and Andreesen 1978). Furthermore, an ornithine decarboxylase (Paulin and Pösö 1983) and an NADPH-dependent alcohol-ketone oxidoreductase (Lamed et al. 1981) were also isolated from *C. thermohydrosulfuricum*; the former enzyme is involved in the cleavage of C-C bonds whereas the latter affects the reduction of methyl ketones.

Some stereospecific conversions, including reductions, are presently carried out in industrial processes by yeast and filamentous fungi, but it is reasonable to expect that the thermophilic clostridia, able to generate an excess of reducing power at low Eh values, will also be considered as suitable catalyzers for applied anaerobic bioreductions.

A larger exploitation of these microorganisms is expected in the future. Some species of mesophilic clostridia have been reported to reduce C═C double bonds and to carry out electrochemical reductions and bioconversions of bile acids and steroids. Similar metabolic features may also be found in thermophilic clostridia and more efforts are required in basic research to elucidate this potential.

16.5 WASTE MANAGEMENT

Waste treatment is one of the main areas in which anaerobic thermophilic microorganisms are being used at a large scale in technical processes. A variety of organisms, including several clostridia, acting as mixed anaerobic cultures of different complexity are involved in the breakdown of polymers to produce CO_2 and CH_4, including several clostridia. The last step in the anaerobic breakdown of polymers is methane formation, a reaction mediated by specialized organisms, the methanogens. The formation of the specific methanogenic substrates such as CO_2, H_2, acetate, formate, methanol, isopropanol, and methylamines from waste material depends on many interactions among different groups of microorganisms (Wiegel and Ljungdahl 1986): the biopolymer degraders (using cellulose, hemicellulose, lignin, starch, pectin, and proteins), which produce sugars, alcohols, and organic acids; the intermetabolic organisms (ethanologenic and acido- and acetogenic organisms), converting the above products into the methanogenic substrates and the methanogenic bacteria that remove inhibitory substrates such as H_2. A major emphasis has been given to interspecies H_2 transfer as the most critical reaction between acid- and alcohol-utilizers and H_2-utilizing methanogens. For example, the degradation of benzoic acids and fatty acids such as butyrate and propionate are thermodynamically only possible at low H_2 concentrations. Since several reviews and reports have been published on the subject of the anaerobic thermophilic waste treatment and the role of thermophilic clostridia that might be important for the development of an effective sewage sludge fermentation system at elevated temperatures (Chen 1983; Marchaim et al. 1983; Matsunaga et al. 1983; Varel 1983; Wiegel and Ljungdahl 1986; Archer and Kirsop 1990; Wohlt et al. 1990), we give here just a general overview. *C. thermocellum*, *C. thermosaccharolyticum*, and *C. thermohydrosulfuricum* are widespread organisms and have been isolated from sewage sludge. They are able to use different polymers such as cellulose (*C. thermocellum* only), starch, (*C. thermosaccharolyticum* and *C. thermohydrosulfuricum*), and hemicellulose (*C. thermocellum* and *C. thermohydrosulfuricum*). These species are thus considered to be potentially involved in the first step of anaerobic waste degradation. However, other, so far unidentified, clostridia are expected to be present and also important (Wiegel 1992). *C. thermosaccharolyticum* and *C. thermohydrosulfuricum* may also, along with *C. thermobutyricum*, *C. thermoaceticum*, and *C. thermoautotrophicum*, play a role in the second stage of the process: the degradation of sugars, alcohols, and organic acids; all of them, in fact, use hexoses and pentoses, and, moreover, *C. thermosaccharolyticum* and *C. thermohydrosulfuricum* are also able to grow on cellobiose. Little is known about thermophilic peptidolytic and lipolytic clostridia.

The anaerobic thermophilic waste degradation is a complex process and thermophilic clostridia naturally interact with other microorganisms: starch- and cellulose-degraders, sulfate reducers, and methanogens. Such a microbial community is large and difficult to characterize due to the number of species and metabolic interactions involved. It is expected, anyway, that the isolation of novel thermophilic clostridia with unique and specific metabolic features will lead to a better knowledge of anaerobic waste-degradation processes at elevated temperatures and, also, to a larger exploitation of these microorganisms in biotechnology.

16.6 FOOD INDUSTRY

Thermophilic clostridia are presently not important for any food-processing or beverage-producing industry. Currently, the only thermophilic clostridia potentially useful for the production of food additives are *C. thermobutyricum* (e.g., butyric acid to synthesize banana aroma), *C. thermolacticum* (lactic acid), and *C. thermoaceticum*/*C. thermoautotrophicum* (acetic acid as a preserving agent in grain storage). Furthermore, a group of thermophilic clostridia (see Table 16–3) can be considered potentially useful for the fruit juice industry (juice clarification). *C. thermosaccharolyticum*, the "swelled can" organism, can be regarded as involved in the food industry, but only as a disturbing contaminant with no potential for a useful application.

16.7 GENETIC STUDIES

In the past and to a great extent still today, the isolation of novel species or strains is the primary approach to selecting a microorganism with specific metabolic and physiological features; nevertheless, the potential use of such a microorganism in industrial processes will probably depend on the future application of traditional genetic manipulations and on the development of molecular techniques suitable for each specific isolate.

Few genetic studies have been carried out to date with thermophilic clostridia and no transformation system for any thermophilic clostridium has been found to be effective. The recent interest in biopolymer degradation (especially cellulose) for the production of valuable compounds at industrial level led to a better knowledge of the genetics of *C. thermocellum* (see Chapter 13 on *C. thermocellum*).

A significant number of mutants have been selected in several species by different methods such as UV light, gamma rays, ethyl methanesulfonate (EMS), and *N*-methyl-*N'*-nitro-*N*-nitrosoguanidine (NNG) (Gomez et al. 1980; Schwartz and Keller 1982a; Duong et al. 1983; Peteranderl 1990). In general, the most effective mutagens have been those acting by a direct mechanism such as EMS. This result, and also similar observations reported in the past for other species of clostridia, suggested that these microorganisms might be generally deficient in effective error-prone DNA repair processes (Walker 1983; Bowring and Morris 1985). Various mutants were obtained in *C. thermocellum* (Gomez et al. 1980; Herrero and Gomez 1980; Méndez and Gomez 1982; Duong et al. 1983), *C. thermoaceticum* (Schwartz and Keller 1982a, 1982b); *C. thermosaccharolyticum* (Avgerinos et al. 1981), and *C. thermohydrosulfuricum* (Peteranderl 1990).

At present, no transducing phages have been found in thermophilic clostridia and the only report on extrachromosomal elements is the isolation of a 58 MDa plasmid from *C. thermoaceticum* (Morton et al. 1992). The plasmid DNA in *C. thermohydrosulfuricum* (Xu and Wiegel 1987) turned out to be protected chromosomal DNA with structures resembling DNA:DNA binding protein complexes (G.-W. Kohring and J. Wiegel, unpublished observations). A procedure for the induction of autoplasts in *C. thermohydrosulfuricum* was independently developed by Kim et al. (1989) and Peteranderl (1990). The latter reported an improved method for autoplast formation and regeneration of walled cells also usable for other anaerobic thermophiles (J. Wiegel, unpublished observations). Attempts, however, to obtain cell-fusion products with protoplasts of *C. perfringens* strain 3624 Rif^{res}/Str^{res} (Solberg et al. 1981) and *C. thermohydrosulfuricum* strain JW 102 were not successful (F. Canganella and J. Wiegel, unpublished observations). The only successful transformation event to date was accomplished in *C. thermohydrosulfuricum* (Soutchek-Bauer and Staudenbauer 1985), but the plasmid used in the experiment could not be re-isolated from the microorganism. In spite of such a failure in transformation techniques, many genes from thermophilic clostridia have been characterized after being cloned into *E. coli*. An updated list of cloned genes from clostridia was reported by Young et al. (1989) and, more recently, other genes have been cloned into *E. coli* from *C. thermosulfurogenes* (Haeckel and Bahl 1989; Burchhardt et al. 1991), *C. stercorarium* (Sakka et al. 1990), *C. thermoaceticum* (Wood and Ljungdahl 1991), *C. thermohydrosulfuricum* 39E (*T. ethanolicus* 39E) (Mathupala et al. 1990), and *C. thermocellum* (Grepinet et al. 1988; MacKenzie et al. 1989).

Further insights into the genetics of thermophilic clostridia will be available only if successful methods of transformation are found, either with whole or walless cells. The ability to transfer specific genes among thermophilic microorganisms and, hopefully, genes modified in *E. coli* and *Bacillus subtilis* back into thermophilic species needs to be achieved. If these attempts are successful, the potential of thermophilic clostridia for exploitation in industrial and biotechnological processes will be greater.

16.8 CONCLUSION

Much more research is needed, especially at a molecular level, to make these interesting thermophilic clostridia suitable for large-scale technical processes. However, the authors feel that the given examples clearly demonstrate the high potential of thermophiles for applications in the near future in biotechnological processes.

REFERENCES

Abdullah, M., Catley, B.J., Lee, E.Y.C., et al. (1966) *Cereal Chem.* 43, 111–118.
Antranikian, G. (1989) *Appl. Biochem. Biotechnol.* 20/21, 267–279.

Antranikian, G., Herzberg, C., and Gottschalk, G. (1987a) *Appl. Environ. Microbiol.* 53, 1668–1673.
Antranikian, G., Herzberg, C., Mayer, F., and Gottschalk, G. (1987b) *FEMS Lett.* 41, 193–197.
Antranikian, G., Zablowski, P., and Gottschalk, G. (1987c) *Appl. Microbiol. Biotechnol.* 27, 75–81.
Archer, D.B., and Kirsop, B.H. (1990) in *Waste Treatment Technology* (Whealthy, A. ed.), pp. 43–91, Elsevier Applied Science, New York.
Avgerinos, G.C., Fang, H.Y., Biocic, I., and Wang, D.I.C. (1981) in *Advances in Biotechnology*, vol. 2 (Moo-Young, M., and Robinson, C.W., eds.), pp. 119–124, Pergamon Press, Elmsford, NY.
Baronowsky, J.J., Schreurs, W.J.A., and Kaskhet, E.R. (1984) *Appl. Environ. Microbiol.* 48, 1134–1139.
Bayer, E.A., Kenig, R., and Lamed, R. (1983) *J. Bacteriol.* 156, 818–827.
Bayer, E.A., Setter, E., and Lamed, R. (1985) *J. Bacteriol.* 163(2), 552–559.
Berenger, J.F., Frixon, C., Bigiarid, J., and Creuzet, N. (1985) *Can. J. Microbiol.* 31, 635–643.
Bowring, S.N., and Morris, J.G. (1985) *J. Appl. Bacteriol.* 58, 577–584.
Braun, M., Schoberth, S., and Gottschalk, G. (1979) *Arch. Microbiol.* 120, 201–204.
Brener, D., and Johnson, B.F. (1984) *Appl. Env. Microbiol.* 47(5), 1126–1129.
Brodel, B., Samain, E., and Debeire, P. (1990) *Biotechnol. Lett.* 12, 65–70.
Brownell, J.E., and Nakas, J.P. (1991) *J. Ind. Microbiol.* 7, 1–6.
Bryant, F.O., Wiegel, J., and Ljungdahl, L.G. (1988) *Appl. Environ. Microbiol.* 54, 460–465.
Bucke, C., and Wiseman, A. (1981) *Chem. Indust.* (April 4), 234–240.
Burchhardt, G., Wienecke, A., and Bahl, H. (1991) *Curr. Microbiol.* 22, 91–95.
Busche, R.M. (1983) *Biotech. Bioeng. Symp.* 13, 597.
Busche, R.M. (1985a) in *Biotechnology Applications and Research* (Cheremisinoff, P.N., and Oulette, R.P., eds.), pp. 88–102, Technicon, Lancaster, PA.
Busche, R.M. (1985b) *Biotechnol. Progr.* 1, 165–180.
Cameron, D.C., and Cooney, C.L. (1986) *Biotechnology* 4, 651–654.
Carreira, L.H., and Ljungdahl, L.G. (1983) in *Liquid Fuel Developments* (Wise, D.L., ed.), pp. 1–29, CRC Press, Boca Raton, FL.
Chen, Y.R. (1983) in *Fuel Gas Systems* (Wise, D.L., ed.), pp. 23–59, CRC Press, Boca Raton, FL.
Chollar, B.H. (1984) *Public Roads* 47, 113–118.
Clarck, J.E., Ragsdale, S.W., Ljungdahl, L.G., and Wiegel, J. (1982) *J. Bacteriol.* 151, 507–509.
Clyde, R.A. (1983) U.S. patent no. 4,407,954.
Cook, G.M., Janssen, P.H., and Morgan, H.W. (1991) *System. Appl. Microbiol.* 14, 240–244.
Debeire, P., Priem, B., Strecker, G., and Vignon, M. (1990) *Biotechnol. Lett.* 12, 65–70.
De Blois, S., and Wiegel, J. (1991) Abstr. O-46, Annual Meeting, Dallas, TX, p. 266, American Society for Microbiology, Washington, DC.
DeLey, J., and Swings, J. (1984) in *Bergey's Manual of Systematic Bacteriology* (Krieg, N.R., and Holt, J.G., eds.), vol. 1, p. 275, The Williams & Wilkins Co., Baltimore.
Drent, W.J., Lahpor, G.A., Wiegart, W.M., and Gottschal, J.C. (1991) *Appl. Environ. Microbiol.* 57(2), 455–462.

Duong, T.-V.C., Johnson, E.A., and Demain, A.L. (1983) in *Topics in Enzyme and Fermentation Biotechnology*, vol. 7 (Wiseman, A., ed.), pp. 156–195, John Wiley and Sons, New York.

Ennis, B.M., Qureshi, N., and Maddox, I.S. (1987) *Enzyme Microb. Technol.* 9, 672–675.

Fogarty, W.M. (1983) in *Microbial Enzymes and Biotechnology* (Fogarty, W.M., ed.), pp. 1–92, Applied Science Publishers, London.

Fogarty, W.M., and Kelly, C.T. (1983) in *Microbial Enzymes and Biotechnology* (Fogarty, W.M., ed.), pp. 131–182, Applied Science Publishers, London.

Freier, D. (1981) M.D. thesis, University of Göttingen, Göttingen, Germany.

Freier, D., Mothershed, Ch.P., and Wiegel, J. (1988) *Appl. Environ. Microbiol.* 54, 104–111.

Freier, D., Wiegel, J., and Gottschalk, G. (1989) *Biotechnol. Lett.* 11, 831–836.

Freitag, R., Scheper, T., Spreinat, A., and Antranikian, G. (1991) *Appl. Microbiol. Biotechnol.* 35, 471–476.

Fuchs, A., de Bruyn, J.M., and Niedeveld, C.J. (1985) *Antonie van Leeuwenhoek J. Microbiol.* 51, 333–351.

Germain, P., Toukourou, F., and Donaduzzi, L. (1986) *Appl. Microbiol. Biotechnol.* 24, 300–305.

Gomez, R.F., Snedecor, B., and Mendez, B. (1980) *Dev. Ind. Microbiol.* 22, 87–96.

Gottschalk, G., Andreesen, J.R., and Hippe, H. (1981) in *The Prokaryotes* (Starr, M.P., Stolp, H., Truper, H.G., Balows, A., and Schlegel, H.G., eds.), pp. 1767–1803, Springer-Verlag, Berlin.

Grepinet, O., Chebron, M.-C., and Beguin, P. (1988) *J. Bacteriol.* 170, 4582–4588.

Haeckel, K., and Bahl, H. (1989) *FEMS Microbiol. Lett.* 60, 333–337.

Hamilton, B.H. (1984) in *Organic Chemicals from Biomass* (Wise, D.L., ed.), pp. 109–143, Benjamin/Cummings, Menlo Park, CA.

Harden, A. (1901) *J. Chem. Soc.* 79, 612–628.

Herrero, A.A. (1983) *Trends Biotechnol.* 1(2), 49–53.

Herrero, A.A., and Gomez, R.F. (1980) *Appl. Environ. Microbiol.* 40, 571–577.

Herrero, A.A., and Gomez, R.F. (1981) in *Advances in Biotechnology*, vol. 2 (Moo-Young, M., and Robinson, C.W., eds.), pp. 213–218, Pergamon Press, Elmsford, NY.

Herrero, A.A., Gomez, R.F., and Roberts, M.F. (1982) *Biochem. Biophys. Acta* 693, 195–204.

Herrero, A.A., Gomez, R.F., and Roberts, M.F. (1985) *J. Biol. Chem.* 260, 7442–7451.

Holdemann, L.V., Cato, E.P., and Moore, W.E.C. (1977) in *Anaerobe Laboratory Manual*, 4th ed. (Holdemann, L.V., Cato, E.P., and Moore, W.E.C., eds.), Virginia Polytechnic Institute, Blacksburg.

Hollaus, F., and Sleytr, U. (1972) *Arch. Microbiol.* 86, 129–146.

Hon-nami, K., Coughlan, M.P., Hon-nami, H., and Ljungdahl, L.G. (1986) *Arch. Microbiol.* 145, 13–19.

Hsu, E.J. (1976) in *Spore Research* (Gould, G.H., and Wolf, J., eds.), pp. 223–242, Academic Press, New York.

Hsu, E.J., and Ordal, Z.J. (1970) *J. Bacteriol.* 102(2), 369–376.

Hugenholtz, J., and Ljungdahl, L.G. (1988) Abstr. K153, Annual Meeting, Miami Beach, FL, p. 232, American Society for Microbiology, Washington, DC.

Hyun, H.H., Shen, J.-G., and Zeikus, J.G. (1985) *J. Bacteriol.* 164, 1153–1161.

Hyun, H.H., and Zeikus, J.G. (1985a) *Appl. Environ. Microbiol.* 49, 1162–1167.

Hyun, H.H., and Zeikus, J.G. (1985b) *J. Bacteriol.* 164, 1162–1170.
Hyun, H.H., and Zeikus, J.G. (1985c) *Appl. Environ. Microbiol.* 49, 1168–1173.
Hyun, H.H., and Zeikus, J.G. (1985d) *Appl. Environ. Microbiol.* 49, 1174–1181.
Ingram, L.O. (1986) *TIBTECH.* Feb., 40–44.
Ingram, L.O., and Buttke, T.M. (1984) *Adv. Microb. Physiol.* 25, 256–300.
Ivey, D.M. (1987) Ph.D. thesis, Univ. of Georgia, Athens.
Ivey, D.M., and Ljungdahl, L.G. (1986) *J. Bacteriol.* 165, 252–257.
Jin, F., Yamasato, K., and Toda, K. (1988) *Int. J. Syst. Bacteriol.* 38(3), 279–281.
Johnson, E.A., Madia, A., and Demain, A. (1981) *Appl. Environ. Microbiol.* 41, 1060–1062.
Jones, D.T., and Woods, D.R. (1989) in *Clostridia* (Minton, N.P., and Clarcke, D.J., eds.), pp. 105–144, Plenum Press, New York.
Katkocin, D., Word, N.S., and Yang, S.S. (1985) U.S. patent no. 4,536,477.
Keller, F.A., Jr., Ganoung, J.S., and Luenser, S.J. (1985) U.S. patent no. 4,513,084.
Kerby, R., and Zeikus, J.G. (1983) *Curr. Microbiol.* 8, 27–30.
Kim, U-H., Park, D-C., Chung, K-T., and Lee, Y-H. (1989) *Kor. J. Microbiol.* 24, 357–365.
Klaushofer, H., and Parkkinen, E. (1965) *Z. Zuckerind. Boehm.* 15, 445–449.
Klingerberg, M., Vorlop, K.D., and Antranikian, G. (1990) *Appl. Microbiol. Biotechnol.* 33, 494–500.
Koch, R., and Antranikian, G. (1990) *Starch* 397–403.
Koch, R., Zablowski, P., and Antranikian, G. (1987) *Appl. Microbiol. Biotechnol.* 27, 192–198.
Kundu, S., Ghose, T.K., and Mukhopadhyay, S.N. (1983) *Biotechnol. Bioeng.* 25, 1109–1126.
Kuo, Y., and Gregor, H.P. (1983) *Sep. Sci. Technol.* 18, 421–440.
Lamed, R., Keinan, E., and Zeikus, J.G. (1981) *Enzyme Microb. Technol.* 3, 144–148.
Lamed, R., Setter, E., and Bayer, E.A. (1983) *J. Bacteriol.* 156(2), 828–836.
Lamed, R., and Zeikus, J.G. (1981) *Biochem. J.* 195, 183–190.
Landuyt, S.L., Hsu, E.J., and Lu, M. (1983) *Ann. N.Y. Acad. Sci.* 413, 473–478.
Larrayoz, M.A., and Puigjaner, L. (1987) *Biotechnol. Bioeng.* 30, 692–696.
Lee, C.K., and Ordal, Z.J. (1967) *J. Bacteriol.* 94, 530–536.
Lee, Y.E., and Zeikus, J.G. (1991) *Biochem. J.* 273, 565–571.
Le Ruyet, P., Dubourguier, H.C., Albagnac, G., and Prensier, G. (1985) *System. Appl. Microbiol.* 6, 196–202.
Lipinsky, E.S. (1981) in *Basic Biology of New Developments in Biotechnology* (Hollander, A., Laskin, A.I., and Rogers, P., eds.), pp. 349–376, Plenum Press, New York.
Ljungdahl, L.G. (1983) in *Organic Chemicals from Biomass* (Wise, D.L., ed.), pp. 219–248, Benjamin/Cummings, Menlo Park, CA.
Ljungdahl, L.G. (1986) *Annu. Rev. Microbiol.* 40, 415–450.
Ljungdahl, L.G., and Andreesen, J.R. (1978) *Methods Enzymol.* 53, 360–372.
Ljungdahl, L.G., Bryant, F., Carreira, L., Saiki, T., and Wiegel, J. (1981) in *Trends in the Biology of Fermentations for Fuels and Chemicals* (Hollaender, A., ed.), pp. 397–419, Plenum Press, New York.
Ljungdahl, L.G., Carreira, L.H., Garrison, R.J., et al. (1986) U.S. Dep. of Transportation, Federal Highway Administration Report No. FHWA/RD-86/117, National Technical Information Service, Springfield, VA.
Ljungdahl, L.G., and Wiegel, J. (1981) U.S. patent no. 4,292,407.

Lundie, L.L., and Drake, H.L. (1984) *J. Bacteriol.* 159, 700–703.

MacKenzie, C.R., Yang, R.C.A., Patel, G.P., Bilous, D., and Narang, S.A. (1989) *Arch. Microbiol.* 152, 377–381.

Madden, R.H. (1983) *Int. J. Syst. Bacteriol.* 33, 837–840.

Madi, E., and Antranikian, G. (1989) *Appl. Microbiol. Biotechnol.* 30, 422–425.

Madi, E., Antranikian, G., Ohmiya, K., and Gottschalk, G. (1987) *Appl. Environ. Microbiol.* 53, 1661–1667.

Marchaim, U., Perach, Z., and Kimchie, S. (1983) in *Fuel Gas Systems* (Wise, D.L., ed.), pp. 141–156, CRC Press, Boca Raton, FL.

Martin, D.R., Lundie, L.L., Kellun, R., and Drake, H.L. (1983) *Curr. Microbiol.* 8, 337–340.

Marynowski, C.V., Jones, J.L., Boughton, R.L., et al. (1983) *Report FHWA/RD*, 86/117-22161, pp. 82–145, National Technical Information Service, Springfield, VA.

Mathupala, S.P., Saha, B.C., and Zeikus, J.G. (1990) Abstr. K53, Annual Meeting, Anaheim, CA, p. 228, American Society for Microbiology, Washington, DC.

Matsunaga, T., Karube, I., and Suzuki, S. (1983) *Enzyme Microb. Technol.* 5, 441.

Mayer, F., Ivey, D.M., and Ljungdahl, L.G. (1986) *J. Bacteriol.* 166, 1128–1130.

Melasniemi, H. (1987a) *J. Gen. Microbiol.* 133, 883–890.

Melasniemi, H. (1987b) *Biochem. J.* 246, 193–197.

Melasniemi, H. (1988) *Biochem. J.* 250, 813–818.

Melasniemi, H., and Paloheimo, M. (1989) *J. Gen. Microbiol.* 135, 1755–1762.

Melasniemi, H., Paloheimo, M., and Hemiö, L. (1990) *J. Gen. Microbiol.* 136, 447–454.

Méndez, B.S., and Gomez, R.F. (1982) *Appl. Environ. Microbiol.* 43, 496–498.

Méndez, B.S., Pettinari, M.J., Ivanier, S.E., Ramos, C.A., and Sineriz, F. (1991) *Int. J. Syst. Bacteriol.* 41(2), 281–283.

Morton, T.A., Chin-Fong, Ch., and Ljungdahl, L.G. (1992) in *Genetics and Molecular Biology of Anaerobic Bacteria* (Sebold, M., ed.), pp. 398–406, Springer-Verlag, New York.

Müller, W., Wehnert, G., and Scheper, T. (1988) *Anal. Chem. Acta* 213, 47–53.

Ng, T.K., Ben-Bassat, A., and Zeikus, J.G. (1981) *Appl. Environ. Microbiol.* 41, 1337–1343.

Ng, T.K., Weimer, P.J., and Zeikus, J.G. (1977) *Arch. Microbiol.* 114, 1–7.

Ng, T.K., and Zeikus, J.G. (1982) *J. Bacteriol.* 150, 1391–1399.

Norman, B.E. (1982) *Starch/Starke* 34, 340–346.

Parkkinen, E. (1986) *Appl. Microbiol. Biotechnol.* 25, 213–219.

Patel, B.K.C., Monk, C., Littleworth, H., Morgan, H.W., and Daniel, R.M. (1987) *Int. J. Syst. Bacteriol.* 37, 123–126.

Patni, N.S., and Alexander, J.K. (1971) *J. Bacteriol.* 105, 226–231.

Paulin, L., and Pösö, H. (1983) *Biochem. Biophys. Acta* 742, 197–205.

Peck, H.D., and Odom, M. (1981) in *Trends in Biology of Fermentations for Fuels and Chemicals* (Hollaender, A., ed.), pp. 375–395, Plenum Press, New York.

Peteranderl, R. (1990) M.S. thesis, University of Georgia, Athens.

Plant, A.R., Patel, B.K.C., Morgan, H.W., and Daniel, R.W. (1987) *Syst. Appl. Microbiol.* 9, 158–162.

Poulsen, P.B. (1981) *Enzyme Microb. Technol.* 3, 271–273.

Prescott, S.C., and Dunn, C.G. (1959) in *Industrial Microbiology*, pp. 428–473, McGraw-Hill Book Co., New York.

Reed, W.M., and Bogdan, M.E. (1985) *Biotech. Bioeng. Symp.* 15, 641–647.

Rogers, P. (1986) *Adv. Appl. Microbiol.* 31, 75–88.

Rothstein, D.M. (1986) *J. Bacteriol.* 165, 319–320.

Saha, B.C., Lamed, R., and Zeikus, J.G. (1989) in *Clostridia* (Minton, N.P., and Clarcke, D.J., eds.), pp. 227–263, Plenum Press, New York.

Saha, B.C., Mathupala, S.P., and Zeikus, J.G. (1988) *Biochem. J.* 247, 343–348.

Saha, B.C., and Zeikus, J.G. (1989) *Trends Biotechnol.* 7, 234–239.

Saha, B.C., and Zeikus, J.G. (1991) *Appl. Microbiol. Biotechnol.* 35, 568–571.

Sakka, K., Kojima, Y., Yoshikawa, K., and Shimada, K. (1990) *Agric. Biol. Chem.* 54, 337–342.

Sanchez-Riéra, F., Cameron, D.C., and Cooney, C.L. (1987) *Biotechnol. Lett.* 9, 449–454.

Savage, M.D., and Drake, H.L. (1986) *J. Bacteriol.* 165, 315–318.

Savage, M.D., Steven, Z.W., Daniel, S.L., et al. (1987) *Appl. Environ. Microbiol.* 53, 1902–1906.

Schink, B., and Zeikus, J.G. (1980) *Curr. Microbiol.* 4, 387–389.

Schink, B., and Zeikus, J.G. (1983a) *J. Gen. Microbiol.* 129, 1149–1158.

Schink, B., and Zeikus, J.G. (1983b) *FEMS Microbiol. Lett.* 17, 295–298.

Schlote, D., and Gottschalk, G. (1986) *Appl. Microbiol. Biotechnol.* 24, 1–6.

Schwartz, R.D., and Keller, F.A., Jr. (1982a) *Appl. Environ. Microbiol.* 43, 117–123.

Schwartz, R.D., and Keller, F.A., Jr. (1982b) *Appl. Environ. Microbiol.* 43, 1385–1392.

Shen, G-J., Saha, B.C., Lee, Y.Y., Bhatnagar, L., and Zeikus, J.G. (1988) *Biochem. J.* 254, 835–840.

Shimshick, E.J. (1981) U.S. patent no. 4,250,331.

Sjolander, N.O. (1937) *J. Bacteriol.* 34, 419–428.

Soh, A.L.A., Ralambotiana, H., Ollivier, B., et al. (1991) *Syst. Appl. Microbiol.* 14, 135–139.

Solberg, M., Blaschek, H.P., and Kahn, P. (1981) *J. Food. Safety* 3, 267–289.

Soutschek-Bauer, E., and Staudenbauer, W.L. (1985) *Biotechnol. Lett.* 7, 705–710.

Specka, U., Mayer, F., and Antranikian, G. (1991) *Appl. Environ. Microbiol.* 57, 2317–2323.

Spreinat, A., and Antranikian, G. (1990) *Appl. Microbiol. Biotechnol.* 35, 511–518.

Su, T.M., Lamed, R., and Lobos, G.H. (1981) in *Second World Congress of Chemical Engineering*, vol. 1, Montreal, Canadian Society of Chemical Engineering, Montreal.

Sukhumavasi, J., Ohmiya, K., Shimizu, S., and Ueno, K. (1988) *Int. J. Syst. Bacteriol.* 38, 179–182.

Turunen, M., Parkkinen, E., Londesborough, J., and Korhola, M. (1987) *J. Gen. Microbiol.* 133, 2865–2873.

U.S. International Trade Commission (1986) Synthetic organic chemicals. United States Production and Sales, USITC Publ. 2009, pp. 209–213. U.S. Government Printing Office, Washington, DC.

Vancanneyt, M., De Vos, P., Maras, M., and De Ley, J. (1990) *Syst. Appl. Microbiol.* 13, 382–387.

Vandamme, E.J., and Derycke, D.G. (1983) *Adv. Appl. Microbiol.* 29, 139–176.

Varel, V.H. (1983) in *Fuel Gas Developments* (Wise, D.L., ed.), vol. 2, pp. 19–47, CRC Series in Bioenergy Systems, CRC Press, Boca Raton, FL.

Vorlop, K.D., and Steinert, H.J. (1987) *Ann. New York Acad. Sci.* 501, 339–342.

Walker, G.C. (1983) in *Basic Biology of New Developments in Biotechnology* (Hollaender, A., Laskin, A.I., and Rogers, P., eds.), pp. 349–376, Plenum Press, New York.

Wang, D.I.C., Biocic, I., Fang, H-Y., and Wang, S-D. (1979) *Third Annu. Biomass Energy Systems Conference Proceedings*, SERI/TP-33-285, Golden, CO, p. 61, U.S. Government Printing Office, Springfield, VA.

Wang, G., and Wang, D. (1984) *Appl. Environ. Microbiol.* 47, 294–298.

Weimar, P.S., and Zeikus, J.G. (1977) *Appl. Environ. Microbiol.* 33, 289–297.

White, H., Lebertz, H., Thanos, I., and Simon, H. (1987) *FEMS Microbiol. Lett.* 43, 173–176.

Wiegel, J. (1980) *Experientia* 36, 1434–1446.

Wiegel, J. (1982) in *New Trends in Research and Utilization of Solar Energy* (Mislin, H., ed.), p. 79, Birkhauser Verlag, Basel.

Wiegel, J. (1992) in *The Bacterial Thermophiles* (Kristjansson, J.K., ed.), pp. 105–184, CRC Press, Boca Raton, FL.

Wiegel, J., Braun, M., and Gottschalk, G. (1981) *Curr. Microbiol.* 5, 255–260.

Wiegel, J., Carreira, L.H., Garrison, R., Rabek, N., and Ljungdahl, L.G. (1991) in *Calcium Magnesium Acetate. An Emerging Bulk Chemical for Environmental Applications* (Wise, D.L., Levendis, Y.A., and Metghalchi, M., eds.), pp. 359–418, Elsevier, Amsterdam.

Wiegel, J., Carreira, L.H., Ljungdahl, L.G., and Puls, J. (1984) in *Proceeding Seventh International FPRS Industrial Wood Energy Forum 83*, Abstr. 7, Nashville, TN, p. 25, Forest Products Research Society, Madison, WI.

Wiegel, J., Carreira, L.H., Mothershed, Ch.P., Ljungdahl, L.G., and Puls, J. (1986) in *Industrial wood energy forum 83*, vol. II, pp. 482–489, Forest Product Research Society, Madison, WI.

Wiegel, J., Carreira, L.H., Mothershed, Ch.P., and Puls, J. (1983) *Biotechnol. Bioeng. Symp.* 13, 193–205.

Wiegel, J., and Dykstra, M. (1984) *Eur. J. Appl. Microbiol. Biotechnol.* 9, 59–65.

Wiegel, J., and Garrison, R. (1985) Abstr. I166, Annual Meeting, Las Vegas, NV, p. 165, American Society for Microbiology, Washington, DC.

Wiegel, J., Kuk, S., and Kohring, G.W. (1989) *Int. J. Syst. Bacteriol.* 39, 199–204.

Wiegel, J., and Ljungdahl, L.G. (1979) in *Technische Mikrobiologie* (M. Dellweg, ed.), Verlag Versuchs- und Lehranstalt für Spiritusfabrikation und Fermentationstechnologie im Institut für Gärungsgeäwerbe und Biotechnologie, Berlin.

Wiegel, J., and Ljungdahl, L.G. (1986) *Cri. Rev. Biotechnol.* 3, 39–107.

Wiegel, J., Ljungdahl, L.G., and Rawson, J.R. (1979) *J. Bacteriol.* 139, 800–810.

Wiegel, J., Mothershed, C.P., and Puls, J. (1985) *Appl. Environ. Microbiol.* 49, 656–659.

Wise, D.L. (1983a) in *CRC Series in Bioenergy Systems: Fuel Gas Systems*, Vol. 1, *Fuel Gas Developments*, vol. 2, *Liquid Fuel Systems*, Vol. 3, *Liquid Fuel Developments*, Vol. 4, *Bioconversion Systems*, Vol. 5 (Wise, D.L., ed.), CRC Press, Boca Raton, FL.

Wise, D.L., (1983b) in *Organic Chemicals from Biomass* (Wise, D.L., ed.), Benjamin/Cummings, Menlo Park, CA.

Wohlt, J.E., Frohish, R.A., Davis, C.I., Bryant, M.P., and Mackie, R.I. (1990) *Biol. Wastes* 32, 193–207.

Wood, H.G. (1982) in *Of Oxygen, Fuels, and Living Matter* (Semenza, G., ed.), Part 2, pp. 173–250, John Wiley and Sons, New York.

Wood, H.G., and Ljungdahl, L.G. (1991) in *Variations in Autotrophic Life* (Shiveley, J.M., and Barton, L.L., eds.), pp. 201–250, Academic Press, London.
Xu, J-G., and Wiegel, J. (1987) Abstr. I-80, Annual Meeting, Atlanta, American Society for Microbiology, Washington, DC.
Yates, R.A. (1981) U.S. patent no. 4,282,323.
Young, M., Staudenbauer, L.H., and Minton, N.P. (1989) in *Clostridia* (Minton, N.P., and Clarcke, D.J., eds.), p. 63, Plenum Press, New York.
Zeikus, J.G. (1979) *Enzyme Microb. Technol.* 1, 243–252.
Zeikus, J.G. (1980) *Annu. Rev. Microbiol.* 34, 423–464.
Zeikus, J.G., Ben-Bassat, A., and Hegge, P.W. (1980) *J. Bacteriol.* 143, 432–440.
Zeikus, J.G., Ben-Bassat, A., Ng, T.K., and Lamed, R.J. (1981) in *Trends in the Biology of Fermentations for Fuels and Chemicals* (Hollaender, A., ed.), pp. 441–461, Plenum Press, New York.
Zeikus, J.G., Dawson, M.A., Thomson, T.E., Ingvorsen, K., and Hatchikian, E.C. (1983) *J. Gen. Microbiol.* 129, 1159–1169.
Zeikus, J.G., and Lamed, T.J. (1982) U.S. patent no. 4,352,885.
Zeikus, J.G., and Saha, B.C. (1989) U.S. patent no. 4,814,267.
Zertuche, C., and Zall, R.R. (1982) *Biotechnol. Bioeng.* 24, 57–68.

INDEX